Horticulture for Health, Livelihood and Economy

THE EDITORS

Dr K. L. Chadha (*Padma Shri Awardee*) is currently President of the Horticultural Society of India and also Adjunct Professor (Hort.) at IARI. He received M.Sc. (Hort.) degree from Punjab University, Chandigarh and Ph.D. from IARI, New Delhi. He had served in various capacities at IARI, New Delhi, PAU, Ludhiana and Central Mango Research Station, Lucknow. He was the Director, IIHR, Bengaluru, Horticulture Commissioner and Executive Director, National Horticulture Board, Govt. of India; DDG (Hort.), ICAR and ICAR-National Professor (Hort.) at IARI, New Delhi.

Dr V. B. Patel is presently Assistant Director General (Fruits & Plantation Crops), ICAR, New Delhi. He received M.Sc. (Ag.) Hort. from GBPUA&T, Pantnagar and Ph.D. from IARI, New Delhi. He also served as University Professor and Chairman of Hort. Dept. as well as Associate Director of Bihar Agricultural University, Sabour (Bihar). He is associated with development and release of five varieties and several production technologies in fruit crops.

Dr S. K. Singh is presently the Director, ICAR-IIHR,Bengaluru. He received M.Sc. and Ph.D. degrees in Horticulture from Banaras Hindu University, Varanasi and joined Agricultural Research Service in 1994. He is involved in P.G. teaching and improvement of mango & grape. He has developed three extra-early grape varieties and production technologies of mango and grape. Dr Singh is working on biotechnological application & including Genomics and marker-assisted breeding in fruit crops.

Dr Jai Prakash is presently the Principal Scientist at the Division of Fruits and Horticultural Technology, IARI, New Delhi. He received M.Sc. (Hort.) from CSAUA&T, Kanpur and Ph.D. degree in Horticulture from Banaras Hindu University, Varanasi and joined Agricultural Research Service of ICAR in 2003. He is serving ICAR for the last 20 years and is involved in research & teaching and has developed improved papaya, grape and mango varieties.

Dr. Shyam Sundar Dey is presently working as Senior Scientist at Division of Vegetable Science, ICAR-Indian Agricultural Research Institute, New Delhi. He obtained his MSc and PhD degree at ICAR-Indian Agricultural Research Institute, New Delhi. He is involved with development of 14 hybrids varieties in different vegetable crops and registered 05 germplasm. Presently, he is working on marker assisted back-cross breeding, development of genomic resources using the indigenous cucumber.

Horticulture for Health, Livelihood and Economy

— *Editors* —

K.L. Chadha

V.B. Patel

S.K. Singh

Jai Prakash

S.S. Dey

Kruger Brentt

Publishers

2024

Kruger Brentt Publishers UK. LTD.
Company Number 9728962

Regd. Office: 68 St Margarets Road, Edgware, Middlesex HA8 9UU

ISBN 978-1-78715-256-4 (Hardback)

For information on all our publications visit our website at http://krugerbrentt.com/

Preface

The vast and ever-evolving landscape of agriculture, horticulture stands as a resilient and promising field that has the potential to address some of the most critical challenges of our times. "Horticulture for Health, Livelihood, and Economy" is a comprehensive compilation of knowledge and insights that celebrates the immense potential of horticulture in improving human well-being, providing sustainable livelihoods, and contributing to economic development. This book brings together a diverse array of topics that shed light on the multifaceted contributions of horticulture to our society. Horticulture encompasses a wide spectrum of plant cultivation, from fruits and vegetables to spices, nuts, and ornamental plants. This diversity not only adds color and flavor to our lives but also plays a pivotal role in promoting health and nutrition. Chapter by chapter, this book unravels the nutritional and health benefits of minor, underutilized, and exotic fruits, as well as arid fruits and vegetables, which often go overlooked in the mainstream discussion of agriculture. We delve into the potential of horticulture to combat malnutrition, especially in the underprivileged regions, and explore the concept of biofortification to enhance the nutritional content of horticultural crops.

The book also examines the challenges and prospects of emerging new generation fruit crops, spotlighting underutilized gems like apricots in the Trans-Himalayan region. Furthermore, it explores the biotechnological and genomic advancements that are transforming the improvement of fruit crops and enhancing the production of coconut, arecanut, and spices. Papaya breeding strategies, the development of new grape rootstocks, and innovative production technologies in vegetable crops are other essential areas of focus. Horticulture is not just about nutrition and health but also offers sustainable economic opportunities. The chapters dedicated to coconut-based cropping systems, floriculture, protected cultivation, and quality pomegranate production underscore the potential for horticulture to drive economic development, even on degraded lands.

As the world grapples with environmental challenges, the book introduces the concept of "cosmic farming" and explores soil and water conservation techniques tailored for horticultural production. It emphasizes how horticulture can transform degraded lands into opportunities for livelihood security and economic growth. Moreover, the book does not shy away from addressing the emerging threats to horticulture, such as diseases affecting mango production, and provides insights into new developments in prolonging the storage of fruits and vegetables. It highlights the importance of India's participation in the global fruit and vegetable markets and delves into the physiology, quality, and safety of fresh-cut produce.

This book extends its focus to the challenges and opportunities in the export of seed spices, as well as the establishment of farmers' producer organizations for quality production and export promotion in horticultural commodities. The culmination of this journey takes us to the empowering of tribal communities in Jharkhand through horticultural enterprises, showcasing how horticulture can be a catalyst for social and economic transformation.

We hope that this book serves as a valuable resource for scholars, researchers, policymakers, and practitioners, offering insights and solutions for harnessing the full potential of horticulture. It is our aspiration that the knowledge contained within these pages inspires and motivates readers to further explore, innovate, and invest in horticulture for the betterment of society, individual well-being, and economic prosperity. The editors are grateful to all the contributors for their cooperation, support and timely submission of their manuscripts for bringing out this publication.

K.L. Chadha

V.B. Patel

S.K. Singh

Jai Prakash

S.S. Dey

Contents

Chapter 1

Know your Minor, Underutilized and Exotic Fruits: Nutritional and Health Benefits

Sanjay Kumar Singh[1*], V.B. Patel[2], Jai Prakash[3], M.K. Verma[4], Chavlesh Kumar[3] and Satyabrata Pradhan[3]

[1]*ICAR–Indian Institute of Horticultural Research, Bengaluru 560089, Karnataka*

[2]*Horticultural Science Division, ICAR–Indian Agricultural Research Institute, New Delhi 110012*

[3]*Division of Fruits and Horticultural Technology, ICAR-Indian Agricultural Research Institute, New Delhi 110012*

[4]*ICAR–Central Institute of Temperate Horticulture, Srinagar 190007, J&K*

e-mail: sanjaydr2@gmail.com

ABSTRACT

India enjoys a prominent position in under-exploited fruit crops of the world. The varied geographical and suitable environment conditions favor growing a wide variety of fruit crops. These fruits are available in abundance and also in different seasons. Most of the minor and underutilized fruits are often available only in the local markets. A large number of these fruits can grow under adverse conditions and are also known for their therapeutic and nutritive value and possess demands of the health-conscious consumers. However, some of these fruits are not acceptable in the market in fresh form due to their acidic/astringent taste. Hence, there is a need to diversify and popularize of such underutilized fruit crops. Since ages, several minor and underutilized fruit crops having medicinal properties are utilized in various indigenous medicinal systems like Ayurveda, Unani, and Homoeopathy. Most of these fruit crops are a rich source of vitamin C, which has been proved as a boon in enhancing immunity in this COVID-19 pandemic times. These crops can even cure insomnia, scurvy, constipation hemorrhage, leucorrhea, anemia, stomach ache, and can be used as a cooling agent to reduce sun strokes ill effects. There is large variability of underutilized and wild fruits present in the entire Western Ghats, western and eastern sub-Himalayan tracts. Similarly, there are several exotic fruits which have acclimatized in the region and acquire good variability due to cross-pollination. A good number of germplasms of underutilized fruits have been collected but most of the species yet to be characterized and conserved. Apart from these, there is a need to develop high yielding varieties, production and plant protection technologies, and post-harvest management practices for these crops. Better coordination among all the agencies involved in research, development, and promotion will help to popularize these fruit crops.

***Keywords**: Minor fruits, Underexploited, Cultivation, Nutrition, Human health.*

Horticulture has emerged as the major driver of growth in the Agriculture Sector in India. However, there was a major boost to this sector with higher allocation of funds by the Government of India during the VII FYP, which resulted in development of sound infrastructure for R&D, besides implementation of several programmes and pro-horticulture policies. At present the area under horticulture crops has increased from 1.95 million ha in 1947 to 26.2 million ha in 2019 resulting in cumulative increase of 506 per cent during the last 60 years. During the same period, the total horticulture production has increased from 31.8 million MT in 1961 to 319.5 million MT in 2019-20 a clear cumulative increase of 904 per cent. Recently, in 2020-21, the estimated production is poised at 329.86 million tonnes (second adv. estimates released on 15.07.2021; Source MoAFW, GoI).

India is a paradise for growing large number of fruit crops due to prevailing diverse agro-ecological conditions. About 40 fruits are grown in India of which several are commercial. Besides major fruit crops, there are several minor, underutilized and exotic which are now being grown commercially. Many of these fruit crops were earlier neglected due to various reasons (due to lack of supply chains, poor shelf-life, unknown for their nutraceutical values, *etc.*). However, these fruit species are well recognized for their tolerance/resistance to several stresses and resilience to climate change. These minor and underutilized fruits are realized for their high nutraceutical values and their use in traditional foods. These fruits are being well adapted in diverse edaphic and climatic conditions, thus have potential to contribute in food security, nutrition, dietary and health food, culinary use and income generation. As consequences, several neglected fruit crops are now of global significantly, dragon fruit, kiwifruit, blue berries *etc.*

Malnutrition is one the biggest challenge among the Indian population and India's child malnutrition is still alarming in the world. Subsequently, child and maternal malnutrition are accounting for the 15 percent human diseases and disorders in India. It is estimated that due to the child malnutrition, India losses about its four per cent of the gross domestic product and eight per cent of its productivity. This challenging situation need to be addressed with nutrient rich diversified diet especially the food rich in iron, zinc and vitamins. This challenging situation need to be addressed with nutrient rich diversified diet especially the food rich with iron, zinc and vitamins along with essential bioactive compounds. The diversified diet always needs to be included fruits because they are rich source of vitamins, minerals and essential bioactive compounds. Fruit crops are considered as protective food because of their richness in various vitamins, minerals and essential bioactive compounds. They are rich sources of certain soluble dietary fiber which reduces erratic bowel movements, aids in cholesterol and fats reduction from the body and ultimately helps in boosting the body immune system. Thus, utilization of minor, underutilized and exotic fruits are essential to combat the challenge of malnutrition because they are inherent source of micro nutrient The present chapter have compiled the information for nutritional and health benefits of minor, underutilized and exotic fruit crop along with their essential bioactive compounds. The diversified diet always needs to be included fruits because they are rich source of vitamins, minerals and essential bioactive compounds. Fruit crops are considered as protective food because of their richness in various vitamins, minerals and essential bioactive compounds. They are rich sources of certain soluble dietary fiber which reduces erratic bowel movements, aids in cholesterol and fats reduction from the body and ultimately helps in boosting the body immune system. Thus, utilization of minor, underutilized and exotic fruits are essential to combat the challenge of malnutrition because they are inherent source of micro nutrient.-

Minor and Underutilized Fruit Crops

Underutilized crops may be termed as the crops that are neither grown on a commercial or large scale nor are traded widely around the world. These crops fall under lesser recognized plant species in terms of research and marketing but are well adapted to the wild and stress like conditions. Neglected and underutilized crops are domesticated plant species that have been used for centuries by locals for their food, fibre, fodder, oil or medicinal properties, but have been reduced in importance due to commercialization of major fruit crops having organized trade. Similarly, these crops are also described as orphan, abandoned, lost, underutilized, local, minor, traditional, alternative, niche, or underdeveloped; recently as neglected, forgotten or smart food. Reductions in use may pertain to, among other things: supply or consumption constraints, poor shelf life, unrecognized nutritional value, poor consumer awareness, and reputational problems.

As the demand for plant and crop attributes changes (reappraisal or discovery of nutritional traits, culinary value, adaptation to climate change, *etc.*), neglected crops can overcome the constraints to the wider production and use. Although the options for scaling up neglected crops for large-scale cultivation appear to be increasingly limited, many species have the potential to contribute to food security, nutrition, dietary, culinary and processing use, health and income generation. Besides, being environmental friendly are also nutritious, higher medicinal properties and multiple uses. Various underutilized fruit crops available in India included; *Aegle marmelos, Annona muricata, Artocarpus heterophyllus, Averrhoa carambola, Bactris gasipaes, Canarium indicum, Carissa edulis, Carya cathayensis, Casimiroa edulis, Cornus mas, Crataegus monogyna, Cydonia oblonga, Dimocarpus longan, Diospyros kaki, Durio zibethinus, Emblica officinalis, Eriobotrya japonica, Garcinia mangostana, Hippophae rhamnoides, Hovenia dulcis, Irvingia gabonensis, Juglans regia, Lagenaria sphaerica, Nephelium lappaceum, Passiflora edulis, Pistacia lentiscus etc.* During present day wild and underutilized fruits, ethnic fruits are gaining popularity along with the major fruits, and thus diversifying the fruit basket on the dining table.

Arid Region Fruits

Arid zone is characterized by extremes of temperature, low and erratic rainfall, high wind velocity, high evapotranspirational losses of water and light sandy soil with very low water holding capacity and poor fertility status. Hence, the crops suited in the region have xerophytic traits and some do have summer dormancy.

Drought-hardy crops especially perennial fruits with deep root systems are capable of surviving extreme radiation and temperatures and provide income security, nutritional security and food security. Some of the fruit crops suited for the region are;

Aonla or Indian Gooseberry (*Emblica officinalis* Gaertn.)

It belongs to the family Euphorbiaceae, is an ancient, indigenous fruit of India. All parts of the plant including the fruits are medicinally rich and are used in the preparation of various Ayurvedic medicines. Fruits are commercially used for preparation of *Chayanprash* and *triphala*. Being a rich source of vit. C, is helpful in curing scurvy, problems of teeth, gums, eye and stomach. The pellets of dried *aonla* powder is being served to soldiers during war and sea voyage for the treatment of scurvy. Being a nutritious and medicinal, *aonla* proves the proverb, "*an aonla a day keeps the doctor away*". It contains 500-1500 mg vit. C, 0.2 mg nicotinic acid, 1.2 mg iron, 0.5 g protein, 0.l g fat, 0.7 g minerals, 2.0-3.4 g fibre, 14-21.8 g carbohydrates, 0.02 g phosphorus and 0.05 g calcium in 100 g pulp. The total sugars content in *aonla* fruit varies from 7 to 9.6 per cent, reducing sugars from 1.04 to4.09 per cent and non-reducing sugars from 3.05 to 7.23 per cent, among the various varieties. Fresh fruits are also used for the preparation of products such as preserve (*murabba*), pickle, chutney, shreds, *etc.* The fruits act as diuretic, laxative and useful in anemia, atherosclerosis, hemorrhages, leucorrhea, bronchitis, liver ailment, diarrhea *etc.*

Bael (*Aegle marmelos* Corr.)

It is a native of India and belongs to the family Rutaceae. The pulp is mucilaginous. The fruit and leaves are used for curing stomach diseases and to improve digestive system. All parts of the tree *i.e.*, root, bark, leaves, flowers, fruits and seed oil are used in different Ayurvedic medicines. *Bael* fruit contains 31.8 g carbohydrates, 1.8 g protein, 2.9 g fibre, 1.7 g mineral (mainly Ca and Fe), 8-18 mg vitamin C, 0.03 mg riboflavin per 100 g pulp. Fruits can be processed into various beverages and preserves. Its wood is used to prepare agricultural implements, besides the trifoliate leaves is grown in temple premises of Lord Shiva. All the parts of plant are economical and possess different medicinal values, *viz.* leaves, roots, seed, bark and fruit *etc.* contain a large amount of coumarins, alkaloids, sterols and essential oils hence, possess analgesic, anti-inflammatory, antipyretic, anti-microfilaria, antifungal, hypoglycemic, anti-dyslipidemic, immunomodulatory, anti-proliferative, wound healing, anti-fertility, and insecticidal abilities. The fruit is considered as one of the richest source of riboflavin. Marmelosin in fruit has therapeutically active factor which is remedy of the stomach ailments. The fresh leaf juice in doses of 8 to 16 g is given with honey as a mild laxative in fever, catarrh and asthama. Fresh leaves and fruits are used as a remedy for beriberi. It is also used in the treatment of chronic diarrhea, dysentery, and peptic ulcers, as a laxative and to recuperate from respiratory affections.

Ber (*Ziziphus mauritiana* Larnk.)

It belongs to the family Rhamnaceae and is indigenous to India. *Ziziphus nummularia* (syn. *Z. rotundifolia* Lamk.), *Z. oenoplia* Mill., *Z. rugosa* Lam. and *Z. sativa* Gaertn are also native of India and yield edible fruits. *Ziziphus jujuba* Mill. (Chinese jujube) produces large fruits but is not common in India. The fruits of *ber* have been in use in India since ancient times. The fruits are rich in sugars, vitamins C, A and B complex. Fully ripe fruits of *ber* contain 0.8-0.9 g starch, 16-23oBrix TSS, 4.9-12.4 g total sugars, 2.0-5.8 g reducing sugar, 0.1-0.5 per cent acidity apart from 0.3-0.5 g ash, 0.03-0.04 g calcium, 0.01- 0.02 g phosphorus, 0.5-1.0 mg iron and 66-133 mg vitamin C per 100 g of pulp. *Ber* leaves contain 5.6 per cent digestible crude protein and 49.7 per cent total digestible nutrients and are thus a nutritive fodder for animals. The decoction from root and bark is good for dysentery and diarrhea and leaf decoction is useful as gargle in sore throat and in bleeding gums. The powder of roots has medicinal properties for curing ulcer, fever and wounds. It is also rich in carotenes and phenolics. The secondary metabolites were present in ber fruits like flavonoids, glycosides, saponins, lignins, sterols and phenols are very effective function against pathogens. The different parts of plant like root, bark, leaves, flowers, seeds are used for treatment and also used as blood purifier and appetizer.

Lasora (*Cordia myxa* Roxb.)

It is also known as *gonda* or Indian cherry or *leusua* is a drought hardy fruit tree of the family Boraginaceae. Cordia is native of India and occurs throughout the warmer regions of the world. A related species *C. gharaf* Forsk. (*gondi*) is found distributed in arid sub-tropical region. Cordia plants are sensitive to frost but can tolerate the extremely high temperatures. Unripe fruits are used as pickle and vegetable. The ripe fruits are eaten as fresh and are rich in sugar and minerals. It has anthelmintic, diuretic, demulcent and expectorant properties. Other parts of the plant also have medicinal properties. Leaves are used for curing ulcer and headache. The bark is used for the treatment of fevers.

Karonda or Christ's Thorn (*Carissa carandas* L.)

It belongs to the family Apocynaceae, is a hardy evergreen spiny shrub of Indian origin. It is a multi- branched bushy tree and is generally planted as a hedge. It can tolerate the hot and dry conditions and intense solar radiation in the arid and plateau region. *Karonda* fruits are rich in Fe (39.1 per cent) and also contain 2.2 per cent protein, 96 per cent fat, 2.8 per cent minerals, 0.10 per cent Ca, 0.06 per cent P and 201-555 mg/100 g vit. C. The fruits are astringent, anti-scorbutic and are used as a remedy for biliousness. The major bioactive constituents, which impart medicinal value to the herb, are alkaloids, flavonoids, saponins and large amounts of cardiac glycosides, triterpenoids, phenolic compounds and tannins. Fruits have been reported to contain carisol, β-caryophyllene, carissone, carissic acid, carindone, carinol, ascorbic acid, lupeol, and β-sitosterol. The fruits are traditionally used for medicinal treatments of malaria, epilepsy, nerve disorder, relieve of pain and headache, fever, blood purifier, itches and leprosy. These chemicals are very effective in the treatment of scabies, intestinal worms, pruritus, biliousness and also used as antiscorbutic, anthelmintic.

Ker [*Capparis decidua* (Forsk.) Edgew.]

It is a native of India a prominent fruit plant found wild in the desert. It belongs to the family Capparidaceae. Almost all the plant parts have economical and medicinal value. The tender branches and leaves are used as a plaster on fractures and swellings. The foliage and stem are chewed to relieve toothache. The root and root bark are pungent and bitter and are used to treat intermittent fevers and rheumatism. The stem bark is laxative, diaphoretic and anthelmintic and is useful in the treatment of cough, asthma and bronchial inflammation. Immature flower buds, flowers and unripe green, immature acrid fruits are used as vegetable. Unripe fruits contain 14.9 g protein, 7.4 g fat, 59.4 g carbohydrate, 12.3 g fibre, 90 mg calcium, 779 mg phosphorus, 3.5 mg iron, 9000 IU vitamin A and 120.7 mg vitamin C in 100g edible matter. These contents are 17.0 g, 5.0g, 71.0 g, 1.0g, 210 mg, 360 mg, 6.0 mg 35000 IU and 119 mg respectively in ripe fruits. It has numerous medicinal properties as its different plant parts have pharmacological activities like hyper- cholesterolemic, anti-inflamatory and analgesic, anti-diabetic, anti-microbial, anti-plaquer, anti-hypertensive, anti-helmintic and purgative activities.

Pilu (*Salvadora oleoides* Decne.)

It is an underexploited multipurpose tree species. Its bushy trees are tolerant to drought and salinity grows successfully in the ravines and degraded lands. Ripe *pilu* fruits are eaten fresh. The fruits contain 6.0 g protein, 2.0 g fat, 76.0 g carbohydrate, 2.0 g fibre, 630 mg Ca, 167 mg P, 8.0 mg Fe and 2.0 mg ascorbic acid in 100 g pulp and the seeds contain 45-46 per cent fat, 18.9 per cent albuminoides, 28.5 per cent carbohydrates, 5.8 per cent fibre and 3.5 per cent ash. Many chemical constituents such as carbohydrates, alkaloids, steroids, glycosides, saponins, tannins, triterpenes, mucilage, fats and oils have been reported from its leaves and stems extracts. Seeds contain 40-45 per cent non-edible greenish-yellow fat (oil), which is a potential industrial substitute for coconut oil. The fat is rich in lamic and myristic acids and is used in medicines. Because of the presence of these active chemical constituents it possesses anti- inflammatory, analgesic, anti-ulcer, anthelmintic, antibacterial, antifungal and diuretic activities.

Cactus Pear [*Opuntia ficus-indica* (L.) Mill.]

Commonly known as prickly pear or Opuntia pear, belongs to the family Cactaceae. Being a succulent xerophytic plant, it is ideally suited to arid and semi-arid regions. Cactus pear is cultivated in several countries for fruits, vegetable, and fodder. The plant has thick succulent spiny or spineless pads called 'cladodes'. The fruit is known as 'tuna'. These are used fresh and for processing as well. The tender cladodes are used as green vegetable and salad. It is a traditional vegetable in Mexico. It also has medicinal value for the treatment of diabetes and is also used as cattle feed during drought periods. *Opuntia* species have been drawn attention of international scientific community and FAO has promoted its cultivation. It cladodes have been used for the treatment for gastritis, hyperglycaemia, hypercholesterolaemic, arteriosclerosis, diabetes, prostatic hypertrophy and they also have an hypolipidemic action and immune-regulation function in gastro-intestinal tract.

Tropical and Sub-tropical Fruits

Custard Apple or *Sitaphal* (*Annona squamosa* L.)

It is a sweet and delicious dessert fruit. It belongs to the family Annonaceae and is native of South America. It has adapted so well in India that a considerable variability is found in Aravali hills in north India and in Telangana and Andhra Pradesh. Some related species such as *A. reticulata* (Ramphal), *A. cherimola* (Laxmanphal), *A. atemoya* (Hanumanphal), *A. glabra* (pond apple) are also important. Custard apple fruits are rich in carbohydrates (23.9 per cent), protein (1.6 per cent), calcium (0.02 per cent), iron (1.0 mg/100 g) and mineral matter (0.7 per cent). The leaves of the plants have been used as insecticide, anti-helmintic and in the healing of bleeding wounds. Unripe and dried fruits are used as anti-dysenteric. Bark act as powerful astringent, anti- dysenteric and vermifuge. Root, bark, leaves and stems are sources of many isoquinoline alkaloids. Annonaceous acetogenins isolated from the seeds were found to possess potent cytotoxicity against different cancer cell lines. The crude extracts as well as isolated compounds from different plant parts exhibited pharmacological activities such as antidiabetic, antimicrobial, antidepressant, anxiolytic, antiviral, antitumor and cytoprotective.

Chironji (*Buchanania lanzan* Spreng.)

It is a tropical evergreen, underutilized nut fruit and is native to India. It is a member of the Anacardiaceae family. *Chironji* plays an important role in lives of tribal people in southern and northern tropical dry deciduous forests in both medicinal and as income generating with fruit kernels for better livelihoods. Fruit kernels are eaten raw or roasted and also used in making dessert, which is rich source of protein, fat, starch, vitamins and minerals and yields sweet oil, which can be used as substitute to olive and almond oil. The tree provides food, fuel, fodder, timber, lac and has medicinal properties. It contains 3.0 g moisture, 19.0 g protein, 59.1 g fat, 3.8 g fibre, 12.1 g carbohydrates, 279 mg Ca, 528.0 mg P, 8.5 mg Fe, 0.69 mg thiamine, 0.53 mg riboflavin, 1.5 mg niacin and 5.0 mg vit. C per 100 g of kernel and provides 656 calories. The polyphenols have been proven for its disease fighting power and possess good antioxidant activities. The fatty acids, polyphenols, phytosterols and stigmasterol in seed extract has both medicinal and nutritional value.

Fig (*Ficus carica* L.)

It belongs to the family Moraceae and is closely related with species, *F. glomerata* (gular) is common in deciduous forests in the subtropical region. It is a native of India and is grown in arid and semi-arid tracts since antiquity. Fruits are valued for their laxative and medicinal properties. The fruit contains 11.5 g sugar, 1.5 mg iron, 270 IU vit. A per 100 g pulp. Fig fruits preserved in vinegar are used to cure liver and spleen enlargement in children and are effective in the treatment of chronic cough and piles. Phytochemistry of the fruit revealed the presence of total phenolics, flavonoids, alkaloids and saponins and other secondary metabolites that contribute to its high antioxidant activity which was evident from ABTS and FRAP assays. Volatile components of fig fruit were identified through GC-MS and showed the presence of vit. E, β-amyrin, stigmasterol, campesterol, oleic acid, isoamyl laurate and / tocopherols majorly. The extract was also screened for antibacterial activity and showed zone of inhibition against *Proteus mirabilis* and *Bacillus subtilis*.

Pommelo (*Citrus maxima* (Burm.) Merr. or *Citrus grandis* L.)

It is the largest citrus fruit of the family Rutaceae and the ancestor of the grapefruit. It is found growing in all Eastern, north-eastern and southern states. The fruit is large, 15–25 cm in diameter, usually weighing 1–2 kg. It has a thicker rind than a grapefruit, and is divided into 11 to 18 segments. Aril juice has good TSS (8- 12 per cent), low acidity (0.2-0.8 per cent), vit. C (41-60 mg/100 g FW), total anthocyanins (0.011- 0.030 μmol ml-1) with high fibre content. Pummelo contains higher antioxidants thus beneficial for health. It contains several vitamins, minerals, and antioxidants. Pomelo is also rich in potassium, which helps regulate fluid balance and blood pressure. It helps in promote weight loss. Pommelos contain high levels of antioxidants, including vitamin C, naringenin, naringin, and lycopene, which may several health benefits. It has anti-aging and heart- healthy properties, and boosts heart health by reducing levels of cholesterol and triglycerides. Thus, reduce these blood fats by preventing the cholesterol. Extract from pomelo peels and leaves have been shown to kill cancer cells and prevent the spread of cancer.

Pomegranate (*Punica granatum* L.)

It belongs to the family Punicaceae. It is a delicious table fruit rich in vitamins (0.06 mg thiamine, 0.1 mg riboflavin, 0.3 mg niacin and 16.0 mg vitamin C per 100 g pulp), minerals (10.0 mg calcium, 70 mg phosphorus and 0.3 mg iron per 100 g pulp), carbohydrate (14.5 per cent) and protein (1.6 per cent). The fruit is used as a special diet for sick and aged persons. Several *in vitro* and *in vivo* studies have revealed its beneficial physiological activities, especially its antioxidative, antimicrobial and anti-inflammatory properties. The health-promoting properties of the fruit are considered to mainly be due to the presence of punicalagin and, to a lesser extent, to other metabolites, such as flavonols and anthocyanins. Several studies have brought up the potential contribution of pomegranate in the treatment of cancer, diabetes and heart disease.

Tamarind (*Tamarindus indica* L.)

It is an important fruit crop of semi-arid, tropical and subtropical regions experiencing extended spell of dry weather. It is a frost susceptible but drought hardy tree species of the family Fabaceae and can be grown in degraded and problematic soils but performs well in deep, well drained, red loam soils. Tamarind pulp is a good source of calcium, tartaric acid (8-18 per cent), malic acid, protein (3.1 per cent), invert sugar (30-40 per cent), vitamins and other minerals. Young leaves, flowers, seed kernels and bark of the tamarind tree are also used for various purposes. It has various pharmacological activities like hypolipidemic, weight reducing, antimicrobial, hepatoprotective, anthelmintic, antioxidant, analgesic and anti-inflammatory *etc.* Tamarind fruit pulp contain proteins, carbohydrates and minerals especially Na, Zn and Fe and can be eaten as fresh or processed into different food products and also the principal culinary souring agent for dishes, squashes, chutney, beverages, *etc.* The fruit has medicinal value and treated as a number of diseases. Antioxidant activity present in tamarind indicate that fruits contain biologically important mineral elements and have high antioxidant capacity associated with high phenolic content appetizer, stomach ailments, *etc.*

Jamun [*Syzygium cuminii* (L.) Skeels]

It belongs to family Myrtaceae and is indigenous to India and grows well under humid tropical conditions. It is also found growing in Thailand, Philippines, Myanmar, which are probably the secondary centre of its origin. *Jamun* is also cultivated in drier parts of Israel, Algeria, Malagassy, West Indies and South Africa. In deciduous forests of subtropical region, it sheds leaves during the spring season. Its fruits are nutritious and are rich in minerals, fibre, carbohydrates and vitamins A and C. The fruit also contains 1.0 g iron, 0.02 g calcium, 0.01 g phosphorus, 0.7 g protein, 0.1 g fat, 0.9 g fibre and 19.7 g carbohydrates in 100 g pulp. *Jamun* fruit is rich in anthocyanin pigments. Fruits, seeds and bark have medicinal value. The gallic acid and ellagic acid content present in seed play an important in role in conversion of starch into sugar thus minimizing blood glucose level. It is also effective in the treatment of inflammation, ulcers and diarrhea. Fruit pulp contains very high anthocyanin content and can be a good source of natural food colourants for the food processing industries. The pigment is known for their strong antioxidant capacity and health-protecting effects and reduces the risk of various diseases.

Wood Apple [*Feronia limonia* (L.) Swingle]

It belongs to the family Rutaceae. It is a native of India and Sri Lanka and is found throughout the tropical and subtropical plains. It grows wild in drier regions of Myanmar and Indochina. Fruits contain 7.3 per cent protein, 15.5 per cent carbohydrates, 170 mg riboflavin, 2 mg vitamin C per 100 g pulp and minerals, especially Ca and P. The fruits are highly acidic and contain 7.6 per cent acid when unripe and 2.3 per cent when ripe. The fruits also contain 3-5 per cent pectin making it suitable for making good quality jelly. The fruits are also used in various ayurvedic preparations for the treatment of diarrhoea, dysentery, gum and throat problems. The fruits contain phytochemicals (polyphenols, phytosterols, saponins, tannins, coumarins, triterpenoids), vitamins, amino acids *etc.* It has curative value for various diseases of bones and joints, bilious diseases, prevention of capillary bleeding, cold, influenza, piles, dysentery, habitual constipation and scurvy. The fruit pulp is used as a liver and cardiac tonic, diarrhoea and dysentery.

Phalsa [*Grewia subinaequalis* DC.) syn. *Grewia asiatica* L.]

It belongs to the family Tiliaceae, is a native of India. It is widely cultivated in tropical and subtropical parts of India, Pakistan, Bangladesh, Thailand and Philippines. Phalsa is a drought hardy fruit and is grown successfully in the hot and dry plains having distinct summer and winter and is, thus, most suited to the arid parts of the country for the production of quality fruits even under extreme summer conditions. Capable of thriving in the humid tracts of coastal India. Its fruit are rich in carbohydrates (14.7), Ca (12.9 mg/100 g), P (39 mg/100 g), Fe (3.1 mg/100 g), b-carotene (419 µg/100 g) and fibre (1.2 per cent). *Phalsa* fruits are cooling aphrodisiac and tonic and alley thirst and burning sensation, remove biliousness, cure inflammation, heart and blood disorders and fever. The extracts from various parts of *phalsa* plants (especially the fruit) possess strong antioxidant, radioprotective, antimicrobial, antidiabetic, anti-inflammatory, anticancer, and cardio-protective properties.

Minor Temperate Fruits

Pistachio (*Pistacia vera* L.)

It belongs to family Anacardiaceae. The earliest attempt to cultivate pistachio nut in India was made as early as 1985 at the Regional Horticultural Research Station, Sharbo-Kinnaur of

H.P. In India, it is mainly grown in Western Himalayan states like J&K, Himachal Pradesh, Uttarakhand and have potential to grow even in Eastern Himalayas (Arunachal Pradesh). It is a nutrient-dense nut with a heart-healthy fatty-acid profile as well as protein, dietary fibre, potassium, magnesium, vitamin K, γ-tocopherol and a number of phytochemicals. The pistachio's unique green and purple kernel colour is a result of its lutein and anthocyanin contents. Among nuts, pistachios contain the highest levels of potassium, γ-tocopherol, vit. K, phytosterols and xanthophyll carotenoids. It is mainly used as roasted and salted for snacks purpose and rest in the form of flavouring agent in ice-cream, confectionary and non-vegetarian dishes. Exploratory clinical studies suggest that pistachios help maintain healthy antioxidant and anti-inflammatory activity, glycemic control, and endothelial function. Pistachios help in controlling body weight because of their satiety and satiation effects and their reduced net metabolizable energy content.

Chilgoza (*Pinus gerardiana* Wallichex. D. Don.)

It is commonly known as the chilgoza pine. It is a pine native to the north-western Himalayas in eastern Afghanistan, Pakistan, and northwest India, growing at elevations between 1800 and 3350 m. In India, it grows naturally in the inner Himalayan zone comprising Kinnaur and Pangi in H.P and Kishtwar and Astur areas of J&K. The Chilgoza pine is well known for its edible pine nuts, rich in carbohydrates and proteins. The nut is eaten as raw, roasted or used in confectionary. It contains 51.3 per cent fat, 8.7 per cent water, 13.6 per cent protein, 22.5 per cent starch, 0.9 per cent fibre and 3 per cent mineral and ash; it possesses a tremendous wound healing property. The plant shows various biological activities such as antioxidant activity, antithrombotic and anti- platelet activity, antidiabetic activity, anti-inflammatory activity, antifungal activity, antibacterial activity *etc.*

Seabuckthorn [*Elaeagnus rhamnoides* (L.) A. Nelson]

It belongs to family Elaeagnaceae, originated from Eurasia. Commonly, it is also known as seaberry, Leh berry, sand thorn, sallow thorn, wonders berry and Ladakh gold, is one of the most nutritious fruit. The leaf extracts are rich in flavonoids, tannins, and triterpenes like isorhamnetin-3-O-glucoside, kaempferol-3-O-β-D-(6-O- coumaryl) glycoside, 1-feruloyl-β-D-glucopyranoside *etc.* Ripe berries are rich in vitamins (A, C, E, and K), organic acids, carotenoids and flavonoids. Sea buckthorn seed is also a good source of vit. C, vit. E, carotenoids, flavonoids, fatty acids, organic acids, triacylglycerol, phytosterols, pro-anthocyanidins, phenolic compounds *etc.* Oil extracted from berries is not only rich in monounsaturated fatty acids (MUFA) and tocopherols but also in various vitamins including E and K, carotenoids, β-70 sitosterol and other compounds. Its seed and root extracts have higher antioxidant activities. Gallic acid was observed as the most important antioxidant in sea buckthorn. Sea buckthorn leaves can be used for the treatment of gastrointestinal disorders, diarrhoea and dermatologic diseases. Leaf tea has proven to play a role in reducing the body weight through lowering the cholesterol, also in controlling blood pressure and blood sugar and platelet aggregation. In Indian and Tibetan medicine, fruit pulp extract, pulp and seed oil have been used to treat a number of diseases including skin disorders, gastric ulcers, coronary heart disease, platelet aggregation *etc.* Apart from this, the seed and fruit oils are also used for curing of lupus erythematosus, inflammatory diseases, trachoma, eczema, chronic wounds, burns *etc.*

Spondias (*Spondias mombin* L.)

Known as hog plum (*S. pinnata*) and great hog plum (*Spondias dulcis*) is an underexploited members of the family Anacardiaceae, are growing widely in India for its edible fruits. These deciduous species are tolerant to drought and growing abundantly as fence side trees in West Bengal. Fruits can be eaten fresh, cooked or made into value-added products. Unripe fruits can be made into jelly, pickle, chutney or relishes or used for flavouring sauces, soups and stews. In some places young leaves are eaten raw or steamed as vegetable for eating with salted fish and rice. Fruits are reported to be rich in vit. A, vit. C and Fe contents. The fruits are used to treat heart ailments and urinary troubles and are used on wounds, sores and burns and for digestion. The bark is considered useful in dysentery and diarrhea and is also given to prevent vomiting. The root is considered useful in menstruation. The plant is reported to have anti-tubercular properties. So, there is great scope of increasing socio-economic status of rural and tribal people by exploiting these two species, *viz.* (hog plum) and *Spondias dulcis* (great hog plum).

Minor Exotic Fruits

Rambutan (*Nephelium lappaceum* L.)

It is a native of Malaysian archipelago and belongs to the family Sapindaceae. The attractive colour and hairy fruits attracted the visitors enthusiastic farmers who introduced in Kerala and coastal Karnataka about half a century ago. Chemical composition of the rambutan fruit on the basis of 100 g pulp: water content (83 g), caloric value (63 cal), proteins (0.8 g), carbohydrates (14.5 g), calcium (25 mg), vit. C (20–45 mg) and iron (3 mg). The pulp has antioxidant and anti-inflammatory activities. However, rambutan seed extract has a high content of α-glucosidase inhibitory activity, with an IC_{50} value of 9.92 μg/mL relevant to low hypoglycemic activity (antidiabetic). Apart from the above, its anti-inflammatory, antiproliferative, antibacterial and antioxidant activities are also well proven.

Longan (*Dimocarpus longan*)

It is also another member of Sapindaceae family. Longan is originated in the mountainous region from Myanmar to Southern China. Fruits are fleshy, very sweet, translucent, white aril which surrounds a red brown to black seed from which it separates easily. Fruit can be eaten fresh, frozen, canned or dried. There are some reports that it is also found wild form in western Ghats region. Longan contains several vitamins and minerals, including iron, magnesium, phosphorus and potassium and large amounts of vitamins A and C. Its pulp has abundant nutritional phytochemicals such as protein, carbohydrates, polysaccharides, polyphenols, *etc.* which show multiple biological activities including antioxidant, immunomodulatory and antitumor effects. Pulp pericarp also demonstrates biological activities because of its rich content of polysaccharides and polyphenols. Longan seed extract is known for antioxidative, antiproliferative, anti-inflammatory, hypoglycemic and hypouremic effects. It has been reported to contain gallic acid, corilagin (an ellagitannin), ellagic acid, soyacerebrosides I and II, 1-O-β-D-glucopyranosyl-(2S,3R,4E,8E)- 2-(2/-lignoceroyl amino)-4, 8-octadecadiene-1,3-diol(longan cerebroside I) and its 8 Zisomer (longan cerebroside II), momor-cerebroside I and phytolacca cerebroside. The most important benefits of longan are skin care, antiaging, boosts libido, flights anxiety, treats insomnia, it may use for blood tonic, promotes weight loss, increase energy, control blood pressure, strengthens immunity, it may use for neuroprotection, speeds-up healing, prevents chronic diseases, aids in digestion, improves memory, it may use for vision health, treats snake bites, dental care *etc.*

Avocado (*Persia americana* Mill.)

It is a native of Central America from where it reached different parts of the world during the sixteenth century. The nutritional composition of avocado pulp has been reported to have a moisture content ranging from 67 to 78 per cent, lipid content ranging from 12 to 24 per cent, carbohydrate content ranging from 0.8 to 4.8 per cent, protein content ranging from 1.0 to 3.0 per cent, ash content ranging from 0.8 to 1.5 per cent, fibre content ranging from 1.4 to 3.0 per cent, and energy between 140 and 228 kcal per g pulp. It is rich in monounsaturated fatty acids, which are effective in reducing the blood levels of undesirable low-density lipoprotein (LDL) and increasing the levels of the beneficial high-density lipoprotein (HDL). Compared to other vegetable oils, avocado oil is known to contain high levels of monounsaturated fatty acids (oleic and palmitoleic acids), low quantities of polyunsaturated fatty acids (linoleic acid) and a significant quantity of saturated fatty acids (palmitic and stearic acids). Another relevant advantage of avocado is the presence of vitamins, such as β-carotene, vitamin E, retinol, ascorbic acid, thiamine, riboflavin, niacin, pyridoxine and folic acid, which are of great importance for overall health and wellbeing. Larvicidal, antifungal, antimicrobial, antioxidant, antiprotozoal, antidiabetic, antihypertensive, hypocholesterolemic, and antimycobacterial activities as well as inhibition of lipid and protein oxidation are the numerous biological activities reported to be related to avocado fruit. Studies have also reported anticancer activity *via* induction of apoptosis of MDA- MB-231 cells by methanol extracts of avocado seeds. Avocatin B, a lipid derived from avocado fruit, is a novel compound with cytotoxic activity in acute myeloid leukemia and has been shown to inhibit fatty acid oxidation and decrease NADPH levels, resulting in ROS-dependent leukemia cell death.

Malayan Apple or Red Jambo (*Syzygium malaccense*)

It is native to Malaysia, Indonesia, Vietnam and Thailand. Malayan apple first came to the shores of Goa by way of the Portuguese and is grown throughout the humid tropical regions of Western Ghats. All portions of *S. malaccense* exhibit significant concentrations of bioactive compounds with biological activity proven to promote human health. The fruit is rich in minerals (Ca, P. and Fe), iron) and vitamins (A and B). The high fiber content allows the fruit to be classified as a food source of these components; this gives it great functional potential, considering the important role of fiber intake, both in aiding digestion and in combating health problems that have emerged among the population, including weight gain and several chronic diseases, such as cardiovascular disease, Type 2 diabetes and some types of cancer. Among the phenolic, anthocyanins are the main pigments responsible for most of the colors of fruits and exhibit anti-inflammatory, antimutagenic and cancer chemo-preventative activities *in vivo.*

Wax Apple [*Syzygium samarangense* (Blume) Merr. and L.M. Perry]

It is a non-climacteric tropical fruit with its origin in the Malay Archipelago. This fruit was introduced in India mainly in the Western Ghats regions of Karnataka, Kerla and Tamil Nadu. The fruit pulp is a rich source of phenolics, flavonoids and several antioxidant compounds. The extracts of wax fruits, flower and bark have potent free radical scavenging, antimutation and anticancer activities. The leaves of wax apple used as tea and is proposed as a possible supplement for type II diabetes patients. Wax apple studied for its numerous pharmacological properties such as antioxidant and antidiabetic properties, anti-inflammation and antinociceptive activity, wound healing activity, antiulcerogenic effect, antibacterial, anticancer and also, it's potential as an uterotonic agent. The fruit can be used to treat high blood pressure and several inflammatory conditions, including sore throat, and can also be used as an antimicrobial, antiscorbutic, carminative, diuretic, and astringent.

Rose Apple [*Syzygium jambos* (L.) Alston]

It is a native of Malay Archipelago, distributed in south India especially in the Western Ghats. The fruits pall yellow colour. Rose apple has a long history of being used in traditional and folk medicine in various cultures. The fruit and root bark are believed to be of use as a blood coolant. The fruit has been used as a diuretic and as a tonic for better health of the brain and liver. With regard to rose apple, reports showed that the ethanol extract was effective as a free radical scavenger in the 1,1-diphenyl- 2-picrylhydrazyl free radical scavenging assay, reducing total phenol capacity, total flavonoid content, and total antioxidant capacity. It health benefits could be antimicrobial, anti-inflammatory, analgesic, antiviral, anti-dermatophytic, anticancer, and hepatoprotective effects.

Durian [*Durio zibethinus* (L.) Murray]

It is a native to Southeast Asia. The unique durian flavour is attributable to the presence of fat, sugar, and volatile compounds such as esters and sulphur-containing compounds such as thioacetals, thioesters, and thiolanes, as well as alcohols. A significant amount of fiber (7.5–9.1 g/100 g DM), carbohydrate (62.9–70.7 g/100 g DM), and sugar (47.9–56.4 g/100 g DM) are found in different varieties. Durian is also rich in polyphenols such as flavonoids (flavanones, flavonols, flavones, flavanols, anthocyanins), phenolic acids (cinnamic acid and hydroxybenzoic acid), tannins, and other bioactive components such as carotenoids and ascorbic acid. Current epidemiological studies have suggested that polyphenols decrease the risk of chronic diseases (*e.g.*, cardiovascular diseases, cancers and diabetes).

Star Fruit or Carambola (*Averrhoa carambola* L.)

It is a native of South Asia and is grown in all parts of India except Arid and temperate regions. There are two distinct classes of star fruit, the smaller, very sour type, richly flavored and the larger, sweet type, mild-flavored, with less acid. The star-fruit is a good source of various minerals and vitamins. Star-fruits are also a rich source of natural antioxidants such as L-ascorbic acid (Vitamin C) and Gallic acid, which aid in scavenging reactive oxidative species (ROS). Star-fruits are a good source of Mg, K, P, as well as β-carotene, and vit. C, which are common antioxidants. The presence of antioxidants like Fe, Zn and Mn in the fruits aid in strengthening the immune system. In addition, the presence of high amounts of fibres in fruits aids in absorbing glucose and retarding the glucose diffusion into the blood stream; thus helps in controlling blood glucose concentration. Its consumption exhibits hypo-cholesterolemic and hypolipidaemic effect as it enhances the removal of cholesterol, lipid, and bile acid through the excrement. It is widely used in Ayurvedic and traditional Chinese Medicine preparations as remedy for fever, sore throat, cough, asthma, chronic headache, and skin inflammations. The phytochemical and pharmacological studies suggest that the extracts of Star- fruit plant leaves, fruits and roots contains saponins, flavonoids, alkaloids and tannins which are known to confer antioxidant and specific healing properties.

Mangosteen (*Garcinia mangostana* L.)

It is an ever-green tree of family Clusiaceae and a native of south Eastern Asian countries. The fruit was introduced more than a century ago in India, but currently it is successfully grown only in selected places on slopes of Nilgiris (Tamil Nadu), Malabar and Kanyakumari (Kerala). The natural molecules α-mangostin, xanthones and other bioactive substances are believed to be responsible for its medicinal activities. It is famous for its anti-inflammatory properties and is used in the treatment of skin infections and wounds. Other applications include the therapy of various conditions such as dysentery, different urinary disorders, cystitis and gonorrhoea. Products containing its fruits are now sold widely as 'liquid botanical supplements'. The biological actions of mangosteen are anti-inflammatory, anti-carcinogenic, antioxidant, anti-allergic, antibacterial, antifungal, antiviral, cardioprotective, anti-depression, and anti-deterioration. It is very effective in treating conditions such as obesity, Alzheimer disease, diarrhea, urinary tract infections, gonorrhea, thrush, tuberculosis, and osteoarthritis.

Passion Fruit (*Passiflora edulis* Sims.)

It belongs to the family Passifloraceae. The edible commercial species of passion fruit originated in the South American rain forests in the Amazon region of Brazil. Passion fruit has two distinct forms, the standard yellow (*Passiflora edulis* f. *flavicarpa* Deg.) and the purple (*Passiflora edulis* f. *edulis*). The yellow is more acidic and less starchy, while the purple is less acidic and starchier. Purple passion fruit (*P. edulis*) and yellow passion fruit (*P. edulis* var. *flavicarpa*) are of commercial importance. Phytochemical analysis of *P. edulis* revealed the presence of carbohydrates, glycosides, flavonoids, resins, alkaloids and phenolic compounds. Tannins were present in the leaf and fruit, saponins were present in the leaf and stem. Organic extract (methanol, ethanol) of *P. edulis* leaves were reported to possess tannins, flavonoids, terpenoid, steroida and saponins. It has anti-inflammatory, anticonvulsant, antimicrobial, anticancer, anti-diabetic, antihypertensive, anti- sedative, antioxidant properties and various remedial measures for treating conditions like osteoarthritis, asthma and act as colon cleanser. Besides, also used for treatment of ulcers, haemorrhoids, as sedatives, remedy for insomnia, digestive stimulant and remedy for gastri carcinoma.

Mulberry (*Morus alba* L.)

It has a long history of use as an edible fruit and traditional medicine. A great diversity of nutritive compounds such as fatty acids, amino acids, vitamins, minerals, and bioactive compounds, including anthocyanins, rutin, quercetin, chlorogenic acid, and polysaccharides have been found in mulberry fruit depending on the cultivars and maturity stages. The fruit extracts have demonstrated numerous biological activities, including antioxidant, neuroprotective, antiatherosclerosis, immune modulative, antitumor, antihyperglycemic, and hypolipidemic activities.

Kiwifruit (*Actinidia chinensis* (Planch.)

It is member of Family Actinidiaceae) and is derived from a deciduous woody, fruiting vine. It is composed of different species and cultivars that exhibit a variety of characteristics and sensory attributes. In India Kiwi is being cultivated in the sub-Himalayan tracts of Himachal Pradesh, in North Eastern states including Arunachal Pradesh *etc.* Kiwifruit is contains vit. C, actinidin, fiber, vit. E, and for selected cultivars, the persistence of chlorophyll in the mature fruit. Kiwifruit are widely known to contain high concentrations of phytochemicals such as carotenoids, polyphenolics, and fiber that may provide health benefits beyond basic nutrition. A limited number of human intervention trials validate some of these health benefits for kiwifruit in the areas of "natural protection" (protection from oxidative stress and DNA damage associated with mutation and cancer), gut health (laxation and healthy bowel habits), recovery from symptoms of cold and flu, and cardiovascular health (reduction in platelet aggregation). The fruit has low glycemic index which makes it suitable for the individuals with diabetes, while being fibre rich maintains blood sugar levels under control.

Kokum (*Garcinia indica* Choisy)

One of the native underexploited tree spices found in Western Ghats of India and some parts of north-eastern India. In spite of its incredible medicinal and nutritive properties, it is generally not cultivated systematically on orchard scale. It is commonly known as *kokum* (Hindi), kokum butter tree, Goa butter tree and mangosteen oil tree. Ripe fruits are harvested during April-May. Fruit rind is widely used in refreshing drinks and curries. The fruit is anti-helminthic, appetizer, cardio-tonic, useful against piles and dysentery. Hydroxyl citric acid (HCA) extracted from kokum is used against obesity, which is available in the form of tablets. The fruit is rich in anti-oxidants that bind with free radicals and prevent oxidative damage to body cells. The anthocyanin pigments obtained from it are used as natural colouring agents for food preservation. Advanced processed products like HCA, garcinol, wine and purified pigments would create more domestic and international demand for *kokum*. Chemical studies have shown that the rind contains protein, tannin, pectin, sugars, fat, organic acids like hydroxycitric acid, hydroxycitric acid lactone and citric acid; the anthocyanins, cyanidin-3-glucoside and cyanidin-3-sambubioside; and the polyisoprenylated phenolics garcinol and isogarcinol. Preclinical studies have shown that *kokum* possess antibacterial, antifungal, anti-ulcerogenic, cardioprotective, anticancer, chemo-preventive, free radical scavenging, antioxidant and anti-obesity effects.

Other Underutilized Fruit crops

Chalta (*Dillenia indica* Linn., Family: Dilleniaceae)

It is an evergreen large shrub or small to medium-sized tree that grows all over the NE states, especially Mizoram and Tripura. Named after Johann Jacob Dillenius, a German botanist, *Dillenia* is a genus of about 100 species of flowering plants growing in tropical and subtropical regions. However, up to now, only eight *Dillenia* species have been reported to be used traditionally for different medical purposes including treatment of cancerous growth. Medicinal properties of the fruit juices of *D. indica* are manifold, as a cooling beverage to treat fever and to relieve fatigue. Although mainly a source of food for elephants and monkeys, the ripped fruits are taken orally to increase appetite and overcome weakness, as laxative and medication for abdominal pain. Native communities used the fruits as a remedy for jaundice. All the plant parts *viz*., leaves, bark, fruit, fruit peel, *etc*. all have proven medicinal properties against most of the common ailments including anti-cancer and anti- microbial properties. Some of the major compounds with medicinal and nutritive value are lupeol, betulinaldehyde, betulinic acid and stigmasterol.

Indian Olive [*Elaeagnus rhamnoides* (L.) A. Nelson]

It is one of the minor crops grown in *terai* region of West Bengal. Fruits are the good source of vitamins and minerals. Fruit had the total soluble solids of 9.98 °Brix, ascorbic acid of 50 mg/100 g pulp, food energy value of 97.23 kcal/g, Zn of 46.31 µg/g DM, Mn of 53.86 µg/g DM, and Cu of 7.82 µg/g DM.

Kendu (*Diospyros melanoxylon* Roxb.)

It is an underexploited fruit species, is grown as natural wild in the forests and marginal lands of West Bengal, Odisha, Madhya Pradesh, Bihar, Jharkhand, Chhattisgarh and Andhra Pradesh. Fruits are a good source of carbohydrates, Ca, P and carotene and can prevent malnutrition. Dried powdered fruit is used as carminative and dried flowers are reported to be useful in urinary, skin and blood diseases. Seeds are prescribed for curing mental disorders, palpitation of heart and nervous breakdown.

Wild Edible Fruits of North-East India

The northeastern region of India, contains more than one third of the country's total fruit diversity. In addition to commercial fruits, the region is also known for its rich genetic resources of underutilized and underexploited fruits. These fruits are neither grown commercially on large scale nor traded widely. They are scattered only in wild forest or semi-wild in homestead gardens and remained uncared for albeit with large variability and are traded and consumed locally. Considerable diversity exists among these species in plant type, morphological and physiological characteristics, adaptability and distribution. They can be grown under stress and adverse conditions and are also known for their medicinal, therapeutic and nutritive values. These fruits, from time immemorial are known to play an important role in food and nutritional supplement of rural community. Most of them are very rich sources of vitamins, minerals along with carbohydrates, proteins and fats. This region harbors a good number of underutilized and under exploited species, *viz.*, *Aegle marmelos*, *Artocarpus heterophyllus*, *Artocarpus lackoocha*, *Averrhoea carambola*, *Citrus medica*, *C. macroptera*, *C. grandis*, *Dillenia indica*, *Elaegnus latifolia*, *Emblica officinalis*, *Garcinia cowa*, *Passiflora edulis*, *Phylanthus acidus*, *Spondias pinnata*, *Syzigium jambos*, *Tamarindus indica*, *Terminalia chebula* and many more. The Nutritional and health benefits of some of these fruits have been discussed under the previous heads and sub- heads.

Sohiong [*Prunus nepalensis* Ser. (Steud)]

Its belongs to the family Rosaceae, is an important underutilized fruit of the tribal population in Meghalaya. The study was conducted to investigate the physical and biochemical contents of two different genotypes of *Sohiong* fruits. Result indicated that the big fruit size type had higher values for various physic-chemical characters except for pulp recovery (74.71 per cent), pulp: stone ratio (2.95) and dry recovery (34.72 per cent), TSS (20.15 per cent), β-carotene (2.76 mg/100 g), anthocyanin (358.86 mg/100 g) and fibre (2.5 per cent). Similarly, the mineral contents of big-sized fruits were comparatively higher than the small fruit type except for potassium and ash content (1.51 per cent). The study indicated that the availability of *sohiong* fruit ranging from August to November. In addition, fruits are rich in vitamin, minerals and bio-chemical properties.

Conclusion

India is bestowed with the three major climatic conditions *i.e.* tropical, subtropical and temperate. Because of this, a majority chunk of the minor and underutilized fruit crops are being grown in many pockets across the nation. This makes the abundance availability of these fruits in different seasons throughout the year. Apart, India is the home for various minor and underutilized fruit crops like *aonla, bael, ber, lasora, karonda. ker, chironji jamum,* wood apple, *phalsa, kokum etc.* There is large variability of these fruits in India especially in the Western Ghats, eastern and western sub-Himalayan tracts including the north eastern states. This is a need to identity, characterize and conserve the new improved genotypes for specific characters including high yield, biotic/abiotic stress tolerance, higher bioactive compounds *etc.* Many of these minor and underutilized fruits although not originated but introduced to India in the recent past, *viz.,* rambutan, longan, avocado, star fruit, passion fruit, kiwi; however, are gaining the momentum in Indian fruit market due to increasing popularity among the consumers owing to multifaceted use. A large number of these fruits can grow under adverse conditions and are also known for their therapeutic and nutritive value and can satisfy demands of the health-conscious consumers. Hence, there is a need to concentrate on research efforts in diversification and popularization of such underutilized fruit crops. This in some extent can be achieved through developing suitable processing and value additives to prolong their utilizable life span. Worldwide researches have shown shreds of evidence of underutilized wild edible plants possessing great valuable nutritional value. Tropical fruits, which are at present underutilized, have an important role to play in satisfying the demand for nutritious, delicately flavored and attractive natural foods of high therapeutic value. These crops can even cure insomnia, scurvy, constipation hemorrhage, leucorrhea, anemia, stomach ache, and can be used as a cooling agent to reduce sun strokes ill effects. Underutilized fruit crops can even adapt to low input agriculture. More research and development efforts in these crops will certainly add substantially to nutritional and food security. Development of better coordination among all the agencies involved in research, development, and promotion will help popularize these fruit crops. There is an urgent need to identify/develop high yielding genotypes, standardization of production and plant protection technologies, and post-harvest management practices for these fruit crops.

Chapter 2

Underexploited Arid Fruits and Vegetables and their Role in Human Nutrition

P.L. Saroj*, R. Kumar and B.R. Choudhary

ICAR–Central Institute for Arid Horticulture, Beechwal, Bikaner 334 006, Rajashtan

**e-mail: plsaroj@yahoo.co.in*

Current Address: ICAR–CISH, Lucknow, Uttar Pradesh

ABSTRACT

Fruits and vegetables play important role in nutritional security by providing nutrients, vitamins and minerals through balanced human diet. This impressive growth in fruits and vegetable production has increased the availability of nutritive food and this was possible due to development of improved varieties and production technologies coupled with large scale adoption by the farmers. However, this remarkable production was contributed by only few major fruit and vegetable crops. The fruit and vegetable crops which are not grown commercially on large scale are termed as underexploited crops. The possible reasons for the low utilization of underexploited fruits and vegetables, in spite of their well known significance are owing to lack of availability of planting material, lack of awareness on nutritional and medicinal aspects and lack of information on production technologies of these crops. Underexploited arid fruit crops such as bael, phalsa, wood apple, fig, tamarind, mulberry, lasoda, kair and in arid vegetables such as kacri, mateera, bitter apple, arya, round melon, spine gourd, jhar karela, cluster bean, moth bean, phog, khimp, ivy gourd, Indian aloe and khejri are rich sources of energy and nutrients particularly micronutrients and vitamins. The micronutrients present are minerals (like iron and calcium) and vitamins (like vitamin B complex, C, folic acid, and carotenoids) while, the macronutrients present are complex carbohydrates/fiber. They contain plentiful amounts of iron, calcium, vitamin C, folic acid, carotenoids (vitamin A precursors) and phytochemicals. In this perspective, there is an imperative need to take up programme on genetic resources exploration, conservation, utilization and improvement of underexploited fruit and vegetable crops to provide nutritional security. The agro-climatic conditions of India are favourable for the production of various underexploited fruits and vegetables. Therefore, the underexploited fruits and vegetables production will meet the shortage of per capita availability of fruits and vegetables and also provide economic security to the rural populations.

Keywords: *Fruits, Vegetables, Nutritional security, Phytochemicals and Sustainable production.*

Underexploited arid fruit crops such as bael (*Aegle marmelos*), phalsa (*Grewia subinaequalis*), wood apple (*Feronia limonia*), karonda (*Carissa carandus* L.), fig (*Ficus carica* L.), tamarind (*Tamarindus indica*), mulberry (*Morus sp.*), kair (*Capparis deciduas*), pilu (*Salvadora oleoides*), lasoda (*Cordia myxa*) *etc.* and in arid and semi-arid vegetables such as kachri (*Cucumis callosus*), snap melon (*Cucumis melo var. momordica*), round melon (*Citrullus vulgaris* var. *fistulolus*), arya (*Cucumis melo* var. *chate*), jhar karela (*Momordica balsamina*), bitter apple (*Citrullus colocynthis*), cluster bean (*Cyamopsis tetragonoloba*), phog (*Calligonum polygonoides*), khimp (*Leptadenia pyrotechnica*), khejri (*Prosopis cineraria*) *etc.* are rich sources of energy and nutrients particularly micronutrients (like iron and calcium) and vitamins (like vitamin B, C, folic acid, and carotenoids) along with phytochemicals (anthocyanins, carotenoids, phenols and flavonoids) and dietary fibers. Some horticultural crops such as water melon and cucumber provide very low calories, whereas some others such as date palm are rich in starch which provides energy in good amount. Therefore, vegetables and fruits can be used to increase or decrease calories in our diet for malnutrition alleviation and obesity management. Fruits and vegetables can be categorized in two groups on basis of their commercial cultivation and utilization (Table 1).

Table 1. Fruits and vegetables of arid regions

Common Name	*Scientific Name*	*Family*	*Chromosome No. (2n)*	*Origin*
Commercial fruit and vegetable crops				
Date palm	Phoenix dactylifera	Palmaeae	36	Persian Gulf, Iraq
Pomegranate	Punica granatum L.	Punicaceae	16, 18	Iran
Ber	*Ziziphus mauritiana*	Rhamnaceae	24	Indian Sub-continent
Aonla	*Emblica officinalis*	Euphorbaceae	18, 28	Peninsular India
Kinnow	Hybrid of 'King' (*Citrus nobilis*) × 'Willow Leaf' (*Citrus deliciosa*)	Rutacae	18	-
Sweet orange	*Citrus sinensis*	Rutaceae	18	South East Asia
Guava	*Psidium guajava* L.	Myrtaceae	22, 33	Brazil
Custard apple	*Annona Squamosa* L.	Annonaceae	14	Tropical America - Peru
Tomato	*Solanum lycopersicum* L.	Solanaceae	24	South America
Brinjal	*Solanum melangena*	Solanaceae	24	Southern India
Chilli	*Capsicum annuum* L.	Solanaceae	24, 26	Central and South America
Bottle gourd	*Lagenaria siceraria*	Cucurbitaceae	22	Sub-Saharan Africa

Common Name	*Scientific Name*	*Family*	*Chromosome No. (2n)*	*Origin*
Cucumber	*Cucumis sativus* L.	Cucurbitaceae	14	India
Water melon	*Citrullus lanatus*	Cucurbitaceae	22	Africa and India
Peas	*Pisum sativum*	Leguminosae	14	Southern Europe and Western Asia
Onion	*Allium cepa* L.	Alliaceae	16	South East Asia
Underexploited fruit and vegetable crops				
Bael	*Aegle marmelos* L.	Rutaceae	36	Northern India
Jamun	*Syzygium cumini*	Myrtaceae	40	Indian Sub-continent
Tamarind	*Tamarindus indica*	Leguminosae	24	Tropical Africa
Wood apple	*Feronia limonia*	Rutaceae	18	India, Sri Lanka
Phalsa	*Grewia subinaequalis*	Tiliaceae	36	India
Mulberry	*Morus alba* L.	Moraceae	308	Northern China
Karonda	*Carissa carandus* L.	Apocynaceae	12	India
Fig	*Ficus carica* L.	Moraceae	26	South East-Asia
Lasoda	Cordia myxa	Boraginaceae	48	India
Chironji	*Buchanania lanzan*	Anacardiaceae	22	Western Peninsula
Kair	*Capparis decidua*	Capparidaceae	14, 16	Arid region of India
Pilu	*Salvadora oleoides*	Salvadoraceae	24	India/Tropical Africa
Khejri	*Prosopis cineraria*	Leguminosae	28	Thar desert of India
Kachri	*Cucumis callosus*	Cucurbitaceae	14	Thar desert of India
Snapmelon	*Cucumis melo* var. *momordica*	Cucurbitaceae	-24	Tropical regions of Africa and Asia
Mateera	*Citrullus lanatus*	Cucurbitaceae	22	Thar desert of India
Bitter apple	*Citrullus colocynthis*	Cucurbitaceae	22	Thar desert of India
Arya	*Cucumis melo* var. *chate*	Cucurbitaceae	24	Thar desert of India
Round melon	*Citrullus vulgaris* var. *fistulolus*	Cucurbitaceae	24	India
Spine gourd	*Momordica dioica*	Cucurbitaceae	28	India
Jhaar karela	*Momordica balsamina*	Cucurbitaceae	22	Africa
Cluster bean	*Cyamopsis tetragonoloba*	Fabaceae	14	India
Moth bean	*Vigna aconitifolia*	Fabaceae	22	India, Pakistan and Burma
Phog	*Calligonum polygonoides*	Polygonaceae	54	Indian Thar desert
Khimp	*Leptadenia pyrotechnica*	Apocynaceae	22	Africa, Indian Thar desert

Role of Fruits and Vegetables in Human Health

According to the *Food and Agriculture Organization*, 'Food Security' refers to a situation that 'all people have physical, social and economic access to sufficient, safe and nutritious food at all times to meet their dietary and food requirements for an active and healthy life' (World Food Summit 1996). The International Food Policy Research Institute defined 'Nutrition Security' as 'adequate nutritional status in terms of protein, energy, vitamins, and minerals for all household members at all times'. Therefore, nutrition security is broader term than food security since it encompasses biological utilization by human body to ingest and metabolize food which promote health and prevent disease. Underexploited fruits and vegetable crops plays important role in nutritional security as they provides micronutrients (minerals and vitamins) along with macro nutrients with high biological utilization efficiency. Underexploited arid fruits and vegetable crops are rich in phytochemicals such as anthocyanins, carotenoids, phenols and flavonoids, which provides protection against reactive oxygen species and act as health promoter.

A balanced diet provides all the required nutrients in appropriate amounts and proportions which can be achieved through a blend of the four basic food groups *i.e.* carbohydrates, proteins, fat and micro nutrients (vitamins and minerals). The quantities of foods needed to meet the nutrient requirements vary with age, gender, physiological status and physical activity. A balanced diet should provide 50-60 per cent of total calories from carbohydrates, possible from complex carbohydrates, about 10-15 per cent from proteins and 20-30 per cent from both visible and invisible fat. In addition, a balanced diet should provide other non-nutrients for example dietary fiber, antioxidants (vitamins C and E, beta-carotene, riboflavin and selenium) and phytochemicals (anthocyanins, carotenoids, phenols and flavonoids) which protect the human body from free radical damage. The *Indian Council of Medical Research* recommended 300 g of vegetables (green leafy vegetable 50 g, other vegetables 200 g and roots and tubers 50 g) and 100 g fresh fruits per day for a balanced diet (Anonymous, 2011).

It is estimated that per capita availability of fruits and vegetable in India is 126 and 286 g per day, respectively after 30 per cent loss, 5 per cent exports and processing deduction which is far below the recommended quantity. This may be attributable to low production of affordable food, supply limitation, poor purchasing power, lack of awareness, uneven consumption, or a combination of these factors. Fruit and vegetable are rich in micro and phytonutrients which play important role in preventing micronutrient deficiencies and non communicable diseases. The contribution of underexploited arid fruits and vegetable crops to the vitamin, mineral and fiber content in a balanced diet and their role is shown in Table 2.

Health Promoting Phytochemicals in Underexploited Fruit and Vegetable Crops

Phytochemicals are naturally occurring chemicals other than nutrients found in fruits and vegetables, which play important role in the human metabolism leading to improvement in human health. Fruits and vegetables have functional food properties to develop nutritional ingredients or supplements. The perception of underexploited fruit and vegetable crops as food is now shifting to nutritious food. The horticultural crops are gaining importance for their metabolite capabilities (Table 3) in preventive health care (Khanuja and Shukla, 2011).

Table 2. Nutrient rich underexploited fruits and vegetables for arid and semi arid region

Crop	*Nutrients Status (per 100 g)*
Bael	Water 61.06 g, crude protein 3.64 g, fat 0.43 g, ash 2.85 g, potassium 603 mg, calcium 78 mg, and phosphorous 51.60 mg
Ber	Moisture 82 g, protein 0.8 g, fat 0.07 g, fiber 0.60 g, carbohydrates 17.0 g, total sugars 10.5 g, ash 0.4 g, calcium 25 mg, phosphorus 26 mg, iron 1.2 mg, carotene 0.021 mg, thiamine 0.021 mg, riboflavin 0.02 mg, miacin 0.7 mg, citric acid 0.6 mg, ascorbic acid 65 mg, fluoride 0.1 ppm, pectin 2.2-3.4 per cent on dry basis.
Jamun	Moisture 82.10 per cent, fat 0.31 per cent, crude fibre 0.25 per cent, total ash 0.36 per cent, vitamin C 19.4 mg, total sugars 12.6 per cent, reducing sugars 8.9 per cent, non-reducing sugars 3.7 per cent, anthocyanins 157 mg, antioxidant activity 415 mg GAE of total phenolics.
Tamarind	Moisture 35.29 g, carbohydrates 50.07 g, dietary fiber 4.13 g, titratable acidity 18.52 g tartaric acid, soluble solids 44.00 °Brix, vitamin C 4.79 mg, folates 59.35 µg, vitamin E 108.78 µg.
Wood apple	Water 64.06 g, carbohydrates 18 g, protein 7 g, fat 4 g, fibre g, calcium 130 mg, and phosphorous 110 mg,
Phalsa	Juice contain protein 1.57 g, dietry, fiber 5.50 g, carbohydrates 21.0 g, phosphorus 24 mg, iron 1.08 mg, potassium 372 mg, sodium 17.3 mg, vitamin A 16.11 ug, thiamine 0.02 mg, riboflavin 0.22 mg, vitamin C 4.38 mg
Mulberry	Moisture 78 g, total soluble solids 28 g, total sugar 17, sucrose 4.2 g, reducing sugar 15 g, protein 0.8 g, ash 2.6 g, pH, titratable acidity 5.7 g, vitamin C 68 mg, total phenolic 21.50 mg, antioxidant activity 19.24 µg gallic acid equivalent/mg.
Karonda	TSS 7.8 per cent, acidity 3.94 per cent, ascorbic acid 6.98 mg, total sugars 5.28 per cent, reducing sugars 4.15 per cent, protein 2.96 per cent, iron 6.88 mg, calcium 16.33 mg and phosphorus 24 mg.
Fig	Moisture 67.82 per cent, ash 0.89 per cent, fibers 2.88 per cent, protein 2.17 per cent, lipids 1.12 per cent, total carbohydrate, sodium 17.33 mg, potassium 208.67 mg, calcium 65.00 mg, magnesium 37.67 mg, iron 3.13 mg, zinc 0.14 mg, copper 0.37 mg and phosphorus 32.76 mg.
Lasoda	Ash 6.7 g, crude protein 8.32 g, lipid 2.2 g, crude fibre 25.7 g, carbohydrates 57.08 g, ascorbic acid 40 mg of dry weigh basis.
Chironji	Moisture 3.0 g, protein 19.0 g, fat 59.1 g, carbohydrate 12.1 g, calcium 279 mg, phosphorus 528.0 mg, iron 8.5 mg, thiamine 0.69 mg, riboflavin 0.53 mg and niacin 1.5 mg.
Kachri	Moisture 88.2 per cent, fibre 1.21 per cent, copper 0.0046 mg, carbohydrate 7.45 per cent, calcium 0.09 mg, zinc 0.052 mg, protein 0.28 per cent, phosphorus 0.0029 mg, manganese 0.058 mg, fat 1.28 per cent, iron 0.182 mg, vitamin C, 29.81 mg.
Snapmelon	Moisture 95.7 per cent, fat 0.1 per cent, carbohydrate 3.0 per cent, vitamin A 265 IU, protein 0.3 per cent and minerals 0.4 per cent.
Round melon	Protein 1.4 per cent, fat 1.4 per cent, carbohydrates 3.4 per cent, fibre 1 per cent, minerals 0.5 per cent, carotene 13 mg and vitamin C 18 mg.
Spine gourd	Moisture 84.1 g, phosphorus 42 mg, protein 3.1 g, carotene 2700 IU, carbohydrate 7.7 g, thiamine 45.2 µg, fibre 2.97 g, riboflavin 176.1 µg, iron 4.6 g, niacin 0.59 mg, calcium 33 mg and ascorbic acid 275.1 mg.
Cluster bean	Moisture 81.0 g, phosphorus 50.0 mg, protein 3.2 g, carotene 198 mg, Carbohydrate 10.8 g, thiamine 0.09 mg, fibre 3.2 g, riboflavin 0.03 mg, iron 1.08 mg, niacin 0.60 mg, calcium 130 mg, ascorbic acid and 49.0 mg.

Crop	*Nutrients Status (per 100 g)*
Moth bean	Moisture 9.7 g, protein 22.9 g, fat 1.6 g, carbohydrate 61.5 g, calcium 150 mg, magnesium 381 mg, phosphorus 489 mg, iron 10.9 mg, zinc 1.9 mg, vitamin A 32 IU, thiamin 0.56 mg, riboflavin 0.09 mg, niacin 2.8 mg, vitamin B_6 0.37 mg, folate 649 µg, and ascorbic acid 4.0 mg.
Phog	Moisture 59.7 g, ash 11.14 per cent, dietary fibre 44.98 per cent, fat 1.27 per cent, sugar 9.75 per cent, protein 8.02 per cent and vitamin C 66.88 mg/100g.
Ivy gourd	Moisture 93.5 g, thiamine 0.07 mg, protein 1.2 g, calcium 40 mg, riboflavin 0.08 mg, fat 0.1 g, phosphorus 30 mg, niacin 0.7 mg, mineral 0.5 g iron 0.38 mg, folic acid 18 µg, carbohydrate 3.1 g carotene 156 µmg and vitamin C 15 mg
Khejri	Protein 23 g, carbohydrate 55.5 g, fat 2 g, fibre 20 per cent, ascorbic acid 532.0 mg, calcium 414 mg, phosphorus 400 mg and iron 19 mg.

Table 3. Health promoting phytochemicals in fruits and vegetables crops

Crop	*Phytochemicals*	*Function*
Karonda and phalsa	Anthocyanin	Antioxidant activity
Bael	Marmelosin	Antihelminthic and antibacterial activity, gastric ulcer treatment
Jamun	Jambolin	Diabetes treatment
Pumpkin and winter squash	Beta carotene	Anti ageing, anti cancer, help in diabetes and lung problem
Mustard green	Glutathione	Antioxidant
Turnip	Ferulic acid	Antioxidant
Bitter apple	Citrullin, citrullene and citrullnic acid, cucurbitacin B and cucurbitacin I.	Traditional medicines and biopesticide formulations
Indian aloe	Alloin	Antioxidant and medicinal properties
Phog	Furan-2,5-dimethyl, 2,3-dihydro-3, 5-dihydroxy-6-methyl-4H-Pyran-4-one,dehydromevalonic lactone, quinic acid, lauric acid, linolenic acid and squalene	Anti-microbial, anti-inflammatory, anticancer, anti-diabetic, hepatoprotective, cardiovascular, antioxidant, and anti-mutagenic.
Kair	Glucocapparin, methyl isothiocyanate	Antibacterial properties
Jhar karela	Saponins and *Momordin*	Anti-diabetic activity biopharmaceuticals and neutraceuticals properties

Source: Rai, 2012.

Enhanced consumption of carotenoids containing fruits and vegetables was useful than carotenoids dietary supplements in enhancing LDL oxidation resistance, reducing DNA damage, and inducing higher repair activity in human (Southon 2000). Antioxidant (vitamins A, C and E) dietary supplements into the diet of cancer patients, who were eating a balanced diet of fruits and vegetables, negatively impacted their radio and chemotherapies (Seifried *et al.*, 2003). Presently, food security mission include not only the problems of physical availability of food but also nutritional status of food with biological absorption. During green revolution stress was given on production exclusively and not on nutritional security, which is the ultimate goal addressed through fruits and vegetables. Preference should also be given to nutritious underexploited fruits and vegetables. For the same, high-yielding varieties of these crops are being developed by ICAR institutes and SAUs and made available to farmers. Each fruit and vegetable crop contains a distinctive combination of phytochemicals, minerals and vitamins which offers a unique health benefit. More than 3,000 phytochemicals have been identified in fruits and vegetables. Besides health benefits, fruits and vegetables growing play an important role in meeting nutritional security, and ensuring socio-economic development of farmers with environmental protection.

Important Underexploited Fruits

Ber (*Zyziphus mauritiana*)

It belongs to the family Rhamnaceae. It is an ideal fruit tree for arid and semi-arid regions in tropical and subtropical climate where most of the fruit crops cannot be grown. It has a high amount of vitamin C. More vitamin C was found in the fruit flesh near the seed rather than near the skin of the fruit. It is also a rich source of vitamin A and B complex. The fruits can also be used for making several products like chutney, dried ber, murabba, jelly, *etc*. Wines can also be prepared from the fruits. The decoction from root and bark is good for dysentery and diarrhoea and leaf decoction is useful as gargle in sore throat and in bleeding gums. The seed kernels are aphrodisiac. The powder of ber roots has medicinal properties for curing ulcer, fever and wounds.

Wood Apple (*Feronia limonia*)

It is originated in India and belongs to the family Rutaceae. It is a hardy upright tree, producing nutritive acidic fruit and the pulp is eaten raw with or without sugar. The pulp is a rich source of calcium, phosphorus, iron and vitamins like carotene, riboflavin, niacin, thiamine and vitamin C. The fruit is used as a liver and cardiac tonic, and when unripe, for easing diarrhoea and dysentery. Wood apple can be converted into value added products like jam and fruit bar in order to avoid glut and utilize the surplus during the season. Wood apple juice when mixed with other fruits can serve as an excellent beverage.

Bael (*Aegle marmelos*)

It is one of the oldest cultivated fruits in India which is native to India and spread throughout South-East Asia. The leaves are used for worshiping Lord Shiva. This fruit is also commonly found in the hilly regions of India. The fruit is used as ayurvedic remedy against diarrhea, dryness of the eye and common cold. It is also useful in preventing scurvy and strengthens the stomach and promotes its actions. It can be used for jam, jelly, juice, murbba squashes and other beverages.

Jamun (*Syzygium cumini*)

It is an important indigenous fruit belongs to the family Myrtaceae. The tree is ideally suited for windbreak and roadside plantations. It is taken as a dessert fruit and is also used in making beverages, squash, jam, jelly and wine. It is used in blood purification, diabetes, diarrhoea, eczema and as an antidote for strychnine poisoning. It is stomachic, carminative and diuretic and lowers the blood pressure. The jamun seed powder reduces the sugar content in urine.

Karonda (*Carissa carandas*)

It is a hardy, evergreen, spiny and indigenous shrub which thrives well as rainfed crop. The fruit belongs to the family Apocynaceae. Fruits, sour and astringent in taste, are a very rich in iron contains a good amount of vitamin C. They also contains protein, carbohydrates, fat, fibre and calcium. The ripened fruits may be eaten as dessert or used for the preparation of jelly, sauce, cream or jellied salad. Unripe fruits are used for making vegetable, pickles, sauces and chutney. The dried fruits may act as a substitute for raisins. The wine prepared from ripe fruits contains about 14.5 to 15 per cent alcohol and is very much liked by wine fanciers. Fruits can also be used in dyeing and tanning industries. Karonda fruit is considered to be antiscorbutic and is also very useful in curing anaemia, stomach ache and is anthelmintic. Root extracts are used in lumbago, chest complains and venereal diseases.

Lasoda (*Cordia myxa*)

It belongs to family of Boraginaceae. It is a perennial, medium sized tree with crooked stem. Lasoda bears small sized fruits in bunches, used as traditional vegetable and pickles. Fruits are good sources of protein, carbohydrates, K and Na, fiber ascorbic acid and vitamins, which gives essential nutrients for the human health. Unripe green fruits are mostly used as vegetable and pickles. Sometime fruits are dehydrated after blanching for used as vegetable during off season. The sticky pulp from ripe fruits is commonly used to make glue.

Fig (*Ficus carica*)

Fig was an important food crop in ancient civilization. It is a highly nutritious fruit. The fruit contains protein, calcium, iron, vitamin A and thiamine at varying concentrations. Figs are consumed as fresh or dried, preserved, candied or canned. Fresh figs are nutritious and used as dessert or for making jam, jelly, pudding, cakes, *etc*. The fruit is valued for its laxative property. It is applied for boils and other skin infection. The latex is used to coagulate milk and leaves are used medicinally as diuretic, demulcent, emollient and anthelmintic properties.

Tamarind (*Tamarindus indica*)

It is native to Tropical Africa and belongs to the family Fabaceae. Tree is large sized, long-lived evergreen tall with a spreading crown. It is an excellent tree for social forestry and agro forestry. This crop is highly suitable for wastelands due to its multi ferrous uses and capacity to withstand adverse agro-climatic conditions. It also acts as a good wind break. Almost every part of the tree finds some uses but fruit is the most useful which contains the sweet acidic pulp. The pulp has low water content and high levels of proteins, carbohydrates and minerals. The pulp is also the principal souring agent for squashes, chutney, beverages, *etc*. The fruits are used in the medicine as a refrigerant, carminative, antiscorbutic and laxative. The active constituents present are furunone, phenyl acetaldehyde and tartaric acid. The products of the seeds are used mostly for manufacture of sizing powders. Tamarind kernel powder is extensively used for starching of cotton yarns, jute and woolen fabrics.

Phalsa (*Grewia subinaequalis*)

Phalsa belongs to family of Tiliaceae. It is a shrub or small tree of Indian origin. The ripe fruits are consumed fresh or processed into refreshing fruit and soft drinks during summer. Phalsa fruit has a short shelf life suitable for only local marketing. It is grown in the northern and the western states of India. Phalsa is found wildly growing in UP, Rajasthan, Punjab, Haryana, MP, West Bengal and many parts of south India. The cultivation of phalsa is limited to very small scale in Punjab, Haryana, Rajasthan, Gujarat and Uttar Pradesh. The fruits are good sources of citric acid, vitamins A, minerals, flavonoids, carotenoids and anthocyanins. The phalsa fruits are potential sources of potassium which plays crucial important role in energy metabolism and normalizing blood pressure.

Chironji (*Buchanania lanzan*)

It belong to family of Anacardiaceae. Nutritional composition of the seeds shows that it as a potential source of protein, fat, dietary fiber, and energy. It is very good source of phosphorus, calcium, magnesium and iron. The fruit has high socio-economic value providing livelihood to rural population. Chironji is low in calories but extremely such in protein and dietary fibre. The fruit is juicy and sweet in taste and used for preparation of various value added products like squash, ready to serve drinks and nectar after juice extraction.

Important Underexploited Vegetables

Kachri (*Cucumis callosus*)

Kachri is a drought hardy and high temperature tolerant crop (up to 46°C), found in the arid zones during rainy season. Unripe fruits are bitter in taste but at ripening they become edible. The mature fruits are usually cooked with various vegetable preparations, *chutney*, pickles and are also used for garnishing the vegetables or as salad. *Kachri* is one of the components of the delicious vegetable popularly known as *Panchkuta* in the desert districts of north-western India (Pareek and Samadia, 1998). Unripe fruits are bitter in taste but at ripening they become edible (Samadia, 2007). *Kachri* can be cultivated successfully both during rainy and spring-summer season. It requires warm and dry weather with plenty of sunshine for growth and production.

Snap melon (*Cucumis melo* var. *momordica*)

Snap melon is commonly grown as inter crop in maize. Its immature fruits are used as salad, vegetables and other culinary preparations. Fruits at ripening stage develop suture. Its fruits are generally less sweet as compared to muskmelon; hence it is much liked by the people who are suffering sugar related disorders. Snap melon pulp is suitable for preparation of jam by adding equal quantity of sugar to pulp. Good quality wine with excellent aroma and taste can be prepared with its pulp. Snap melon is originated in the tropical regions of Africa and Asia. It is a very popular vegetable of arid zone. It is commonly grown as a rainfed crop in Rajasthan and Gujarat. A great extent of genetic variability exists in India with respect to vegetative growth, quality attributes and resistance to biotic and abiotic stresses.

Mateera (*Citrullus lanatus*)

Mateera is generally known as the poor man's vegetable and the common man's fruit in the desert. It is an indigenous type of drought hardy watermelon and extensively grown on barren sand dunes of western Rajasthan. Watermelon is indigenous to Tropical Africa where wild forms are still available. The immature green fruits at the tender stage are rich in protein, carbohydrate, crude fibre, calcium and phosphorus. The mateera fruits are sweet with refreshing edible flesh and consumed fresh as dessert and have juicy properties. Besides, the tender fruits (*Loiya*) weighing 80-150 g are used extensively as fresh vegetable for making *rayta* and curries. The seeds are protein rich (25-32 per cent) and are roasted and eaten as snacks.

Bitter Apple (*Citrullus colocynthis*)

Bitter apple is the probable ancestor of watermelon which is bitter in taste and also known as *Tumba* or 'bitter apple'. It resembles a common watermelon vine, but bears small, hard fruits with a bitter pulp. Fruit contains 15 per cent pulp, 62 per cent seed and 23 per cent rind. The mesocarp contains glucose (1.3 per cent on flesh basis). The processed mesocarp may be good sources of pectin. The juice of the fruit contains citrullin, citrullene and citrullnic acid. The fruits of bitter apple also contain cucurbitacin B and its glycoside, cucurbitacin I. The peel free flesh of ripe fruits contains yellow bitter oil. The seeds are used for edible purposes as well as to extract oil. Seeds contain 16.7 per cent yellow coloured semi-dry oil rich in linoleic acid. It is one of the most drought cucurbit showing maximum diversity in Thar Desert. The fruits are widely used in traditional medicines and preparation of biopesticide formulations.

Arya (*Cucumis melo* var. *chate*)

Arya is an annual plant and monoecious. Tender fruits are generally used as salad and harvested before maturity. The fruit is climacteric. It is propagated by seeds. Fruit shape is long, skin colour light to dark green and smooth at tender stage. Fruit flesh is light orange at ripening without sugar and aroma. It is cultivated in several parts of arid zone. Fruits are rich in vitamins, minerals and phytochemicals.

Roundmelon (*Citrullus vulgaris* var. *fistulolus*)

Roundmelon is also known as round gourd, Indian squash, squash melon and *Tinda*. It is grown for small, tender fruits that are roughly spherical and about 5-8 cm in diameter. The fruits at cooking stage contain 1.4 per cent protein, 1.4 per cent fat, 3.4 per cent carbohydrates 1 per cent fibre, 0.5 per cent minerals, 13 mg carotene and 18 mg vitamin C/100 g of fresh weight. *Tinda* is extensively cultivated in North India, especially in Punjab, Uttar Pradesh and Rajasthan. It is believed to have originated in India and several landraces are prevalent in arid zone.

Spine Gourd (*Momordica dioica*)

Spine gourd or *kartoli* is a perennial climbing plant. The spiny fruits are used as vegetable in all regions of India. They are quite palatable, rather sweet and entirely free from bitterness. Fruits are good source of protein and iron. It is also rich in ascorbic acid content (275.10mg/100g). It also contains iodine (0.7µg/100g). Fruits, leaves and tuberous roots are used as a folk remedy for diabetes. It has small leaves and small yellow flowers. Fruits are small, dark green, round or oval with spines. Fruits are available from July to October in North India. Momodicausenal (a pentacyclic triterpene) was isolated from the seed (Ali and Srivastava, 1998). Spine gourd is dioecious and propagated by underground tubers and stem cuttings. Wide genetic variability exists in arid zone of Rajasthan particularly in natural habitats.

Jhaar Karela (*Momordica balsamina*)

Jhaar karela is a wild climber containing wide spectrum of medicinal and nutritional value. The fruits are harvested from waste land from July to October in arid zone. It is monoecious and propagated through seeds. Fruits are small and used as vegetable. The leaves, fruits, seeds and bark of the plant contain resins, alkaloids, flavonoids, glycosides, steroids, terpenes, cardiac glycoside, saponins having various medicinal importance. The therapeutic agent is '*Momordin*' which possess very good anti-diabetic activity. The commercial exploitation for biopharmaceuticals and neutraceuticals are some of the prospective future potential of *Jhaar karela*. It is found in natural habitats of arid zone particularly neglected places. It possesses resistance against biotic and abiotic stresses which may prove worth in utilizing in the breeding programmes.

Cluster Bean (*Cyamopsis tetragonoloba*)

Cluster bean commonly known as *Guar* is one of the most drought hardy legume vegetable. Its immature pods are primarily used as vegetable. It is also dehydrated and stored for future use as vegetable. Cluster bean possesses several medicinal properties. Being a legume crop it is much preferred for forage. It fixes nitrogen in soil and preferred in crop rotation cycle in arid zone. It is relatively tolerant to salt and water stress and give high yield under adverse conditions and need less fertilizer. It is very good source of protein and dietary fibers.

Moth Bean (*Vigna aconitifolia*)

Moth bean is an important crop of arid and semi-arid regions. It has higher adaptability to uncongenial ecological environments makes it a perfect choice for areas receiving lesser rainfall. Moth bean seeds are a good and potential reservoir of proteins and other essential minerals and vitamins. The green pods are eaten as a vegetable and the ripe seeds, whole or split, are eaten cooked. Dry seeds of moth bean offers a variety of edible products-vegetable, fodder for animals, whole seed, *Papad, Bhujia, Dal, Mangori, Vada, etc.* It is also assimilated as a flour to prepare south Indian food delights like *Idli* and *Dosa*. India being the origin place has numerous landraces and cultivars of moth bean are available in arid zone.

Phog (*Calligonum polygonoides*)

Phog is an endemic and threatened species reported from Thar Desert. It belongs to the family Polygonaceae and commonly grows on dry sandy soils and sand dunes. Its native habitat is hot arid region of Thar Desert of India. *Phog* is a perennial shrub, usually 4-6 feet in height but occasionally may reach even 15 feet height. It is drought hardy and capable of growing under adverse conditions of soil and moisture. It is not affected by frost. *Phog* has high economic values as its all plant parts are useful in one or other way. Its abortive flowers and succulent fruits are the most important source of food for sustenance during frequently occurring famines and also valued for medicinal properties. Dried tender flower buds are known as *phogla* in Rajasthan and used to prepare *rayta*. The wood is used as raw materials in building huts/shelter and scaffolding of wells and other structures. The aqueous paste of plant acts as an antidote against snake bite. It is used for curing typhoid, asthma, cough and cold.

Khimp (Leptadenia pyrotechnica)

Khimp is a amazing desert plant belongs to the Apocynaceae family. It is leafless, erect, evergreen perennial shrub and widespread in arid zone of India. Being highly drought resistant, it played an important role in afforestation of desert. Roots have a strong soil binding capacity and used in sand dune fixation. *Khimp* possesses antifungal, antibacterial, anticancer, antioxidant, wound healing, anthelmintic, antiatheroscloretic, hypolipidemic, antdiabetic and hepatoprotective activities coupled with other multifarious uses. Almost all plant parts are used in the traditional medicinal system to treat various disorders. It's flowering and fruiting time is August to January. The pods are known as *khimpoli* which ripe in the month of March. Pods are cooked as vegetable and used possess medicinal value. The plant is used in thatching of huts and plant fiber for making ropes.

Ivy Gourd (*Coccinia grandis* syn. *C. indica*)

Ivy gourd or *kundru* is perennial climber and dioecious plant. Being perennial in nature it continues up to 2-3 years in same field. It is native of India. Fruits are smooth, small (5-6 cm long) and cooked as vegetable. It is propagated through stem cuttings. Parthenocarpic germplasm is also found in nature which does not require pollination for fruit set. It is rich source of vitamins and minerals.

Indian Aloe (*Aloe barbadensis*)

Indian aloe or *Guarpatha* is a small, succulent, stemless, herbaceous perennial plant (60-100 cm height) with shallow root system. It contains alloin, a yellow coloured liquid and colourless aloe gel. It is rich in vitamin C (53 mg/100g edible part) with antioxidant and medicinal properties. The tender aloe, free from bitter content, is commonly used for vegetable purpose. The aloe vera is widely used for preparation of juice, vegetable and cosmetic purpose.

Khejri (Prosopis cineraria)

Khejri is a drought hardy perennial plant belonging to family Leguminosae. It is called as *Kalpvriksh* of Thar Desert. It grows luxuriantly under the extremely adverse agroclimate in hot arid regions without much care. *Khejri* have multiple uses *viz.*, nutritious rich pods, fodder and fuel, besides its favourable effects on ecology and soil fertility. It not only tolerates the extreme edapho-climatic conditions of Thar Desert but also has plentiful foliage, bears flowers and fruits during the dry period. Being leguminous species, it is compatible to almost all companion crops grown under the traditional cropping systems. The immature pods called as *sangri* are used as vegetable. It is a major source of leaf fodder (*loong*) in arid zone of country. The pods and fodder is nutritious and high valued. About 10-15 kg tender pods and 25-30 kg *loong* can be harvested annually from a 20 years old tree.

Improved Varieties of Underexploited Fruit and Vegetable Crops

The climatic conditions of arid region are very adverse for growth and production of a crop therefore; selection of a variety/crop for this region is highly important for economic production. The selected crop/variety should complete their vegetative and reproductive phase during maximum moisture availability period. The underexploited arid fruits and vegetables crops meet this requirement. The genetic resource of underexploited arid fruits and vegetables crops are being conserved and maintained in field repository at ICAR-CIAH, Bikaner and its regional station CHES, Godhara, Gujarat. The collected germplasm has been evaluated and characterized for utilization in improvement programmes. The ICAR-Central Institute for Arid Horticulture, Bikaner has released numbers of varieties of arid fruits and vegetable crops (Table 4). A number of varieties have been developed and released through AICRP on Arid Zone Fruits. Bael can be grown successfully in semiarid and arid regions and varieties NB-5 and NB-9 doing well under irrigated hot arid conditions. Goma Yashi bael variety developed by ICAR-CIAH, Bikaner is well accepted among farmers under semi-arid conditions due to optimum fruit size, high yield and quality (Singh *et al.*, 2011). Similarly improved varieties of arid vegetables like khejri, kachri, snap melon, Indian bean and clusterbean have been released for cultivation. AHK-200 and AHK-119 variety of Kachri and Thar Manak of mateera are becoming popular in arid region.

Table 4. Underexploited fruits and vegetable varieties released by ICAR-CIAH, Bikaner

Fruits		*Vegetables*	
Bael	Goma Yashi, Thar Divya, Thar Neelkanth, Thar Shristi	Khejri	Thar Shobha
Jamun	Goma Priyanka, Thar Kranti	Bottle guard	Thar Samridhi
Lasoda	Thar Bold	Mateera	AHW-19, AHW- 65, Thar Manak
Phalsa	Thar Pragati	Kachri	AHK-119, AHK-200
Karonda	Thar Kamal	Snap melon	AHS-10, AHS-82
Mulberry	Thar Lohit, Thar Harit	Kakri	AHC-2, AHC-13
Mahua	Thar Madhu	Sword bean	Thar Mahi
Tamarind	Goma Prateek	Indian bean	Thar Kartiki, Thar Maghi
Chironji	Thar Priya	Ridge gourd	Thar Karni
Khirni	Thar Rituraj	Clusterbean	Goma Manjri, Thar Bhadvi
Ber	Goma Kirti, Thar Sevika, Thar Bhivraj, Thar Malti	Sponge gourd	Thar Tapis

After evaluation of germplasm by AICRPs, ICAR institutes and SAUs, a number of varieties of arid fruits and vegetable crops have been recommended for cultivation in arid and semi regions for income and nutritional security (Table 5).

Table 5. Promising varieties of arid fruit and vegetable crops

Fruit Crops	
Ber	Gola, Seb, Mudia, Kathli, Banarsi Karaka, Umran, Ilaichi, Rashmi
Bael	NB-5, NB 9, NB-16, NB-17, Pant Aparna, Pant Sujata, Pant Shivani, CISH Bael-1, CISH Bael-2
Fig	Poona Fig, Dinkar, Conadria, Excel, Dianna, Brown Turkey
Tamarind	PKM 1, Pratisthan, Yogeshwari, Goma Prateek
Aonla	NA 6, NA 7, NA 10, Kanchan, Krishna, Banarasi, Chakiya, Lakshi 52, BSR 1, Anand 1, Anand 2.

Fruit Crops	
Custard apple	Balanagar, Arka Sahan, Phule Janki, Washington, Pink's Mammoth.
Pomegranate	Ganesh, G 137, Jalore Seedless, Mridula, Ruby, Arakta, Bhagwa, Phule Super Bhagawa, Solapur Lal, CAZRI, Vishal
Vegetables	
Cowpea	Pusa Dofasali, Pusa Phalguni, Pusa Barsati, Pusa Rituraj, Pusa komal, Kashi Nidhi,
Cluster bean	Pusa Sadabahar, Pusa Mausami, Pusa Navbahar, Durga Bahar, AHG-13
Amaranth	CO-1, CO-2, CO-3, Chhoti Chauali, Badi Chaulai

Conclusion

Underexploited fruits and vegetables are well accepted and more sustainable source of minerals, vitamin phytochemicals and antioxidant for nutritional security along with socio-economic development and environment protection. Integration of fruits and vegetables in daily diet not only supply nutrients and vitamins but they also improve nutrient absorption and impart unique flavors and taste of food too. Underexploited fruits and vegetables are cheapest and effective sources of vitamin and dietary fiber. Under exploited fruit and vegetable crops have adaptability to diverse agro-climatic condition and ability to survive in degraded lands. The area and production of underexploited fruits and vegetables crops are increasing fast because of the development of new varieties and agro-techniques for arid and semi-arid regions. However, there is a need for germplasm conservation, evaluation, standardization of region specific technologies, infrastructure development in form of market, storage and processing for sustainable development of under exploited arid fruits and vegetable crops.

REFERENCES

1. Anonymous. 2011. Dietary Guidelines for Indians. 2011. A manual, National institute of nutrition, Hyderabad – 500 007, India.
2. Khanuja S. P. S. and Shukla A. K. 2011. Human health and nutrition: Functional foods. In: *Horticulture to Horti-Business* (Editors: KL Chadha, AK Singh, VB Patel), Proceedings Book of the Fourth Indian Horticulture Congress held at New Delhi during 18-21 November, 2010, Westville Publishing House, New Delhi, pp 433-445.
3. Rai, S. K., Arora N., Pandey N., Meena R. P., Shah K. and Rai S. P. 2012. Nutraceutical enriched vegetables: Molecular approaches for crop improvement. *International Journal of Pharma and Bio Sciences*, 3(2): 363-79.
4. Saroj, P. L., Bhargava, R, Singh, R S and Krishna, H. 2018. *ICAR-CIAH An overview*, Technical Bulletin, ICAR-CIAH, Bikaner, Rajasthan 334 006.
5. Seifried, H E, McDonald S S, Anderson D E, Greenwald P and Milner J. A. 2003. The antioxidant conundrum in cancer. *Cancer Res.* 63: 4295-4298.

6. Southon S. 2000. Increased fruit and vegetable consumption within the EU: Potential health benefits. *Food Research International,* 33: 211-217.
7. World Food Summit 1996, *Rome Declaration on World Food Security.*
8. Ali, M. and Srivastava, V. 1998. Characterization of phyto-constituents of the fruits of *Momordica dioica. Indian J. Pharmaceutical Sci.*, 60(5), 287-285.
9. Pareek, O.P. and Samadia, D.K. 1998. A booklet on beneficial *Kachri.* NRCH, ICAR, Bikaner.
10. Samadia, D.K. 2007. Production and processing of *kachri* (In Hindi). Technical Bulletin, CIAH, Bikaner, pp. 1-20.

Chapter 3

Underexploited Fruits for Combating Malnutrition from North Eastern Region

B.N. Hazarika* and N. Bidyaleima Chanu

Central Agricultural University, College of Horticulture and Forestry, Pasighat 7911102, Arunachal Pradesh

**e-mail: bnhazarika13@yahoo.co.in*

ABSTRACT

The North-East India is a part of both Himalaya as well as Indo-Burma biodiversity hotspots in the world. It has the richest reservoir of plant diversity in India and is one of the 'biodiversity hotspots' of the world supporting about 50 per cent of India's biodiversity. North Eastern region occupy 7.7 per cent of total geographical area of country and harbours 50 per cent of Indian flora (8,000 species) of which about 4 per cent is endemic (2,526 species). The distinct tribes in the region have rich indigenous knowledge system on the use of components of biodiversity for their daily sustenance like food, fodder, shelter and healthcare. The region has several unique features such as fertile land, abundant water resources, evergreen dense forests of about 66 per cent, high rainfall, and agriculture-friendly climate. Its unique phyto-geographical positions, topography and high degree of precipitation are some of the important factors which are mainly responsible for its enormous biological diversity. As a result, an array of unexplored plants are grown across the region ranging from tropical to alpine. A large number of diversity in fruits belonging to the genera Artocarpus, Annona, Averrhoa, Garcinia, Musa, Passiflora, Phyllanthus, etc. are reported from the region besides diverse vegetables, and spices with some unique quality because of their locational advantage. The region has a great ethno-cultural diversity with major and sub-tribes, which explains the wealth of traditional ecological knowledge among farmers. People of region have their own culture, tradition and medicinal system of treatment and knowledge acquired through close observation of nature. Its ethnic people living in the remote forest areas still depend to a greater extent on the forest ecosystems for their livelihood and avoid malnutrition. They collect different wild fruit plants along with other medicinal plants and use them in traditional ways to cure their health-related forms. The minor and wild fruits are mostly used to cure various gastrointestinal disorders, respiratory problems, cardiovascular compliance, muscular illness, bone diseases, gynaecological problem, cancers, snake bite, allergy and malaria etc. by local people of the region besides their use as fruit. These lesser known fruits are rich in minerals like Ca, Fe, P, Mn, organic acid, vitamins and other nutrients like carbohydrates, proteins, and fats. Tribal people used to take lesser known fruits either raw or in the form of beverages, pickles or cooked/boiled with some other diets. This indigenous system of treatment based on such fruits is still an important part in social life and culture of the tribal people. Malnutrition affects badly in countries like India. The hidden potential of these novel fruits need to expore to strengthen the nutritional security. Even though, works on underutilized fruits are done, yet a large number of species being endemic remained underexplored. Hence cultivation, promotion and conservation of these underexploited are very important.

This chapter provides the information on some of the wonder underutilized fruit plants of the region and their ethno medicinal as well as uses in overcoming malnutrition by the tribes of the region.

Keywords: *Unexploited fruits, Nutrition, North Eastern Regions of India, Economy.*

The North-Eastern stretches between 21°50' and 29°34'N latitude and 85°34' and 97°50'E longitude with altitude varying from 100 meters to over 7000 meters above mean seal level. The North Eastern region has a wide range of climatic condition because of its unique position in the Indian subcontinent. It has humid subtropical or nearly tropical plains which receives high rainfall, temperate climate as well as snow covered high mountains. Varying range of climate and soil condition, moderate temperature and plenty of rainfall are conducive for growing diverse crops which may be either of edible or may be of other uses. Many of underutilized species are well adapted to marginal lands either in homestead gardens or growing wild and with low cost inputs, able to withstand adverse soil and climatic condition to a considerable extent, they multiply and grow spontaneously rather than coming to extinction, attack by serious pests and diseases are relatively less. However, lack of attention has made their potential value is under-estimated and under-exploited. The increase in area and production of these crops not only provide nutritional security also helps in economic return. On the other hand underutilized fruits were known to mankind since the time immemorial and they have been the prime source of food for the native people. Underutilized fruits play a vital role in providing food, nutrition and other substances to the native communities and are an additional source of income. These fruits have great potential as a source of food, besides meeting multipurpose needs of local communities, can make a specific contribution to human nutrition (Table 1). Organoleptic behaviour of the fruits belonged to many of the minor fruit crops are excellent, a number of excellent products are also made from many of these fruits such as *morabba* from *aonla,* jam from bael, pickles *etc.* In the recent past, considerable interest has been paid in the production of underutilized fruits worldwide. The importance of these lesser-known fruits is increasing because of medicinal properties and essential nutrients. A number of species are also utilized as rootstocks or in breeding programme to evolve new varieties.

The nutritive value of some unexploited fruits are as follows:

Table 1. Nutrient content in 100 grams of edible portion of the underutilized fruit crops

Crops	*Carbo-hydrate (g)*	*Protein (g)*	*Fat (g)*	*Iron (mg)*	*Vit. C (mg)*	*Vit. A (IU)*	*Ca (mg)*	*P (mg)*	*Fibre (g)*
Rambutan	14	0.9	0.1	1.8	31	4	3	6	1.1
Longan	25.2	1	0.5	0.3	8	28	2	6	0.4
Bael	31.8	1.8	0.39	0.3	8	-	43	25	-
Aonla	13.7	0.5	0.1	1.2	600	-	50	20	3.4
Ber	12.8	0.8	0.1	300	70	-	30	50	-
Jamun	14	0.7	0.1	190	18	48	20	10	0.9
Pomegranate	26.44	1.46	0.46	0.34	16	166	12.07	4.91	0.9
Fig	9.59	4	1.5	4	1	100	200	7	1.5
Tamarind	70.8	3.1	0.4	11	-	-	170	110	3
Passion fruit	13.6	9.4	0.1	0.2	30	717	3.6	12.5	-
Persimmon	24.97	0.47	0.20	0.25	11	2710	8	28	4.9
Carambola	7.83	0.54	0.35	0.26	50	560	4	16	2.7
Delinia	2.9	3	0.1	-	4	-	0.5	26	-
Jackfruit	18.9	1.9	0.1	500	-	540	20	30	1.1
Guava	11.88	0.82	0.6	0.31	183.5	31	20	25	5.4

The region has rich diversity of different minor fruit crops. Some important underutilized fruits are Garcinia, Rambutan, Longan, Bael, Aonla, Ber, Jamun, Pomegranate, Fig, Tamarind, Passion fruit, Persimmon, Carambola, Delinia (Chalta), Jackfruit *etc*. The potentialities some unexploited are discussed below.

Passion Fruit

The passion fruit, belonging to the family Passifloraceae is an important fruit crop grown in the world having economic value. The juice but mainly the leaves of passion fruit contain the alkaloids, which has blood pressure lowering, sedative and antispasmodic action. The flower of passion fruit has a mild sedative and can help to induce sleep. Passion flower has been used in the treatment of nervous and easily excited children, bronchial asthma, insomnia, nervous gastrointestinal disorders and menopausal problems. Passion flower is sometimes used as a mild hallucinogen.

Persimmon

Persimmons belong to the family "Ebenaceae. Perhaps most important is the use of the fruit as an astringent for treating sores in the throat and mouth. They also used that quality of the fruit to treat hemorrhoids, and they chewed the bark to treat heartburn. Truly, persimmon fruit has lot of medicinal properties. The calyx of the persimmon fruit where it is connected to the branch of the tree can be used to make a calyx tea. Drinking calyx tea is good for stopping hiccup, and is also good for bed-wetting. For bee sting, paste a grinded astringent persimmon is very effective.

Carambola

Carombola belonging to the family Oxalidaceae. Carambola contains a relatively high oxalic acid and sour in taste which makes it little commercial value for consumption as a fresh fruit. However, new sweet flavored cultivars with a low oxalic acid concentration are rapidly becoming popular. Fruits are important sources of minerals like potassium, iron, calcium, sodium and phosphorus. The ripe fruit is digestible, tonic and strengthening. The juice of fruit, when used as eye drop, is also considered to be useful for vision. The leaves are antipruritic, antipyretic and anthelmintic. They are useful in scabies, various types of poisoning, pruritus, intermittent fevers and intestinal worms. These fruits also allay thirst. The ripe fruit said to be a good remedy for bleedings piles, particularly the internal one. The dried fruit is given in some parts of India in fevers.

Delinia

Elephant apple belongs to the family Dilleniaceae; is a large, knobby fruit with acidic flavoured. Fruits are consumed as tonic and laxative. The fruit juice is mixed with sugar and used as a cooling beverage in fevers and as a cough mixture. Green fruits are very effective against rheumatism. Durian (*Durio zibethinus* Murr.): Durian belongs to the family Bombacaceae, is one of the popular seasonal fruit in South East Asia. It is popular because of its taste and strong odour. The fruit is usually consumed fresh, but it can be processed into paste, or the flesh can be frozen or powdered or added into other products such as ice cream, cakes and confectionery, seeds roasted and eaten.

Longan

Longan is a subtropical fruit of Sapindaceae family. Longan can be eaten fresh, dried or quick frozen. The fruit can be peeled, pitted and canned. The juice of most cultivars is sufficiently sweet for processing without addition of sugar. The flesh of the fruit is administered as a stomachic, febrifuge and vermifuge, and is regarded as an antidote for poison. A decoction of the dried flesh is taken as a tonic and treatment for insomnia and neurasthenic neurosis. Dried flowers are used for medicinal purposes. The seeds are administered to counteract heavy sweating and the pulverized kernel, which contains saponin, tannin and fat, serves as astyptic.

Rambutan

The rambutan or hairy litchi belongs to Sapindaceae family is a delicious fruit common in Arunachal Pradesh The white semi-translucent, sub-acid sweet flavored aril is the edible flesh of the fruit. Root decoction for fevers. Leaves for poulticing. Bark used as astringent for tongue maladies. Fruit used for dysentery and as warm carminative in dyspepsis. Fruit decoction used for diarrhea.

Ber

Ber is an ideal fruit tree for arid and semi-arid regions in tropical and subtropical climate where most of the fruit crops cannot be grown either due to lack of irrigation facilities or adverse climate and soil condition. The fruits are laxative and invigorating. The leaves are used to treat conjunctivitis. The bark is used for the treatment of diarrhea. Root decoction is used in fever and its powder is applied to ulcers and old wounds.

Jamun

The jamun belongs to Myrtaceae family The ripe fruit is delicious as dessert. The ripe fruit is very useful in curing diarrhoea and diabetes. It is stomachie, carminative and diuretic, apart from having cooling and digestive properties which help to cure diabetes, diarrhoea, and dysentery. Recent studies have shown that it markedly lowers blood pressure. The seed powder of jamun reduces the quantity of sugar in urine very quickly and permanently. The berry is known for its hypoglycemic (lowering blood sugar) properties. The seed is used in various alternative healing methods in Unani, Ayurveda and Chinese medicine for digestive ailments. The leaves and bark are used for gingivitis and controlling blood pressure.

Bael

Bael belonging to family Rutaceae. Bael is known for its high medicinal and nutritional properties. Almost every part of the bael tree is used. The fruit is very rich in sugar, riboflavin (Vitamin B2) and minerals. The ripe fruit is a tonic, restorative, laxative and good for heart and brain. The mature fruit is astringent, digestive and stomachic, and is usually prescribed for diarrhoea and dysentery. Ripe bael fruit could be used as an excellent laxative. A decoction made out of the root and bark of bael tree is used in treating fever. The bael root could also prove a useful home remedy for getting rid of ear problems.

Phalsa

The Phalsa plant belongs to the familyTiliaceae The edible part of fruit varies from 69 to 93 per cent of its fresh weight. Ripe phalsa fruits are sub-acidic and good source of vitamin A and C and are fair sources of phosphorus and iron. The popularity of phalsa fruit is due to its attractive colour ranging from crimson-red to dark purple and its pleasing taste. The ripe fruits are used for making refreshing drink in summer having cooling effect. Several beverages, *viz.*, nectar, concentrate, squash and crushed phalsa packed in glass bottles remain acceptable for up to 8 months in cool storage. The shoots of the plants after pruning can be utilized either for making baskets or supporting vegetable crops.

Fig

Fruits are laxative and are rich in protein digestive enzymes. It is also effective in the treatment of piles, constipation, asthma, corns, and chronic cough. It is an excellent tonic for weak people who suffer cracks in lips, toungh and mouth.

Tamarind

Tamarind pulp is useful in treating fevers. Pulp of the ripe fruit is useful in the treatment of bilious vomiting, flatulence and indigestion. It is also beneficial in constipation. An infusion of the pulp prepared by softening it in water is particularly useful for loss of appetite and lack of inclination for food intake. The tender leaves of tamarind tree are an effective remedy for treating burns. Its leaves prevent oedema and help in the growth of healthy, normal skin. The oils keep the affected part well protected against moisture and entry of harmful germs.

Durian

Durian is one of the popular seasonal fruit in South East Asia. It is popular because of its taste and strong odour. The fruit is usually consumed fresh, but it can be processed into paste, or the flesh can be frozen or powdered or added into other products such as ice cream, cakes and confectionery, seeds roasted and eaten

Karonda

Karonda is an evergreen, spiny small tree belongs to Apocynaceae family. The karonda is a non-traditional fruit crop. Ripe fruits are sub-acidic to sweet in taste with peculiar aroma. The fruits may be eaten as a dessert when ripe or used in the preparation of fruit products such as jelly, squash and chutney. The dried fruits may become a substitute for raisins. The unripe fruits yield milky white latex which can also be used in preparing chewing gum and rubber. Fruits can also be used in dyeing and tanning industries. Karonda fruit is considered to be antiscorbutic. Root extracts are used in lumbago, chest complains and venereal diseases.

There are various factors which favours cultivation of underutilised fruit crops in this region. The SWOC analysis of it will give a thorough understanding of potentiality of this crop.

Strength

- Undulating topography and micro climate favours the growth of the crop
- New soils are highly fertile which can be utilized with low or no input management
- The crop itself highly tolerant for acidity which is the prevalent chemical character of the Northeastern soils
- Every state in the region has the scientific institute working on the problems of this crop with one exclusive college on horticulture at Pasighat Arunachal Pradesh under CAU.

Weakness

- Low penetration of modern technologies
- Undulating topography makes the farmer inaccessible
- Lack of genuine nursery for supply of quality plant material
- Inadequate market facilities
- Poor communication and transport facility
- Lack of storage and processing facilities

Opportunities

- Abundant land for area expansion

- Ample scope for production and productivity increase
- Enormous potentiality for organic production
- Better export opportunity due to international boundaries

Challenges

- Quality planting material production
- Getting inputs at proper time
- Managing pest and diseases under organic production
- Access to the market information
- Sustaining the supply for the storage and processing units

Conclusion

North East is reservoirs of genetic variability and diversity of underutilised fruits which exist in plant types, morphological and physiological variations or reactions to diseases and pests. It is bestowed with the most congenial climatic conditions for the production of minor fruit crops. There are vast untapped lands of different kinds which are fit for supporting cultivation of these fruit crops. Such lands can easily be put to use for growing such crops in order to diversify the present-day agriculture. The increase in area and production of these fruit crops will not only provide nutritional security, it can provide many folds employment

REFERENCES

1. Hazarika, T.K., Lalramchuana, Nautiyal, B.P. 2012. Studies onwild edible fruits of Mizoram, India used as ethno-medicine. *Genet. Resour. Crop Evol.* **59(8)**: 1767– 1776.
2. Myers, N., Mittermeier, R.A., Mittermeier, C.G., da Fonseca, G.A.B., Kent, J. 2000. Biodiversity 'Hotspots' for conservation priorities, *Nature*. **403**: 853-858.

Chapter 4

Biofortification of Horticultural Crops for Nutrition and Health

Tusar Kanti Behera

ICAR-Indian Institute of Vegetable Research, Varanasi 221305, Uttar Pradesh

e-mail: tusar@rediffmail.com

ABSTRACT

Malnutrition because of insufficient intake of vital nutrients is a wide spread phenomenon among world population and women and children are most affected. Development of crop varieties in different horticultural crops through different approaches is one of the prime focus of the scientists associated with production and improvement of vegetable and fruit crops. Enhancement in specific nutrients through specified agronomic practices through supplementation of nutrients in reported in number of crops. However, breeding approaches and application of modern genomics tools are proved to be very important in biofortification of crop plants. Selection and hybridization, mutation and polyploidy approaches were used selectively in crops like tomato, citrus, banana and watermelon with different degree of success. In the last decades, a large number of DNA based markers were identified to be closely associated with genes responsible higher concentration of different nutrients. Marker assisted introgression of these identified genes though marker assisted selection holds great promise in nutritional enhancement of agronomically superior crop varieties. Besides, transgenics were also developed in crops like banana and cauliflower with enhanced beta carotenoid concentration. Application of genome editing technologies like CRISP/Cas9 mediated targeted mutation are extremely useful in enhancement of traits for bi-fortification in large number of horticultural crops

***Keywords**: Nutritional security, Biofortification, Agronomy, Breeding, MAS, Genome editing.*

Nutrient deficiency leads to poor health, mental impairment, low productivity, and even death. Its adverse effects on child health and survival are devastating, particularly the first 1,000 days of a child's life, from conception to the age of two, resulting in serious consequences such as morbidity, mortality, physical, and mental defects. Childhood stunting and wasting rate is highest in India due to chronicity of energy-protein malnutrition, occurring in approximately one-third of all children of the world. As per the National Health and Family Survey (2015-16) by the Indian Government, in rural India about 27 per cent women and 23 per cent men are malnourished (Verma and Kumar, 35). Comprehensive National Nutrition Survey (2016-18) data showed 34.7 per cent of children under 5 years of age are still low height for age (stunted growth) and 33.4 per cent are low weight for age (underweight). The subsequent health and productivity costs of hidden hunger in the adult population also result in severe economic losses; the economic cost of micronutrient deficiency in India is ~2.4 per cent of GDP which is equivalent to $15–46 billion (Ricthie *et al.,* 28).

Fruits and vegetables are a natural source of micronutrients essential for the human body and horticultural production has the potential for attaining nutritional security. Biofortification is a feasible and cost-effective means of delivering micronutrients to populations that may have limited access to diverse diets and other micronutrient interventions. Developing nutrient dense horticultural crops can be achieved by biofortification (through agronomic and breeding practices, transgenic and genome editing approach). Presently, plant breeding programmes primarily focused on high agronomic yields and conventional efforts like diet diversification, supplementation, and industrial fortification are not enough to alleviate the situation. An efficient and complete food system is necessary to deliver all basic nutritional requirements.

Biofortification through Agronomic practices

The agronomic biofortification of vegetables is one of the simplest and easy methods of biofortification. However, this strategy requires a long period and adequate funds, and this technique is useful in the countries where the genetic engineering method of biofortification is not well accepted. In this approach, generally, fertilizer is used either in the form of spray on leaves or the application of fertilizer in soil (Weng *et al.,* 36). The biofortification of Iron (Fe) and Zinc (Zn) was reported to be successful where the foliar application was used to enhance these nutrients in plant tissue and edible part. The agronomic approach for biofortification also includes management practices during the crop growing season. The package of practices like tillage, water management, and nutrient interaction are involved in enhancing micronutrient. Foliar application is the better option for agronomic biofortification, which requires fewer amounts of Fe and Zn fertilizer as compared to soil application.

In agronomic biofortification, application of potassium (K) influences quality of fruit and yield of *Cucumis melo*. Soil application of 11 mM K increased phenol content and antioxidant capacity of *C. melo* to 213.10 µM equivalent Trolox/100 g fresh base compared to treatment at 7 mM K, respectively, which produced 162.90 µM (Preciado-Rangel *et al.*, 26). This is an advantage since fruit with extensive phytonutrients are associated with lower risk of chronic diseases and cancers (Linnewiel-Hermoni *et al.*, 17). Foliar application of silicon at 100 and 200 mg"L–1 increased ascorbic acid (Vitamin C) content of *Cucumis sativus* (Abd-Alkarim *et al.*, 1). Vitamin C helps the body defence system, and boosts immunity, and cell functions in humans

(Linnewiel-Hermoni *et al.*, 17). Besides, biofortified pumpkin varieties, Oleshkivskyi and Sviten, treated with Riverm fertilizer, which contains organic materials, microorganisms, enzymes and growth substances, caused protein to increase by 16.8 and 15.4 per cent, respectively, compared to non-treated plants (Deinychenko and Yudicheva, 5). The results indicated that the application of fertilizer could improve the protein content in pumpkin varieties, thus, improves the quality of the fruits produced.

Biofortification through Breeding and Transgenic Approaches

Biofortification is a practice of adding nutrition value to the crop by breeding, improved agronomy and microbiological interventions, improving micronutrient uptakes and their proper distribution in the edible part, genetic engineering/transgenic approach *etc.* Biofortification of crop plants have potential to reduce the gap between micronutrient requirements and intake by increasing the proportion of dietary vitamins, iron, zinc and selenium which are essential micronutrients of health significance. In developing countries, more than 20 million people are now growing and consuming biofortified crops. Considering the nutritional impact, biofortification program have been carried out in many horticultural crops like Potato, Orange sweet potato, Cow pea, Beans, Pumpkin, Bananas, Cassava, *etc.* These are joint efforts of national and international organizations by conventional and transgenic/genome editing approaches (Prasad *et al.*, 25).

Breeding Approaches

To overcome the malnutrition/hidden hunger, biofortification of vegetable crops can be a feasible and economical approach. Vegetable crops are important part of the daily diet in developing countries of the world. Vegetable crops are often deficient in some of mineral elements. Thus, by increasing the bio-available concentration of micronutrients in vegetable crops *via* biofortification is a promising strategy in modern agriculture, providing more nutritious foods to more people with the use of fewer lands and can effectively contribute in alleviating the micronutrient deficiency. Biofortification of vegetable crops is under the process of development in many parts of the world. Different breeding methods used in developing biofortified crop varieties are explained below:

Selection and Hybridisation

The International Centre for Potato (CIP) developed iron rich potato by conventional biofortification method having 11-30 ppm iron with low phenolic compounds and hence has better bioavailability of iron to the human body. CIP in South Africa, Uganda (Harvest Plus) and National Agriculture Research and Extension System (NARES) jointly developed and released provitamin-A biofortified orange sweet potato in 2002. The biofortified orange sweet potato has upto 7.4 mg/100g fresh weight β-carotene. Donald Danforth Plant Science Center, Nigeria has developed high β-carotene cassava (7.6 ppm) by hybridization and selective breeding methods to fight against vitamin A malnutrition in Nigeria and Kenya as Bio Cassava. International Centre for Tropical Agriculture (CIAT), Colombia developed high iron and zinc biofortified beans. Usually, iron content in common bean is about 50 ppm and biofortified beans developed with 83 ppm iron and 44 ppm zinc, released in the year 2012 and 2014. In India, G.B. Pant University of Agriculture and Technology, Pantnagar, developed two early maturing high iron and zinc fortified varieties namely Pant Lobia-1 (82ppm Fe and 40ppm Zn), Pant Lobia-2 (100ppm Fe and 37 ppm Zn) by conventional plant breeding and released in 2008 and 2010. Pant Lobia-3 (67 ppm Fe and 38 ppm Zn), Pant Lobia-4 (51ppm Fe and 36 ppm Zn). Indian Agricultural research Institute, New Delhi, India has developed biofortified β-carotene rich cauliflower (Pusa Kesari VitA-1: 1000 μg/100g β-carotene). Its curds are yellow/orange coloured and compact with average marketable curd weight is 1.250 Kg and 42.0 to 46.0 t/ha yield. Biofortified crops can have a significant impact on the lives and health of millions of people, especially for micronutrient deficient population, the effectiveness of the biofortification program depends on the farmer's and consumer's acceptance and future policy interventions.

Mutation Breeding

Out of 3000 mutant varieties developed globally, 776 have been induced for nutritional quality (Jain and Suprasanna, 11). In a simple way, mutation is a random or directed change in the structure of DNA or the chromosome which often result in a visible or detectable chance in specific trait. Relatively minor genomic changes (point mutations, single gene insertions) are regularly observed following metabolomic analysis, leading to significant changes in biochemical composition and antioxidants (anthocyanin, lycopene) in the tomato *cv.* Moneymaker (Giliberto *et al.*, 9). In self-pollinated crops, it is well known whereas in cross pollinated crops its application is more difficult and it is tough to identify the origin of the desirable genotypes. Both physical (X-ray, UV) and chemical (Ethyl Ethane Sulphonete, Ethyl Methane Sulphonate, 5-bromouracil) mutagens cause macro-mutation (major, mono or oligogenic qualitative mutation) and/or micro-mutation (minor, polygenic or quantitative changes) in the genotypes under treatment. In tomato, Sapir *et al.* (30) reported that high pigment-1 (*hp-1*) mutation known to increase flavonoids content in fruits. Spontaneous mutation such as 'orange' cauliflower (Li *et al.*, 16) and in sweet potato orange mutants are rich in beta-carotene (30–100 ppm) than white fleshed (2 ppm) and some of the common orange fleshed mutant varieties are Nancy Gold and Murff Bush Porto Rico (LaBonte and Don, 15).

Polyploidy Breeding

Polyploid can be induced due to aberration in the process of cell division. This may occur both in the mitosis as well as in meiosis. This method can be used successfully in vegetable breeding as a means of enhancing nutraceuticals and colours in vegetables. Tetraploids in radish, pumpkin, muskmelon and watermelon are highly productive and have improved quality. Zhang *et al.* (38) developed tetraploid muskmelon which is rich in soluble solid, soluble sugar and vitamin-C contents which were distinctly higher than those in the diploid fruit. Liu *et al.* (18) reported that the range of lycopene content among diploid watermelon fruit was from 33.2 to 54.8 mg/kg and in triploid it was ranged from 41.2 to 61.8 mg/kg. The range of lycopene content among tetraploid was from 38.1 to 59.8 mg/kg. They also reported that the lycopene content in the triploid and tetraploid was higher than that of diploid however; ploidy did not affect lycopene content in variety Fan Zu No. 2. Marzougui *et al.* (22) induced polyploidy in *Trigonella foenum-graecum* L. using a 0.5 per cent colchicine solution and reported that the autotetraploid cultivar has larger leaf area and larger productivity compared to the diploids concerning seed number, pod number and branch number. Its leaves were richer in potassium, sodium, calcium and phosphorus.

Marker Assisted Breeding

Marker assisted breeding strategies were effectively applied in developing biofortified crop varieties. Zhang *et al.* (37) found SCAR markers linked to "*or*" gene inducing β-carotene accumulation in Chinese cabbage. Ripley and Roslinsky (27) identified an ISSR Marker for 2-propenyl glucosinolate content in Brassica. The orange cauliflower was first discovered in Bradford Marsh, Ontario, Canada in 1970. The orange cauliflower resulted from a spontaneous mutation of a single dominant gene designated as '*Or*' for orange gene (Dickson *et al.*, 7). This *Or* mutant was originally found in white curded autumn crop *cv.* extra early snowball. This trait is absent in Indian cauliflower where large population is suffering from carotene deficiency. At ICAR-IARI, New Delhi marker assisted breeding was effectively used to introgress '*Or*' gene into Indian cauliflower with enhanced β-carotene concentration.

Quantitative traits due to genetic variation can be regulated by combining effects of genes, known as quantitative trait loci (QTLs) (Sehgal *et al.*, 31). The QTL analysis using statistical methodology links phenotypic and genotypic data to genetically explain variation in complex traits. Marker assisted breeding can be used to develop gene mapping of carotene content and control of flesh color. The 4 QTLs were identified for total carotenoid content, based on phenotypic data in *Cucurbita moschata* with phenotypic variance explained (PVE) of 13.2 to 28.6 per cent, (Zhong *et al.*, 39). This will serve as a basis for map-based gene cloning, draft genome assembling and molecular breeding. In *Citrullus lanatus*, one QTL at *LCYB4.1* (lycopene β-cyclase) occurs with a high value of trait variation (83.50 per cent) related to lycopene content and mapped on Chromosome 4 between the CAPS markers, WII04E07-33 and WII04E07-40, respectively (Liu *et al.*, 19). The QTLs can facilitate study on determining lycopene content related genes and cloning watermelon genes. Other example for marker assisted breeding was in *C. sativus* with yellow flesh that contain more β-carotene than those with green and white fleshed genotypes (Lu *et al.*, 20). This is because yellow fruit flesh of *C. sativus* inbred line PI200815 was controlled by a single recessive gene, *yf*, and mapped at Chromosome 7 linked by markers yf SSR108 and yf Indel29, at 92.3 and 84.6 per cent, respectively (Lu *et al.*, 20)

Transgenic Approach

Biofortification in fruit crops achieved very limited success so far in spite very high potential. Bananas (*Musa* spp.) are economically important fruit crops grown throughout tropical and sub-tropical regions of the world where Vitamin-A deficiency is most prevalent. Some banana genotypes are rich in provitamin-A carotenoids, providing an opportunity to use bananas as a readily available vehicle for provitamin A delivery. Queensland University of Technology, Australia, developed provitamin-A (β-carotene), alfa-tocopherol and iron rich varieties of banana by transgenic approach as Super banana. The biofortified banana has up to 20 ppm beta carotene, 2.6mg/100 gm iron and higher alfa-tocopherol content.

Three genes, encoding phytoene synthase (*CrtB*), phytoene desaturase (*CrtI*) and lycopene beta-cyclase (*CrtY*) from *Erwinia* have been introduced in potato to produce beta carotene.

Romer *et al.* (29) developed transgenic tomato to enhance the carotenoid content and profile of tomato fruit, we have produced transgenic lines containing a bacterial carotenoid gene (*crtI*) encoding the enzyme phytoene desaturase which converts phytoene into lycopene. Gerjets and Sandmann (8) developed genetically engineered potato enriched with commercially important keto-carotenoids including astaxanthin (3, 32-dihydroxy 4, 42-diketo-β-carotene). Lu *et al.* (21) suggested that transgenic cauliflower with *Or* transgenesis associated with a cellular process that triggers the differentiation of proplastids or other non-coloured plastids into chromoplasts for carotenoids accumulation and also reported *Or* gene can be used as a novel genetic tool to induce carotenoid accumulation in a major staple food crop. The post-transcriptional gene silencing is also used for change in nutrient biosynthesis pathways which improve the nutrient contents.

Another big health challenge is folate deficiency which causes neural tube defects and other human diseases. Folates are synthesized from pteridine, p-aminobenzoate (PABA), and glutamate precursor. Díaz de la Garza *et al.* (6) engineered tomatoes by fruit- specific over-expression of GTP cyclohydrolase I that catalyzes first step of pteridine synthesis, and ami-nodeoxychorismate synthase that catalyzes the first step of PABA synthesis. Vine-ripened fruits contained on average 25-fold more folate than controls by combining PABA- and pteridine overproduction traits through crossbreeding of transgenic tomato plants. This research programme showed way for biofortification of folate in other vegetables for millions of people deprived of adequate folate in their regular diets however adequate attention is required on safety issues.

Biofortification through Genome Editing

CRISPR/Cas (Clustered regularly interspersed short palindromic repeats/CRISPR—associated proteins) is one of the most promising genome editing tool emerging as a revolutionary technology in diverse plant species because of its easy manipulation, high efficiency and its wide application in functional studies of genes and genetic crop improvement (Thakare *et al.*, 32). CRISPR/Cas9 tool contains a non-specific Cas9 nuclease and a single guide RNA which is specific to target gene directs Cas9 to the specific genomic location creating double-strand breaks and subsequent repair process creates insertion or deletion mutations. This is currently the widely adopted tool for reverse genetics, and crop improvement in large number of agricultural crops. Among horticultural crops, 42 per cent genome editing studies were on tomato crops, targeting development related gene edited to manipulate flowering patterns and fruit development and 13 per cent genome editing work was carried out in potato. Among horticultural crops, 72 per cent genome editing was performed in vegetable crops, some medicinal and flowering plants.

Genome Editing in Vegetable Crops

Most of the CRISPR/Cas9 experiments of horticultural crops were done in tomato where 18 different genes have been edited independently. The first CRISPR/Cas9-mediated genome editing in vegetable crops was reported in tomato in 2014, where ARGONAUTE7 (SlAGO7) gene involved in leaf development was targeted (Brooks *et al.*, 2014). Later on, other genes such as *Anthocyanin 1* (*ANT1*) gene of anthocyanin biosynthetic pathways in tomato (Cermak *et al.*, 2015), *Phytoene desaturase* (*SlPDS*), *Phytochrome interacting factor* (*SlPIF4*) (Pan *et al.*, 24), and *Phytoene synthase* (*PSY1*) (Hayut *et al.*, 2017) functioning in carotenoid biosynthesis were mutated by CRISPR/Cas9.

Potato starch quality is an important factor for food as well as many commercial product developments. The "waxy genotype" of potato, producing only amylopectin containing starch was developed by mutating granule bound starch synthase (*GBSS*) gene using CRSIPR/Cas9 (Andersson *et al.,* 2). The starch characterization of genome-edited lines produced only amylopectin and completely lacking amylose confirming the knock-out of all the four alleles of *GBSS* gene.

Parthenocarpy in horticultural crop plants is an important trait having huge demand in processing industry as well as for direct eating quality. Development of parthenocarpic fruits is considered as an important goal especially for maintaining sustainable agriculture in the face of global warming. Keeping these facts in view, Ueta *et al.* (34) have successfully introduced mutations in *SlIAA9*, a key gene controlling parthenocarpy and their results showed that, regenerated mutants exhibited seedless fruit; a characteristic of parthenocarpic tomato. Klap *et al.* (14) developed parthenocarpic tomato through CRISPR/Cas9 gene - SlAGAMOUS-LIKE 6 (SlAGL6) knockout and confirmed that a mutation in *SlAGL6* was responsible for the parthenocarpic phenotype.

Genome Editing in Fruit Crops

CRISPR/Cas9 applications in fruit crops are still at preliminary level but in view of the potential of this technology, and very few reports have appeared so far due to its perennial nature, and mutation was done mostly for carotenoid biosynthesis pathway genes. Jia and Wang (12) reported editing in citrus by mutating *Phytoene desaturase* (*CsPDS*) gene of carotenoid biosynthesis pathway. Other than citrus, genome editing by CRISPR/Cas9 has also been reported in watermelon (Tian *et al.,* 33) and grape (Nakajima *et al.,* 23) by targeting *Phytoene desaturase* gene in both the crops.

Traditional agricultural practices can partly enhance the nutritional content in plant foods but biofortification is a practice of nutrient fortification into food crops using agronomic, conventional and transgenic breeding methods to provide a sustainable and long-term strategy to address vitamin and nutrient deficiencies.

REFERENCES

1. Abd-Alkarim, E., Bayoumi, Y., Metwally, E. and Rakha, M. 2017. Silicon supplements affect yield and fruit quality of cucumber (*Cucumis sativus* L.) grown in net houses. *Afr. J. Agric. Res.* 12(31): 2518–2523.
2. Andersson, M., Turesson, H., Nicolia, A., Fält, A.S., Samuelsson, M. and Hofvander, P. 2017. Efficient targeted multiallelic mutagenesis in tetraploid potato (*Solanum tuberosum*) by transient CRISPR-Cas9 expression in protoplasts. *Plant cell reports.* 36: 117-128.
3. Brooks, C., Nekrasov, V., Lippman, Z.B. and Van Eck, J. 2014. Efficient gene editing in tomato in the first generation using the clustered regularly interspaced short palindromic repeats/CRISPR-associated9 system. *Plant physiology.* 166(3): 1292-1297.
4. Cermak, T., Baltes, N.J., Cegan, R., Zhang, Y. and Voytas, D.F. 2015. High-frequency, precise modification of the tomato genome. *Genome biology.* 16: 1-15.

5. Deinychenko, G. and Yudicheva, O. 2016. Studying the accumulation of nitrogenous substances in biofortified pumpkin vegetables. *East-Eur. J. Enterp. Technol.* 3: 40–46.

6. Diaz de la Garza, R. I., Gregory, J. F. and Hanson, A. D. 2007. Folate biofortification of tomato fruit. *Proc. Natl. Acad. Sci. USA.* 104: 4218- 4222.

7. Dickson, M.H., Lee, C.Y. and B.lamble, A. E. 1988. Orange-curd high carotene cauliflower inbreds, NY 156, NY 163, and NY 165. *Hort. Sci.* 23: 778–779

8. Gerjets, T. and Sandmann, G. 2006. Keto-carotenoid formation in transgenic potato. *J. Exp. Bot.* 57(14): 3639-3645.

9. Giliberto, L., Perrotta, G., Pallara, P., Weller, J.L., Fraser, P.D., Bramley, P.M., Fiore, A., Tavazza, M. and Giuliano, G. 2005. Manipulation of the blue light photoreceptor cryptochrome 2 in tomato affects vegetative development, flowering time, and fruit antioxidant content. *Plant Physiol.* 137(1): 199-208.

10. Hayut, S.F., Bessudo, C.M. and Levy, A.A. 2017. Targeted recombination between homologous chromosomes for precise breeding in tomato. *Nat Commun.* 8: 15605.

11. Jain, S. M. and Suprasanna, P. 2011. Induced mutations for enhancing nutrition and food production. *Gene Conserve.* 40: 201–215.

12. Jia, H. and Wang, N. 2014. Targeted genome editing of sweet orange using Cas9/sgRNA. *PloS one.* 9(4): p.e93806.

13. Karkute, S.G., Singh, A.K., Gupta, O.P., Singh, P.M. and Singh, B. 2017. CRISPR/Cas9 Mediated Genome Engineering for Improvement of Horticultural Crops. *Front. Plant Sci.* 8: 1635.

14. Klap. C., Yeshayahou, E., Bolger, A.M., Arazi, T., Gupta, S.K., Shabtai, S., Usadel, B., Salts, Y. and Barg, R. 2016. Tomato facultative parthenocarpy results from SlAGAMOUS-LIKE 6 loss of function. *Plant Biotechnol J.* doi: 10.1111/pbi.12662.

15. LaBonte. and Don, R. 2012. “Sweet potato Lists 1-26 Combined”. Vegetable cultivar descriptions for North America. Department of Horticulture, Louisiana State University, USA.

16. Li, L., Paolillo, D.J., Parthasarathy, M.V., DiMuzio, E.M. and Garvin, D.F. 2001. A novel gene mutation that confers abnormal patterns of β-carotene accumulation in cauliflower (Brassica oleracea var. botrytis). *The Plant Journal.* 26(1): 59-67.

17. Linnewiel-Hermoni, K., Khanin, M., Danilenko, M., Zango, G., Amosi, Y., Levy, J. and Sharoni, Y. 2015. The anti-cancer effects of carotenoids and other phytonutrients resides in their combined activity. *Arch. Biochem. Biophys.* 572: 28–35.

18. Liu, W., Zhao, S., Cheng, Z., Wan, X., Yan, Z. and King, S. R. 2010. Lycopene and citrulline contents in watermelon (*Citrullus lanatus*) fruit with different ploidy and changes during fruit development. *Acta Horticulturae.* 871: 543-550.

19. Liu, S., Guo, P., Wang, X., Davis, A.R., Baloch, A.M., and Luan, F. 2015. Mapping of quantitative trait loci for lycopene content and fruit traits in *Citrullus lanatus. Euphytica.* 202(3): 411–426

20. Lu, H.M., Miao, H., Tian, G.L., Wehner, T.C., Gu, X.F. and Zhang, S.P. 2015. Molecular mapping and candidate gene analysis for yellow fruit flesh in cucumber. *Mol. Breed.* 35(2): 1–8.

21. Lu, S., Eck, J.V., Zhou, X., Lopez, A.B., Halloran, D. M., Cosman, K. M., *et al.,* 2006. The cauliflower *Or* gene encodes a DnaJ cysteine-rich domain-containing protein that mediates high levels of β-carotene accumulation. *Plant Cell.* 18: 3594-3605.

22. Marzougui, N., Boubaya, A., Elfalleh, W., Ferchichi, A. and Beji, M. 2009. Polyploidy induction in Trigonella foenum-graecum L.: morphological and chemical comparison between diploids and induced autotetraploids cultivars. *Acta Botanica Gallica,* 156(3): 379-389.

23. Nakajima, I., Ban, Y., Azuma, A., Onoue, N., Moriguchi, T., Yamamoto, T., Toki, S. and Endo, M. 2017. CRISPR/Cas9-mediated targeted mutagenesis in grape. *PLoS One.* 12(5): p.e0177966.

24. Pan, C., Ye, L., Qin, L., Liu, X., He, Y., Wang, J., Chen, L. and Lu, G. 2016. CRISPR/Cas9-mediated efficient and heritable targeted mutagenesis in tomato plants in the first and later generations. *Sci. Rep.* 6(1): 24765.

25. Prasad B.V.G., Mohanta S., Rahaman S., Bareily P. (2015). Bio-fortification in Horticultural Crops. *J. Agri Engineer. Food. Technol.* 2(2): 95-99.

26. Preciado-Rangel, P., Salas-Pérez, L., Gallegos-Robles, M.A., Ruiz-Espinoza, F.H., AyalaGaray, A.V., Fortis-Hernández, M. and Murillo-Amador, B. 2018. Increasing doses of potassium increases yield and quality of muskmelon fruits under greenhouse. *Hortic. Bras.* 36 (2): 184–188.

27. Ripley, V. L. and Roslinsky, V. 2005. Identification of an ISSR marker for 2-propenyl glucosinolate content in *Brassica juncea* and conversion to a SCAR marker. *Mol. Breed,* 16 (1): 57-66.

28. Ritchie H., Reay D.S. and Higgins, P. 2018 Quantifying, Projecting, and Addressing India's Hidden Hunger. Front. Sustain. Food Syst. https: //doi.org/10.3389/fsufs.2018.00011

29. Romer, S., Fraser, P. D., Kiano, J. W., Shipton, C. A., Misawa, N., Schuch, W. and Bramley, P.M. 2000. Elevation of the pro-vitamin A content of transgenic tomato plants. *Nat. Biotechnol.* 18: 666–669.

30. Sapir, M., Oren-Shamir, M., Ovadia, R., Reuveni, M., Evenor, D., Tadmor, Y., Nahon, S., Shlomo, H., Chen, L., Meir, A. and Levin, I. 2008. Molecular aspects of Anthocyanin fruit tomato in relation to high pigment-1. *Journal of Heredity.* 99(3): 292-303.

31. Sehgal, D., Singh, R. and Rajpal, V.R. 2016. Quantitative trait loci mapping in plants: Concepts and approaches, p. 31–59. In: V.R. Rajpal, S. Rao, and S. Raina (eds.). Molecular breeding for sustainable crop improvement, sustainable development and biodiversity. Springer International Publishing, Basel, Switzerland.

32. Thakare, S.S., Bansal, N., Vanchinathan, S. Prashat G.R., Krishnana K., Sachdev, A. Praveen, S. and Vinutha T 2019. *GFP* tagging based method to analyze the genome editing efficiency

of CRISPR/Cas9-gRNAs through transient expression in *N. benthamiana*. *J. Plant. Biochem. Biotechnol.* https: //doi.org/10.1007/s13562-019-00540-0

33. Tian, S., Jiang, L., Gao, Q., Zhang, J., Zong, M., Zhang, H., Ren, Y., Guo, S., Gong, G., Liu, F. and Xu, Y. 2017. Efficient CRISPR/Cas9-based gene knockout in watermelon. *Plant Cell Reports*, 36: 399-406.

34. Ueta, R., Abe, C., Watanabe, T. *et al.*, 2017. Rapid breeding of parthenocarpic tomato plants using CRISPR/Cas9. *Sci. Rep.* 7: 507

35. Verma, S. and Kumar, A. 2019. Biofortified staples may hold the key to India's rural malnutrition. International Food Policy Research Institute (IFPRI Blog).

36. Weng, H.X., Weng, J.K., Yan, A.L. *et al.*, 2008b. Increment of iodine content in vegetable plants by applying iodized fertilizer and the residual characteristics of iodine in soil. *Biol. Trace. Elem. Res.* 123: 218–228.

37. Zhang, F., Wang, G., Mang, M., Liu, X., Zhao, X. *et al.*, 2008. Identification of SCAR markers linked to or, a gene inducing beta-carotene accumulation in Chinese cabbage. *Euphytica*, 164(2): 463-471

38. Zhang, W., Hao, H., Ma, L., Zhao, C. and Yu, X. 2010. Tetraploid muskmelon alters morphological characteristics and improves fruit quality. *Sci. Hort.* 125(3): 396-400.

39. Zhong, Y.J., Zhou, Y.Y., Li, J.X., Yu, T., Wu, T.Q., Luo, J.N., Luo, S.B. and Huang H.X. 2017. A high-density linkage map and QTL mapping of fruit-related traits in pumpkin (*Cucurbita moschata* Duch.). *Sci. Rep.* 7(1): 1–12.

Chapter 5

Emerging New Generation Fruit Crops: Associated Challenges and Perspectives

G. Karunakaran[1]*, M. Arivalagan[1], T. Sakthivel[1], P.C. Tripathi[2], V.B. Patel[2], A. Thiruganavel[3] and M. Kavino[4]

[1]ICAR–Indian Institute of Horticultural Research, Bengaluru 560089, Karnataka

[2]Horticultural Science, ICAR, New Delhi 110012

[3] ICAR-Central Citrus Research Institute, Nagpur 440033, Maharashtra

[4]Department of Fruit Crops, HC&RI, Tamil Nadu Agricultural University, Coimbatore 641003, Tamil Nadu

**e-mail: karanstg@gmail.com*

ABSTRACT

Among the new generation fruit crops, dragon fruit, avocado, rambutan, etc. hold high potential for diversification as a commercial fruit in India; and this is mainly due to nutraceutical benefits, consumer preference, present market demand, premium price, and, hence needs due attention. Empirical appraisal of agro-climatic requirements for these crops, profiling of potential tropical humid areas in parts of India through advanced tools for area expansion needs consideration. Such efforts need to be appropriately backed up by introduction of elite genetic resources already available with other countries. On-farm evaluation and grower's experiences showed that the yield and quality of these fruits grown in India are comparable to that in Thailand, Vietnam, Indonesia and Malaysia. At present, the major market demand for these crops is from major metropolitan cities in India, and market prices for these fruits varies between Rs. 150-300 per kg depending upon the grade and quality. Undertaking awareness programme among the growers, consumers and entrepreneurs regarding the potential of these fruits in the local and export markets, development of trait-specific superior types, optimization of plant propagation techniques for rapid and large scale multiplication of quality planting materials, crop husbandry practices including pests and diseases management, post-harvest management and value addition protocols are the critical needs for promoting these crops for diversification and sustainable livelihoods of the growers.

Keywords: *New generation fruit crops, Cultivation, Quality, Value addition, Export.*

India is bestowed with diverse agro-climatic conditions that could favour the production of a wide range of exotic fruit crops having national importance in view of their yield and market potential. Many of these exotic fruits got introduced in to India through the travellers and migrant settlers from the South East Asian countries. Among them, Dragon fruit (*Hylocereus* spp), Avocado (*Persea Americana* Miller), Rambutan (*Nephelium lappaceum* L.), Mangosteen (*Garcinia mangostana* L), Passion fruit (*Passiflora edulis* Sims), Durian (*Durio zibethinus* L.,) Longan (*Dimocarpus longan* Lour) were primarily introduced and raised in quite a few homesteads in India. Although precise documentary information is not available, all these introductions are of seedling origin to a great extent. Even though they got adapted gradually to the tropical humid western peninsular India, in majority, sexual propagation practised by the growers has led to development of some amount of genetic diversity through gene flow, genetic introgression and probably some extent of genetic consolidation which has not been exploited to their full potential despite for their novelty, niche markets and economic value. There exists good amount of genetic diversity in exotic fruit crops, among the seedling population across the agro-ecologies which should offer sufficient genetic base for genetic improvement of this crop through applying selection pressure initially and subsequently for developing useful recombinants. Systematic work on the exotic fruit crops started at ICAR-IIHR, Bengaluru, Karnataka, India during the year 2000s by collecting and assembling the prevailing genetic diversity from their respective growing areas.

Dragon Fruit

Dragon fruit (*Hylocereus* spp.) is an herbaceous perennial climbing cactus (Figure 1), widely known as Pitaya, has recently drawn much attention among the Indian growers due to its economic value as well as numerous health benefits. Dragon fruit is native to Southern Mexico, Guatemala and Costa Rica. It prefers dry tropical and subtropical climate with average temperature of 20-29°C, but can withstand up to 40°C, and as low as 0°C for short periods, well distributed annual rainfall of 800-2500 mm, and sandy loam soil with acidic pH (pH ranging from 5.5 to 6.5). The crop provides fast return with economic production in the first year after planting (1.5 t/acre accommodated 448 poles), and full production is attained in 3-4 years. Although, the initial investment (11.95 lakhs/ha) is relatively high, profit is substantial within 2-3 years. The biggest advantage of this crop is that once planted, it can continuously yield for about 20 years (Karunakaran *et al.*, 7).

Dragon fruit is currently cultivated in Vietnam, China, Mexico, Colombia, Nicaragua, Ecuador, Thailand, Malaysia, Indonesia, Australia and United States. Area under cultivation in most of these countries is expanding every year. For example, in Vietnam the area under cultivation increased from 2250 ha in 1995 to 54,000 hectares in 2018. As on 2017, Vietnam is the leading exporter of dragon fruit in the world, and reached the production of about 1 million metric tons (MT) with a value of US$ 1.1 billion. In India, dragon fruit import was started during 2017 with a quantity of 327 tonnes with the value of Rs 2.1 Crores, which has increased sharply to 9162 tonnes in 2019 with the value of 100 Crores; and the estimated import for 2020 and 2021 is about 11916 and 15491 tonnes, respectively, and the projected import value is about 150 crores for 2021. Presently, dragon fruit is mainly imported from Vietnam, Sri Lanka, Malaysia, and Thailand. Dragon fruit import was noticed mainly in the months of January, April, October, November, and December, *i.e.,* at the non-productive season India either due to higher demand in Indian market or glut production in the importing countries. At present area expansion of dragon fruit cultivation, India has great scope in a near future for self-sustainability.

Figure 1. Dragon fruit.

In India, dragon fruit cultivation was initiated during the 2008's and the area under cultivation is gradually increasing since 2015. Karnataka, Gujarat, Maharashtra, Telangana, Andhra Pradesh, West Bengal, Bihar, Kerala, Tamil Nadu, Orissa, NEHs, and Andaman and Nicobar Islands are the major states cultivating dragon fruit, and now cultivation extended to Rajasthan, Punjab, Haryana, Madhya Pradesh, and Uttar Pradesh. The most widely grown species is the red skin-white pulp (63 per cent) followed by red skin-pink/red pulp (33 per cent) and yellow skin-white pulp (<1 per cent). According to the data collected from MIDH, GOI, and New Delhi from the State Horticulture Mission (SHM) across the country, the total area under cultivation is about 3150 ha as on 2020. The expected area expansion is about 25000 ha in 2025. Out of a total area (3,150 ha), more than 75 per cent area (2,360 ha) is under new cultivation with less than 18 months old plantation. Further, the average productivity of these areas is ranging from 1.5 to 3.0 MT/ha. While remaining 25 per cent cultivation area (790 ha) is already well established and attained its full maturity level with average productivity of 12-18 MT/ha. Indian farmers, who follow proper management practices, realized the yield up to 18-22 MT from third year onwards.

Dragon fruit prefers full sunlight; open areas are very well suitable for planting. Generally, 1.5 m to 2 m tall single pole system is used for planting with a spacing of 10 x8 feet apart in raised bed system. Dragon fruit is a semi epiphytic vine that can climb with their aerial roots to any natural or artificial support. However, proper training of the vines on to suitable support and pruning are important for its successful cultivation. Many training systems are used in India, and single pole trellis system showed better performance in growth and yield than other systems. Pruning is important when vertical and horizontal supports are used, and the stems should be selected in such a way as to force the plant to climb over the entire support. All lateral growth and parts of the plant facing the ground should be removed, while the main stems and branch stems are kept. The removal of tip of main stem to allow new shoots to grow laterally and climb at the ring to form an umbrella like structure of vines where flowers will emanate and develop into fruits induces lateral branching. Dragon fruit normally start flowering 12-15 months after planting. Flowering is dependent on the high temperature and long photoperiod occurring during warmer months of May - September. As many as 7-8 flowering cycles may occur in tropical areas. Fruits were ready to harvest 29-30 days after fruit set for pink pulp while 31-33 days for white pulp cultivar, fruit maturity could be optimised with the change of fruit epicarp colour from green to red. Proper time of harvesting was found 3-5 days of the colour transition. Maximum shelf life is 7 days under ambient conditions and fruits can be stored 15 days under 4 ^{0}C to 6^{0}C with 90 per cent humidity (Karunakaran and Arivalagan, 4; Karunakaran *et al.,* 5, 6, and 7).

The fruit price in the beginning of the season (May- June) is much higher than those in the mid-season (around July- September). This abundant fruit causes a serious problem to the growers due to its nominal price. Whereas, during November- March, there is less number of fruits, since it is off-season and the fruit price is expensive. Therefore, farmers can produce off-seasonal fruits by adopting technology of LED bulbs (7 w 1500 lumens). It will be an alternative way to solve this problem of price fluctuations and year round production possible.

Dragon fruit is considered to be one of the 'super foods' due to its nutrient richness. It is rich in various nutrients and low in calories. The TSS of the fruit is about 9-11°Brix, and sugar content is about 5.0 to 7.0 g/100g. Glucose and fructose are the major sugars present in the fruit. Fruit contains about 1.0 g of protein and dietary fibre. It is good source for essential amino acids such as histidine, lysine, methionine, and phenyl alanine. Malonic acid and citric acid are the major organic acids identified. Presence of organic acids ensures the increased shelf life, stability, and microbial safety. Fruits with red pulp have significantly high quantum of phenolics and antioxidant potential than white pulped ones. 100g of Fruit contained 120-200 mg K, 30-45 mg Mg, 20-45 mg Ca, 20-35 mg P, 0.70-1.5 mg Fe and 0.20-0.40 mg Zn. Vitamin C was found maximum (~6 mg), followed by vitamin E (~150 μg), pantothenic acid (~50 μg), vitamin K1 (~25 μg) and niacin (~15 μg). Due to its low caloric value dragon fruit is ideal option for weight loss treatment, control of diabetes and lowering cholesterol, *etc.* Reports suggest that it's consumption could help in the control of chronic illnesses, improves the health of alimentary canal and boosting the body's immunity. It is also rich in various vitamins, minerals and dietary fibre. All these beneficial factors make dragon fruit a suitable candidate for inclusion in diet aimed at weight loss treatment, control of diabetes, lowering the cholesterol level, *etc.* (Arivalagan *et al.,* 1).

Currently, dragon fruit fresh juice is mainly sold in the markets. Many value added products can be produced on commercial scale with attractive colour without adding any synthetic colour as fruit itself is having bright purplish to pink colour pigment. Different products include ready-to-serve beverage (RTS), squash, fermented beverage, jelly, osmotically dehydrated cubes or slices, frozen cubes or slices, ice cream, flavoured milk and yoghurt can be prepared from the dragon fruit. Dragon fruit peel can be used for pectin extraction, used in the food industry as a thickener, emulsifier, gelling agent, stabilizer, and also as a fat substitute. Betacyanin pigment from peel can be extracted and used as a natural colourant in different food preparations such as ice cream, spreads, and candies.

Dragon fruit provides fast return with economic production in the first year after planting (1.5 t/acre accommodated 448 poles), and full production is attained in 3-4 years. The life expectancy of the crop is about 20 years. Initial investment for dragon fruit cultivation is comparatively high, and it mainly due to training system establishment. Estimated cost of establishment of one acre land for dragon fruit production is Rs 5,90,000, which includes cost of land, single pole training system (about 450 Nos. at Rs 1250/pole) and land preparation *etc.* For the fully established orchard, the annual maintenance cost is about Rs 3,65, 500 per acre. Average economic yield after 2 years of planting is 10 tonnes per acre. At present the market rate is Rs 100 per kg fruit, so the revenue generated by selling fruits is per year is Rs 10,00,000. Benefit Cost Ratio (BCR) is: 2.58

In India, there is an assured market for pink/red pulp dragon fruit when comparatively white pulped ones due to consumer preference. The present trend of high demand is due to the limited number of commercial producers. In Indian market consumer demand red and pink pulp cultivar, juicy and sweetness when compared to white pulp. Due to the awareness among the consumers about the nutritional and medicinal properties of dragon fruit, the demand may increase in near future. Mostly in the Indian retail market, dragon fruit is often sold as fresh produce. In many places, growers could also sell their produce directly to consumers through mobile/digital markets or farmers' markets (Mandi). At present, the demand for dragon fruit is mainly in metropolitan cities like Mumbai, Delhi, Kolkata, Chennai, Pune, Hyderabad, and Bengaluru, and a few tier-two cities/towns. Dragon fruit with red pulp fetches more price (Rs 90- 160/kg) compared to white pulped fruits (Rs 55-160/kg).

Since the dragon fruit is introduced to India, the availability of diverse germplasm is limited, there is a need to enrich varietal wealth by widening the genetic base. Location specific complete package of nutrient management needs to be developed. Post-harvest handling, cold storage, processing and value addition needs to strengthened. Technical know-how for farmers and awareness campaign for consumers about health benefits to be imparted.

Avocado

Avocado is the most nutritive fruit (Figure 2) and is regarded as the most important contribution of new World to human diet (Karunakaran *et al.,* 8). The pulp is rich in fat (up to 30 per cent), but low in carbohydrates. The fat is similar to olive oil in composition and is widely used in the preparation of cosmetics. Avocados has the highest energy value (245 cal/100 g) of any fruit besides being a reservoir of several vitamins and minerals (USDA, 21). Avocado contains a significant amount of mono-saturated fat, which helps to reduce bad cholesterol, and vitamin E also aid in delaying of aging of skin. It is also a great source of potassium which is known to prevent high blood pressure and fluid retention. Avocado is mainly used fresh, in sandwich filling or in salads. It can also be used in ice creams and milk shakes and the pulp may be preserved by freezing.

Figure 2. Avocado (Arka Supreme).

In India, avocado was introduced from Sri Lanka in the early part of the twentieth century. In a very limited scale and in a scattered way it is grown in Tamil Nadu, Kerala, Maharashtra, Karnataka in the south-central India and in eastern Himalayan state of Sikkim (Ghosh, 3). It cannot tolerate the hot dry winds and frosts of northern India. Climatically, it is grown in tropical or semitropical areas experiencing some rainfall in summer, and in humid, subtropical summer rainfall areas.

The agroclimatic conditions prevailing in various parts of the country appear to be favourable for bringing more area under avocado. Presently, plantations are not well organized and they are scattered. Also, quite a good number of improved varieties are now available with higher yield potential. Vegetative propagation techniques have also been standardized. Multiplication of a large number of high-quality nursery plants of selected varieties and their systematic planting, both in tropical southern India and humid semi-tropical areas of the northeastern region of India, could help to place avocado properly on the fruit map of India. The research support for the crop is still very poor, but the available research information from Tamil Nadu and Sikkim amply demonstrate that size, colour and quality of fruits attained in India are comparable with those of avocado fruits grown elsewhere. Avocado is grown successfully in neighbouring Sri Lanka, where good-quality fruits are harvested during May to August and December to January at different regions. Similar agro-climatic conditions to Sri Lanka are available in the Andamans and Nicobar Islands and in the tropical southern India. With proper varietal selection it should be possible to exploit also the possibility of out-of-season production, thus enhancing the availability of fruits for a longer period during the year. Avocado fruits produced in the country can be marketed without much difficulty, particularly to meet the requirement of the growing tourist industry. The mainland India and the Andaman and Nicobar Islands are attracting foreign tourists in a large number of places, where avocado could find a good market access. Avocado has also a good export potential (Karunakaran *et al.,* 8).

All three horticultural races adapted to tropical and sub-tropical conditions, *i.e.* West Indian, Guatemalan and Mexican have been tried in India. The cultivars of West Indian race are grown in localized pockets in Maharashtra, Tamil Nadu and Karnataka. In tropical and near-tropical areas, only West Indian race is well-adapted but its hybrids with Guatemalan (*e.g.* Booth selection) perform well and are considered valuable for extending the harvest season. In less tropical regions, hybrids of Guatemalan with Mexican race are predominate since they combine the cold hardiness of the latter with the superior horticultural traits of both and also bridge the two seasons of maturity. In the eastern Himalayan state of Sikkim, avocado has been introduced successfully in hill ranges with an altitude of 800- 1,600 metres. Both the Mexican and Gautemalan races are grown successfully in Sikkim. The varieties that are cultivated in India go by several names, such as Purple, Green, Fuerte, Pollock, Peradeniya Purple Hybrid, Trapp, Round and Long. Among the several existing varieties, perhaps Fuerte is the most widely grown, but it is regarded as unsuitable for the tropics. The Purple and Green varieties were introduced into India from Ceylon in 1941. In Kodagu, avocado is grown as one of the mixed crops in coffee-based cropping system. Almost each house is maintaining few plants of avocado, and lots of variability of avocado is available in Coorg and adjoining areas.

In India, avocado is commonly propagated through seeds. The seeds taken from mature fruits are sown directly in the nursery or in polyethylene bags. After 6-8 months, the seedlings are ready for transplanting. Vegetative propagation by means of budding or grafting has resulted in establishment of selected varietal clones. Need of *Phytophthora*-resistant rootstocks is recognized as the most important factor for the success of its cultivation. In India, presently, there is many commercial nurseries engaged in vegetative propagation of avocado. It was concluded that cleft grafting in the month of September and March was found most suitable for the vegetative propagation of avocado under humid tropical conditions (Tripathi and Karunakaran, 16).

If an avocado plantation is to be established in a relatively new area, the varieties to be selected for planting should belong to both A and B groups in the ratio of 1:1 or 2:1 and their flowering must overlap. Avocado is planted out to a distance of 6 metres depending on the vigour of variety and its growth habit. In areas prone to excess water, they should be planted on mounds as avocados cannot withstand water logging. The pollination in avocado is mainly performed by insects. The studies conducted revealed that insect species including 7 hymenopterans, 7 dipterans, 3 coleopterans were recorded on avocado flowers. Among all the visitor groups, hymenoptera order constituted the major pollinators (52.66 per cent), followed by dipteran order (39.57 per cent) and coleopteran was the least (7.62 per cent) of the total pollinators in avocado orchard. In the group of hymenoptera it was found that *Apis cerena* as a major pollinator with 17.56 per cent followed by *Apis florea* with 11.3 per cent (Tripathi *et al.,* 15).

Among insect pests, stem borer, scales, mealy bugs and mites are the important ones, and may be controlled by suitable insecticides. Fruit spot disease caused by *Colletotrichum gleosporioides* results in shedding of young fruits. Fruits often become deformed. The infection may remain latent in some fruits. Another strain of the same fungus causes leaf spot. The Fuerte cultivar is reported to be more susceptible to anthracnose (*Glomerella cingulata* var. minor) and stem-end rot (*Dothiorella aromatica*) from fruit set till harvest. The time of infection has been found to vary with the seasons and is related to rainfall. In addition, Cercospora spot (*Cercospora purpurea*) and scab (*Sphaceloma perseae*) have been found to attack both leaves and fruits also. The most serious disease of avocado is the root rot caused by *Phytophthora cinnamoni*, leading to death of plant. The disease situation is aggravated by ill drained and waterlogged conditions.

Avocado plants raised from seeds start bearing four to five years after planting. Mature fruits of purple varieties change their colour from purple to maroon, whereas fruits of green varieties become greenish-yellow. Fruits are ready for harvest when the colour of seed coat within the fruit changes from yellowish white to dark brown. Mature fruits ripen six to ten days after harvesting. The fruits remain hard as long as they remain on the trees, softens only after harvest. The yield ranges from about 100 to 500 fruits per tree. Avocados do not ripen on the tree, and fruits soften only after they are picked. Fruits need to be picked carefully. They should be harvested at the correct stage of maturity, when they are still hard and have a minimum oil content of 12 per cent. In India, fruits of 250 to 300 grams in size are preferred. Presently, there is no organized marketing system for avocado as the production is small and production areas are scattered.

One of the important constraints appears to be due to the consumer preference. In the domestic market the avocado may not be liked by the common people due to the nature of its taste and also due to availability of many other tropical fruits throughout the year which are more palatable than avocado. Nevertheless, due to increasing health consciousness among the educated population and the high nutritive value of avocado, it is expected to find its rightful place in the Indian market in due course. The successful introduction of avocado and its wide acceptance among the tribal population of the hilly state of Sikkim indicate that for household nutrition security avocado should be a potential fruit crop in India. Presently, there is plan to strengthen research and development of avocado in India. Due to the availability of a large number of fruit crops in India and consumer preference for more palatable fruits of sweet taste, avocado has not caught the imagination of the average Indians. With some governmental support it can be popularized in the country and it can provide much needed household nutrition security to the Indian population.

Rambutan

Rambutan, one of the important fruits of humid tropical region, is native to the Malay Archipelago or Malesia (Muhamed *et al.,* 13), and widely cultivated in Thailand, Indonesia, Malaysia and Sri Lanka. In India, it is primarily introduced and grown in backyards of some parts of Kerala, Karnataka and Tamil Nadu (Tripathi *et al.,* 20). Fruit resembles litchi fruit in appearance and hairy outgrowth has eye catching yellow or pink and red colours and it imparts a distinctive exotic appearance to fruits (Figure 3). The translucent pulp is sweet, juicy and delicious with a pleasant flavour. Recently, cultivation of rambutan gained substantial momentum due to steadily increasing market demand and lot of planting material is being produced, and plantings have increased appreciably in southern India. Currently, all the rambutan grown are for intrastate and interstate trade. The major market demand from Bangalore, Chennai, Mumbai and Coimbatore and market prices varied between Rs. 200-300 per kilo depending upon the grade and quality (free stone nature) (Karunakaran *et al.*, 9).

Figure 3. Rambutan.

The crop is adapted to warm tropical and subtropical climate. It requires around 22-30 °C for optimum growth and very sensitive to low temperature. The high (>40 °C) and low (less than 10 °C) temperatures affect the growth and development of plants. The tree grows well up to 700 m above sea-level and requires well-distributed rainfall of 200-500 cm throughout the year, but a short spell of the dry period is essential for initiation of floral bud (Ravishankar *et al.*, 2007). Rambutan seedlings grow into male, female, or hermaphrodite trees which can be identified only after flowering. Hence, rambutan plant should be vegetatively propagated by approach grafting, patch budding and air layering. It can be propagated through seeds, cuttings and grafting or budding. However, patch budding on 8 to 12-month-old rootstocks during May or July give the best results. Several commercial varieties of rambutan are grown in South-Eastern Asian countries. CHES (IIHR) Chettalli has recently released two varieties namely Arka Coorg Arun and Arka Coorg Peetabh. Arka Coorg Arun is an early maturing variety with red coloured fruits (Figure 4). Fruits mature in September - October and average fruit weight is 40 to 45 g. Arka Coorg Peetabh is a high yielding, yellow coloured variety with the average fruit weight of 25-30 g (Tripathi *et al.*, 18).

Generally, rambutan is planted at 10-12 m spacing; however, the vegetatively propagated trees can be planted at the spacing of 8x8 or 8x6 or 6x6 m. The close planting of 6x6 or 5x5 or 4x4 m may need pruning to control the tree size which facilitates easy cultural operations. The trees exhibit strong apical dominance and tend to produce long, upright growth. Early pruning and training to form an open center is recommended, and it is done either by cutting down the height of the plant 3-4' above the ground level after 1-2 year of planting which allows primary branches in different directions. While doing pruning, one must make sure that the plants have at least minimum canopy beneath the cutting portion to protect the plant from dying out. In Kerala, rambutan usually flowers during December-January and rambutan fruits are available in June-August. Pruning time varies depending upon of the location and time of harvest (Karunakaran *et al.,* 9).

Figure 4. Arka Coorg Arun and Arka Coorg Peetabh.

Exact fertilizer requirement of rambutan not worked out and it is also difficult to specify a common dose as soils vary in fertility. For fruiting trees, 200g N, 25g P and 130g K per tree per year of age are recommended and 20 kg FYM along with Fish meal and goat manure each 5 kg annually in two splits. Further, it recommends that three foliar sprays of 19:19: 19 NPK during new flush growth, after fruit set and after pruning. It may be noted that the dose indicates here has a general applicability. Providing shade is desirable in the first two years after planting. In order to induce the flowering of full-grown plants, there is a need to allow the rambutan plant to pass through a period of drought for about 21-30 days. Then irrigation is done to induce floral bud initiation within 10-15 days after water manipulation.

Rambutan is cross-pollinated and can be staminate or pistillate or hermaphrodite and it blooms after a dry period of about 21-30 days (Tripathi *et al.,* 17). The flowering period continues for 30-45 days depends on stress and climatic conditions. Irrigation is necessary for dry seasons. This is insect-pollinated crop, pollinated mostly by *Apis cerana* followed by *A. florea*, *Trigona spp.*, ants, and wasp.

Rambutan gets very less insect pest damage. The most important yield reducing factor in rambutan plantations are birds and bats (flying foxes) and possibly some other non-flying mammals that consume the fruits. Major pest is mealy bug (*Pulvinaria psidii*) and it is noticed on rambutan fruits when they attain pea size and continue to persist till the ripening stage. Application of carbaryl 4 g/L or dimethoate 2 ml/L mixed with neem oil may be helpful in reducing the damage caused by mealy bug infestation. However, recent experiment shows that the mealy bugs can be reasonably well suppressed by applying Lastraw @ 5 ml/L when the fruits are of coffee berry size. Attendant ants that may feed on the honey dew may be prevented reaching the target by drenching soil with Chlorpyriphos (3 ml/L). There are several pathogens that attack the fruit and cause rotting under warm and moist conditions, apply bordeaux mixture (1 per cent) fungicides after fruit set and pruning.

In Kerala and some parts of Karnataka, rambutan usually fruits during June-August. However, in Coorg fruits available from September - October. Well managed rambutan trees begin to fruiting at fourth year onwards. The budded trees start bearing fruits from the fourth year onwards. It may take up to five months for the flowering to harvest. An average tree produces 5,000-6,000 fruits (60–70 kg). Studies indicated that there is ample scope and potential future of successful rambutan production with the area expansion in Kerala, Karnataka and Tamil Nadu and additional areas may also be explored.

Passion Fruit

Passion fruit belongs to the family Passifloraceae and is native to Tropical America (Knight and Winters, 11). There are two botanical forms of edible passion fruit (*P. edulis*) namely purple type (*P. edulis* Sims) and yellow type (*P. edulis* f. *flavicarpa* Deg.). Passion fruits have a tough, smooth, waxy dark purple/yellow coloured rind with faint, fine white specks (Figure 5). Fruit contains orange colored pulpy juice with large number of small, dark brown to black pitted seeds. The fruits are generally not used for table purpose but they possessed unique flavor and aroma and high nutritional and medicinal properties (Chen *et al.*, 2). They are processed to make fruit juice and concentrate. The juice is delicious and has an excellent flavour. It is known for its blending quality. Fruits contain good proportion of reducing and non-reducing sugars and acids. In India, though passion fruit cultivation was limited to few districts of Karnataka, Kerala and Tamil Nadu (Tripathi *et al.,* 19). Since last one decade it's cultivation started in some parts of northern India, especially north eastern states. The area under this crop is rapidly increasing in Mizoram, Nagaland, Manipur and Sikkim because of great potential for its cultivation in these states (Thokchom and Mandal, 14). It is grown in very limited scale and remains in back yard cultivation. Presently, there is a huge demand for passion fruit pulp due to attractive flavour of juice. In Coorg many small-scale industries and NGOs are established for passion fruit pulp processing. There is no released variety of purple and yellow colour passion fruit in India. The farmers grow only local genotypes of purple and yellow varieties in South and Northeast India and some farmers of South India cultivate a hybrid 'Kavery' (Kishore *et al.*, 10). A hybrid 'Kavery' was developed at IIHR- Central Horticultural Experimental Station, Chettalli in 1986. This is a cross between purple and yellow variety. Fruits are ovoid to round, big (85-110 g) and purple dotted colour. It is field tolerant to nematodes, wilt, brown leaf spot and collar rot.

Passion fruit is propagated through seeds, stem cutting as well as grafting. Plants of seedling origin are more vigorous than the plants raised by cuttings. Passion fruit propagated through cutting or grafting starts fruiting much earlier (7-6 months) than those from seeds (10-12 months). In case of grafting on resistant rootstocks (yellow passion fruit) it can avoid damage caused by wilt or root rot. Stem cuttings is most popular method of multiplication of passion fruit. Rooting may be hastened by treatment of 200 ppm NAA for a very short period (3-5second). Planting is done in pits of 45 x 45 x 45 cm at three metre from row to row and two metre plant to plant. Passion fruit is a woody vine it needs support for good growth and fruiting. For commercial cultivation, the vines are trained on a frame of wires and poles 1.5 to 2 meter above the soil surface. Among the different types of trellising, kniffin system is the most economical. Passion fruit vines bear fruits only on current season's growth and systematic prun-ing of vine encourage new growth resulting in regular and higher yield of fruits. Pruning is generally done twice in a year, first in March - April and another in October-November depending upon the harvest of the crop (Karunakaran *et al.,* 9).

Figure 5. Passion fruit (Kaveri hybrid).

A fertilizer dose of 110g N, 60g P_2O_5 and 110g K_2O per vine per annum is recommended for the four-year-old orchards in South India. Prolonged dry spell during January-March may reduce main summer crop and may also affect adversely the development of flowering laterals. Passion fruit is shallow rooted thus deep digging should be avoided. Passion fruit flowers and fruits produces throughout the year under favourable conditions. There are two main periods of fruiting *i.e.* August - December and March - May. The first fruiting is obtained from the ninth month after planting. On an average, yield of 8-10 tonnes per hectare per year can be obtained. The vines are perennial and can produces yield for 10 to 15 years but maximum production can be obtained up to six years after that the yield declines. Root rot diseases cause considerable damage to crop. To control the disease, water logging should be avoided by providing proper drainage. Drenching with Bordeaux mixture (1 per cent) may be done and affected plants should be removed. Among the Insect-Pests, fruit fly, thrips, mites and nematodes are major. Suitable pesticides should be used to control these pests.

Mangosteen

Mangosteen is a broad-leaved, medium-size, ever-green tree. It is considered by many to be the most delicious fruit of the tropics with a universal appeal, 'the finest fruit of the world' or 'queen of fruits' (Moongkarndi *et al.,* 12). The fruit was introduced more than a century ago in India, but currently it is successfully grown only in selected places on slopes of Nilgiris, Malabar and Waynad. The mature tree reaches a height of 10 to 15m with a dense pyramidal crown and glossy bright leaves. Mangosteen fruit resembles small-sized orange with a smooth reddish purple cortex (Figure 6). The pulp is white, soft and juicy that almost melts in mouth with an indescribably sweet and pleasant flavour. Mangosteen requires humid tropics with 180-250 cm rainfall. It is found growing in an altitude of 400-900 m in south India.

Figure 6. Mangosteen.

The occurrence of natural variability is also limited by the fact that 'seeds' are of asexual origin; they are formed from the nucellar tissue in the parthenocarpic fruit. Mangosteen is commonly propagated through seeds. Planting distances vary from 8 to 10 m between trees, depending on soil fertility. It is grown as a component of a mixed crop planting. Mangosteen is a slow growing tree that can grow up to 10 to 25 meters high, but it has a long juvenile period. Mangosteen requires good soil moisture and higher humidity for better growth and yield. Regular irrigation is required in dry season to sustain the growth of the plants. Adequate amount of farmyard manures should be applied. Mangosteen trees produce flowers and fruits on the terminals of new growth on small twigs that develop from the main stem or large branches. A large number of strong healthy and productive branches and terminal shoots should be left on the tree. The trees begin to bear fruits 10 to 12 years after seeding. However, if the plant is well cared, fruiting can begin at 7 to 8 years after seeding. Technology needs to be established to improve the mangosteen production.

REFERENCES

1. Arivalagan, M., Karunakaran, G., Roy, T.K., Dinsha, M., Sindhu, B.C., Shilpashree, V.M., Satisha, G.C., and Shivashankara, K.S. 2021. Biochemical and nutritional characterization of dragon fruit (*Hylocereus* species). *Food Chemistry*. **353**: 2021-129426.

2. Chen, S. L., Wu, J. S. B. and Wu, C. M. 1987. The preparation of concentrated passion fruit juice by methods of flavour recovery and enzyme treatment. *Journal of Chinese Agricultural Chemistry Society*. **25**: 177-187.

3. Ghosh, S. P. 2000. Avocado Production in India. In Avocado production in Asia and the Pacific. FAO Corporate Document Repository. Bangkok, Thailand: FAO Regional Office for Asia and the Pacific.

4. Karunakaran, G., and Arivalagan, M. 2019. Dragon fruit. In: Production Technology of Tropical and Sub Tropical Fruits --A Hand Book. Eds: Satisha J, Kurian RM, Vincent L, and Anushma PL. ICAR-IIHR, Bengaluru. Pp 136-140.

5. Karunakaran, G., Arivalagan, M., and Sriram, S. 2019. Dragon fruit country report from India. In: Dragon Fruit network (DFNet): Marketing and the Whole Value Chain, Vietnam. Pp-105-112.

6. Karunakaran, G., Arivalagan, M., and Sriram, S. 2020. Dragon Fruit - A potentially healthy niche fruit crop in waiting. *Indian Horticulture*. **65(4)**: 41-44

7. Karunakaran, G., Arivalagan, M., and Sriram, S. 2021a. Cultivation and Characterization of Dragon fruits in India. In: Dragon fruit – From Production to market. Eds: Fullerton RA, and Lee WL. Publisher: FFTC, Taiwan. Pp. 131-150

8. Karunakaran, G., Tripathi, P.C., Pritee Singh, Arivalagan, M., and Sriram, S. 2021b. Smart Fruits for 21st Century. In: Current Horticulture: Improvement, Production, Plant Health Management and Value-addition. Pp-423-437

9. Karunakaran, G., Tripathi, P.C., Thiruganavel., and Arivalagan, M. 2021c. Underutilized Fruits of Southern India. In: Underutilized Fruits of India. Eds: Singh, R.S., Singh, A.K., Maheshwari, S.K., Bhargava, R., and Saroj, P.L. Brillion Publishing, New Delhi. Pp. 39-54.

10. Kishore, K. K., Pathak, A., Yadav, D.S., Bujarbaruah, K.M., Bharali, R. and Shukla, R. 2006. Passion Fruit. *Tech Bull.* pp. 2-4.

11. Knight, R.J. and Winters, H.F. 1962. Pollination and fruit set of yellow passion fruit in southern Florida. Proc. *Florida Hort. Soc.* **75**: 412–418

12. Moongkarndi, P., Jaisupa, N., Samer, J., Kosem, N., Konlata, J., Rodpai, E., and Pongpan, N. 2014. Comparison of the biological activity of two different isolates from mangosteen. *Journal of Pharmacy and Pharmacology*. **66**: 1171–9.

13. Muhamed, S., Kurien, S., Iyer, K.S., Remzeena, A. and Thomas, S. 2019. Natural diversity of rambutan (*Nephelium lappaceum* L.) in Kerala, India. *Genetic Resources and Crop Evolution*. **66(5)**: 1073-1090.

14. Thokchom, R., and Mandal, G. 2017. Production preference and importance of passion fruit (*Passiflora edulis*): A review. *Journal of Agriculture Engineering and Food Technology*. **4(1)**: 27-30.

15. Tripathi, P. C., G. Karunakaran, T. Sakthivel, V. Sankar, and R. Senthilkumar. 2014a. Avocado Cultivation in India. Technical Bulletin 2/2014, ICAR-IIHR, Central Horticultural Experiment Station, Chettalli, Kodagu, Karnataka, Kodagu, Karnataka

16. Tripathi, P.C., and Karunakaran, G., 2019. Standardization of time and method of propagation in avocado. *Journal of Applied Horticulture*. **21(1)**

17. Tripathi, P.C., and Karunakaran.G, 2012. Butter Fruit: Cultivation Practices (In Kannada). Sujath: *Progressive farmers Magazine*. **18(12)**: 10-12.

18. Tripathi, P.C., G. Karunakaran and Sakthivel, T. 2011. Promising lines of under-utilized fruits at CHES, Chettalli, pp 8.

19. Tripathi, P.C., Karunakaran, G., Sakthivel, T., Sankar, V. and Senthil kumar, R. 2014c. Passion fruit cultivation in India. Tech Bull., ICAR-IIHR, Central Horticultural Experiment Station, Chettalli, Kodagu, Karnataka, Kodagu, Karnataka
20. Tripathi, P.C., Karunakaran, G., Sakthivel, T., Sankar, V. and Senthil kumar, R. 2014b. Rambutan Cultivation in India. Tech Bull., pp. 1-2.
21. USDA. 2011. Avocado, almond, pistachio and walnut composition. Nutrient Data Laboratory. United States Department of Agriculture (USDA) National Nutrient Database for Standard Reference, Release 24. Washington, DC: USDA.

Chapter 6

Apricot (*Prunus armeniaca* L.): An Underutilized Fruit Crop in Trans-Himalayan Ladakh, India

Tsering Stobdan[1]*, Deldan Namgial[2], O.P. Chaurasia[1], Mushtaq Wani[3], Tsewang Phunchok[4] and Mohmad Zaffar[5]

[1]*Defence Institute of High-Altitude Research, Defence R&D Organisation, Leh-Ladakh 194101, Union Territory of Ladakh*

[2]*High Mountain Arid Agriculture Research Institute, SKUAST-K, Stakna, Leh Ladakh*

[3]*Sher-e-Kashmir University of Agricultural Sciences and Technology of Kashmir, Shalimar, Srinagar, J&K*

[4]*Horticulture Department, Leh Ladakh 194101, Union Territory of Ladakh*

[5]*Horticulture Department, Kargil Ladakh, Union Territory of Ladakh*

**e-mail: stobdan.dihar@gov.in*

ABSTRACT

Ladakh is the biggest apricot producer in India. The total fruit production is 15,789 MT (Leh: 5,133; Kargil:10,656 MT). The fruit crop is a major source of livelihood and is also deeply associated with the tradition and culture of the region. The major part of apricot produced is consumed locally. The fruit is being sold in various forms viz. fresh, dried, kernels, oils etc. Historically, the premium quality dried apricots, locally known as Phating, with high sugar and dry matter content, were one of the main trading commodities. Raktsey Karpo, apricots with white seed coat, is unique to Ladakh. It is the sweetest apricot and has been ranked as the most preferred cultivar for fresh consumption. However, of today, the popularity of Ladakhi apricots remains restricted to the region due to limited production, incidence of codling moth (Cydia pomonella), and lack of organized marketing system. The efforts to organize apricot marketing and drying have not been initiated in the region, and thus there is a need to organise fruit collection, processing, value-addition and marketing. The disorganized nature of the sector and adherence to certain inefficient production and processing techniques implies that 40-50 percent of the apricot production is wasted. Critical areas that need focused attention include promotion of premium quality cultivars, plantation in orchard system, adoption of standard cultural practices, conservation of the rich genetic resource, organic certification, adoption of integrated post-harvest management, management of insect-pests, and establishment of marketing infrastructure. In view of the uniqueness and premium quality of apricots of Ladakh there is immense scope for Ladakh to emerge on the world map for apricot production. Export of Ladakh fresh apricots started in year 2021.

***Keywords**: Underutilized fruit, Ladakh, Kargil, Leh, Organic, Export.*

Apricot (*Prunus armeniaca* L.) is one of the most popular temperate fruit tree species. It is highly appreciated by consumers, as the fruit exhibits a perfect balance between sugars and organic acids, combined with a strong and rich aroma. The common apricot grows in geographically diverse areas ranging from the cold winters of Siberia to the subtropical climate of North Africa and from the deserts of Central Asia to the humid areas of Japan and eastern China. However, areas under commercial production are still very limited (Mehlenbacher *et al.*, 4). Traditionally, apricot has been one of the few temperate fruit trees not affected by overproduction and often premium prices are reported for both fresh and processed fruits (Hormaza *et al.*, 2). Some apricot cultivars are particularly prone to irregular productions that have been associated to the narrow adaptability range of the species. Indeed, while in other fruit tree species a few cultivars are grown all around the world, in apricot each cultivar is usually restricted to a particular geographical area with certain ecological conditions (Layne *et al.*, 3).

World Scenario

Global apricot production is estimated at 4.5 million tonnes in 2016-18. The world's largest producers of apricots are Turkey and Uzbekistan, accounting for 17.3 per cent and 17.2 per cent of world apricot production, respectively during 2016-18.

The largest volumes of apricot exports are traded fresh. Global fresh apricot export is estimated at 406,300 tonnes in 2017. Fresh apricot exports have increased by 59 per cent from 2010 to 2017, while dry apricot exports have increase by 18 per cent during the same period. Although Spain is only the world's seventh largest producer of apricots, in 2017 it exported 52.7 per cent of its harvest, which is 22 per cent of total world exports, making it the global leader in fresh apricots. Correspondingly, fresh apricot imports have also increased. Germany is the world's leading importer, importing 17.6 per cent of all fresh apricot traded in 2017 (FAOSTAT).

Global dry apricot export is estimated at 139,900 tonnes in 2017. Turkey is the leading exporter of dried apricots, exporting 67.9 per cent of all world exports in 2017 followed by Uzbekistan (6.8 per cent) and Tajikistan (6.6 per cent). Kazakhstan is the world's leading importer, importing 13 per cent of all dried apricot traded in 2017. USA is the second leading importer, importing 11.3 per cent of dried apricot in 2017 (FAOSTAT).

Indian Scenario

Apricot is grown commercially in Ladakh, Jammu and Kashmir, Himachal Pradesh,

Uttaranchal and to a limited extent in the north eastern hills. India produces approximately 15,279 tonnes of fresh apricot annually during 2015-2018, which is only 0.35 per cent of the global apricot production. Apricot is cultivated on 5389 ha, which represents only 0.95 per cent of the total area under apricot cultivation in the world. India is the sixth largest importer of dried apricot, importing 4.2 per cent of all world import in 2017 (5457 tonnes). Import value of dried apricot was 24.8 million USD in 2017. However, import of fresh apricot was only 113 tonnes in 2017, which is valued at 0.1 million USD. India also exported small quantity of fresh (36 tonnes) and dried apricots (42 tonnes) in 2017 (FAOSTAT).

Apricot In Ladakh: Current Status

Ladakh is the biggest apricot producer in India with a total production of 15,789 tonnes (Leh: 5133; Kargil:10,656 tonnes). The fruit crop is a major source of livelihood and is also deeply associated with the tradition and culture of the region. Apricots of Ladakh are known for their quality and the region has not witnessed introduction of cultivars from outside the region except for trial purposes. The gene pool is maintained due to geographical isolation and natural high mountain barrier. Historically, the premium quality dried apricots, locally known as *Phating*, with high sugar and dry matter content, were one of the main trading commodities. *Raktsey Karpo*, apricots with white seed coat, is unique to Ladakh. It is the sweetest apricot and has been ranked as the most preferred cultivar for fresh consumption (Naryal *et al.*, 6). However, of today, the popularity of Ladakhi apricots remains restricted to the region due to limited production and incidence of codling moth.

Apricot, locally known as *Chuli*, is classified in Ladakh into two broad categories based on kernel taste. Fruit with a bitter kernel are called *Khante*, meaning bitter, while those with a sweet kernel are called *Ngarmo*, meaning sweet (Targais *et al.*, 9). The *Ngarmo* is further divided into two subgroups based on seed coat color. Fruits with white seed coats are called *Raktsey Karpo*, while those with brown seed coats are called *Nyarmo*. Large fruits are called *Chenmo*, meaning large, while small ones are called *Chuun*, meaning small. Fruit weight ranged from 7.1-53.7 g with mean weight of 21.6±9.3 g (Naryal *et al.*, 7).

Area under Cultivation

The total area under apricot cultivation in Ladakh is 2,303 ha (Kargil: 1,645 ha; Leh: 658 ha) in 2019. The area under apricot cultivation has increased by 9.4 per cent in Kargil district and decreased by 16.9 per cent in Leh district from 2015 to 2019. The decrease in area under apricot cultivation in Leh district is due to dying of plants in flood affected areas.

Production

Ladakh produces approximately 15,789 tonnes of fresh apricot, making it the largest producer of apricot in the country. The fresh apricot production in Kargil and Leh district are shown at Table 1. The fresh apricot production has increased by 3.8-fold in Kargil district and 1.6-fold in Leh district during 2015 - 2019.

Table 1. Fresh apricot production in Ladakh (tonnes) (2017-19) (Stobdan *et al.*, 8)

District	*Zone*	*2017*	*2018*	*2019*
Kargil	Kargil	4045	5075	6375
	Batalik	2455	2555	2750
	Chicktan	1276	1403	1531
	Total	**7776**	**9033**	**10656**
Leh	Khalsi	542	1850	2910
	Saspol	195	481	384
	Leh	603	640	451
	Kharu	118	260	107
	Nubra	466	1068	1281
	Total	**1924**	**4299**	**5133**
Total		**9,700**	**13,332**	**15,789**

Dried Apricots

The dried apricot production in Ladakh is shown at Table 2. Ladakh produced approximately 1,999 tonnes of dried apricot in 2019, making it the largest producer of dried apricot in the country. The dried apricot production has increased by 6.1-fold in Kargil district and no significant increase was observed in Leh district from 2015 to 2019.

Table 2: Dried apricot production in Ladakh (tonnes) (2017-19) (Stobdan *et al.*, 8)

District	*Zone*	*2017*	*2018*	*2019*
Kargil	Kargil	501.6	628.1	790.5
	Batalik	303.4	316.8	341.0
	Chicktan	158.4	174.0	189.7
	Total	**963.4**	**1118.9**	**1321.2**
Leh	Khalsi	70.5	240.5	384.1
	Saspol	70.6	63.5	51.0
	Leh	78.6	84.5	59.7
	Kharu	13.5	34.2	14.2
	Nubra	61.5	141.0	169.0
	Total	**294.7**	**563.7**	**678.0**
Total		**1258.1**	**1682.6**	**1999.2**

Not all parts of Ladakh produce premium quality dried apricots. Micro-climate plays an important role in production of apricots that possess characteristics of quality dried fruits. In general, villages that are located in narrow valley below 9,500 feet elevation produce the best quality dried apricots. Villages that are known for producing premium quality dried apricots (Halman/*Phating*) in Kargil district are Hardaz, Karkichu, Majidas, Kharul, Hundurmal,

Shilikchay, Gongma Kargil, Manjee, Batalik, Chulichan, Gurgurdo, Sanatsey, Hordas, Garkhon, Darchik, Sanjak and Dargo. Similarly, in Leh district the villages known for producing premium quality dried apricots are Dha, Biama, Hanu Thang, Achinathang, Lehdo, Skurbuchan, Domkhar-Dho, Takmachik, Turtuk, Bokdang and Thang (Stobdan *et al.*, 8).

Value Added Products

A small quantity of processed products such as jam, juice and seed oil are produced by few privately-owned firms. However, less than 50 tonnes of apricot pulp are being used for product development, and are locally consumed. There is ample scope to increase the market share of locally produced apricot-based value-added products (Stobdan *et al.*, 8).

Area Expansion

Worldwide fresh apricot exports have increased by 59 per cent from 2010 to 2017, while dry apricot exports have increased by 18 per cent during the same period. India is the sixth largest importer of dried apricot, importing 4.2 per cent of all world import in 2017 (5457 tonnes). Therefore, there is immense scope for Ladakh to supply high quality apricots not only for the domestic market but also has the potential to produce enough premium quality fruit for exports.

Agriculture production in Ladakh is entirely based on irrigation. The main constraints in area expansion under fruit tree plantation is the assured water source for irrigation. There is immense scope to expend area under apricot cultivation from the existing 2303 hectares. The vast area of barren land remains unused. With technological advancement, it is now feasible to lift water from the river for irrigation. The glacial water that drains into the rivers can be conserved and judiciously used to bring more area under cultivation. There is a range of technologies available for improved operation, better management and efficient use of irrigation water – ranging from simple mulching technology to sophisticated canal automation and telemetry.

Characteristic Features of Ladakhi Apricots

Apricots of Ladakh are known for its quality. Historically, the premium quality dried apricots were one of the main trading commodities with neighbouring countries. *Phating*, the premium quality dried apricots of Ladakh, are popularly known as *Nyari Khambu* in Tibet (Stobdan *et al.*, 2021). Uniqueness of the apricots of Ladakh was recently proven by Defence Institute of High-Altitude Research (DIHAR). The same have been recognized by the scientific community as evident from publications of their studies in reputed peer-reviewed international journals.

Raktsey Karpo is Unique to Ladakh

Raktsey Karpo, apricots with white seed stone, is unique to Ladakh. It has not been reported from anywhere else in the world. White seed stone phenotype is associated with sweet kernel, brightly coloured fruit with high TSS content (Angmo *et al.*, 1). *Raktsey Karpo* is most preferred by the consumers for fresh consumption. It ranked first in terms of sweetness, juiciness, aroma, flesh colour and overall appreciation (Naryal *et al.*, 6).

Raktsey Karpo contains significantly higher sorbitol (53.6±26.5 mg/g DW) followed by fruits with brown coat with sweet kernel (34.6±27.0 mg/g DW) and fruits with brown coat with bitter kernel (23.8±36.1 mg/g DW) (Naryal *et al.*, 5). Sorbitol is a sugar alcohol that is used as a sweetening agent in various food products such as sugar-free sweets and chewing gum. The white seed coat phenotype of *Raktsey Karpo* can be considered an easily identifiable marker for quality fruit. The phenotype can be explored as a distinguishing feature of high-quality apricots of Ladakh. It can be used to develop consumer's loyalty and strong image of apricots of Ladakh (Angmo *et al.,* 1).

Fruit TSS Increase with Increasing Altitude

Apricots of Ladakh are sweeter. The sugar content in terms of total soluble solids (TSS) ranged from 11.0-37.9^{0}Brix with mean value of 23.9±5.7^{0} Brix. Altitude showed linear relationship with fruit TSS content. For every 100 m increase in elevation, the fruit TSS increased by 1.2^{0} Brix. *Raktsey Karpo* has significantly higher TSS (28.1±3.8^{0} Brix) than cultivars with brown seed coat (Angmo *et al.,* 1). In comparison, the value ranged from 11-27^{0} Brix among 128 apricot cultivars in Turkey, 15.7-18.9^{0} Brix in 14 genotypes grown in Central Serbia, 10.6-16.3^{0} Brix in 43 cultivars grown in Spain, 9.3-18.7^{0} Brix in 55 cultivars in Spain, 8.7-22.4^{0}Brix in 51 genotypes from germplasm collection in France, 12.3-15.8^{0} Brix in 18 genotypes belonging to the Italian and international germplasm, and 12.7-20.0^{0} Brix in six cultivars in Pakistan.

Fruit Sugar Increase with Increasing Altitude

High altitude environmental conditions contribute to the sweetness of the apricot of Ladakh. Linear relationship between sugar contents and increasing altitude has been observed. For every 100 m increase in altitude, total sugar contents increased by 64.8 mg/g DW. For every 100 m increase in altitude, sucrose contents increased by 49.1 mg/g DW. Fruit sorbitol contents showed linear relationship with increasing altitude. Sorbitol contents increased by 11.9 mg/g DW for every 100 m increase in altitude. Therefore, high altitude climatic conditions are favourable for production of apricots with high sugar contents (Naryal *et al.,* 5).

Small Size Fruit

Apricots of Ladakh are small. Fruit weight ranged from 7.1-53.7g with mean weight of 21.6 ± 9.3 g. For every 100 m increase in elevation the fruit weight decreases by 0.5 gm (Naryal *et al.,* 7). In comparison, fruit weight among the promising genotypes of the Lake Van region ranged from 24.2-48.3 g, 36.5-105.3 g in Greek and American origin apricots, 49.1-81.5 g in promising apricot resources in Central Serbia, 28.1-77.7 g with mean weight of 42.44 g from Canada, Czech Republic, Ukraine and USA.

Late Fruit Ripening

Apricots of Ladakh are late maturing as compared to apricots from other parts of the world. The fruits are harvested between mid-July and early September. For every 100 m increase in elevation, fruit ripening is delayed by 7.1 days. In comparison, apricots in Anatolia, Turkey are harvested in late June and early July while those from Lake Van Region, Turkey are harvested in late July to early August. Cultivars and selections grown in Spain attain maturity in mid-May and late June and those grown in Greece and America are harvested in May to June. Therefore, apricots of Ladakh region have comparative advantage as it does not coincide with main apricot season in the market (Naryal *et al.*, 7).

Extended Fruit Ripening

The harvesting season of apricots of Ladakh remained for 39 days in an orchard at DIHAR Leh. In contrast, only 14-15-day difference in harvest season has been reported for 15 apricot genotypes grown in Central Serbia. Harvesting spanning more than a month is an important element in view of short shelf life of fresh apricot fruit. Therefore, apricots of Ladakh can be marketed over an extended period.

Apricot Growing and Management: Constraints and Suggested Solutions

The leading apricot producing countries are upgrading their orchard models and management techniques. Cultural improvements include the introduction of improved cultivars and rootstock, and a trend towards intensive, irrigated and mechanized orchards compatible with the fresh market and processing industry. Apricot demands specific geographical and climatic requirements. Traditional techniques have given way to more innovative management approaches, resulting in reduced risk of alternate bearing and loss of yield thus stabilizing orchard returns.

In Ladakh, apricots are being grown either as individual trees or small groups of trees in traditional way without the use of chemical fertilizer and pesticides. Standard cultural practices such as training and pruning are not followed. The fruit trees are hardly applied chemical fertilizers and manures resulting in low productivity and small fruit size. Majority of the trees are seedling origin. There is a need to grow improved cultivars and follow standard cultural practices in the region.

Cultivars

Majority of the trees are seedling origin. However, in selected villages grafting is widely practiced on seedling rootstock to grow premium quality cultivars especially *Halman* and *Raktsey Karpo*. However, within *Halman* and *Raktsey Karpo* there exist wide variations. The variation observed within *Halman* or *Raktsey Karpo* is largely ignored resulting in difference in size and quality of the fruit within a cultivar in a village. At DIHAR there are seven variants of *Raktsey Karpo* which differ significantly in size, sweetness and maturity.

Trees bearing high quality *Halman, Raktsey Karpo* or any other cultivars in each village need to be selected. The selected trees will serve as mother plant for obtaining scion for grafting on seedling rootstocks. Incentives may be provided to growers for maintaining the mother trees and for allowing unrestricted access to other growers for taking scion.

Mass Propagation

To get uniformity in products, fresh as well as dried, it is important to grow known cultivars on large scale. Mass propagation of a selected mother trees is laborious and time consuming by the traditional grafting technique. Adoption of tissue culture technology for multiplication of selected mother trees on large scale will bring quick results. Several privately-owned tissue culture labs undertake mass multiplication of fruit trees as per requirement.

A project needs to be undertaken for a large-scale multiplication of selected mother plant using tissue culture technology. However, instead of setting up a tissue culture lab in Ladakh, the job can be outsourced to public or privately-owned tissue culture labs which are located in a temperate zone of the country. The firm can undertake multiplication and supply of one- or two-years old plants in polybags.

Fruit Nursery

Supply of quality nursery plants of known cultivar is vital for promotion of apricot in the region. So far, the majority of the apricot plant is procured through informal farmer-to-farmer contact, while a small number of them have been obtained from government-owned nurseries. However, majority of the nursery plants which have been raised in government-owned nurseries are unbudded due to lack of skilled manpower.

There is need to promote government or private-owned nurseries to raise only budded or grafted apricot nursery plants. The sale price of such budded plants needs to be regulated. Measures need to be put in place to restrict the sale of unbudded plants from nurseries.

Orchard System

Apricot trees are planted around the house or around the field boundaries. The orchard is poorly managed and scattered over a large area. Well managed apricot orchards are hard to find in the region.

There is a need for plantation of apricot trees in orchard system. World over the tradition of 6m × 6m spacing between the plants have been largely abandoned. At DIHAR Leh apricots are planted in 4m × 4m spacing. Studies suggested that increased density reduces yield per tree but augment yield per hectare. Therefore, orchard management practices are needed to be identified as one of the key areas for improvement.

Incentives for Establishment of New Orchard

Establishment of new orchard involves planting material, land preparation, pit digging, nutrient management, irrigation and fencing. High mortality of young trees is observed when young trees are planted in orchard without protection from domesticated and stray animals. In view of the high cost of land preparation and fencing growers are finding it difficult to establish new orchards with the existing financial assistance.

Incentives may be provided to the interested growers keeping in view the actual cost for establishment of new orchard in the region.

Creation of Water Resources

Assured irrigation is a must for fruit tree cultivation in Ladakh. Hot and dry conditions result in high evaporation rates in Ladakh. The soil is mostly sandy loam with very less moisture holding capacity. Hence, the crops need more water with higher frequency. Creation of water resources is a must in the region.

There is need to create water resources in the region for apricot plantation. There are provisions for creation of water resources both at community and individual level under MIDH scheme.

Micro-irrigation

Assured irrigation is a vital need for fruit tree cultivation in Ladakh. During summer, the evaporation rates are high in Ladakh owing to hot and dry conditions. Moreover, the soil is mostly sandy loam with very less moisture holding capacity due to which the crops need more water with higher frequency. Adoption of micro-irrigation system is a viable option for Ladakh region. Studies conducted at SKUAST-K Stakna has found that drip irrigation system results in 40-50 per cent water saving. There is no leaching loss under drip irrigation since water is applied directly to root zone of plant. Since weed growth is less in drip method, the yield of crop increases apart from saving the labour cost in weeding. Drip irrigation system also checks the erosion of sloppy lands.

Adoption of micro-irrigation system may be made mandatory for plantation on barren lands. In view of the high initial cost of installation of drip-irrigation system, incentives may be given to the farmers for adoption of the micro-irrigation system. Privately-owned agencies may be encouraged to open sale and service outlets in Kargil and Leh.

Training and Pruning

Standard cultural practices such as pruning and training of trees are not performed in Ladakh. Pruning is opposed by majority of the growers as it is viewed as an unnecessary cutting of branches. Lack of training and pruning resulted in trees reaching heights that hinder fruit harvesting. Besides, the fruits remain small and low quality.

There is a need for an intense awareness campaign on the need of training and pruning. Farmers need to be taken on exposure tours to show orchards where trees are regularly trained and pruned. One such orchard is at DIHAR, Leh. Growers lack the skill in scientific training and pruning methods. Thus, it is important to train farmers in pruning and tree management to produce higher yield and better-quality fruit. Front line demonstrations need to be conducted in every village.

Application of Manure and Fertilizer

In Ladakh apricots are being grown either as individual trees or small groups of trees. The fruit trees are hardly applied any manures and fertilizers resulting in low productivity and small fruit size.

Importance and method of application of manure to apricot trees need to be communicated to the growers. Front line demonstrations on application of fertilizers need to be conducted in every village. Benefits of application of manure for enhancing crop productivity and increased fruit size can be demonstrated to the farmers.

Slow Plant Growth

The growth of plants, especially at early stage, is slow in Ladakh due to low temperature. Growers have shown reservations on establishment of new orchard since it takes 8-10 years for trees to come into bearing. Besides, frequent irrigation is required at initial stages. However, recent studies at DIHAR has shown that use of black plastic mulch (100 microns) reduces the time for fruit trees to reach fruit bearing stage almost by half. Besides, use of mulching significantly reduced the number of irrigations. Black plastic mulch increases soil temperature, checks evaporation from soil surface and does not allow weeds to grow around the tree trunk.

Promote application of black plastic mulch while establishing new orchards. Growers should be demonstrated the effect on plastic mulching on growth and fruiting. Distribution of black plastic mulch sheet (more than 100 microns) need be a part of regular plan of the Directorate of Horticulture. Front line demonstrations on application of mulching technology need to be conducted in every village.

Planting Season

Apricot nurseries are planted in Mar and April when the plants are in dormant stage. Planting after leaf emergence results in high mortality rates due to transplanting shock. Besides, direct transplanting in the field often results in high mortality especially in the initial two year. DIHAR has demonstrated a method for extending planting season by growing the fruit nursery plant in polybags. The method is highly successful in extending the planting season and achieving over 95 per cent success rate in field condition.

Promote use of new methods for extending the planting season and for achieving higher success rate. Selected Departmental nurseries may raise plants in polybags that can be made available to growers for planting even during summer months. Rates for nursery plant in polybags need to be regulated to enable the growers and private nursery to benefit from the new technology.

Input Supply

The supply of inputs such as farm tools and agrochemicals is being done primarily through the Horticulture Department in Kargil and Leh. However, due to several factors it is not always possible to cater to the actual needs of the growers. It is, therefore, important to make the alternative sources of input suppliers available in the form of privately-owned firms. Few privately-owned input suppliers have opened shops in Kargil and Leh. However, they are finding it difficult to sustain the business due to heavy subsidy on input items by the Agriculture and Horticulture Departments.

There is a need to provide incentives to encourage the privately-owned suppliers to open outlets in Kargil and Leh. Incentives may also be provided to private enterprises for setting up of workshops for repair of farm tools and machineries.

Gaps and Interventions in Value Chain of Apricots in Ladakh

Organic Certification

Apricots are being grown in the region either as individual trees or a small cluster of trees in traditional way. The fruit is not certified as organic even though no chemical fertilizer or pesticides have ever been used on them. However, additions of chemical preservatives after primary processing turn the pulp into a non-organic produce.

Certify and promote apricots of Ladakh as organic. Develop facilities at apricot processing sites to maintain organic integrity and the vital natural qualities of the fruit.

Geographical Indication (GI) Certification

Raktsey Karpo, apricot with white seed coat, is unique to Ladakh and its quality attributes are intrinsically due to the place of origin. However, lack of GI tag for the cultivar is a major disadvantage to the growers and traders to realize the full commercial potential.

Initiate GI tagging of apricots of Ladakh and its products.

High Wastage Rates

The disorganized nature of the sector and adherence to certain inefficient production and processing techniques currently result in wastage of approximately 40-50 percent of the apricot produced in Ladakh.

(a) Adopt integrated post-harvest management; (b) establish marketing infrastructure for apricots; (c) provide incentives to existing fruit processing units for up-scaling production capacity; (d) invite potential buyers for apricot pulp and seed

Harvesting Tools

Currently apricot fruit are harvested by shaking or beating the branches. The fruit fall on the ground which is then handpicked and collected in baskets. When ripe fruits fall from a height and strike the hard ground surface, the fruit get damaged and results in reduced shelf life and quality. Handpicking each fruit from the ground is cumbersome and laborious.

(a) Provide incentives to the farmers for adoption of 'Harvest Net'; (b) conduct in-situ demonstration and training at farmer's location (c) Directorate of Horticulture may include 'Harvest Net' in its regular plan.

Input Supply

The supply of inputs such as farm tools and agrochemicals is being done primarily through the Horticulture Department in Kargil and Leh. However, due to several factors it is not always possible to cater to the actual needs of the growers. It is, therefore, important to make the alternative sources of input suppliers available in the form of privately-owned firms. Few privately-owned input suppliers have opened shops in Kargil and Leh. However, they are finding it difficult to sustain the business due to heavy subsidy on input items by the Agriculture and Horticulture Departments.

There is a need to provide incentives to encourage the privately-owned suppliers to open outlets in Kargil and Leh. Incentives may also be provided to private enterprises for setting up of workshops for repair of farm tools and machineries.

Integrated Post-harvest Management

Ripe apricot fruit is delicate and needs to be processed as soon as possible. Several steps are involved in post-harvest management of the fruits for fresh as well for development of value-added products. Integrated post-harvest management system is lacking in Ladakh.

Identify sites in major apricot growing areas where integrated post-harvest management projects can be undertaken. There is a need for integrated post-harvest management of apricots in major apricot growing belts in Kargil and Leh. There are several provisions under MIDH schemes for post-harvest management of fruits. It includes pack house; integrated pack house with facilities for conveyor belt, sorting, grading units, washing, drying and weighing; pre-cooling unit; cold room; mobile pre-cooling unit; cold storage; refrigerated transport vehicle; minimal processing unit; ripening chamber.

Establishment of Marketing Infrastructure for Apricots

Apricot is locally consumed and lacks a marketing system for trading outside the region. There are opportunities for fresh as well as processed products from Ladakh to be sold in down-country supermarkets.

There is a need for establishment of marketing infrastructure for fresh and processed apricots of Ladakh. There are several provisions under MIDH schemes for establishment of marketing infrastructure for horticultural crops. It includes terminal markets; wholesale market; Rural Markets/Apni Mandies/Direct Markets; Retail markets; mobile vending cart; collection, sorting, grading, packing units; quality control/analysis lab.

Cold-chain Infrastructure for Fresh Fruit Trading

Fresh apricot is entirely consumed locally. There is no marketing system for trading outside the region although there exists immense potential for the fresh apricot from Ladakh to be sold in down-country supermarkets. About 20 MT of fresh apricots from Ladakh was sold for the first time in Mumbai, Hyderabad, Ahmedabad and Delhi/NCR markets in 2021. The maiden export (150 kg) of Ladakh apricot to Dubai also took place in 2021.

With the setting up of a cold chain a facility for washing, grading and packing of the fruits will also be required for the producers. A central cold store in Kargil and Leh will be required to chill the fruit before transporting down-country in refrigerated vans. A substantial capital investment is required to develop cold-chain infrastructure. There is a need to develop a business model for sale of quality fresh apricots to supermarket and retail chains.

Hygienic Drying

Majority of the apricot fruit is dried in open sun, which is prone to spoilage due to incorrect picking methods, unhygienic and non-controlled drying procedures. The fruits are generally sun-dried on the roof of houses or on open fields before it is sold off in local markets with very little value addition. The major disadvantage of traditional drying is the visual appearance of the fruits as the apricots dried with this method can be susceptible to rapid darkening and decomposition of chemical structure that leads to objectionable changes in colour, texture, and flavour that fetches very low prices in the market. It is due to the suboptimal drying practices that the dried fruits of Ladakh have a limited access to other markets. Of late, some small-scale solar apricot dryers have been installed by several agencies in Ladakh. However, due to their small size, growers are finding it difficult to achieve quality dried fruits at commercial scale. No efforts have been made to bring innovations in apricot drying and marketing in the region.

There is a need to develop a systematic fruit collection, processing, drying and marketing in the region, which will largely solve the problem of unhygienic drying conditions in the region. Technologies for processing and drying of apricots are well known and need to be adopted.

Seed Oil Extraction

Apricot oil is extracted from the kernel, which contains 40-50 per cent oil. Oil extracted from the sweet kernel is used for edible purpose while that of bitter kernel has religious, cosmetic and medicinal values. The oil is popular as body oil or massage oil, and known to relieve backache and joints ache. The oil extracted by the traditional method is still being preferred due to its distinct smell. However, the traditional method of oil extraction is cumbersome and is slowly declining. Currently oil is extracted using mechanical method that generates high heat. Many temperature sensitive compounds are lost in the process which reduces the overall quality of the oil.

Centralized facilities for oil extraction using cold pressed method may be created at major apricot growing region. Privately-owned units involved in extraction of apricot oil may be given financial incentives for adoption of cold-pressed or better technology for oil extraction. The apricot oil from Ladakh may be marketed as certified organic cold press oil.

Value Added Products

Approximately 40-50 percent of the apricot produced is wasted. There is a need for development of value-added products to reduce wastage.

Provide incentives to develop value added products for Ladakhi apricots. Government support is required for infrastructural development, development of skill and training for value addition. There is a need for promotion of apricot as a unique product of Ladakh.

Strengthening Market Linkages

Market linkage for apricot is poor. There are many small players in the value chain. Processors find it difficult to engage major buyers due to their small size and distance from market.

Set up a single window online system for providing information and services related to apricot trade. Develop market linkages with emphasis on helping farmers to identify profitable markets even at overseas markets.

Strengthen Local Entrepreneurs

There are many small local traders and processors in apricot value chain. Their risk-taking capacity is low.

Government support in strengthening the local entrepreneurs. Need to form ‘Ladakh Fruit Processor Association’.

Conclusion

Ladakh is the biggest apricot producer in India. The total fruit production is 15,789 MT (Leh: 5,133; Kargil:10,656 MT). However, of today, the popularity of Ladakhi apricots remains restricted to the region, and major part of apricot produced is consumed locally. The disorganized nature of the sector and adherence to certain inefficient production and processing techniques implies that 40-50 percent of the apricot production is wasted. In view of the uniqueness and premium quality of apricots of Ladakh there is immense scope for Ladakh to emerge on the world map for apricot production. Export of Ladakh fresh apricots started in year 2021.

REFERENCES

1. Angmo, P., Angmo, S., Upadhyay, S.S., Targais, K., Kumar, B. and Stobdan, T. 2017. Apricots (*Prunus armeniaca* L.) of trans-Himalayan Ladakh: Potential candidate for fruit quality breeding programs. *Sci. Hortic.* **218:** 187-192.
2. Hormaza, J.I., Yamane, H. and Rodrigo, J. 2007. Apricot. **In:** *Genome Mapping and Molecular Breeding in Plants*, Vol 4: Fruits and Nuts. C. Kole (ed.), Springer-Verlag Berlin Heidelberg, pp 171-187.
3. Layne, R.E.C., Bailey, C.H. and Hough, L.F. 1996. Apricots. **In:** *Fruit Breeding*, Vol 1: Tree and Tropical Fruits. J. Janick. and J.N. Moore (ed.), John Wiley and Sons, New York, USA, pp 79–111.
4. Mehlenbacher, S.A., Cociu, V. and Hough, L.F. 1990. Apricots (*Prunus*). *Acta Hortic.,* **290:** 65-107.
5. Naryal, A., Acharya, S., Bhardwaj, A.K., Kant, A., Chaurasia, O.P. and Stobdan, T. 2019b. Altitudinal effect on sugar contents and sugar profiles in dried apricot (*Prunus armeniaca* L.) fruit. *J. Food Compos. Anal.,* **76:** 27-32.
6. Naryal, A., Angmo, S., Angmo, P., Kant, A., Chaurasia, O.P. and Stobdan, T. 2019a. Sensory attributes and consumer appreciation of fresh apricots with white seed coats. *Hort. Environ. Biotechnol,* **60:** 603-610.
7. Naryal, A., Dolkar, D., Bhardwaj, A.K., Kant, A., Chaurasia, O.P. and Stobdan, T. 2020. Effect of altitude on the phenology and fruit quality attributes of apricot (*Prunus armeniaca* L.) fruits. *Def. Life Sci. J.,* **5:** 18-24.
8. Stobdan, T., Namgial, D., Chaurasia, O.P., Wani, M., Phunchok, T. and Zaffar, M. 2021. Apricot (*Prunus armeniaca* L.) in trans- Himalayan Ladakh, India: Current Status and Future Directions. *J. Food Agric. Res.* **1:** 86-105.
9. Targais, K., Stobdan, T., Yadav, A. and Singh, S.B. 2011. Extraction of apricot kernel oil in cold desert Ladakh, India. *Indian J. Tradit. Know.,* **10:** 304-306.
10. www.fao.org/faostat/en

Chapter 7

Biotechnological and Genomic Advancements in Improvement of Fruit Crops

Malarvizhi M. and Ravishankar K.V.*

Division of Basic Sciences, ICAR–Indian Institute of Horticultural Research, Hessaraghatta Lake Post, Bengaluru 560089, Karnataka

**e-mail: ravishankar.kv@icar.gov.in*

ABSTRACT

Crop improvement strategies in fruit crops are faced with challenges due to their heterozygous nature, long juvenile phase, polyploidy and parthenocarpy. The application of genomic approaches can mitigate the problems and expedite the conventional breeding methods. At present, fruit crop research is progressing towards a new era due to the availability of sequence information of various model and non-model crops like apple, mango, banana, citrus, papaya etc. Molecular markers have aided in diversity analysis of germplasms and screening of mapping populations for fruit quality traits with commercial importance. With the advance of next generation sequencing technologies like GBS and RNA Seq helped the construction of high-density linkage maps and identification of QTLs and crop specific SNP and SSR markers linked to important quantitative traits in fruit crops. These predictive markers for fruit traits have augmented the breeding efforts through Marker- Assisted Selection (MAS) of progenies at seedling stage, Marker-Assisted Backcrossing (MAB) and Marker-Assisted Introgression (MAI) of desired traits in fruit species. Integration of 'omics' approaches like genomics, transcriptomics, metabolomics, proteomics and phenomics can be informative and assist in understanding the molecular mechanisms underlying fruit quality traits, bearing nature, shelf life, biotic and abiotic stress tolerance in fruit crops. Further, identification of Single Nucleotide Polymorphisms have led to successful application of genomic assisted breeding strategy like Genome Wide Association Studies (GWAS) and model based predictions like Genomic Selection (GS) in model crops like apple, peach and banana etc. In addition, CRISPR/ Cas based genetic modification is gaining more significance in fruit crops with the success of various studies in banana, apple and papayas a proof of concept. Improvement in tissue culture techniques for fruit crops can ensure the success of these transgenic and genomic modification efforts. Some of the new sequencing innovations like pan-genomes, pan-transcriptomes, super pan-genomes and single cell transcriptomics can exploit the potential of the vast genetic diversity of fruit crops. Overall, the advancements and success of genomic assisted breeding strategies can help improve the fruit quality traits, reduce the time and need for vast orchards and mapping populations.

Keywords: *Fruit crops, Genomics, Molecular markers, QTLs, CRISPR/Cas9.*

The global fresh fruit production had increased from 576.65 million metric tons in 2000 to 883.42 million metric tons in 2019. With the increase in world population and wide prevalence of hunger and malnutrition in the developing and under-developed countries; fruits can serve as a source of nutritional and financial security to the people (FAO, 23). Some of the fruits with high worldwide production are from both temperate and tropical regions of the world (Shahbandeh, 76). Fruit crops across the globe, grown under different climatic conditions are facing new challenges due to climate change in addition to their high perishability. Production of high quality and high yielding varieties are the major goals of fruit breeding effort. In order to achieve this goal amidst the various biotic and abiotic stress conditions we need an integrated approach. The advancements in sequencing technologies are a major factor that influences the breeding efforts of various fruit species. Availability of sequence information on the economically important fruits like apple, peach, banana, citrus, mango, papaya *etc.* have contributed in the development of improved breeding strategies (Luo *et al.,* 48; Eduardo *et al.,* 20; Rawandoozi *et al.,* 70; Nyine *et al.,* 62; Zhang *et al.,* 101).

The progress in next generation sequencing (NGS) has revolutionized development of markers that can accurately predict the presence of a particular gene locus and efficiently screen the germplasms and progenies for specific trait linked to it. NGS has helped in development of genomics assisted breeding strategies like genomic selection (GS) and marker assisted selection (MAS) for fruit crop improvement (Endo *et al.,* 22; Nyine *et al.,* 62). Genotyping by Sequencing (GBS) combine with high throughput phenotyping has paved the way for high density linkage mapping and genome wide association studies (GWAS) for identification of genes and QTLs involved in fruit quality traits, tolerance to biotic and abiotic stresses *etc.* (Kuhn *et al.,* 40). Further, availability of whole genome sequencing (WGS) data has facilitated accurate assembly and mapping of related species and genotypes and led to identification of genotypic variants, epigenetic variations and evolutionary changes in different fruit crop species (Feng *et al.,* 24; Jian-ye *et al.,* 36; Cao *et al.,* 7; Li *et al.,* 44).

The main aim of these techniques is to develop predictive markers which can be applied in marker assisted selection (Zhang *et al.,* 101), marker assisted introgression (Luo *et al.,* 48) or marker assisted backcrossing (Kim *et al.,* 39). High throughput phenotyping and/or genotyping combined with genomic selection can accurately predict the breeding values of the individuals in a population and their suitability as parents in breeding pipeline (Roth *et al.,* 73; Cazenave *et al.,* 8). All these strategies can reduce the length of the breeding cycle and expedite progeny selection with desired traits at the seedling stage of a fruit crop (Figure 1). Further advances in transcriptomes studies using RNA seq (Rahim *et al.,* 69) and single cell transcriptomics (Rich-

Griffin *et al.*, 72) can expedite the process of identification of the genes governing important horticultural traits and biotic and abiotic stress tolerance in fruit species.

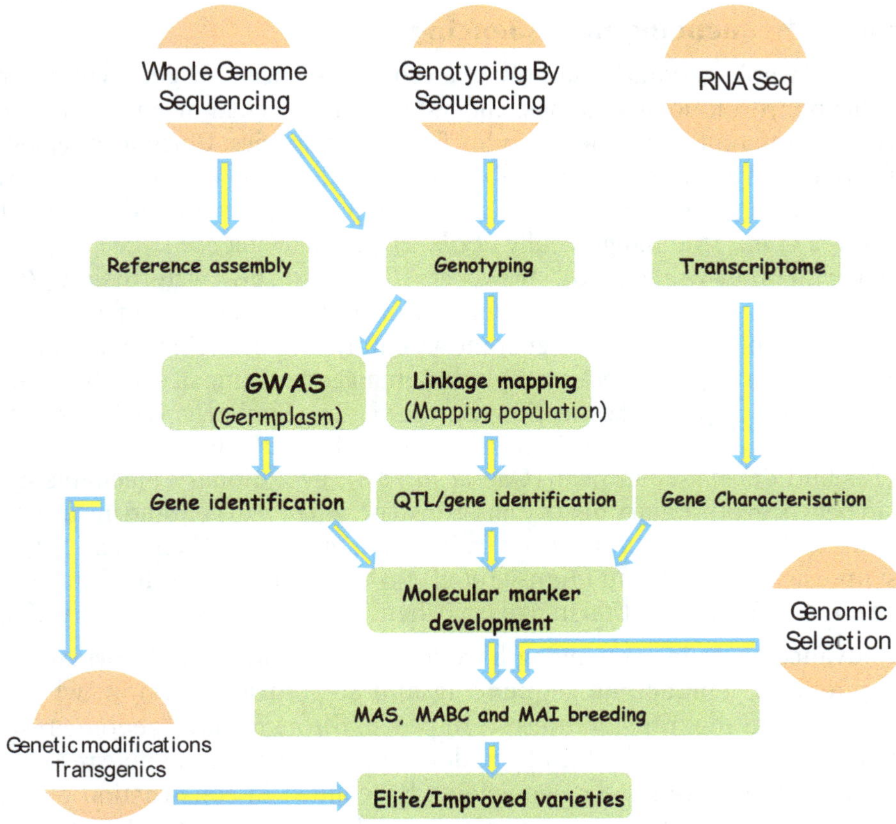

Figure 1. Genomics assisted breeding strategies in fruit crops.

Genetic modifications in crops have achieved a new boost in the wake of CRISPR/Cas success. Enormous opportunities have been created due to its ease and efficiency in genetic manipulations in fruit crops. Even though tomato has been used extensively as a model crop for CRISPR/Cas studies, it has now been extended to other fruit crops like apple (Zhang *et al.*, 102), orange (Peng *et al.*, 67), banana (Tripathi *et al.*, 88), strawberry (Wilson *et al.*, 94) *etc.* It has been employed for improvement of fruit quality and resilience towards biotic and abiotic factors. The current trends in fruit crop improvement is very promising and seeing a surge in techniques that can complement and hasten the breeding programmes.

Next Generation Sequencing Techniques in Fruit Crops

Next generation sequencing has revolutionised crop science due to the availability of sequence information for wide range of plant species with precision, ease and at reduced cost. NGS based techniques have been adapted efficiently in fruit crops for studying epigenetic changes, phylogenetic relationships, discovery of non coding regions, fine mapping of genes and study of intricate molecular mechanisms behind the traits.

Whole Genome Sequencing/Resequencing

Availability of whole genome assembly of a crop is essential for genotyping for variants, marker identification, evolutionary studies, understanding molecular mechanism of a character/ trait and further as reference genome assembly for related species. Chromosome level genome assembly of different fruit crops are available in public domain. Whole genome assembly of apple (Daccord *et al.*, 13), pear (Shirasawa *et al.*, 80), plum (Liu *et al.*, 46), peach (Cao *et al.*, 7), banana (Davey *et al.*, 15), mango (Bally *et al.*, 2), guava (Feng *et al.*, 24), papaya (Ming *et al.*, 54), dragon fruit (Jian-ye *et al.*, 36), strawberry (Qiao *et al.*, 68), watermelon (Renner *et al.*, 71), ground cherry (Woehner *et al.*, 95), longan (Wang *et al.*, 90), passion fruit (Ma *et al.*, 49), citrus (Zhu *et al.*, 105) *etc.*, High quality genome assembly of apple helped in understanding the abundance of repetitive regions, DNA methylation changes resulting in early fruit development (Daccord *et al.*, 13) and apple domestication (Sun *et al.*, 84). Similarly, WGS has facilitated the understanding of aroma biosynthesis in passion fruit (Ma *et al.*, 49); epigenetic modification responsible for bud dormancy in pear (Gao *et al.*, 28); evolutionary mechanism of volatile compounds in peach; evolutionary history and construction of pan genome in strawberry (Qiao *et al.*, 68); accumulation of ascorbic acid and fruit softening in guava (Feng *et al.*, 24); regulation of betalain synthesis in dragon fruit (Jian-ye *et al.*, 36), candidate genes for fruit size in mango (Bally *et al.*, 2) and sex determination in papaya (Chen *et al.*, 11) and red bayberry (Jia *et al.*, 31).

Further advancements in sequencing have led to development of information on whole genome resequencing, epigenome sequencing and structural variation, which can help researchers understand the traits of interest. In peach, through whole genome resequencing, molecular markers linked to flat fruit trait was developed, which is a valuable resource for MAS (Zhang *et al.*, 101). Whole genome resequencing have aided in construction of high density linkage maps and mapping of firmness genes in grapes (Jiang *et al.*, 34), SNP mining, diversity analysis in plum (Wei *et al.*, 93) and epigenetic studies on variation in skin colour in apples (Naibin *et al.*, 56). Whole genome bisulphide sequencing (BS) had uncovered the whole genome methylation patterns helpful in studying the regulatory mechanisms behind floral induction in apple (Zhou *et al.*, 104). Likewise, BS had revealed epigenetic regulation in response to drought stress in mulberry (Li *et al.*, 44) and flowering in citrus (Agusti *et al.*, 1). Epigenetic modifications play a major role in regulation of fruit development, ripening and response to stress conditions in fruit crops. Hence these studies can have a greater impact on breeding efforts undertaken for trait improvement. Skim sequencing is one of the recent advances of NGS, which is low coverage whole genome sequencing with reduced cost. This can be used for genetic linkage mapping, QTL mapping and GWAS (Kumar *et al.*, 41).

Genotyping by Sequencing

GBS is a reduced representation sequencing carried out at a low cost. It is helpful in genotyping and identification of SNP markers throughout the genome. Unlike the PCR based markers, SNPs provides a greater coverage of the genome with abundant markers. Restriction site associated genomic DNA (RAD) sequencing is a slightly modified version of GBS which has also been utilized in genotyping. GBS-based SNPs and SSR markers, facilitates construction of high density linkage maps and mapping of QTLs and genes with better precision than conventional markers (Hu *et al.,* 30; Ollitrault *et al.,* 63). Mapping populations (F_1) generated through crossing of heterozygous cultivars are usually utilized for linkage mapping in fruit crops (Yang *et al.,* 97). QTLs and markers closely related to important horticultural traits were fine mapped in various fruit crops facilitating marker assisted selection and introgression and genomic modifications in target crops (Nuñez-Lillo *et al.,* 60; Branham *et al.,* 5). Numerous SNP and SSR markers used in linkage map have a higher probability to be related to a particular trait of interest. Hence QTLs and markers developed using high throughput genotyping opens up avenues for genomic assisted breeding in fruit crops.

RNA Sequencing

RNA seq or transcriptome sequencing provides information on the functional genomics of fruit crops and helps characterise genes expressed during different conditions or among different tissues. Various studies have been taken up over the years to unravel mechanisms involved in pulp and peel colouration, shelf life, biotic and abiotic response, fruit development and ripening *etc.* in fruit species. But the application of RNA Seq has immensely contributed towards understanding transcriptional control of genes (SampangiRamaiah *et al.,* 74) and comparative gene expression profiling (Zhang *et al.,* 100; Kaur *et al.,* 38). The major advantage of RNA seq is that, it is applicable in crops with reference genome or without it (*de novo*). Hence, in fruit crops, where majority of them does not have a reference genome, RNA seq has transformed the molecular level studies by facilitating studies on splice variants (Sun *et al.,* 85) and non- coding RNAs (SampangiRamaiah *et al.,* 74). The application of transcriptome analysis varied from studies on fruit development (Chen *et al.,* 10), ripening (Nawaz *et al.,* 57), fruit peel and pulp colour formation during ripening (Jiang *et al.,* 35, Wu *et al.,* 96), regulatory role of non coding RNAs in ripening and stress mechanisms (Ma *et al.,* 50; Sun *et al.,* 85), transcription factors (Deng *et al.,* 18), shelf life (Onik *et al.,* 64; Dwivany *et al.,* 19), response to diseases (Tian *et al.,* 87; Li *et al.,* 45); drought (Gao *et al.,* 27; Santos *et al.,* 75); temperature (Yuan *et al.,* 99; Vidya *et al.,* 89) *etc.* An integrated approach using different 'omics' techniques like transcriptomics, metabolomics, proteomics and phenomics have the potential for detailed analysis of molecular mechanisms and the functional role of each gene under study (Sun *et al.,* 83; Milyaev *et al.,* 51). RNA sequencing had also been successfully applied to study the alternate bearing characters in fruit crops like apple (Gottschalk *et al.,* 29), mango (Sharma *et al.,* 77), citrus (Agusti *et al.,* 1) *etc.* which gave insights into the positive and negative controllers of fruit bearing. Further, validation of RNA seq data through qPCR studies facilitated the corroboration of transcriptome data and quantification of gene expression (Patil *et al.,* 65; Karanjalker *et al.,* 37). Recently, RNA seq has been integrated with new sequencing technologies like CHiP seq (Muhammad *et al.,* 55) to study the molecular architecture and interaction of proteins and transcription factors with DNA in a better way. Likewise, introduction of single cell transcriptomics (scRNA seq) and Full length RNA profiling at a single-nucleus level (FlsnRNA-seq) in crop science can have a greater impact in understanding of gene expression and cell to cell interactions in fruit crops related to fruit quality and tolerance to stress conditions (Shaw *et al.,* 78; Long *et al.,* 47).

Genomics Assisted Breeding

Genomics assisted breeding (GAB) requires precise and high throughput genotyping and phenotyping to identify candidate genes of the desired trait. The major advantage of GWAS is that it does not require a huge mapping population for analysis. The wide variety of germplasm available with the breeder can be utilised for trait- marker association studies. Digital phenotyping platforms and genome-wide high density marker coverage aids in fine mapping of QTLs and genes of interest. GWAS facilitates association of quantitative traits to markers that can be used as predictive markers of a specific trait in conventional breeding. Association of genotyping by GBS and the SNP markers identified enables association mapping and genomic prediction in fruit crops for different traits of interest. Genotyping by GBS facilitated GWAS in the identification of genes related to marssonina Blotch disease resistance, fruit shape, flesh browning *etc.* in apples (Noh *et al.*, 58; Kunihisa *et al.*, 43). Further integration of *de novo* genome assembly, resequencing and GWAS had identified genes related to fruit shape, fruit development period, and floral morphology in Prunus species (Tan *et al.*, 86). Similar genome wide association studies have been conducted in other model fruit crops like peach (Fu *et al.*, 26), citrus (Minamikawa *et al.*, 52) and banana (Nyine *et al.*, 61). GWAS and genomic prediction models utilizing parental and mapping population had been applied to study fruit quality traits in citrus (Minamikawa *et al.*, 52) and pear (Minamikawa *et al.*, 53). Genomic prediction models predict the genomic estimated breeding values (GEBV) of population that are suitable for utilizing in breeding processes. Genomic selection reduces the time and lists out progenies with superior character. Genomic selection had been applied in to predict fruit texture in apple (Roth *et al.*, 73), fruit weight and quality in peach (Biscarini *et al.*, 3), fruit yield and quality in strawberry (Cockerton *et al.*, 12), autotetraploidy in blueberry (de Bem Oliveira *et al.*, 16), fruit characters in pear (Kumar *et al.*, 42) *etc.* Extending association studies and genomic selection to fruit crops can enhance the speed of conventional breeding programs along with strategies like marker assisted selection and marker assisted introgression (Endo *et al.*, 22; Sheikh *et al.*, 79; Zhu *et al.*, 106; Noh *et al.*, 59).

Genetic Modifications

Trangenics cisgenics, RNAi, and CRISPR/Cas are some of the tools used for genome modifications in fruit crops. Availability of sequence information for various fruits has increased prototypes for genetic modifications. Transgenics has been successfully applied in apple (Zhao *et al.*, 103), banana (Paul *et al.*, 66), papaya (Blas *et al.*, 4), orange (De Francesco *et al.*, 17), plum (Sidorova *et al.*, 81), citrus (Jia *et al.*, 31) *etc.* RNAi technology had been utilized to produce transgenic papaya (Cabanos *et al.*, 6) and banana (Elitzur *et al.*, 21) with delayed ripening and extended shelf life and transgenic papaya with PRSV-resistance (Jia *et al.*, 33). Cisgenetic approach had been applied in Cavendish banana to increase the resistance to Panama disease and Provitamin A levels (Dale *et al.*, 14). Some of the commercialized transgenic fruit crops are 'SunUp', PRSV resistant papaya (Ye and Li, 98); 'Arctic', non-browning apple (Stowe and Dhingra, 82) and 'Pinkglow', Pink-fleshed pineapple (Firoozbady and Young, 25). Many transgenic trials are underway in fruit crops that focus on disease resistance, fruit quality and abiotic stress tolerance. CRISPR/Cas has become an inevitable alternative approach for editing the fruit crops for improving fruit quality, tolerance to stresses and domestication of wild species and also to overcome regulatory bottlenecks faced by transgenic crops. It has been utilized for inducing targeted mutagenesis in crops like apple (Zhang *et al.*, 102), citrus (Peng *et al.*, 67), pear (Charrier *et al.*, 9), banana (Tripathi *et al.*, 88), grapes (Wang *et al.*, 91), kiwifruit (Wang *et al.*,

92) and strawberry (Wilson *et al.,* 94) as proof of concept studies. These genetic modification approaches are a boon to fruit crop improvement programs, as it reduces the hurdles of long juvenile phase, large mapping population and time taken for breeding cycles. However, the tissue cultural techniques for their regeneration and multiplication need to be standardised.

Conclusion

Development of sequencing technologies and modern breeding techniques supported by predictive markers for specific traits will assist breeders in identifying superior progenies and develop improved cultivars. Application of integrated 'omics' approaches, marker-trait association studies, genomic selection and genetic modifications in fruit crop improvement will be powerful tools in the coming years.

REFERENCES

1. Agustí, M., Mesejo, C., MuñozFambuena, N., VeraSirera, F., de Lucas, M., MartínezFuentes, A., Reig, C., Iglesias, D.J., PrimoMillo, E. and Blázquez, M.A. 2020. Fruitdependent epigenetic regulation of flowering in Citrus. *New Phytol.* **225**(1): 376-384.
2. Bally, I.S., Bombarely, A., Chambers, A.H., Cohen, Y., Dillon, N.L., Innes, D.J., Islas-Osuna, M.A., Kuhn, D.N., Mueller, L.A., Ophir, R. and Rambani, A. 2021. The 'Tommy Atkins' mango genome reveals candidate genes for fruit quality. *BMC Plant Biol.* **21**(1): 1-18.
3. Biscarini, F., Nazzicari, N., Bink, M., Arús, P., Aranzana, M.J., Verde, I., Micali, S., Pascal, T., Quilot-Turion, B., Lambert, P. and da Silva Linge, C. 2017. Genome-enabled predictions for fruit weight and quality from repeated records in European peach progenies. *BMC genomics.* **18**(1): 1-15.
4. Blas, A. L., Ming, R., Liu, Z., Veatch, O. J., Paull, R. E., Moore, P. H., and Yu, Q. 2010. Cloning of the Papaya Chromoplast-Specific Lycopene β-Cyclase, CpCYC-b, Controlling Fruit Flesh Color Reveals Conserved Microsynteny and a Recombination Hot Spot. *Plant Physiol.* **152** (4): 2013–2022.
5. Branham, S.E., Wechter, W.P., Ling, K.S., Chanda, B., Massey, L., Zhao, G., Guner, N., Bello, M., Kabelka, E., Fei, Z. and Levi, A. 2020. QTL mapping of resistance to *Fusarium oxysporum* f. sp. niveum race 2 and Papaya ringspot virus in *Citrullus amarus*. *Theor. Appl. Genet.* **133**(2): 677-687.
6. Cabanos, C.S., Sajise, A.G., Garcia, R.N., Siar, S.V; Mae, E. and Mendoza, T. 2013. Compositional analysis of transgenic papaya with delayed ripening trait. *Philipp. Agric. Sci.* **96** (4): 331–339.
7. Cao, K., Yang, X., Li, Y., Zhu, G., Fang, W., Chen, C., Wang, X., Wu, J. and Wang, L. 2021. New high-quality peach (*Prunus persica* L. Batsch) genome assembly to analyze the molecular evolutionary mechanism of volatile compounds in peach fruits. *Plant J.* **108**(1): 281-295.
8. Cazenave, X., Petit, B., Laurens, F., Durel, C.E. and Muranty, H. 2021. Combining genetic resources and elite material populations to improve the accuracy of genomic prediction in apple. *bioRxiv*.

9. Charrier, A., Vergne, E., Dousset, N., Richer, A., Petiteau, A. and Chevreau, E. 2019. Efficient targeted mutagenesis in apple and first-time edition of pear using the CRISPR-Cas9 system. *Front plant sci.* ***10***: 40.

10. Chen, C., Guo, J., Cao, K., Zhu, G., Fang, W., Wang, X., Li, Y., Wu, J., Xu, Q. and Wang, L. 2021. Identification of candidate genes associated with slow-melting flesh trait in peach using bulked segregant analysis and RNA-seq. *Sci Hortic.* **286**.

11. Chen, J.R., Ueno, H., Matsumura, H., Urasaki, N., Lee, C.Y., Chen, F.C., Chin, S.W., Liu, C.C., Chiu, C.T., Tarora, K. and Li, J.Y. 2021. Genomic characterization of a rare *Carica papaya* X chromosome mutant reveals a candidate monodehydroascorbate reductase 4 gene involved in all-hermaphrodite phenomenon. *Mol. Genetics Genomics*. 1-13.

12. Cockerton, H.M., Karlstrom, A., Johnson, A., Li, B.M., Stavridou, E., Hopson, K., Whitehouse, A. and Harrison, R.J. 2021. The Genetic Architecture of Strawberry Yield and Fruit Quality Traits. *bioRxiv.*

13. Daccord, N., Celton, J.M., Linsmith, G., Becker, C., Choisne, N., Schijlen, E., Van de Geest, H., Bianco, L., Micheletti, D., Velasco, R. and Di Pierro, E.A. 2017. High-quality de novo assembly of the apple genome and methylome dynamics of early fruit development. *Nat Genet.* **49**(7): 1099-1106.

14. Dale, J., Paul, J., Dugdale, B. and Harding, R. 2017. Modifying Bananas/: From transgenics to organics/? *Sustainability*. **9**(3): 333.

15. Davey, M.W., Gudimella, R., Harikrishna, J.A., Sin, L.W., Khalid, N. and Keulemans, J. 2013. A draft *Musa balbisiana* genome sequence for molecular genetics in polyploid, inter- and intra-specific Musa hybrids. BMC genomics. **14**(1): 1-20.

16. de Bem Oliveira, I., Amadeu, R.R., Ferrão, L.F.V. and Muñoz, P.R. 2020. Optimizing whole-genomic prediction for autotetraploid blueberry breeding. *Heredity*. **125**(6): 437-448.

17. De Francesco, A., Simeone, M., Gómez, C., Costa, N. and Garcia, M.L. 2020. Transgenic Sweet Orange expressing hairpin CP-mRNA in the interstock confers tolerance to citrus psorosis virus in the non-transgenic scion. *Transgenic Res.* **29**(2): 215-228.

18. Deng, G.M., Zhang, S., Yang, Q.S., Gao, H.J., Sheng, O., Bi, F.C., Li, C.Y., Dong, T., Yi, G.J., He, W.D. and Hu, C.H. 2021. MaMYB4, an R2R3-MYB repressor transcription factor, negatively regulates the biosynthesis of anthocyanin in banana. *Front. Plant Sci.* **11**. 2023.

19. Dwivany, F.M., Nugrahapraja, H., Sari, L.D.N., Putri, R.R. and Novianti, C. 2021. Transcriptome dataset of ethylene-treated Klutuk Wulung banana. *Data Brief.* **38.**

20. Eduardo, I., de Tomás, C., Alexiou, K.G., Giovannini, D., Pietrella, M., Carpenedo, S., Raseira, M.B., Batlle, I., Cantín, C.M., Aranzana, M.J. and Arús, P. 2020. Fine mapping of the peach pollen sterility gene (Ps/ps) and detection of markers for marker-assisted selection. *Mol. Breed.* **40**(6): 1-10.

21. Elitzur, T., Yakir, E., Quansah, L., Zhangjun, F., Vrebalov, J., Khayat, E., Giovannoni, J. J. and Friedman, H. 2016. Banana MaMADS transcription factors are necessary for fruit ripening and molecular tools to promote shelf-life and food security. *Plant Physiol.* **171** (1): 380–391.

22. Endo, T., Fujii, H., Omura, M. and Shimada, T. 2020. Fast-track breeding system to introduce CTV resistance of trifoliate orange into citrus germplasms, by integrating early flowering transgenic plants with marker-assisted selection. *BMC Plant Biol.* **20:** 1-16.

23. FAO. 2020. World Food and Agriculture - Statistical Yearbook 2020. Rome. https: //doi. org/10.4060/cb1329en.

24. Feng, C., Feng, C., Lin, X., Liu, S., Li, Y. and Kang, M. 2021. A chromosome level genome assembly provides insights into ascorbic acid accumulation and fruit softening in guava (*Psidium guajava*). *Plant Biotech. J.* **19**(4): 717-730.

25. Firoozbady, E. and Young, T.R. 2015. Del Monte Fresh Produce Co. Pineapple plant named 'Rosé'. U.S. Patent Application 13/507,101.

26. Fu, W., da Silva Linge, C. and Gasic, K. 2021. Genome-Wide Association Study of Brown Rot (*Monilinia* spp.) Tolerance in Peach. *Front. Plant Sci.* **12**: 354.

27. Gao, T., Zhang, Z., Liu, X., Wu, Q., Chen, Q., Liu, Q., van Nocker, S., Ma, F. and Li, C. 2020. Physiological and transcriptome analyses of the effects of exogenous dopamine on drought tolerance in apple. *Plant Physiol.Biochem.* **148**: 260-272.

28. Gao, Y., Yang, Q., Yan, X., Wu, X., Yang, F., Li, J., Wei, J., Ni, J., Ahmad, M., Bai, S. and Teng, Y. 2021. High-quality genome assembly of 'Cuiguan'pear (*Pyrus pyrifolia*) as a reference genome for identifying regulatory genes and epigenetic modifications responsible for bud dormancy. Hortic Res. **8**(1): 1-16.

29. Gottschalk, C., Zhang, S., Schwallier, P., Rogers, S., Bukovac, M.J. and van Nocker, S. 2021. Genetic mechanisms associated with floral initiation and the repressive effect of fruit on flowering in apple (*Malus* x *domestica* Borkh). *PloS one*, **16**(2): p.e0245487.

30. Hu, F.C., Zhe, C., Wang, X.H., Wang, J.B., Fan, H.Y., Qin, Y.H., Zhao, J.T. and Hu, G.B. 2021. Construction of high-density SNP genetic maps and QTL mapping for dwarf-related traits in *Litchi chinensis* Sonn. *J. Integr. Agric.* **20**(11): 2900-2913.

31. Jia, H.M., Jia, H.J., Cai, Q.L., Wang, Y., Zhao, H.B., Yang, W.F., Wang, G.Y., Li, Y.H., Zhan, D.L., Shen, Y.T. and Niu, Q.F. 2019. The red bayberry genome and genetic basis of sex determination. Plant Biotechnol. J. **17**(2): 397-409.

32. Jia, H. and Wang, N. 2014. Xcc-Facilitated agroinfiltration of citrus leaves: a tool for rapid functional analysis of transgenes in citrus leaves. *Plant Cell Rep.* **33** (12): 1993–2001.

33. Jia, R., Zhao, H., Huang, J., Kong, H., Zhang, Y., Guo, J., Huang, Q., Guo, Y., Wei, Q., Zuo, J. and Zhu, Y.J. 2017. Use of RNAi technology to develop a PRSV-resistant transgenic papaya. *Sci. Rep.* **7**(1): 1-9.

34. Jiang, J., Fan, X., Zhang, Y., Tang, X., Li, X., Liu, C. and Zhang, Z. 2020. Construction of a high-density genetic map and mapping of firmness in grapes (*Vitis vinifera* L.) based on whole-genome resequencing. *Int. J. Mol. Sci.* **21**(3): 797.

35. Jiang, S.H., Sun, Q.G., Chen, M., Wang, N., Xu, H.F., Fang, H.C., Wang, Y.C., Zhang, Z.Y. and Chen, X.S. 2019. Methylome and transcriptome analyses of apple fruit somatic mutations reveal the difference of red phenotype. *BMC Genomics.* **20**(1): 1-13.

36. Jian-ye, C., Fang-fang, X., Yan-ze, C., Can-bin, C., Wang-jin, L., Xiao-di, H., Qing-zhu, H., Zhao, J., Zhi-jiang, W., Gao, D. and Zhi-ke, Z. 2021. A chromosome-scale genome sequence of pitaya (*Hylocereus undatus*) provides novel insights into the genome evolution and regulation of betalain biosynthesis. *Hortic Res.* **8**(1).

37. Karanjalker, G.R., Ravishankar, K.V., Shivashankara, K.S., Dinesh, M.R., Roy, T.K. and Sudhakar Rao, D.V. 2018. A study on the expression of genes involved in carotenoids and anthocyanins during ripening in fruit peel of green, yellow, and red colored mango cultivars. *Appl. Biochem. Biotechnol.* **184** (1): 140–154.

38. Kaur, K., Awasthi, P. and Tiwari, S. 2021. Comparative transcriptome analysis of unripe and ripe banana (cv. Nendran) unraveling genes involved in ripening and other related processes. *Plos one.* **16**(7): p.e0254709.

39. Kim, J.H., Jung, Y.J., Seo, H.K., Kim, M.K., Nou, I.S. and Kang, K.K. 2019. Recurrent parent genome (RPG) recovery analysis in a marker-assisted backcross breeding based on the genotyping-by-sequencing in tomato (*Solanum lycopersicum* L.). *J. Plant Biotechnol.* **46**(3): 165-171.

40. Kuhn, D.N., Livingstone III, D.S., Richards, J.H., Manosalva, P., Van den Berg, N. and Chambers, A.H. 2019. Application of genomic tools to avocado (*Persea americana*) breeding: SNP discovery for genotyping and germplasm characterization. *Sci. Hortic.* **246**: 1-11.

41. Kumar, P., Choudhary, M., Jat, B.S., Kumar, B., Singh, V., Kumar, V., Singla, D. and Rakshit, S. 2021. Skim sequencing: an advanced NGS technology for crop improvement. *J. Genetics.* **100**(2): 1-10.

42. Kumar, S., Kirk, C., Deng, C.H., Shirtliff, A., Wiedow, C., Qin, M., Wu, J. and Brewer, L. 2019. Marker-trait associations and genomic predictions of interspecific pear (Pyrus) fruit characteristics. *Sci. Rep.* **9**(1): 1-10.

43. Kunihisa, M., Hayashi, T., Hatsuyama, Y., Fukasawa-Akada, T., Uenishi, H., Matsumoto, T., Kon, T., Kasai, S., Kudo, T., Oshino, H. and Yamamoto, T. 2021. Genome-wide association study for apple flesh browning: detection, validation, and physiological roles of QTLs. *Tree Genet. Genomes.* **17**(1): 1-16.

44. Li, R., Hu, F., Li, B., Zhang, Y., Chen, M., Fan, T. and Wang, T. 2020. Whole genome bisulfite sequencing methylome analysis of mulberry (*Morus alba*) reveals epigenome modifications in response to drought stress. Sci. Rep. **10**(1): 1-17.

45. Li, W.M., Dita, M., Rouard, M., Wu, W., Roux, N., Xie, J.H. and Ge, X. J. 2020. Deep RNA-seq analysis reveals key responding aspects of wild banana relative resistance to *Fusarium oxysporum* f. sp. cubense tropical race 4. *Funct. Integr Genomics.* 1-12.

46. Liu, C., Feng, C., Peng, W., Hao, J., Wang, J., Pan, J. and He, Y. 2020. Chromosome-level draft genome of a diploid plum (*Prunus salicina*). *GigaScience.* **9**(12).

47. Long, Y., Liu, Z., Jia, J., Mo, W., Fang, L., Lu, D., Liu, B., Zhang, H., Chen, W. and Zhai, J. 2021. FlsnRNA-seq: protoplasting-free full-length single-nucleus RNA profiling in plants. *Genome Biol.* **22**(1): 1-14.

48. Luo, F., Norelli, J.L., Howard, N.P., Wisniewski, M., Flachowsky, H., Hanke, M.V. and Peace, C. 2020. Introgressing blue mold resistance into elite apple germplasm by rapid cycle breeding and foreground and background DNA-informed selection. *Tree Genet. Genomes.* **16**(2): 1-15.

49. Ma, D., Dong, S., Zhang, S., Wei, X., Xie, Q., Ding, Q., Xia, R. and Zhang, X. 2021. Chromosome level reference genome assembly provides insights into aroma biosynthesis in passion fruit (*Passiflora edulis*). *Mol. Ecol. Resour.* **21**(3): 955-968.

50. Ma, Y., Xue, H., Zhang, F., Jiang, Q., Yang, S., Yue, P., Wang, F., Zhang, Y., Li, L., He, P. and Zhang, Z. 2021. The miR156/SPL module regulates apple salt stress tolerance by activating *MdWRKY100* expression. *Plant Biotechnol. J.* **19**(2): 311.

51. Milyaev, A., Kofler, J., Klaiber, I., Czemmel, S., Pfannstiel, J., Flachowsky, H., Stefanelli, D., Hanke, M.V. and Wünsche, J.N. 2021. Toward Systematic Understanding of Flower Bud Induction in Apple: A Multi-Omics Approach. *Front. Plant Sci.* **12**: 286.

52. Minamikawa, M.F., Nonaka, K., Kaminuma, E., Kajiya-Kanegae, H., Onogi, A., Goto, S., Yoshioka, T., Imai, A., Hamada, H., Hayashi, T. and Matsumoto, S. 2017. Genome-wide association study and genomic prediction in citrus: potential of genomics-assisted breeding for fruit quality traits. *Sci. Rep.* **7**(1): 1-13.

53. Minamikawa, M.F., Takada, N., Terakami, S., Saito, T., Onogi, A., Kajiya-Kanegae, H., Hayashi, T., Yamamoto, T. and Iwata, H. 2018. Genome-wide association study and genomic prediction using parental and breeding populations of Japanese pear (*Pyrus pyrifolia* Nakai). *Sci. Rep.* **8**(1): 1-12.

54. Ming, R., Hou, S., Feng, Y., Yu, Q., Dionne-Laporte, A., Saw, J.H., Senin, P., Wang, W., Ly, B.V., Lewis, K.L. and Salzberg, S.L. 2008. The draft genome of the transgenic tropical fruit tree papaya (*Carica papaya* Linnaeus). *Nature.* **452**(7190): 991-996.

55. Muhammad, I.I., Kong, S.L., Akmar Abdullah, S.N. and Munusamy, U. 2020. RNA-seq and ChIP-seq as complementary approaches for comprehension of plant transcriptional regulatory mechanism. *Int. J. Mol. Sci.* **21**(1): 167.

56. Naibin, D.N.D., Yumin, M.Y.M., Kun, W.K.W., Xiaomu, W.X.W., Kun, X.K.X., Jing, B.J.B., Yongyi, Y.Y.Y., Yanyan, P.Y.P. and Yongchao, G.Y.G. 2021. SNP Mining by Genome Resequencing of 30 Apple Varieties in Shandong Province. *Mol. Plant Breed.* **12**.

57. Nawaz, I., Tariq, R., Nazir, T., Khan, I., Basit, A., Gul, H., Anwar, T., Awan, S.A., Bacha, S.A.S., Zhang, L. and Zhang, C. 2021. RNA-Seq profiling reveals the plant hormones and molecular mechanisms stimulating the early ripening in apple. *Genomics.* **113**(1): 493-502.

58. Noh, J., Do, Y.S., Kim, G.H. and Choi, C. 2020. A genome-wide association study for the detection of genes related to apple Marssonina Blotch disease resistance in apples. *Sci. Hortic.* **262**.

59. Noh, Y.H., Lee, S., Whitaker, V.M., Cearley, K.R. and Cha, J.S. 2017. A high-throughput marker-assisted selection system combining rapid DNA extraction high-resolution melting and simple sequence repeat analysis: strawberry as a model for fruit crops. *J. Berry Res.* **7**(1): 23-31.

60. Nuñez-Lillo, G., Balladares, C., Pavez, C., Urra, C., Sanhueza, D., Vendramin, E., Dettori, M.T., Arús, P., Verde, I., Blanco-Herrera, F. and Campos-Vargas, R. 2019. High-density genetic map and QTL analysis of soluble solid content, maturity date, and mealiness in peach using genotyping by sequencing. *Sci. Hortic.* **257**.

61. Nyine, M., Uwimana, B., Akech, V., Brown, A., Ortiz, R., Doležel, J., Lorenzen, J. and Swennen, R. 2019. Association genetics of bunch weight and its component traits in East African highland banana (*Musa* spp. AAA group). *Theor. App. Genetics.* **132**(12): 3295-3308.

62. Nyine, M., Uwimana, B., Blavet, N., Høibová, E., Vanrespaille, H., Batte, M., Akech, V., Brown, A., Lorenzen, J.H., Swennen, R.L. and Doležel, J. 2018. Genomic prediction in a multiploid crop: genotype by environment interaction and allele dosage effects on predictive ability in banana. *Plant Genome.* **11**(2).

63. Ollitrault, P., Ahmed, D., Costantino, G., Evrard, J.C., Cardi, C., Mournet, P., Perdereau, A. and Froelicher, Y. 2021. Segregation distortion for male parents in high density genetic maps from reciprocal crosses between two self-incompatible cultivars confirms a gametophytic system for self-incompatibility in citrus. *Agriculture.* **11**(5): 379.

64. Onik, J.C., Wai, S.C., Li, A., Lin, Q., Sun, Q., Wang, Z. and Duan, Y. 2021. Melatonin treatment reduces ethylene production and maintains fruit quality in apple during postharvest storage. *Food Chem.* **337**.

65. Patil, S.I., Vyavahare, S.N., Krishna, B. and Sane, P.V. 2021. Studies on the expression patterns of the circadian rhythm regulated genes in mango. *Physiol. Mol. Bio. Plants.* **27**(9): 2009-2025.

66. Paul, J., Khanna, H., Kleidon, J., Hoang, P., Geijskes, J., Daniells, J., Zaplin, E., Rosenberg, Y., James, A., Mlalazi, B., Deo, P., Arinaitwe, G., Namanya, P., Becker, D., Tindamanyire, J., Tushemereirwe, W., Harding, R. and Dale, J. 2017. Golden bananas in the field: elevated fruit pro-vitamin A from the expression of a single banana transgene. *Plant Biotechnol. J.* **15** (4): 520–532.

67. Peng, A., Chen, S., Lei, T., Xu, L., He, Y., Wu, L., Yao, L. and Zou, X. 2017. Engineering canker resistant plants through CRISPR/Cas9 targeted editing of the susceptibility gene Cs LOB 1 promoter in citrus. *Plant Biotechnol. J.* **15**(12): 1509-1519.

68. Qiao, Q., Edger, P.P., Xue, L., Qiong, L., Lu, J., Zhang, Y., Cao, Q., Yocca, A.E., Platts, A.E., Knapp, S.J. and Van Montagu, M. 2021. Evolutionary history and pan-genome dynamics of strawberry (*Fragaria* spp.). *PNAS.* **118**(45).

69. Rahim, M.A., Afrin, K.S., Jung, H.J., Kim, H.T., Park, J.I. and Nou, I.S. 2021. Identification and characterization of carotenoid biosynthesis related genes in a novel dark skinned citrus mutant cultivar 'Suneat'. Hortic. Environ. Biotechnol. **62**(1): 99-111.

70. Rawandoozi, Z.J., Hartmann, T.P., Carpenedo, S., Gasic, K., da Silva Linge, C., Cai, L., Van de Weg, E. and Byrne, D.H. 2020. Identification and characterization of QTLs for fruit quality traits in peach through a multi-family approach. *BMC Genomics.* **21**(1): 1-18.

71. Renner, S.S., Wu, S., Pérez-Escobar, O.A., Silber, M.V., Fei, Z. and Chomicki, G. 2021. A chromosome-level genome of a Kordofan melon illuminates the origin of domesticated watermelons. *PNAS.* **118**(23).

72. Rich-Griffin, C., Stechemesser, A., Finch, J., Lucas, E., Ott, S. and Schäfer, P. 2020. Single-cell transcriptomics: a high-resolution avenue for plant functional genomics. *Trends Plant Sci.* **25**(2): 186-197.

73. Roth, M., Muranty, H., Di Guardo, M., Guerra, W., Patocchi, A. and Costa, F. 2020. Genomic prediction of fruit texture and training population optimization towards the application of genomic selection in apple. *Hortic. Res.* **7**(1): 1-14.

74. SampangiRamaiah, M.H., Ravishankar, K.V. and Ajitha, R. 2019. Long non-coding RNAs in banana/: prediction, mapping and their comparative studies using *Musa balbisiana* and *Musa acuminata* transcriptome. *Trees.* 359–369.

75. Santos, A.S., de Andrade, E.M., de Oliveira Sousa, A.R., do Carmo Costa, M.M., Togawa, R., Grynberg, P., Pirovani, C.P., Oliveira, S.A.S., dos Santos Soares Filho, W., Coelho Filho, M.A. and da Silva Gesteira, A. 2021. Transcriptome profile of drought responsive candidate genes in varieties of citrus rootstocks with different tolerance strategies. *Sci. Hortic.* **277**.

76. Shahbandeh. 2021. Fresh fruit production worldwide 1990-2019. https: //www.statista.com/statistics/262266/global-production-of-fresh-fruit/

77. Sharma, N., Singh, A.K., Singh, S.K., Mahato, A.K., Srivastav, M. and Singh, N.K. 2020. Comparative RNA sequencing based transcriptome profiling of regular bearing and alternate bearing mango (*Mangifera indica* L.) varieties reveals novel insights into the regulatory mechanisms underlying alternate bearing. *Biotechnol. Letters.* **42**(6).

78. Shaw, R., Tian, X. and Xu, J. 2021. Single-cell transcriptome analysis in plants: advances and challenges. *Mol. Plant.* **14**(1): 115-126.

79. Sheikh, M.A., Mushtaq, K., Mir, J.I., Amin, M. and Nabi, S.U. 2020. Introgression of Scab Resistance Gene *Vf* (Rvi6) in commercially grown Susceptible Cultivar Fuji Azitec of apple (*Malus domestica*) using Marker Assisted Selection. *Res. J. Biotechnol.* **15**. 9.

80. Shirasawa, K., Itai, A. and Isobe, S. 2021. Chromosome-scale genome assembly of Japanese pear (*Pyrus pyrifolia*) variety 'Nijisseiki'. *DNA Res.* **28**(2).

81. Sidorova, T., Mikhailov, R., Pushin, A., Miroshnichenko, D. and Dolgov, S. 2019. Agrobacterium-mediated transformation of Russian commercial plum cv. "Startovaya" (*Prunus domestica* L.) with virus-derived hairpin RNA construct confers durable resistance to PPV infection in mature plants. *Front. Plant Sci.* **10**. 286.

82. Stowe, E. and Dhingra, A. 2021. Development of the Arctic® Apple. *Plant Breed. Reviews.* **44:** 273-296.

83. Sun, T., Zhang, J., Zhang, Q., Li, X., Li, M., Yang, Y., Zhou, J., Wei, Q. and Zhou, B. 2021. Transcriptome and metabolome analyses revealed the response mechanism of apple to different phosphorus stresses. *Plant Physiol. Biochem.* **167**: 639-650.

84. Sun, X., Jiao, C., Schwaninger, H., Chao, C.T., Ma, Y., Duan, N., Khan, A., Ban, S., Xu, K., Cheng, L. and Zhong, G.Y. 2020. Phased diploid genome assemblies and pan-genomes provide insights into the genetic history of apple domestication. *Nature Genetics*. **52**(12): 1423-1432.

85. Sun, Y., Hao, P., Lv, X., Tian, J., Wang, Y., Zhang, X., Xu, X., Han, Z. and Wu, T. 2020. A long noncoding apple RNA, MSTRG. 85814.11, acts as a transcriptional enhancer of SAUR32 and contributes to the Fedeficiency response. *Plant J.* **103**(1): 53-67.

86. Tan, Q., Li, S., Zhang, Y., Chen, M., Wen, B., Jiang, S., Chen, X., Fu, X., Li, D., Wu, H. and Wang, Y. 2021. Chromosome-level genome assemblies of five Prunus species and genome-wide association studies for key agronomic traits in peach. *Hortic. Res.* **8**(1).

87. Tian, X., Zhang, L., Feng, S., Zhao, Z., Wang, X. and Gao, H. 2019. Transcriptome analysis of apple leaves in response to powdery mildew (*Podosphaera leucotricha*) infection. *Int. J Mol. Sci.* **20**(9): 2326.

88. Tripathi, J.N., Ntui, V.O., Shah, T. and Tripathi, L. 2021. CRISPR/Cas9 mediated editing of DMR6 orthologue in banana (*Musa* spp.) confers enhanced resistance to bacterial disease. *Plant Biotechnol. J.*

89. Vidya, S. M., Kumar, H. V., Bhatt, R. M., Laxman, R. H., and Ravishankar, K. V. 2018. Transcriptional profiling and genes involved in acquired thermotolerance in Banana: a non-model crop. *Sci. Rep.* **8**(1): 1-10.

90. Wang, J., Li, J., Li, Z., Liu, B., Zhang, L., Guo, D., Huang, S., Qian, W. and Guo, L. 2021. Genomic insights into longan evolution from a chromosome-level genome assembly and population analysis of longan accessions. *BioRxiv*.

91. Wang, Y., Liu, X., Ren, C., Zhong, G.Y., Yang, L., Li, S. and Liang, Z. 2016. Identification of genomic sites for CRISPR/Cas9-based genome editing in the *Vitis vinifera* genome. *BMC Plant Bio.* **16**(1).

92. Wang, Z., Wang, S., Li, D., Zhang, Q., Li, L., Zhong, C., Liu, Y. and Huang, H. 2018. Optimized pairedsgRNA/Cas9 cloning and expression cassette triggers highefficiency multiplex genome editing in kiwifruit. *Plant Biotechnol. J.* **16**(8): 1424-1433.

93. Wei, X., Shen, F., Zhang, Q., Liu, N., Zhang, Y., Xu, M., Liu, S., Zhang, Y., Ma, X. and Liu, W. 2021. Genetic diversity analysis of Chinese plum (*Prunus salicina* L.) based on whole-genome resequencing. *Tree Genet. Genomes*. **17**(3): 1-10.

94. Wilson, F. M., Harrison, K., Armitage, A. D., Simkin, A. J. and Harrison, R. J. 2019. CRISPR/Cas9-mediated mutagenesis of phytoene desaturase in diploid and octoploid strawberry. *Plant Methods*. **15**(1).

95. Woehner, T.W., Emeriewen, O.F., Wittenberg, A., Schneiders, H., Vrijenhoek, I., Halasz, J., Hrotko, K., Hoff, K.J., Gabriel, L., Keilwagen, J. and Berner, T. 2021. The draft chromosome-level genome assembly of tetraploid ground cherry (*Prunus fruticosa* Pall.) from long reads. *BioRxiv*.

96. Wu, Y., Xu, J., Han, X., Qiao, G., Yang, K., Wen, Z. and Wen, X. 2020. Comparative transcriptome analysis combining SMRT-and Illumina-based RNA-Seq identifies potential

candidate genes involved in betalain biosynthesis in pitaya fruit. *Int. J Mol. Sci.* **21**(9): 3288.

97. Yang, C., Sha, G., Wei, T., Ma, B., Li, C., Li, P., Zou, Y., Xu, L. and Ma, F. 2021. Linkage map and QTL mapping of red flesh locus in apple using a R1R1× R6R6 population. *Hortic. Plant J.* 7(5): 393-400.

98. Ye, C. and Li, H. 2010. 20 Years of transgenic research in China for resistance to Papaya ringspot virus. *Transgenic Plant J.* **4**: 58-63.

99. Yuan, W., Liu, J., Takáè, T., Chen, H., Li, X., Meng, J., Tan, Y., Ning, T., He, Z., Yi, G. and Xu, C. 2021. Genome-Wide Identification of Banana *Csl* Gene Family and Their Different Responses to Low Temperature between Chilling-Sensitive and Tolerant Cultivars. *Plants.* **10**(1): 122.

100. Zhang, H., Li, M., Kong, M., Dunwell, J.M., Zhang, Y., Yue, C., Wu, J. and Zhang, S. 2021. Study on the differences of gene expression between pear and apple wild cultivation materials based on RNA-seq technique. *BMC plant biology.* **21**(1).

101. Zhang, Y., López-Girona, E. and Aranzana, M. J. 2017. March. Region-wide association analysis and high-throughput resequencing strategies in peach to develop molecular markers for flat fruit marker-assisted selection. *In*: IV International Symposium on Molecular Markers in Horticulture. **1203**: 79-84.

102. Zhang, Y., Zhou, P., Bozorov, T.A. and Zhang, D. 2021. Application of CRISPR/Cas9 technology in wild apple (*Malus sieverii*) for paired sites gene editing. *Plant Methods.* **17**(1): 1-9.

103. Zhao, S., Gao, H., Jia, X., Wang, H., Ke, M. and Ma, F. 2020. The HD-Zip I transcription factor *MdHB-7* regulates drought tolerance in transgenic apple (*Malus domestica*). Environ. Exp. Botany. **180**.

104. Zhou, H., Zhang, C., Wang, M., Zhang, W., Ma, J., Qi, S., Wang, S. and Xing, L. 2021. Epigenomic Regulatory Mechanism of Flowering in Apple Demonstrated in Two Varieties with Contrasted Flowering Behaviors. *Research square.*

105. Zhu, C., Zheng, X., Huang, Y., Ye, J., Chen, P., Zhang, C., Zhao, F., Xie, Z., Zhang, S., Wang, N. and Li, H. 2019. Genome sequencing and CRISPR/Cas9 gene editing of an early flowering Mini Citrus (*Fortunella hindsii*). *Plant Biotechnol. J.* **17**(11): 2199-2210.

106. Zhu, P., Gu, B., Li, P., Shu, X., Zhang, X. and Zhang, J. 2020. New cold-resistant, seedless grapes developed using embryo rescue and marker-assisted selection. Plant *Cell Tissue Organ Cult.* **140**(3): 551-562.

Chapter 8

Biotechnological Advancements in the Improvement of Coconut and Arecanut

M.K. Rajesh[1,2*], S.V. Ramesh[1], K.S. Muralikrishna[1], M. Shareefa[3] and V.B. Patel[4]

[1]*ICAR–Central Plantation Crops Research Institute, Kasaragod 671124, Kerala*

[2]*ICAR–Central Plantation Crops Research Institute, Regional Station, Vittal 574243, Karnataka*

[3]*ICAR–Central Plantation Crops Research Institute, Regional Station, Kayamkulam, Kerala*

[4]*Horticultural Science, Indian Council of Agricultural Research, New Delhi 110012*

**e-mail: rajesh.mk@icar.gov.in*

ABSTRACT

Coconut (Cocos nucifera L.) and arecanut (Areca catechu L.) are perennial palms with a wide distribution range across the tropics. The multitude of uses of coconut and arecanut has a significant bearing on the socio-economic framework among rural smallholders in producing countries. Adopting biotechnological tools has greatly aided in comprehending complex living systems in an integrated manner. In general, the utilisation of biotechnology for the advancements of plantation crops such as coconut and arecanut has been trailing compared to the significant accomplishments made elsewhere in cereals, legumes or other commercial crops. Adopting genomic, proteomic, transcriptomic, epigenomic or such multi-omics approaches in improving these plantation crops has been hampered due to their perennial nature, long juvenile period, and high degree of genetic heterozygosity from cross-pollination.

This review briefly describes the significant progress made in the field of in vitro culture, molecular markers and multi-omics of coconut and arecanut at ICAR-CPCRI. A perspective on the way forward is also provided.

Keywords: *Coconut, Areca nut, Molecular markers, In vitro culture, Somatic embryogenesis, Cryopreservation, Genetic resources, Plantation crops, Multi-omics.*

Coconut and arecanut are predominant plantation crops cultivated in the tropics for their myriad economic uses. These two palms are particularly vulnerable to a range of biotic and abiotic stresses that significantly restrict their productivity. The demand for high-yielding, improved varieties of coconut and arecanut, coupled with the necessity for enhanced resistance to biotic and abiotic stresses, has steadily increased. Additionally, it is anticipated that palms must adapt to climate-change-induced fluctuations (Ramesh *et al.,* 26). Crop improvement programs for coconut and arecanut relying heavily on conventional breeding approaches have largely proven inadequate for various reasons, including their long juvenile period, perennial nature, and genetic heterozygosity. Progress in the field of crop biotechnology has empowered palm researchers to embrace and apply innovative techniques for the enhancement of perennial crops. This article provides an overview of the advancements made in using tissue culture and cryopreservation, applying molecular marker techniques, and utilising next-generation sequencing (NGS) to enhance the cultivation of coconut and arecanut.

Using coconut embryo culture has significantly facilitated the unrestricted collection and exchange of germplasm across the countries. For coconut accessions that do not naturally germinate and sterile inter-specific hybrids of arecanut, effortless embryo rescue techniques have been successfully employed for regeneration. A primary objective in palm tissue culture is the multiplication of elite accessions through *in vitro* culture techniques. Despite significant progress in the tissue culture of other palms, achieving success in direct or callus-mediated regeneration in coconut has remained a challenge. However, a recently standardised protocol for direct regeneration from coconut immature inflorescence explants appears promising.

On the other hand, arecanut has witnessed considerable progress in the regeneration of superior genotypes using immature inflorescence tissues as explants. Establishing and adopting effective cryopreservation techniques for zygotic embryos and pollen of both coconut and arecanut have enabled the safe and long-term preservation of genetic resources, making it a feasible approach. Diverse molecular marker technologies have been deployed to uncover the genetic diversity of coconut and arecanut accessions. Furthermore, the application of molecular markers has played a crucial role in constructing linkage maps and identifying QTLs that govern yield and yield-attributing traits. The recent progress in genomics and the widespread availability of whole-genome sequence data are expected to unlock even more of the genomic potential in these crops. This is especially promising in the realms of breeding and the development of economically valuable, climate-resilient, and specialised cultivars. This article offers a brief glimpse into the remarkable accomplishments in the biotechnology of coconut and arecanut.

Embryo Culture and Embryo Rescue

Culturing Coconut Zygotic Embryos for International Germplasm Exchange

At ICAR-CPCRI, an effective protocol has been established for *in vitro* culturing of coconut zygotic embryos (Figure 1). This protocol has demonstrated success in numerous germplasm expeditions. The method relies on specially formulated artificial media and carefully controlled *in vitro* micro-climatic conditions to facilitate optimal growth and development of the embryos into healthy plantlets. Additionally, a field collection technique for zygotic embryos has been devised, aiding in germplasm collection during expeditions to regions with limited access to sophisticated laboratories or equipments. Researchers at ICAR-CPCRI have also devised a simple yet efficient method for storing zygotic embryos for up to two months. This technique involves utilising sterile water as a medium, proving useful in germplasm collection initiatives. The coconut embryo culture protocol pioneered at ICAR-CPCRI has gained international acceptance and could serve as a valuable germplasm collection and exchange technology.

Figure 1. Embryo culture of coconut: ICAR-CPCRI protocol (a) Germinating embryo (b) Development of complete plantlet (c) *In vitro* hardened plantlet developed from a zygotic embryo (d) Establishment of embryo culture derived plantlet in the field and (e) Embryo culture derived palm bearing nuts.

The embryo culture protocol developed by ICAR-CPCRI (Karun *et al.,* 7) was adopted in collecting six accessions from the Pacific Ocean collections housed at the World Coconut Germplasm Center (WCGC) in the Andaman and Nicobar Islands, India. The process involved the collection of mature zygotic embryos, their transportation to the mainland, *in vitro* retrieval, and subsequent acclimatisation of the resulting plantlets. These plantlets were then field planted in 1996 at the International Coconut Gene Bank for South Asia and the Middle East (ICG-SAME) in Kidu, Karnataka State, India. Following this success, the protocol was effectively utilised in five international germplasm collection missions conducted by ICAR-CPCRI between 1997 and 2001. The primary objective of these missions was to enhance coconut genetic diversity in India. During these missions, zygotic embryos were collected from 45 accessions from eight countries: Bangladesh, Comoros, Madagascar, the Maldives, Mauritius, Réunion, Seychelles, and Sri Lanka. These embryos were retrieved *in vitro,* and the exotic collections obtained through the ICAR-

CPCRI embryo culture protocol were planted in ICG-SAME, India. Subsequent performance evaluations highlighted the promise of certain exotic collections. Consequently, these promising exotic collections have been integrated into coconut breeding programs. Insights gained from these endeavours indicated that approximately 300 to 400 embryos are essential for successfully establishing 100 palms in a field gene bank.

Embryo Rescue in Coconut

In vitro embryo rescue entails the cultivation of plantlets from embryos that either fail to germinate or exhibit very slow germination in their natural environment. One such coconut variant is *Mohachao Narel*, discovered in the Guhaghar taluk, Ratnagiri district, Maharashtra, India. The kernels of this variant are characterised by a sweet and soft texture and a relatively low fibre content. Due to the sweet endosperm nuts and the comparatively low embryo weight, these accessions struggle to germinate under natural conditions. Applying the embryo culture protocol developed at ICAR-CPCRI, plantlets were successfully regenerated through the embryo rescue technique for these sweet endosperm nuts (Karun *et al.*, 8) (Figure 2).

Figure 2. Embryo rescue in special sweet endosperm type of coconut: (a) *In vitro* culture initiated for the rescued and surface sterilised embryo from sweet endosperm coconut (b) Embryo germination (c) Development of plantlet (d) Plantlets in greenhouse post hardening process and (e) Field planting of embryo-rescued plantlet.

In another instance, Horned Coconuts, exclusive to the Andaman Islands, present a unique feature where multiple ovaries result in horn-like structures covering the mature nuts, causing a delay in the natural germination process. The ICAR-CPCRI embryo culture technique was effectively employed to culture Horned Coconut embryos *in vitro*. Upon field planting these embryo-cultured palms derived from horned coconuts, flowering was observed within six to eight years.

Cryopreservation: A Boon for Perennials

The genetic resources of coconut and arecanut have conventionally been conserved *ex-situ* by preserving entire palms in field gene banks. However, this method is space-intensive and requires substantial maintenance costs. Cryopreservation, on the other hand, offers a long-term conservation strategy that ensures the safe and efficient preservation of genetic diversity and overcomes the limitations of traditional conservation methods. Cryopreservation of coconut zygotic embryos and coconut and arecanut pollen emerges as a highly effective approach for the enduring conservation of valuable genetic resources. Moreover, this technique can act as a practical backup for the genetic resources of these crops held in field gene banks, adding an extra layer of security and preservation for these crucial genetic materials.

Vitrification-based cryopreservation of Coconut Zygotic Embryos

A crucial step for the successful cryopreservation of mature zygotic embryos involves pre-treating them before subjecting them to liquid nitrogen treatment at −196°C. Several pre-treatments have been identified to be effective, including simple desiccation (utilising a laminar air current and silica gel), high sucrose concentrations, encapsulation with 3 per cent sodium alginate, and cryoprotectants such as plant vitrification solution (PVS).

In 2005, a standardised cryopreservation technique was developed for the West Coast Tall cultivar, utilising desiccation pre-treatments for mature embryos. Normal plantlets were successfully derived from zygotic embryos that had undergone desiccation for 18 hours with silica gel or 24 hours with laminar airflow, followed by cryopreservation. However, irreversible damage to the shoot meristem was observed when the moisture content of the embryo was reduced to less than 20 per cent. By implementing a pre-growth desiccation method using a high sucrose concentration, the moisture content of the embryos could be reduced to 27-30 per cent, respectively, with a final recovery of 20.8-29 per cent after cryopreservation.

Subsequent research has focused on refining the vitrification technique for cryopreservation of mature coconut zygotic embryos. The PVS_3 solution, comprising equal parts of glycerol and sucrose, was identified as the most suitable for regenerating cryopreserved embryos. Thus, a standardised protocol involved pre-culturing embryos for three days on a medium containing 0.6 M sucrose, PVS3 treatment for 16 hours, rapid cooling in liquid nitrogen, rewarming, and final unloading in a 1.2 M sucrose liquid medium for 90 minutes. This refined protocol achieved survival rates of 70-80 per cent (corresponding to size enlargement and weight gain) (Figure 3) and successfully established 20-25 per cent of the regenerated plants (displaying normal shoot and root growth) from cryopreserved embryos in pots (Sajini *et al.,* 30).

Cryopreservation of Coconut Pollen

Pollen, a valuable source of diverse alleles within a gene pool, is crucial for germplasm conservation, hybrid seed production, and assisted pollination. Therefore, efficient pollen storage methods are imperative. The cryopreservation of coconut pollen is a noteworthy advancement in ensuring its long-term viability and utility. Interestingly, storing coconut pollen in liquid nitrogen for 24 hours does not adversely affect its *in vitro* germination.

The process of pollen cryopreservation involves several key steps: (a) Extraction of Pollen: pollen is extracted by sieving male flowers, typically using a mesh size of 0.2 mm. The male flowers are placed in an oven at 40°C for 24 hours to facilitate extraction. (b) Packaging Pollen: The extracted pollen is carefully wrapped in strips of aluminium foil. (c) Storage in Cryovials: The foil-wrapped pollen is inserted into cryovials, ensuring a secure and protective storage environment. (d) Plunging into Liquid Nitrogen: The cryovials containing the wrapped pollen are immersed in liquid nitrogen, providing a suitable cryogenic storage condition. Pollen viability and fertility are assessed in vitro and in vivo. *In vitro* assessment involves observing the growth of pollen in artificial media, while *in vivo* assessment is conducted through hand pollination in the field (Figure 4). Remarkably, even after seven years of storage in liquid nitrogen, the

Figure 3. Vitrification-based cryopreservation of zygotic embryos of coconut: (a) Pre-culturing of embryo in high sucrose medium (b) PVS3 treatment (c) Cryovial with dehydrated embryos (d) Embryo regeneration in retrieval medium following the cryopreservation process (e) Development of complete plantlet (f) *In vitro* hardened plantlet (g) *ex vitro* hardened plantlet and (h) Establishment of plantlet in the field and (i) Flowering and bearing palm.

coconut pollen retained its viability and fertility. Hand pollination using cryopreserved pollen for seven years resulted in a normal nut set. Furthermore, embryos extracted from hybrid nuts produced with cryostored pollen displayed a remarkable 100 per cent germination rate and exhibited normal plantlet development. This study conclusively proves the viability and feasibility of establishing a pollen cryobank in coconut for long-term germplasm preservation and utilisation (Veluru *et al.,* 34; Karun *et al.,* 5).

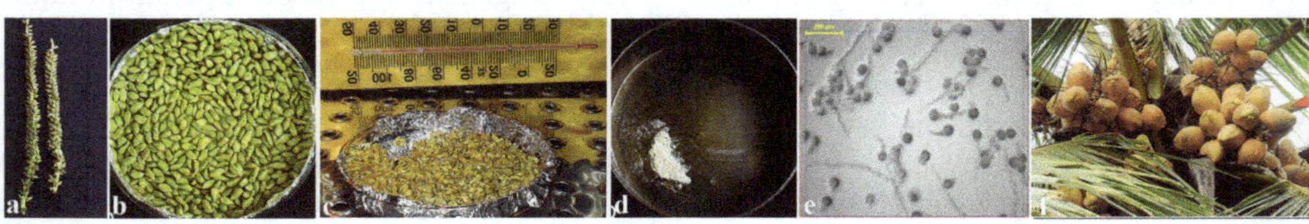

Figure 4. Simple desiccation-based pollen cryopreservation in coconut (a) Coconut spikelets with mature male flowers (b) and (c) Separation of male flowers and desiccation (d) Extraction of pollen grains from desiccated flowers (e) Assay for in vitro pollen germination and (f) Fertility assessment of cryopreserved pollen in the field.

Cryopreservation of Arecanut Pollen

Different aspects of arecanut pollen cryopreservation have been successfully standardised, including desiccation, pollen collection, in vitro germination, and viability and fecundity studies. In the case of *in vitro* viability tests, both fresh and desiccated pollen from various arecanut genotypes were utilised. The desiccated pollen was cryopreserved by immersing it directly in liquid nitrogen for varying durations, ranging from 24 hours to two years. Subsequent viability and fertility studies were conducted using the cryopreserved pollen (Figure 5). Conspicuously, the use of cryopreserved pollen resulted in a normal nut set, affirming the effectiveness and viability of cryopreservation in maintaining the fertility of arecanut pollen (Karun *et al.*, 9; Karun *et al.*, 5).

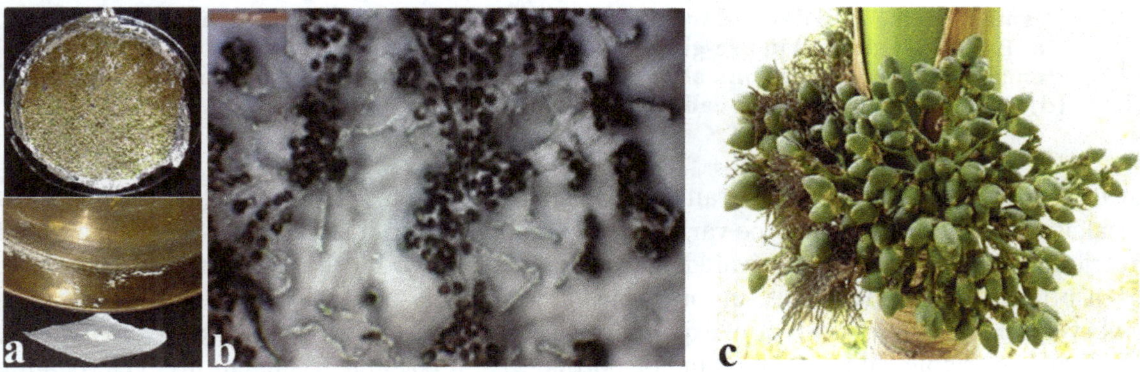

Figure 5. Simple desiccation-based pollen cryopreservation in arecanut (a) Desiccated male flowers and extracted pollen (b) Assay for *in vitro* pollen germination and (c) Fertility assessment of cryopreserved pollen in the field.

Cryopreservation of Arecanut Embryogenic Calli

The embryogenic calli underwent a pre-growth stage in Y3 medium, supplemented with sucrose, for three days. Subsequently, they were desiccated using PVS3, as illustrated in Figure 6. The results of this process indicated a successful recovery rate of 8-10 per cent of the embryogenic calli, ultimately leading to the normal production of plantlets (Muralikrishna *et al.*, 11).

Start codon targeted (SCoT) markers were employed to assess clonal fidelity. The results of this genetic evaluation demonstrated no discernible genetic variation between the cryopreserved calli and the original calli. This study showcased the effective utilisation of the V-cryoplate technique for the long-term cryopreservation of embryogenic calli of arecanut, highlighting its potential and efficiency in preserving crucial genetic materials (Muralikrishna *et al.*, 11).

In vitro Culture

The multiplication of elite coconut palms, known for their yielding nature and resilience to various stresses, holds immense significance for multiple reasons. The need for a technique enabling large-scale coconut multiplication *in vitro* is paramount due to the inability of conventional methods to meet the demand for seedlings. Unfortunately, this demand is hindered

Figure 6. Vitrification-based cryoplate (V-cryoplate) technique for cryopreservation of embryogenic calli derived from immature inflorescence culture in arecanut (a) Embryogenic calli in pre-growth medium (b) Calli placed in V-cryoplate and encapsulation with sodium alginate solution (c) PVS3-based dehydration and (d) Growth of embryogenic calli in retrieval medium following cryopreservation.

by the substantial *in vitro* recalcitrance of coconuts. Several factors contribute to this *in vitro* recalcitrance, including genotype variations, levels of explant maturity, and the absorption of nutrients and hormones by activated charcoal. Various coconut tissue culture explants have been utilised, including leaves, inflorescence, plumular tissues, ovaries, anthers, roots, and embryos. Irrespective of the type of explants, the presence of abnormal tissues and the challenge of obtaining friable calli persist as major bottlenecks in coconut tissue culture. Overcoming these barriers is essential for efficient large-scale coconut multiplication vital for meeting the demand for quality coconut seedlings.

Coconut Immature Inflorescence Culture

Recently, the utilisation of immature inflorescence culture has emerged as a promising technique for the mass multiplication of elite coconut genotypes. Immature inflorescences are abundant in meristematic points, making them an excellent source of explants. The success of regeneration depends on selecting immature inflorescences at the appropriate maturity stage. The procedure involves collecting immature inflorescence explants with spathe lengths ranging from 2-12 cm. Under aseptic conditions, the outer and inner spathes are carefully removed. The rachillae are then sliced into 1-1.5 mm long sections, inoculated in Y3 media with a low concentration of 2,4-D, and incubated in the dark for 16 weeks. Within one or two weeks of incubation, the rachillae bits develop tiny swellings that eventually transform into vegetative buds. After eight months of dark incubation and regular sub-culturing in Y3 medium, with low 2,4-D concentrations, structures resembling shoots are observed (Shareefa *et al.*, 32).

These shoot-like structures are then transferred to 1/2 MS media supplemented with NAA and BAP and incubated for one month under diffused light. Upon transitioning from diffused to direct light, the cultures exhibit elongation and formation of vegetative shoots. At the end of this incubation cycle, distinct shoots are observed, with multiple shoots also forming. Individual shoots can be separated from the multiples and transferred to media containing NAA and 2IP to induce shoot elongation. The individual shoots are further cultured in Y3 media with

NAA and BAP to stimulate rooting and promote normal shoot growth. Particularly, immature inflorescences with an outer spathe length of 4-7 cm display the highest shoot regeneration. Plantlets developed through this technique can be successfully hardened (Figure 7) (Shareefa *et al.*, 32).

Figure 7. Mass multiplication of coconut palm using immature inflorescence *in vitro* cultures (a) Immature inflorescence explants of varying length from elite palm (b) *In vitro* culture of rachillae bits in Y3 medium (c) and (d) Multiple shoot initiation in rachillae segments (e) Development of plantlet (f) In vitro hardened plantlet and (g) Field planting.

Coconut Plumule Culture

When utilised as explants, plumular tissues have consistently demonstrated exceptional results among various explants cultured *in vitro*. These tissues exhibit higher frequencies of embryogenic calli, somatic embryoids, and meristemoids. When Y3 media is supplemented with an auxin (2,4-D) combined with a cytokinin (TDZ), calli can be induced from plumular tissues. These calli are sub-cultured in media with lower levels of 2,4-D and a constant level of either cytokinin (BA and TDZ) or polyamines (spermine and putrescine) at regular intervals to encourage plantlet formation (Figure 8). Plantlets with well-balanced shoots and roots can be potted and transplanted to the greenhouse (Rajesh *et al.*, 23).

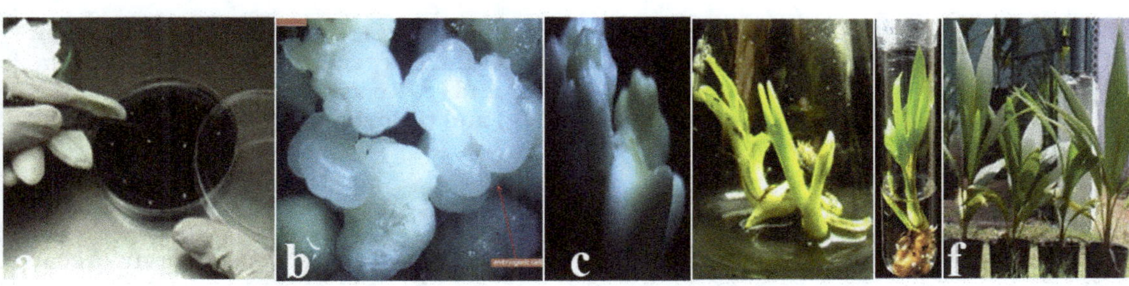

Figure 8. Regeneration of coconut from plumular explants via *in vitro* somatic embryogenesis (a) *In vitro* growth of plumule in callusing medium (b) Embryogenic calli (c) Somatic embryo (d) and (e) Plantlet development and (f) Hardened plantlets.

Histological studies have confirmed the development of shoot buds (organogenesis) and typical embryoids (bipolar). Despite successful regeneration and establishment of plantlets in the field, a commercial-scale protocol is yet to be fully developed. Forming friable calli and converting somatic embryos into plantlets present significant challenges, necessitating further refinement of the protocol. Recent studies have demonstrated that a maximum of seven plantlets could be obtained from a single plumular explant by supplementing a novel aromatic cytokinin (meta-topolin) in the tissue culture media. This finding suggests a potential enhancement for more efficient plantlet production in tissue culture (Rajesh *et al.,* 23). Efforts are also being made to generate cell suspension cultures from embryogenic calli derived from plumular explants (Bhavyashree *et al.,* 2).

Arecanut Immature Inflorescence Culture

Using immature inflorescence explants, a highly reproducible protocol has been standardised at ICAR-CPCRI for arecanut somatic embryogenesis and subsequent plantlet regeneration. This protocol allows for the rapid multiplication of elite genotypes, including Yellow Leaf Disease (YLD) resistant palms, Hirehalli Dwarf (HD) mutants characterised by short stature, and dwarf hybrids like VTLAH–1 and VTLAH–2. To achieve mass multiplication, immature inflorescence rachillae are sliced into 1-2 cm-sized explants and introduced into the Y3 basal medium to initiate somatic embryogenesis. Calli is induced through subsequent subcultures, wherein the auxin concentrations are gradually reduced. Somatic embryo formation is successfully attained in a hormone-free Y3 medium. For germination, the medium is supplemented with cytokinins. Following this, plantlet development is initiated in a regeneration medium under light room conditions, maintaining a photoperiod of 16 hours (Figure 9). This protocol serves as an efficient and reliable means for the multiplication of desired arecanut genotypes, providing a valuable tool for further research and crop improvement efforts (Karun *et al.,* 10; Radha *et al.,* 16).

Figure 9. Mass multiplication of elite palms through Immature inflorescence in arecanut (a) Explants: immature inflorescence (b) In vitro culture of rachillae bits in Y3-medium (c) Induction of callus and somatic embryos (d) and (e) Regeneration of plantlet and (f) Plantlet after hardening and (g) and (i) Field planting.

In vitro Assay to Screen for Phytophthora Resistance

A simple in vitro assay was developed to screen Phytophthora resistance in Areca sp. Challenge inoculation was done either directly on the in vitro plantlets derived from embryo culture or on the leaf segments of adult field-grown plants cultured in Y3 media. Both these in vitro methods prove to be effective in screening the resistance to disease-causing fungi in arecanut as the method is rapid and easy to execute (Muralikrishna *et al.,* 12).

Molecular Markers

Understanding the genetic diversity of a crop is a fundamental aspect of the success of any breeding program. Molecular markers have emerged as highly effective tools for assessing the genetic diversity of germplasm collections and comprehending the population structure of various crops. Their efficient utilisation enables the characterisation and management of germplasm, facilitates genetic diversity studies, aids in linkage mapping, and plays a vital role in identifying quantitative trait loci (QTLs) for marker-assisted selection (MAS). By employing molecular markers, breeders and researchers can assess the genetic makeup of crops at the molecular level. It allows for identifying desirable traits, understanding the inheritance patterns of these traits, and accelerating the breeding of improved cultivars with enhanced traits such as disease resistance, higher yield, better nutritional value, and adaptability to specific environmental conditions. In brief, molecular markers are powerful tools that significantly contribute to advancing breeding programs, ensuring the sustainable improvement and development of crops to meet the growing demands of agriculture and addressing global challenges such as food security and climate change.

Application of Molecular markers: Potent Tool for Genetic Diversity Analysis and Marker-Assisted Selection (MAS) in Coconut

Enhancing breeding efficiency is crucial for improving crops, and this can be achieved through various molecular techniques that assess genetic diversity, germplasm characterisation, management, linkage mapping, and the identification of quantitative trait loci (QTLs) for marker-assisted selection (MAS). In the realm of coconut research, several molecular markers have been instrumental in these efforts: Restriction Fragment Length Polymorphism (RFLP), Random Amplified Polymorphic DNA (RAPD), Amplified Fragment Length Polymorphism (AFLP), Inverse Sequence-Tagged Repeats (ISTR), Simple Sequence Repeats (SSR), Expressed Sequence Tag-Simple Sequence Repeats (EST-SSRs), Inter Simple Sequence Repeats (ISSR), Start Codon Targeted (SCoT) Markers, Single Nucleotide Polymorphisms (SNPs), These molecular markers have facilitated the categorisation of coconut accessions into genetic groups, aiding in the identification of redundant collections. Studies utilising these markers have revealed significant genetic differences between tall and dwarf coconut accessions in accordance with their distinct breeding behaviours. For instance, self-pollinating dwarf accessions exhibit low genetic diversity and high homozygosity compared to cross-pollinated tall accessions (Rajesh *et al.*, 19; Regi *et al.*, 27; Preethi *et al.*, 14; Jerard *et al.*, 4; Sabana *et al.*, 28).

Using molecular markers associated with important traits in perennial crops like coconut can significantly increase breeding efficiency and reduce breeding cycles. Specific markers distinguishing tall and dwarf coconut cultivars have been identified (Rajesh *et al.*, 21). Moreover, these markers have been instrumental in confirming the authenticity of dwarf (D) x tall (T) hybrids at an early seedling stage, assuring farmers regarding the authenticity of the hybrids they are planting (Rajesh *et al.*, 20; Sudha *et al.*, 33). These advancements in marker-based analyses have opened up new avenues for phylogenetic analysis and have greatly improved the efficient conservation and utilisation of coconut genetic resources.

Molecular Markers: A Potent Tool for Genetic Diversity Analysis and Evaluation of Clonal Fidelity in Arecanut

Traditionally, the characterisation and evaluation of arecanut germplasm have relied on morphological and yield parameters. However, the genetic diversity information obtained from morphological characters is limited due to potential influences of environmental, genetic, and physiological factors on these parameters. Molecular techniques have been employed to address this limitation and gain a deeper understanding of genetic relationships among arecanut germplasm, including indigenous and exotic accessions.

In particular, techniques like Random Amplified Polymorphic DNA (RAPD) and Simple Sequence Repeats (SSR) primers have been utilised to assess the genetic relationships among arecanut germplasm. The clustering of genotypes observed through these molecular markers often corresponded to the geographic origins of the accessions (Bharath *et al.*, 1; Rajesh *et al.*, 18, 19; Nagaraja *et al.*, 13). This molecular approach offers a more precise and unbiased understanding of genetic diversity, providing insights that might not be apparent through morphological parameters alone. Moreover, molecular markers have proven valuable tools in confirming the genetic fidelity of tissue-cultured derived plantlets (Karun *et al.*, 6). The use of tissue culture for large-scale multiplication of elite arecanut palms is promising, especially given the low level of variability observed among tissue culture-derived plantlets. This reaffirms the potential and efficiency of tissue culture techniques for the mass production of high-quality arecanut palms.

Molecular Markers: A Tool to Screen Inter-specific Hybrids in Arecanut

Areca catechu L. and its wild relatives, including *Areca concinna* and *Areca triandra*, have been characterised using Start Codon Targeted (SCoT) markers. This approach has enabled the identification of markers that are distinctive to both *A. catechu* and *A. triandra*. Subsequently, these SCoT markers were transformed into Sequence Characterised Amplification Markers (SCAR) specific to *A. catechu* and *A. triandra*. These SCAR markers have proven to be highly effective in authenticating inter-specific hybrids between *A. catechu* and *A. triandra* (Rajesh *et al.*, 24). This demonstrates the utility and specificity of SCAR markers in distinguishing and confirming genetic relationships among different Areca species, contributing to a deeper understanding of their genetic makeup and facilitating hybrid authentication.

Integrating Multi-omics Data

Integrating genomics, transcriptomics, proteomics, and metabolomics approaches offers a powerful means to understand the intricate mechanisms that govern the complex traits of agricultural significance. Numerous initiatives have significantly contributed to developing genomic resources in the context of coconuts, a few of which are briefly highlighted here.

RNA-sequencing (RNA-seq) is a pivotal tool for characterising genes, including those differentially expressed. This technique is particularly valuable for functional genomics studies, especially when the available genome information is limited. In coconut, transcriptomic studies employing RNA-seq have elucidated the molecular basis of critical processes such as somatic embryogenesis (Rajesh *et al.*, 25). Furthermore, these studies have shed light on the differential responses of coconut genotypes to root (wilt) disease (Rajesh *et al.*, 24; Rachana *et al.*, 15), bud rot disease (Gangaraj *et al.*, 3), as well as water deficit stress (Ramesh *et al.*, 26). Leaf transcriptome profiles of Kalpa Sree and Kalpatharu coconut seedlings subjected to water deficit stress identified upregulation of PHLOEM PROTEIN 2-LIKE A1-like, WRKY transcription factor 40 isoform X1 and downregulation of glycerol-3-phosphate in Kalpasree.

On the other hand, upregulation of transcripts encoding polyamine oxidase, and arabinose 5-phosphate isomerase, among others and downregulation of aquaporin PIP1-2 were documented in Kalpatharu leaves. Besides, long non-coding RNA and genic SSRs were also identified from the transcriptome data, which would pave the way for its utilisation in developing climate-smart coconut. The insights gained from these resources encompass spatio-temporal expression patterns of key genes, their regulatory mechanisms, and their involvement in a wide array of biological processes. RNA-seq data has facilitated the annotation of the nuclear genome and revealed an abundance of genic molecular markers crucial for genetic research and breeding applications in coconuts.

Notably, the nuclear and organellar genomes of the dwarf cultivar Chowghat Green Dwarf (CGD) have been successfully characterised using a hybrid sequencing strategy (Rajesh *et al.*, 17). The availability of the CGD genome, known for its enhanced resistance to root (wilt) disease, represents a valuable genetic resource that significantly accelerates genomics-assisted breeding efforts in the domain of coconuts. In conclusion, these advanced genomic technologies and resources are instrumental in unravelling the complexities of coconut biology and are pivotal for driving advancements in coconut breeding and cultivation practices. Genome-wide exploration of auxin response factors (ARFs) and their expression dynamics in response to abiotic stresses and growth regulators in coconut has also been undertaken towards developing functional markers for abiotic stress tolerance (Santhi *et al.*, 31).

To gain a better comprehension of the molecular basis of *in vitro* recalcitrance in coconut and to shed insights into the intricate regulatory processes, comparative genome-wide profiling of small RNAs from embryogenic and non-embryogenic calli has been undertaken using the NGS platform. Conserved and novel miRNAs (microRNAs) were identified. The expression patterns of miRNAs and their putative targets were analysed using qRT-PCR, and most of the analysed miRNA-target pairs showed inverse correlation during somatic embryogenesis. The results point towards the role of a few conserved miRNAs in regulating the process of somatic embryogenesis in coconut (Sabana *et al.*, 29).

Way Forward

Moving forward, biotechnological tools present promising opportunities for the improvement of coconut and arecanut crops. Key techniques such as cryopreservation, embryo culture, and embryo rescue have played critical roles in collecting and preserving valuable germplasm for these crops. Clonal propagation appears to be a viable solution for scaling up production to meet the increasing demand for high-quality planting materials. Successful implementations have been observed in arecanut. However, in the case of coconut, challenges such as genotypic variations in response to in vitro culture, low rates of somatic embryo formation, difficulties in converting somatic embryos into plantlets, and abnormal somatic embryo formation persist.

Addressing these challenges requires a focused approach. Optimising media combinations with precise plant growth regulators, incorporating additives, and utilising bioreactor systems are essential steps toward standardising a commercial-level protocol for coconut. Furthermore, enhancing omics strategies can provide valuable insights into the molecular mechanisms underlying in vitro recalcitrance in coconut. By leveraging these approaches, we can overcome hurdles and expedite progress in the biotechnological enhancement of coconut and arecanut crops, ensuring sustainable and efficient agricultural practices.

While conventional breeding techniques have successfully improved coconut and arecanut, integrating multiple-omics technologies is still in its early stages for both crops. Notably, the genome assembly of a dwarf coconut cultivar has been accomplished, serving as a foundational step. However, efforts must be intensified to re-sequence elite or pivotal genotypes, strengthening genomics-assisted breeding. This involves identifying and characterising Single Nucleotide Polymorphisms (SNPs), quantitative trait loci (QTLs) associated with economically significant traits, and advancing towards genotyping by sequencing (GBS) to conduct genome-wide association analysis (GWAS).

Incorporating cutting-edge breeding tools like genomic selection (GS) holds significant promise and can potentially enhance genetic gains in coconut and arecanut crops. By leveraging these initiatives, we can pave the way for the widespread adoption of biotechnology programs globally. This adoption will play a critical role in developing climate-smart and disease-resistant phenotypes, ultimately contributing to sustainable agricultural practices and addressing the challenges posed by evolving climates and diseases. Further implementation of novel approaches like speed breeding, high-throughput phenotyping, and multi-omics tools could accelerate the process of genetic gain in coconut and arecanut.

REFERENCES

1. Bharath, B.G., Ananda, K.S., Nagaraja, N.R. and Rajesh, M.K. 2021. Microsatellite markers-based assessment of genetic diversity and population structure in arecanut (*Areca catechu* L.). *Indian J. Genet. Plant Breed.* **81**(04): 107–110.
2. Bhavyashree, U., Lakshmi Jayaraj, K., Muralikrishna, K. S., Sajini, K. K., Rajesh, M.K. and Anitha Karun. 2016. Initiation of coconut cell suspension culture from shoot meristem derived embryogenic calli: A preliminary study. *Journal of Phytology*. **8**(1): 13-16. doi: 10.19071/jp.2016.v8.2979.
3. Gangaraj, K.P., Muralikrishna, K.S., Antony, G., Sahu, B.B., Hegde, V. and Rajesh, M.K. 2021. A rapid *in vitro* leaf inoculation assay to investigate *Phytophthora palmivora*–coconut interactions. *J. Phytopathol.* **169**(5): 316-328.
4. Jerard, B.A., Rajesh, M.K., Regi Jacob Thomas, Niral, V. and Samsudeen K. 2017. Island ecosystems host rich diversity in coconut (*Cocos nucifera*): Evidences from Minicoy Island, India. *Agric. Res.* **6**(3): 214-226.
5. Karun, A., Muralikrishna, K.S., Sajini, K.K. and Rajesh, M.K. 2023. Cryopreservation of Coconut and Arecanut Pollen. **In**: *Pollen Cryopreservation Protocols*. New York, NY: Springer US. pp. 527-540.
6. Karun, A., Radha, E., Sangeetha Vijayan, P., Jiji George, Rajesh, M.K. and Ananda, K.S. 2008. Assessment of genetic fidelity of arecanut plantlets derived through direct somatic embryogenesis by RAPD markers. *J. Plant. Crops*. **36**: 228-231.
7. Karun, A., Sajini, K.K. and Shivashankar, S. 1999. Embryo culture of coconut: the CPCRI protocol. *Indian J. Hortic.* **56**(4): 348-353.

8. Karun, A., Sajini, K.K., Muralikrishna, K.S., Rajesh, M.K. and Engelmann, F. 2017. Cryopreservation of arecanut (*Areca catechu* L.) pollen. *CryoLett.* **38**(6): 463-470.

9. Karun, A., Sajini, K.K., Rajesh, M.K., Muralikrishna, K.S., Samsudeen, K., Ajith Kumar, P. and Nagwekar, D.D. 2017. ***In vitro* retrieval via embryo rescue of 'Mohachao Narel', a sweet endosperm coconut from Maharashtra.** *Int. J. Innov. Hortic.* **6**(2): 147-150.

10. Karun, A., Siril, E.A., Radha, E. and Parthasarathy, V.A. 2004. Somatic embryogenesis and plantlet regeneration from leaf and inflorescence explants of arecanut (*Areca catechu* L.). *Curr. Sci.* **86**(12): 1623-1628.

11. Muralikrishna, K.S., Sajini, K.K., Kavya, P., Prakash, K., Sabana, A.A., Rajesh, M.K. and Karun, A. 2021. Cryopreservation of arecanut *(Areca catechu* L.) embryogenic calli by V cryo-plate method. *Curr. J. Appl. Sci. Technol.* **40**(20): 76-82.

12. Muralikrishna, K.S., Sharadraj, K.M., Gangaraj, K.P., Nagaraja, N.R., Anitha Karun, Rajesh, M.K. and Chowdappa, P. 2018. *In vitro* assay for screening of *Areca* spp. for *Phytophthora* resistance. *Int. J. Innov. Horti.* 7(2): 139-142.

13. Nagaraja, N.R., Ananda, K.S. and Rajesh, M.K. 2022. *Genetic variation among varieties, wild species and related genera of arecanut (Areca catechu L.) as revealed by microsatellite markers. Int. J. Plant Soil Sci.* **34**(22): 846-855.

14. Preethi, P., Rahman, S., Naganeeswaran, S., Sabana, A.A., Gangaraj, K.P., Jerard, B.A., Niral, V. and Rajesh, M.K. 2020. Development of EST-SSR markers for genetic diversity analysis in coconut (*Cocos nucifera* L.). *Mol. Biol. Rep.* **47**(12): 9385-9397.

15. Rachana, K.E., Naganeeswaran S., Fayas T.P., Thomas, R. J. and Rajesh, M.K. 2016. Cloning, characterization and expression analysis of NBS-LRR-type resistance gene analogues (RGAs) in coconut. *Acta Bot. Croat.* **75**(1): 1-10.

16. Radha, E., Karun, A., Rajesh, M.K. and Nancy J. Mariya. 2008. Effect of cytokinins on growth and maturation of direct somatic embryos in arecanut (*Areca catechu* L.). *J. Plant. Crops.* **36**: 243-245

17. Rajesh, M.K., Chowdappa, P., Behera, S.K., Kasaragod, S., Gangaraj, K.P., Kotimoole, C.N., Nekrakalaya, B., Mohanty, V., Sampgod, R.B., Banerjee, G., Das, A.J., Niral, V., Karun, A., Mahato, A.K., Gaikwad, K., Singh, N.K. and Keshava Prasad, T.S. 2020. Assembly and annotation of the nuclear and organellar genomes of a dwarf coconut (Chowghat Green Dwarf) possessing enhanced disease resistance. *OMICS J. Integr. Biol.* **24**(12): 726-742.

18. Rajesh, M.K., Fayas, T.P., Naganeeswaran, S., Rachana, K.E., Bhavyashree, U., Sajini, K.K. and Karun, A. 2016. *De novo* assembly and characterization of global transcriptome of coconut palm (*Cocos nucifera* L.) embryogenic calli using Illumina paired-end sequencing. *Protoplasma.* **253**(3): 913-928.

19. Rajesh, M.K., Gangurde, S.S., Pandey, M.K., Niral, V., Sudha, R., Jerard, B.A., Kadke, G.N., Sabana, A.A., Muralikrishna, K.S., Samsudeen, K., Karun, A. and Keshava Prasad, T.S. 2021. Insights on genetic diversity, population structure, and linkage disequilibrium

in globally diverse coconut accessions using genotyping-by-sequencing. *OMICS J. Integr. Biol.* **25**(12): 796-809.

20. Rajesh, M.K., Jerard, B.A., Preethi, P., Regi Jacob Thomas and Karun, A. 2014. Application of RAPD markers in hybrid verification in coconut. *Crop Breed. Appl. Biotechnol.* **14**: 36-41.
21. Rajesh, M.K., Jerard, B.A., Preethi, P., Regi Jacob Thomas, Fayas, T.P., Rachana, K.E. and Karun, A. 2013. Development of a RAPD-derived SCAR marker associated with tall-type palm trait in coconut. *Sci. Hortic.* **150**(4): 312-316.
22. Rajesh, M.K., Rachana, K.E., Kulkarni, K., Sahu, B.B., Thomas, R.J. and Karun, A. 2018. Comparative transcriptome profiling of healthy and diseased Chowghat Green Dwarf coconut palms from root (wilt) disease hot spots. *Eur. J. Plant Pathol.* **151**: 173-193.
23. Rajesh, M.K., Radha, E., Sajini K.K. and Karun, A. 2014. Polyamine-induced somatic embryogenesis and plantlet regeneration *in vitro* from plumular explants of dwarf cultivars of coconut. *Indian J. Agric. Sci.* **84**(4): 527-530.
24. Rajesh, M.K., Sabana A.A., Rachana K.E., Shafeeq Rahman, Ananda K.S. and Karun, A. 2016. Development of a SCoT-derived SCAR marker associated with tall-type palm trait in arecanut and its utilization in hybrid (dwarf x tall) authentication. *Indian J. Genet. Plant Breed.* **76**(1): 119-122.
25. Rajesh, M.K., Sabana, A.A., Rachana, K.E., Ananda, K.S. and Karun, A. 2016. Potential of start codon targeted (SCoT) markers for assessment of genetic diversity of arecanut (*Areca catechu* L.). *Indian J. Hortic.* **73**: 423-426.
26. Ramesh, S.V., Hebbar, K.B., Rajesh, M.K., Abhin Sukumar, P., Gangaraj, K.P. and Athul Bobby. 2020. Transcriptome analysis of *Cocos nucifera* L. seedlings having contrasting water-use efficiency (WUE) under water-deficit stress: Molecular insights and genetic markers for drought tolerance. *Biol. Life Sci. Forum.* **4**(1): 73.
27. Regi J. Thomas, Rajesh, M.K., Jacob, P.M., Mejosh Jose and Nair, R.V. 2015. Studies on genetic uniformity of Chowghat Green Dwarf and Malayan Green Dwarf varieties of coconut using molecular and morphometric methods. *J. Plant. Crops.* **43**(2): 89-96.
28. Sabana, A.A., Antony, G., Rajesh, M.K., Gangaraj, K.P., Niral, V., Sudha, R. and Jerard, B.A. 2022. Development and characterization of non-coding RNA-derived simple sequence repeat markers in coconut (*Cocos nucifera* L.). *Funct. Integr. Genomics.* **22**: 1243–1251.
29. Sabana, A.A., Rajesh, M.K. and Antony, G. 2020. Dynamic changes in the expression pattern of miRNAs and associated target genes during coconut somatic embryogenesis. *Planta.* **251**(4): 1-18.
30. Sajini, K.K., Karun, A., Amarnath, C.H. and Engelmann, F. 2011. Cryopreservation of coconut (*Cocos nucifera* L.) zygotic embryos by vitrification. *Cryo-Lett.* **32**(4): 317-328.
31. Santhi, CKV, Rajesh, M.K., Ramesh, S.V., Muralikrishna, K.S., Gangaraj, K.P. and Payal, G. 2021. Genome-wide exploration of auxin response factors (ARFs) and their expression dynamics in response to abiotic stresses and growth regulators in coconut (*Cocos nucifera* L.). *Plant Gene.* **28**: 100344.

32. Shareefa, M., Regi J. Thomas, Sreelekshmi J.S., Rajesh, M.K. and Karun, A. 2019. *In vitro* regeneration of coconut plantlets from immature inflorescence. *Curr. Sci.* **117**(5): 813-820.

33. Sudha, R., Samsudeen, K., Rajesh, M.K. and Niral, V. 2022. Molecular marker assisted confirmation of hybrids in coconut (*Cocos nucifera* L.). *Indian J. Genet. Plant Breed.* **82**(03): 369–372.

34. Veluru, A., Prakash, K., Neema, M., Muralikrishna, K.S., Samsudeen, K., Chandran, K.P., Rajesh, M.K. and Karun, A. 2021. Pollen storage of coconut dwarf accession Chowghat Orange Dwarf at low temperature. *Indian J. Agric. Sci.* **91**(2): 321-324.

Colour Plates

Figure 1. Dragon fruit (pg.55)

Figure 2. Avocado (Arka Supreme) (pg.58)

Figure 3. Rambutan (pg.61)

Figure 4. Arka Coorg Arun and Arka Coorg Peetabh (pg.62)

Figure 5. Passion fruit (Kaveri hybrid) (pg.64)

Figure 6. Mangosteen. (pg.65)

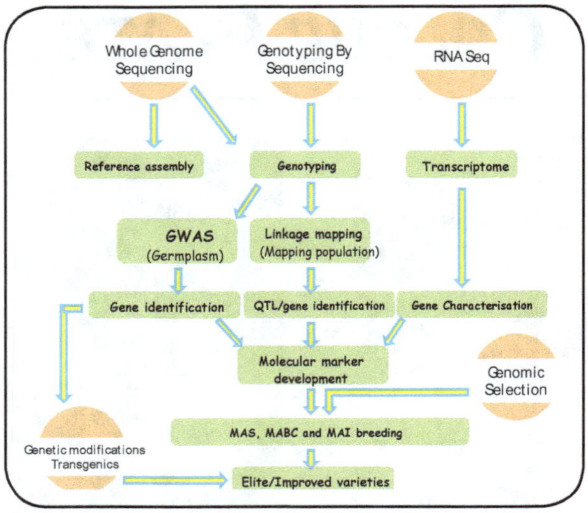

Figure 1. Genomics assisted breeding strategies in fruit crops. (pg.85)

Figure 1. Embryo culture of coconut: ICAR-CPCRI protocol (a) Germinating embryo (b) Development of complete plantlet (c) In vitro hardened plantlet developed from a zygotic embryo (d) Establishment of embryo culture derived plantlet in the field and (e) Embryo culture derived palm bearing nuts. (pg.101)

Figure 2. Embryo rescue in special sweet endosperm type of coconut: (a) In vitro culture initiated for the rescued and surface sterilised embryo from sweet endosperm coconut (b) Embryo germination (c) Development of plantlet (d) Plantlets in greenhouse post hardening process and (e) Field planting of embryo-rescued plantlet. (pg.102)

Figure 3. Vitrification-based cryopreservation of zygotic embryos of coconut: (a) Pre-culturing of embryo in high sucrose medium (b) PVS3 treatment (c) Cryovial with dehydrated embryos (d) Embryo regeneration in retrieval medium following the cryopreservation process (e) Development of complete plantlet (f) In vitro hardened plantlet (g) ex vitro hardened plantlet and (h) Establishment of plantlet in the field and (i) Flowering and bearing palm. (pg.104)

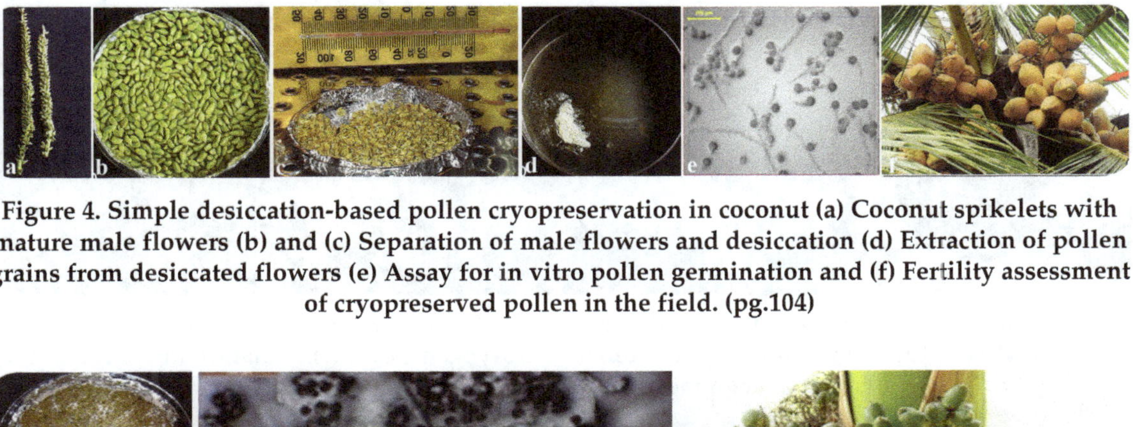

Figure 4. Simple desiccation-based pollen cryopreservation in coconut (a) Coconut spikelets with mature male flowers (b) and (c) Separation of male flowers and desiccation (d) Extraction of pollen grains from desiccated flowers (e) Assay for in vitro pollen germination and (f) Fertility assessment of cryopreserved pollen in the field. (pg.104)

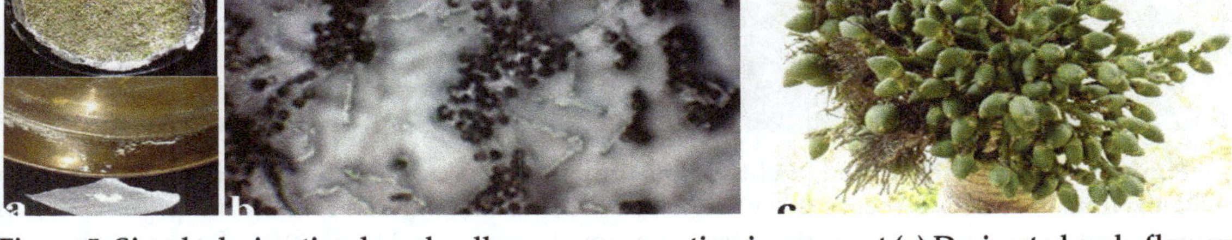

Figure 5. Simple desiccation-based pollen cryopreservation in arecanut (a) Desiccated male flowers and extracted pollen (b) Assay for in vitro pollen germination and (c) Fertility assessment of cryopreserved pollen in the field. (pg.105)

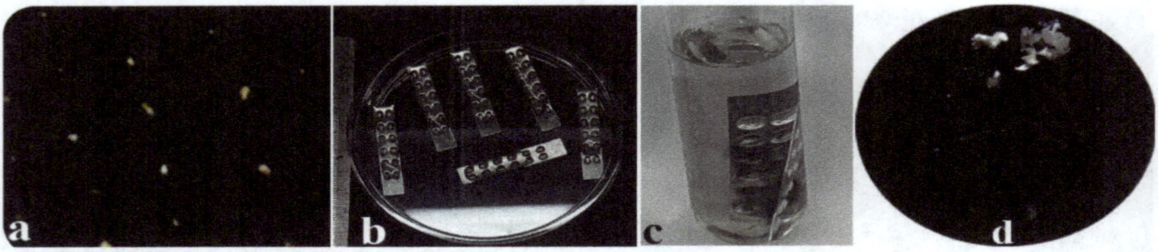

Figure 6. Vitrification-based cryoplate (V-cryoplate) technique for cryopreservation of embryogenic calli derived from immature inflorescence culture in arecanut (a) Embryogenic calli in pre-growth medium (b) Calli placed in V-cryoplate and encapsulation with sodium alginate solution (c) PVS3-based dehydration and (d) Growth of embryogenic calli in retrieval medium following cryopreservation. (pg.106)

Figure 7. Mass multiplication of coconut palm using immature inflorescence in vitro cultures (a) Immature inflorescence explants of varying length from elite palm (b) In vitro culture of rachillae bits in Y3 medium (c) and (d) Multiple shoot initiation in rachillae segments (e) Development of plantlet (f) In vitro hardened plantlet and (g) Field planting. (pg.107)

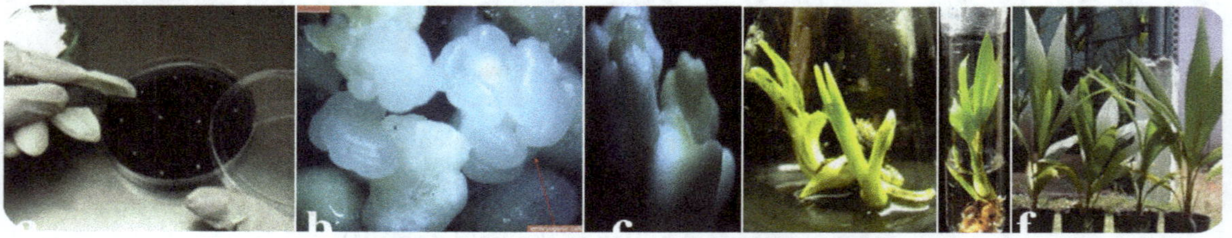

Figure 8. Regeneration of coconut from plumular explants via in vitro somatic embryogenesis (a) In vitro growth of plumule in callusing medium (b) Embryogenic calli (c) Somatic embryo (d) and (e) Plantlet development and (f) Hardened plantlets. (pg.107)

Figure 9. Mass multiplication of elite palms through Immature inflorescence in arecanut (a) Explants: immature inflorescence (b) In vitro culture of rachillae bits in Y3-medium (c) Induction of callus and somatic embryos (d) and (e) Regeneration of plantlet and (f) Plantlet after hardening and (g) and (i) Field planting.
(pg.108)

Chapter 9

Breeding Next Generation Spices for Nutraceuticals and Industrial Use

D. Prasath*, M.S. Shivakumar, R. Sivaranjani, and S. Aarthi

ICAR–Indian Institute of Spices Research, Kozhikode, Kerala

e-mail: dprasath@gmail.com

ABSTRACT

The spices face highly dynamic situations arising from requirements of nutraceutical industries for high quality spices varieties. The proactive nature of response to this factor will translate to the release of new cultivars meeting the requirements of spiceceuticals. Development of improved cultivars by spice crop breeders may require many breeding cycles and dozens of years because of perennial nature, polyploidy, incompatibility, and sterility. The scientific and technological progresses are undoubtedly key factors to obtain quality improved spices derived from the new plant breeding technologies, which have remarkable potentialities. Spices are high value and low volume, export-oriented commodities, commonly used for flavouring and seasoning of food and beverages. India is the world's largest producer, consumer, and exporter of spices; the country produces about 75 of the 109 varieties listed by the International Organization for Standardization (ISO) and accounts for half of the global trading in spices. Growing demand from the emerging segment of nutraceuticals is driving the global consumption of Indian spices further at a time when the country is straining to meet the needs of the traditional food sector. Non-traditional use of spices including nutraceuticals now accounts for nearly 15 per cent of spice production in India, estimated at 50 lakh tonne a year. The spice nutraceutical sector growth rate is projected as 10 per cent-12 per cent per year. Also, the demand of spice oils and oleoresins is increasing mostly due to enhanced use in food and beverage industries. Curcumin, in turmeric; capsaicin, in red pepper; piperine, in black pepper; [6]-gingerol, in ginger, saponin in fenugreek are immensely valuable in health care with their multiple physiological effects. Since these parameters are genetically controlled, genotypes can be successfully improved through breeding techniques. Classical breeding techniques along with recent biotechnological intervention can facilitate the improvement in quality parameters in spices. It requires a continuous flow of genes and allelic diversity into different crop gene pool. The primary, secondary gene pool and pre-breeding plays a vital role. The vast chemo-diversity available in India can be exploited for the understanding of biosynthetic pathway modifications or activity of different isoforms of enzymes which leads to accumulation of secondary metabolites of interest. Also, development of mapping population and extensive genotyping and phenotyping is very crucial to identify genomic regions controlling neutracitical components. The NGS based whole genome and transcriptome approaches have clearly demonstrated their advantages over previously developed methods and are becoming the new standard for NGS studies in spices. The functional genomics/metabolomics approach help to elucidate the biochemical pathways and key genes underlying specific neutracitical compounds.

The collaborations between public spices research institutions and human health research centers, with spices-based industries would play a key role in the success of this endeavour.

Spices are high value and low volume, export-oriented commodities, commonly used for flavouring and seasoning of food and beverages. India is the world's largest producer, consumer, and exporter of spices; the country produces about 75 of the 109 varieties listed by the International Organization for Standardization (ISO) and accounts for half of the global trading in spices. These spices impart characteristic flavours, aromas, or piquancy and colours to foods. Apart from their role in flavouring, spices often are used for preservation, medicinally, pharmaceutically, and for health maintenance purposes.

Growing demand from the emerging segment of nutraceuticals is driving the global consumption of Indian spices further at a time when the country is straining to meet the needs of the traditional food sector. Though spices contained diverse array of phytochemicals, the nutraceutical preparation often revolve around few major bioactive metabolites. The term spiceutical is gaining momentum among major spice processing industries of the country. The term roughly translates into "the nutraceutical products derived specifically from spices with proven health benefits". Non-traditional use of spices including nutraceuticals now accounts for nearly 15 per cent of spice production in India, estimated at 50 lakh tonne a year. The spice nutraceutical sector growth rate is projected as 10 per cent-12 per cent per year.

The scientific and technological progresses are undoubtedly key factors to obtain quality improved spices derived from the new technologies, which have remarkable potentialities.

Area and Production

Every state/Union Territory in India grows at least a few spice crops. Rajasthan is the leading state in spices area of cultivation while Madhya Pradesh is the leading state in spices production. The cultivation of spices in India is in an area of around 4.5 million ha with a production of 10.5 million tonnes during 2021-22. The global spice exports amount to 3.5 million metric tons, accounting for US$ 8.7 billion in value. India's share in spices export at Global level is 1.5 million metric tons, accounting for US $ 4.102 billion *i.e.,* 43 per cent in volume and 47 per cent in value (Source: Production - DASD and Export- Spices Board).

The spices sector has witnessed a period of robust growth and development during the last two decades. Area and production of spices in the country has increased to the tune of 4.4 per cent and 7 per cent per annum respectively during the period from 2005-06 to 2020-21. Productivity of spices has also increased substantially during the period from 1.63 tonnes/ha to 2.5 tonnes/ha. at the same time imports of spices into India stood at 2,24 million tonnes valued at 1,087.46 million USD (Figures 1, 2 and 3, Tables 1). This indicates a technology driven productivity enhancement across spice crops during the last two decades.

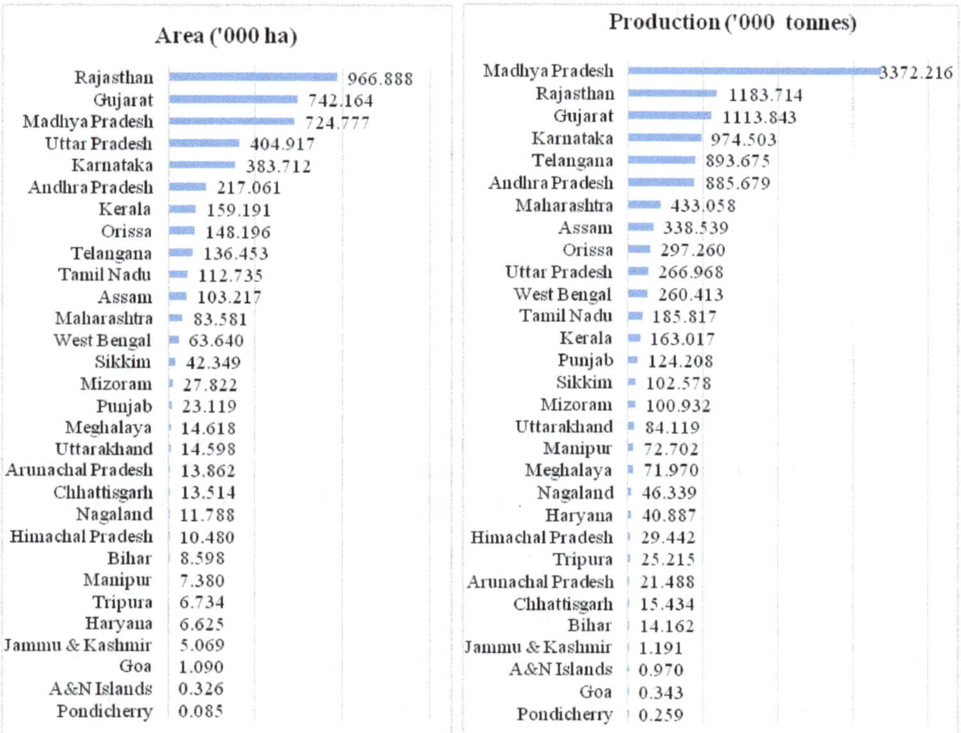

Figure 1. State wise area and production of total spices.
***Source*: Directorate of Arecanut and Spices Development (DASD), Ministry of Agriculture and Farmers Welfare, Government of India.**

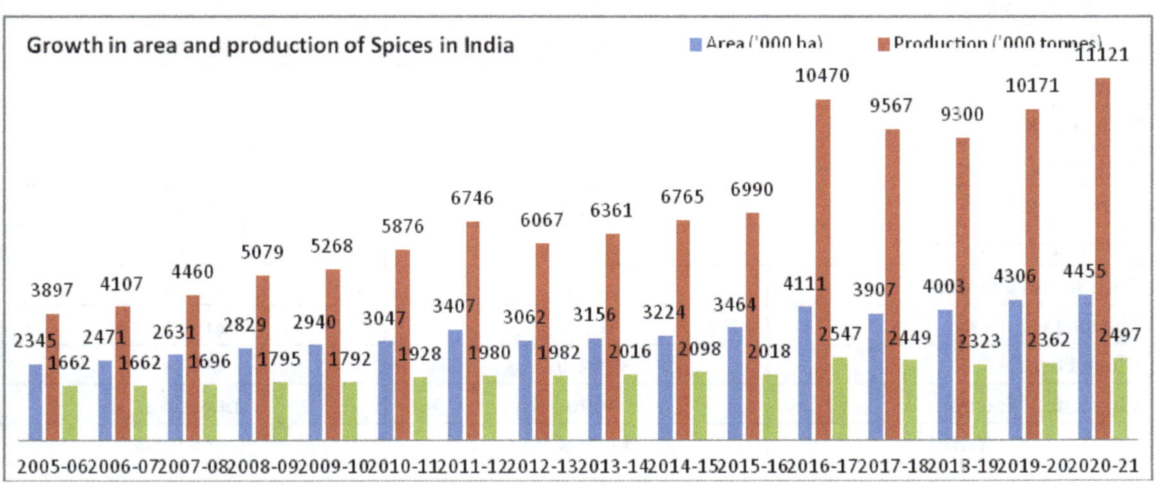

Figure 2. Growth in area and production of spices in India during 2005- 2020.
***Source*: Directorate of Arecanut and Spices Development (DASD), Ministry of Agriculture and Farmers Welfare, Government of India**

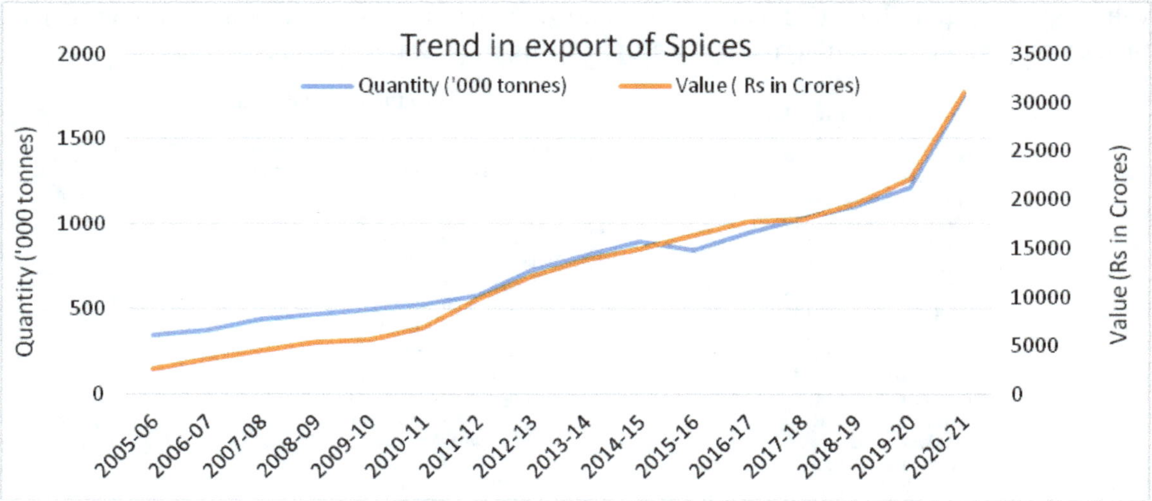

Figure 3. Growth in export of spices from India.
***Source*: Directorate of Arecanut and Spices Development (DASD), Ministry of Agriculture and Farmers Welfare, Government of India.**

Table 1. Export of selected spice value added commodities (2021-22)

Product	*Quantity (tonnes)*	*Value (Million INR)*
Pepper oleoresin	1490.09	3796.46
Ginger Oleoresin	359.74	950.64
Cassia Oil	4.05	5.66
Cinnamon Bark Oil	12.74	34.35
Cinnamon Leaf Oil	14.24	23.98
Clove Leaf/Stem Oil	101.94	131.44
Ginger Oil	73.76	577.81
Clove Bud Oil	24.98	48.53
Nutmeg Oil	72.61	258.44
Pepper Oil	133.39	430.84
Turmeric Oil	25.67	83.86
Turmeric Oleoresin	2251.40	9273.89
Cardamom Oleoresin	29.41	125.24
Nutmeg Oleoresin	231.43	416.75
Clove Oleoresin	23.16	69.30
Capsicum Oleoresin	6930.18	14495.62
Total	**11,778.79**	**30,722.81**

Source: Export-Import data bank, Department of commerce, Ministry of Commerce and Industry, Government of India.

Demand and Production Targets

As a traditional hub of spice trade and a major exporter of spices, the performance of the spice export sector is critical for the spice economy of the country. The projected growth rate for spices demand in the world is around 3.2 per cent. The projected population in India in 2050 is 1619 million, with increased GDP and per capita food spending. The per capita demand for spices is expected to increase many folds by 2050.

India should plan to double its production of spices, rich in intrinsic quality, to about 20 million tonnes by the next decade by adopting high yielding varieties, precision production technologies and environmentally safe agriculture in a multispecies cropping systems mode and conservation of resources. Indian spices are facing stiff competition from other spices producing countries especially Vietnam in black pepper, Guatemala in small cardamom, China in ginger Russia in coriander and Iran in saffron essentially due to their production volume and lower prices. This is further enhanced by poor nutrient management, not so efficient post harvest management, value addition and lack of aggressive marketing in India. Presently India exports about 14.93 per cent of its total production accounting for US $ 4.102 billion. Spices which are having great export potential are Chilli, Cumin, Turmeric, Ginger, Cardamom, Mint, Coriander, fenugreek, fennel, pepper, Nutmeg etc. Targeted production using precision technologies and high-quality varieties, value addition and aggressive marketing of spices would double our trade and progressively increase to 10 billion USD by next decade and around 25 billion USD by 2050. To achieve this, India must change its strategy from major exporter of spices mostly in raw form to exporter of value-added products like food flavours, aroma therapy products, natural food colours, nutraceuticals, cosmeceuticals and novel high value phytochemicals.

The processed food industry and pharmaceutical industry, both of which has recorded robust growth rates in the recent past, has played a key role in maintain a global demand growth rate of 3.2 per cent. Many high value commodities and value-added products from spices are further integrated to the consumer value chain to produce a range of products that are intended to meet the growing needs of the consumers across the world.

Nutraceuticals in Spices

In addition to nutrients such as vitamins, minerals, amino acids, peptides, and some fatty acids, there are other specific bioactive compounds that can be considered as nutraceuticals: enzymes, pre- and probiotics, coenzyme Q10, glucosamine, chondroitin, etc., as well as many phytochemicals (polyphenols, sulfur-containing compounds, alkaloids, terpenoids, etc.) (Nicoletti, 37). Additionally, nutraceuticals can be classified into potential and/or established nutraceuticals according to their proven beneficial effects as demonstrated in either preclinical or clinical studies (Pandey, 38). In addition to their sensory properties, the spices in our diet are rich sources of different phytochemicals with putative beneficial effects *e.g.*, antioxidative, anti-inflammatory, chemo preventive, antimutagenic, immune-modulatory properties (Kurian, 32).

Spices like turmeric, ginger, black pepper, fenugreek, coriander, etc., contains considerable quantity of biologically active specific metabolites with numerous biological functions. Curcumin, in turmeric; capsaicin, in red pepper; piperine, in black pepper; [6]-gingerol, in ginger, saponin in fenugreek are immensely valuable in health care with their multiple physiological effects.

Table 2. Active ingredients and functional properties of major spices

Spice	*Active Ingredients*	*Functional Properties*
Black pepper	Piperine, β-caryophyllene, β-pinenes, sabinene	Bioenhancer, antioxidant
Cardamom	1, 8-cineole, α-terpinyl and linalyl acetates, limonene, sabinene	Antiseptic, aromatic, stomachic, carminative, and stimulant
Ginger	Gingerols, shogaols, zingiberene, sesqui-phellandrene, limonene, α-terpineol and citral.	Carminative, Anti-inflammatory, Antioxidant, anti-ulcer principles, analgesic and beneficial for cardiovascular issues.
Turmeric	Curcumin, ar-turmerone, α-turmerone, ar-curcumene, **turmerin** and zingiberene	Anti-inflammatory, anti-mutagenic, antioxidant, anti-cancerous, antidiabetic, hepatoprotective, antifungal etc.
Tamarind	Furfural, Glycosides, palmitic acid and phenylacetaldehyde	Digestive, anti-inflammatory, antioxidant and hepatotonic
Nutmeg	Myristic (trimyristin), petroselinic and palmitic acids, myristicin, elemicin, sabinene, safrole	Carminative, aphrodisiac, anti-carcinogenic, astringent and hallucinating
Cinnamon and cassia	Cinnamaldehdyde, cinnamyl acetate, caryophyllene, linalool and eugenol	Anti-diabetic, antioxidant, food flavorant
Clove	Eugenol, eugenyl acetate and β-caryophyllene	Beneficial for respiratory and digestive ailments, antibacterial, antifungal, antispasmodic, expectorant and aphrodisiac.
Allspice	Eugenol, methyl eugenol, myrcene, limonene, α-phellandrene, pimentol, quercetin, gallic acid, galloylglucosides, ampelopsin.	To treat indigestion, gastritis, dysenteryand uterine infections.
Curry leaf	β-caryophyllene, β-gurjunene, β-elemene, β-phellandrene and β-thujene. Carbazole alkaloids, murrayanine isomahanimbine and koenimbidine	Analgesic, astringent, antidysenteric, antioxidant, febrifuge, hypolipidemic, hypoglycemic.
Garcinia	Xanthones, phenolic compounds Garcinol, Hydroxy Citric acid	Anti-bacterial agent, Anti-obesity factor, Veterinary medicine, anti-inflammatory
Kokum	Garcinol, HCA, Adscorbic acid	Anti cancerous, anti-obesity, antioxidant
Coriander	d-linalool, α-and-β-pinenes, dipentene (limonene), p-cymene, α-and γ-terpinenes, n-decanal, geraniol and l-borneol -.	Antioxidant, carminative, digestive and antispasmodic agent
Fenugreek	Galactomannan including n-alkanes, diosgenin, trigonelline, saponins, alkaloids, steroidal saponins and flavonoids	Hypoglycemic, Hypocholesterolaemic, anticarcinogenic, antipyretic, Immunomodulatory, Nematicidal, Larvicidal wound healing and antidiabetic..
Fennel	*trans*-anethole, fenchone, estragol, limonene, camphene, α-pinene	Antioxidant, antimicrobial, hepatoprotective, acricidal
Nigella	Rich in alkaloids and phytosterols-	Reduces trglycerides, and blood pressure

Spice	*Active Ingredients*	*Functional Properties*
Cumin	Cuminaldehyde (4-isopropylbenzaldehyde), cumin alcohol, α-pinene, β-pinene, dipentene, *p*-cymene pyrazines and cuminoside A and B	Hypogglyemic, hypolipidemic, antioxidant and antimicrobial.
Ajowan	Thymol, paracymene, γ-terpinene, β-pinene, dipentene, camphene and myrcene	Strong germicide, anti-spasmodic, fungicide, in toothpaste and perfumery
Garlic	Diallyl disulfide, and diallyl trisulfide followed by allyl methyl trisulfide, allyl methyl disulfide and allicin	Broad-spectrum therapeutic effect with minimal toxicity, treatment of cardiovascular disease, cancer, diabetes, blood pressure, atherosclerosis, and hyperlipidaemia
Saffron	Picrocrocin, trimethyl hydroxy carboxaldehydecyclohexene, kaempherol, cis/trans-crocins and safranal.	Antioxidant, anti-depressant, aid weight loss and anti-inflammatory.

Source: Parthasarathy *et al.* (39)

Curcumin

The bioactive compound curcuminoids extracted from turmeric has proven nutraceutical properties including antioxidant, antidiabetic, anti-inflammatory, antimicrobial and anticancer and hypolipidemic properties (Ahmed *et al.,* 3). Curcuminoids is a complex of three components namely bisdemethoxycurcumin, demethoxycurcumin and curcumin which are synthesized in the phenylpropanoid pathway (Kita *et al.,* 29). The ratio of each of these compounds in the turmeric rhizome is characteristics of genotypes and environment it grows (Aarthi *et al.,* 1). Over 100 different clinical trials have been completed with curcumin, which clearly show its safety, tolerability, and its effectiveness against various chronic diseases in humans (Kunnumakkara *et al.,* 31).

Piperine

It is known that piperine present in black pepper possesses several properties. In its interaction with other drugs, it can act as a bioavailability enhancer; this effect is also manifested in combination with other nutraceuticals, e.g. with curcumin, i.e. piperine can modify curcumin's antioxidant, anti-inflammatory, antimicrobial and anticancer effects. Piperine displays significant immune modulating, antioxidant, chemopreventive and anticancer activity; these effects have been shown to be dose-dependent and tissue-specific (Smilkov *et al.,* 51).

Gingerol/Shogaol

The spice ginger has many terpenoids and phenolic constituents in its rhizome. Among them, vanilloid compounds namely gingerol, paradol and shogaols and other compounds such as zingerone, and galanals A and B showed therapeutics properties by modulation of enzymes involved in the disease pathway (Aggarwal and Shishodia, 2). The compounds gingerol and shogaols found to have beneficial effect in prevention and reduction in colon cancer incidents in humans, proved through several *in vitro* and *in vivo* studies (Prasad and Tyagi, 40; Chen *et al.,* 14; Shukla and Singh, 46).

Capsaicin

Capsaicin is the primary bioactive substance in red chili peppers, which produces the pungent flavor. During the past few decades, pharmacological benefits of capsaicin and its underlying mechanisms have been examined extensively. Major biological efficacies of capsaicin are analgesic, antioxidant, anti-inflammatory, anti-cancer, anti-obesity, cardio-protective, and metabolic modulation effects (Lu *et al.,* 34).

Saponin

Fenugreek is a commonly used spice in Indian cuisine. The nutraceutical properties of this spice are attributed to different type of both primary and secondary metabolites namely lipids, polysaccharides, amino acids, steroids, saponins, phenolics and flavonoids. Numerous clinical investigations using fenugreek seed showed that it has antidiabetic, antiobesity, antioxidant, anti-inflammatory, anticancer and has numerous benefits in menstruating women (Yao *et al.,* 57). Saponin inhibits the absorption of cholesterol from the small intestine or the reabsorption of bile acids thereby lowers the cholesterol in circulatory pathway (Matsuura, 35)

Problems in Spices Breeding for Nutraceuticals

Unlike other sectors of horticulture, the spices face highly dynamic situations arising from requirements of nutraceutical industries for high quality spices varieties. The proactive nature of response to this factor will translate to the release of new cultivars meeting the requirements of spiceuticals. Development of improved cultivars by spice crop breeders may require many breeding cycles and dozens of years because of perennial nature (black pepper, cardamom, and tree spices). Other major issues in spices breeding are polyploidy (turmeric), incompatibility, and sterility (ginger).

In a typical plant biochemical pathway, a single precursor synthesizes multiple metabolites. So the presence of a precursor does not assure the synthesis of any specific metabolite (Dixon *et al.,* 18). Several plant biochemical pathways crosstalk for effective regulation of cellular processes in an organism. A product of one pathway becomes the precursor or intermediate substrate for another pathway. Furthermore, a single enzyme affects several biochemical pathways. Therefore, a single point mutation could lead to significant changes in a metabolome (Newell-McGloughlin, 36). Alteration in one pathway could potentially change the flux in non-target metabolites (Davies, 16). Quality-enhancing metabolites are often governed by quantitative trait loci, and epistatic interactions make a major contribution to these phenotypic variations (Fernie and Schauer, 19; Rommens *et al.,* 42). The diversity in quality- enhancing metabolites in plants, not only comes from genes, but also from divergent, substrate- specific enzymes and sub-cellular, non-enzymatic activities (Hall *et al.,* 25).

GxE Interaction on Secondary Metabolites

The individual climate change components, like high temperature, elevated carbon dioxide, drought stress, enhanced ultraviolet-B radiation influences on secondary metabolites, such as phenolics, terpenes, and alkaloids. The understanding of these aspects on secondary metabolites plays important role in spices. Terpenes are the largest class of secondary metabolites, accounting for at least 25,000 compounds (Taiz *et al.,* 54), which are derived from the five-carbon isoprene unit. Plants lose approximately 10–20 per cent of assimilated carbon to isoprene, even under carbon-limited environments (Brunetti *et al.*, 11). Terpenes have a variety of important functions, such as plant communication and protection against abiotic stress and some are medically important (Krauss and Nies, 30; Singsaas, 49). Variation in environmental, soil nutrients and agroclimatic conditions in different zones results in varied essential oil, total curcuminoids and its composition, as well yield in turmeric (Singh *et al.*, 48; Anandaraj *et al.*, 6; Aarthi *et al.*, 1). GxE interaction was evident for oil content and t-anethole content in fennel (Lal, 33; Samira Shojaiefar, 43).

Breeding for nutraceuticals

Present status

Little efforts have been made in this area of quality upgradation of spices. Most of the released varieties aimed for higher yield and disease tolerance. Following are the some of the varieties released in India with specific quality attributes.

Table 3. Spice varieties for high quality attributes

Crop	*Varieties*
Black pepper	Quality of black pepper is measured in terms of piperine, oil and oleoresin content. PLD 2 variety contains oleoresin 15.45 per cent and essential oil 4.8 per cent. Sreekara and Subhakara are rich in essential oil with 6.0 and 7.0 per cent respectively. IISR Malabar Excel is rich in oleoresin 13.5 per cent.
Small cardamom	Market value of cardamom is decided by its appearance and size. Malabar types are characterized by globose-oblong shaped capsules and Mysore types are of ovoid, bold and dark green capsules. IISR Avinash has bold dark green capsules and ICRI 8 has essential oil of 8 per cent
Ginger	Varieties with low fibre *viz.,* Suprabha, Himgiri and IISR Varada are suitable for fresh ginger. Essential oil is one of the major quality characters of ginger and varieties such as Athira, Karthika and Aswathy are rich in essential oil (>3 per cent). IISR Vajra has desirable flavor owing to its high zingiberene content (29.83 per cent).
Turmeric	Curcumin is one of the important quality parameters and varieties rich in curcumin content are Roma (6.1 per cent), Suroma (6.1 per cent), IISR Prathiba (6.2 per cent), IISR Prabha (6.5 per cent), MeghaTurmeric-1 (6.8 per cent), IISR Alleppey Supreme (5.5 per cent) and IISR Pragati (5 per cent).
Coriander	High quality lines with high essential oil content: RCr-480 (0.425 per cent), Sudha (0.40 per cent) Sindhu (0.40 per cent), RD 385 (0.45 per cent)
Cumin	Guj. Cumin 3 (4.4 per cent), Gujarat Cumin 2 (4.0 per cent) and RZ-341 (3.87 per cent) are the varieties with high volatile oil. RZ-223 is a variety with long bold attractive seeds. Ac-01-167 is another bold seeded variety
Fennel	RF- 205 (2.48 per cent), HF 143 (2.4 per cent), RF 281 (2.58), Gujarat Fennel 2(2.4 per cent) and RF-178 (2.13 per cent) are the varieties with high essential oil. RF-157 is a variety with long, attractive, bold seeds and a volatile oil content of 1.95 per cent, Sweet seeded type: Pant Madhurika is a variety with sweet seeds and green fine ridges

Crop	*Varieties*
Fenugreek	HM 444/ Hisar Manohar has unique green seed colour. Gujarat Methi-2 (Gujarat Fenugreek 244) and RMt-351(UM-351) are varieties with bolder lustrous grains and uniform in size. Ajmer fenugreek 3 has 1.79 per cent diosgenin and 0.97 per cent 4-hydroxy isoleucine content in seed. Ajmer fenugreek 5 has 3.98 per cent volatile oil and high antioxidant content (66.428mg/BHTE/ppm).
Ajwain	Lam ajwain 3 (7.8 per cent) and Ajmer ajwain 73 (6.38 per cent) are rich in volatile oil content, suitable for essential oil producing industries
Nigella	Ajmer Nigella 1 has high essential oil (19.7 per cent) and high oleic acid (3.32 per cent) content
Nutmeg	Konkan Sanyukta possess high nut oil (27 per cent) and mace oil (17.75 per cent)
Cassia	IISR Cassia (D1) is the first cassia variety with low coumarin content 19.9 mg/kg, oil content in bark 24 per cent.
Clove	PPI(CL) 1, the first clove variety has oil content of 6 per cent

Chemo-diversity in Spices

Cultivar diversity is one of the principal components of diversity in the genus, Piper. Evaluation of the black pepper germplasm collections has resulted in identifying high piperine, oil and oleoresin types. Essential oil is found to vary from 0.4–7 per cent, while piperine varies from 2–7.4 per cent among the cultivars. Some of the high piperine cultivars are Ceylon, Kaniyakkadan, Kottanadan, Kumbhakkodi, Kuthiravally, Munda, Perumunda, and Karimunda (Gopalam and Ravindran, 27). Among *Piper longum* collections, the genotype collected from Pandur, Maharashtra recorded maximum (11.27 ppm) piperine content followed by Kolkewadi, Maharashtra (9.92 ppm) (Gupta *et al.*, 23).

Major phyto-constituents of turmeric are diarylheptanoids, which occur in a mixture termed curcuminoids that generally make up approximately 2–8 per cent of turmeric by dry weight (Prasath *et al.*, 41). Variation in curcumin content of the same variety occurs in different locations (Anandaraj *et al.*, 6). This variation in phyto-constituents (curcumin, oleoresin and essential oils) of turmeric can likely be attributed to differences in climate and soil conditions that occur in different agroclimatic zones. The variation in curcumin content among genotypes under similar climatic conditions may be due to genetic factors (Sandeep *et al.*, 44) indicating that phyto-constituents are attributes of biosynthetic gene expression pathways. Among the Indian traditional cultivars of ginger, Kuruppampady, Wayanad, Maran and Rio-de-Janeiro are superior in total gingerol from oleoresin (Zachariah, 58). Meghalaya Local contained the highest concentration of total gingerol (3.436 µg/g). The cultivar "Jamaican" contained the highest concentrations of all three gingerols and was the most pungent of the clones analysed (Wohlmuth *et al.*, 56).

Fenugreek genotypes rich in saponins will be useful for pharmaceutical industries. The germplasm was screened for high saponins and it ranged from 0.92 g to 1.68 g with the mean value of 1.34 g /100 g DW (Arivalagan *et al.*, 9).

Pre-breeding and Breeding

But because of the inherent variability already available, both selective breeding and cross breeding strategies may be fruitful in developing high quality lines. The development of hybrid cultivars instead of open- pollinated cultivars could potentially increase certain metabolites without any yield decrease. Recent estimates of metabolite heritability are 25- 35 per cent, encouraging future breeding efforts with metabolomics (Fernie and Schauer, 19).

Gamma-ray irradiation is an effective method to increase the quality of spices for a long time. M2 offspring from a parent ginger rhizome irradiated with 20 Gy was found to have a high 6-gingerol content (38.4 ± 0.01 mg/g methanol extract in comparison to 22.1 ± 0.03 mg/g in non-irradiated control samples) (Asmaa *et al.*, 10). Another strategy to accelerate breeding is polyploid induction, which could improve yield and quality.

Applied Genomics

Marker Assisted Breeding

The use of molecular markers in traditional plant breeding is increasing rapidly. The identification of new molecular markers for quality traits requires a thorough understanding of the biochemical basis of intricate pathways for metabolites. The integration of metabolomics with genomics will provide new molecular markers of quality traits for marker assisted selection in existing breeding programs. Transcription factors play a major role in secondary metabolite biosynthesis and offer new hope for nutraceutical breeding through their integration into marker assisted breeding (Newell-McGloughlin, 36). In spices, linkage maps have been developed for few agronomically important traits using different types of markers and mapping populations. RNA-Seq based association mapping reported association of SNPs with three yield traits in garlic (Zhu *et al.*, 59).

Transcriptomics

High performance de novo transcriptome assembly based on in-depth sequencing provides an opportunity for genome-wide gene expression analysis of secondary metabolites pathway genes, which is particularly useful to study spices. George *et al.* (20) identified the entire terpene synthase family responsible for the biosynthesis of the flavor-imparting volatiles in black pepper berries using a combinatorial approach, the berry hybrid transcriptome assembly of llumina and nanopore sequencing. Also, identified the transcripts related to biosynthetic pathways of several anti-cancer compounds like taxol, curcumin, and vinblastine in addition to anti-malarial compounds like artemisinin and acridone alkaloids, emphasizing turmeric's importance as a highly potent phytochemical (Annadurai *et al.*, 8). Transcriptome of four turmeric cultivars (Suroma, Roma, Lakadong and Megha) revealed functional basis of differentially expressed pigment synthesis genes among the four cultivars (Ambika *et al.*, 4). In ginger, five genes associated with gingerol biosynthesis were identified as being significantly differentially expressed in the rhizome at an early developmental stage (Jiang *et al.*, 28). In ajwain (*Trachyspermum ammi*), transcriptomic study by Soltani Howyzeh *et al.* (52) revealed key candidate genes along with the molecular mechanism of terpene synthase and thymol biosynthesis.

Metabolomics Assisted Breeding

Breeding for nutraceuticals needs metabolic screening and therefore, integration of metabolomics with traditional breeding is needed for releasing new cultivars with nutraceutical properties. Modified breeding programs could reveal the hidden biochemical variation for targeted crop improvement and better-quality products.

Metabolomics-assisted breeding, in conjunction with genomics and proteomics, offer such tools for nutraceutical breeding. Techniques, such as, mass spectrometry, nuclear magnetic resonance spectroscopy, chromatography, etc., have been utilized to study the impact of time, stress, and environmental conditions and to analyse hundreds of metabolites simultaneously (Dixon *et al.,* 18). Instead of monitoring a single putative bioactive metabolite, hundreds of biosynthetically related metabolites and enzymes could be analysed by metabolic profiling, which in turn could be used to develop stable genotypes for nutraceutical cultivars.

Genomics

Spices being non-model crops have been schemed out of sequencing projects earlier, due to high running cost of Sanger sequencing. However, the advent of fast and cost-effective next generation sequencing (NGS) platforms in the recent past has enabled the unearthing of certain characteristic gene structures unique to these species. It has also aided in gaining insight about mechanisms underlying processes of gene expression and secondary metabolism.

All orthologous gene sequences involved in alkaloid and piperine biosynthesis in black pepper were captured from comparative genomics analyses, and corresponding gene functions were verified by searching the published literature and the NCBI and UniProt databases (Hu *et al.,* 27). Based on the results of ginger genome sequencing, transcriptome analysis, and metabolomic analysis, backbone biosynthetic pathway of gingerol analogues consisted of 12 enzymatic gene families, PAL, C4H, 4CL, CST, C3'H, C3OMT, CCOMT, CSE, PKS, AOR, DHN, and DHT were identified (Li *et al.*, 2021). A chromosome-scale Vanilla planifolia genome (Hasing *et al.,* 26) revealed gene variants affecting the vanillin pathway, bean quality and productivity.

Small RNA sequencing in black pepper identified 110 known miRNAs, 18 novel miRNAs, and 1007 unique targets in secondary metabolism, particularly piperine biosynthesis (Ding *et al.,* 17). The ginger transcriptomic studies reported 16 miRNA families for the production of a bioactive component, gingerol (Singh *et al.,* 47). High-throughput sequencing-based identification of miRNAs from the curry leaf and their functional implications in secondary metabolite biosynthesis open a new avenue for miRNAs based regulation underlying spice accumulation (Gutierrez-Garcia *et al.,* 24).

Interventions to Enhance Nutraceuticals

Plant Secondary Metabolite (PSM) which is the basis for developing preventive and therapeutic nutraceuticals has wide role in commercial and industrial application. As the secondary metabolite is influenced by environment factors, controlled system may enhance the production of PSM. For curcumin content, most sensitive factors were; the altitude, soil pH, nitrogen and Potassium content. In contrast, the temperature and phosphorous content were found critical for leaf essential oil while, nitrogen, phosphorous and potassium contents were the key quantitative and qualitative determinants for rhizome essential oil. Managing varying environmental parameters of the spices cultivation, can enhance quality of curcumin, rhizome and leaf essential oil in turmeric (Sriram *et al.,* 53).

Elicitor treatments has role in increasing PSM, which helps in accumulation of secondary metabolites. Understanding signal transduction paths underlying elicitor-induced production of secondary metabolite is important for optimizing their commercial production. Sivaranjani *et al.* (50) reported on influence of elicitor high essential oil production in turmeric. The elicitor effect was observed at 0.25 mM of Salicylic Acid, resulting in greater economic value of the biomass due to the higher production of secondary compounds, such as phenolic compounds and flavonoids in fennel (Gorni *et al.,* 22). Application of paclobutrazol at 50 ppm recorded the maximum fixed oil content of 38.74 per cent, followed by chitosan at 100 ppm (37.75 per cent) in black cumin (Anil Kumar *et al.,* 7). MeJA 0.01 per cent increased diosgenin levels, in 12 days old seedlings, from 0.5 per cent–0.9 per cent to 1.1 per cent–1.8 per cent in fenugreek (Chaudhary *et al.,* 13).

Arbuscular mycorrhizal fungi (AMF) establish mutualistic symbiotic associations with plant roots and act as biofertilizers by enhancing plant nutrient and water uptake. Moreover, AMF colonization may influence the biosynthesis of plant bioactive compounds in medicinal and aromatic plants (Colla *et al.,* 15). In saffron, using AMF, high quality (ISO category) and had high contents of bioactive compounds was produced in saffron, with very high total polyphenol content and elevated antioxidant activity (Caser *et al.,* 12).

Tissue culture plays important role in enhancing PSM. In many cases rigourously controlled plant in vitro cultures can generate valuable natural products (Anand, 5). Cell culture technology source of PSM production could be used for the large-scale culture of plant cells from which this metabolite can be extracted. Plant cell and tissue cultures hold great promise for large scale-controlled production of useful secondary metabolites on demand (Vijaya Sree *et al.,* 55).

Conclusion

Since quality parameters of spices are genetically controlled, genotypes can be successfully improved through breeding techniques. Classical breeding techniques along with recent biotechnological intervention can facilitate the improvement in quality parameters in spices. It requires a continuous flow of genes and allelic diversity into different crop gene pool. The primary, secondary gene pool and pre-breeding plays a vital role. The vast chemo-diversity available in India can be exploited for the understanding of biosynthetic pathway modifications or activity of different isoforms of enzymes which leads to accumulation of secondary metabolites of interest. Also, development of mapping population and extensive genotyping and phenotyping is very crucial to identify genomic regions controlling nutraceutical components. The NGS based whole genome and transcriptome approaches have clearly demonstrated their advantages over previously developed methods and are becoming the new standard for NGS studies in spices. The functional genomics/metabolomics approach help to elucidate the biochemical pathways and key genes underlying specific nutraceutical compounds.

The future of nutraceuticals is bright for most horticultural crops, including spices. Scientists are gradually appreciating the integrated approach of 'system biology', where genomics and metabolomics are coupled to study the organism at a cellular and organ levels and describe networks of biological regulation in a most inclusive way (Saxena and Cramer, 45). Plant breeders will benefit from a system biology approach to develop nutraceutical cultivars for improvement in overall human health. The collaborations between public and private crop breeding institutions, with human health research centers would play a key role in the success of this endeavour. The agenda set forth, in tune with the needs and leads, for the coming years are destined to turn a new leaf in spices research in the days to come.

REFERENCES

1. Aarthi, S., Suresh, J., Leela, N.K. and Prasath, D. 2020. Multi environment testing reveals genotype-environment interaction for curcuminoids in turmeric (*Curcuma longa* L.). *Industrial Crops and Products.* **145**: 112090.
2. Aggarwal, B.B. and Shishodia, S. 2006. Molecular targets of dietary agents for prevention and therapy of cancer. *BiochemPharmacol.* **71(10)**: 1397-421.
3. Ahmed, R.S., Hussain, M.B., Sultan, M.T., Arshad, M.S., Waheed, M., Shariati, M.A., Plygun, S. and Hashempur, M.H. 2020. Biochemistry, Safety, Pharmacological Activities, and Clinical applications of turmeric: A Mechanistic Review. Evidence based complementary and Alternative Medicine, Article ID: 7656919.https://doi.org/10.1155/2020/7656919.
4. Ambika, S., Basudeba, K., Suprava, S., Sudipta, J., Ananya, K., Sriram, P., Jeetendranath, P. and Sanghamitra, N. 2019. De Novo transcriptome sequencing explored cultivar specific sequence variation and differential expression of pigment synthesis genes in turmeric (*Curcuma longa* L.). *Industrial Crops and Products.* **134**:388-402. https://doi.org/10.1016/j.indcrop.2019.04.021.
5. Anand. S. 2010. Various approaches for secondary metabolite production through plant tissue culture. *Pharmacia.* **1(1)**: 1-7.
6. Anandaraj, M., Prasath, D., Kandiannan, K., Zachariah, T.J., Srinivasan, V., Jha, A.K., Singh, B.K., Singh, A.K., Pandey, V.P., Singh, S.P., Shoba, N., Jana, J.C., Ravindra Kumar, K. and Uma Maheswari, K. 2014. Genotype by environment interaction effects on yield and curcumin in turmeric (*Curcuma longa* L.). *Ind. Crops Prod.* **53**:358-364.
7. Anil Kumar, G.S., Umesha, K., Vishnuvardhana., Shivanna, M. and Shankarappa, T.H. 2021. Influence of elicitors in enhancing the fixed oil content and yield of black cumin (*Nigella Sativa* L.). *The Pharma Innovation Journal.* **10(1)**: 559-562
8. Annadurai, R.S., Neethiraj, R., Jayakumar, V., Damodaran, A.C., Rao, S.N., Katta MAVSK., *et al.,* 2013. De Novo transcriptome assembly (NGS) of Curcuma longa L. rhizome reveals novel transcripts related to anticancer and antimalarial terpenoids. *PLoS ONE.* **8(2)**: e56217. https://doi.org/10.1371/journal.pone.0056217
9. Arivalagan, M., Gangopadhyay K. K., and Kuma G. 2013. Determination of Steroidal Saponins and Fixed Oil Content in Fenugreek (*Trigonella foenum-graecum*) Genotypes. *Indian J Pharm Sci.* **75(1)**:110-113.
10. Asmaa, M., Magdy., Eman, M., Fahmy., Abd EL-Rahman, M.F., AL-Ansary. and Gamal Awad. 2020. Improvement of 6-gingerol production in ginger rhizomes (*Zingiber officinale*

Roscoe) plants by mutation breeding using gamma irradiation. *Applied Radiation and Isotopes*. 162. 109193, https://doi.org/10.1016/j.apradiso.2020.109193.

11. Brunetti, C., Guidi, L., Sebastiani, F. and Tattini, M. 2015. Isoprenoids and phenylpropanoids are key components of the antioxidant defense system of plants facing severe excess light stress. *Environ. Exp. Bot.* **119**: 54–62.
12. Caser, M., Íris Marisa Maxaieie Victorino., Sonia Demasi., Andrea Berruti., Dario Donno., Erica Lumini., Valeria Bianciotto. and Valentina Scariot. 2018. Saffron Cultivation in Marginal Alpine Environments: How AMF Inoculation Modulates Yield and Bioactive Compounds. Agronomy. **9**: 12; doi:10.3390
13. Chaudhary, S., Chikara, S.K., Sharma, M.C., Chaudhary, A., Syed, B.A., Chaudhary, P.S., Aditya Mehta, Maulik Patel, Arpita Ghosh and Marcello Iriti. 2015. Elicitation of Diosgenin Production in *Trigonella foenum-graecum* (Fenugreek) Seedlings by Methyl Jasmonate. *Int. J. Mol. Sci.* **16(12)**: 29889–29899.
14. Chen, H., Fu, J., Chen, H., Hu, Y., Soroka, D.N., Prigge, J.R., Schmidt, E.E., Yan, F., Major, M.B., Chen, X. and Sang, S. 2014. Ginger compound [6]-shogaol and its cysteine-conjugated metabolite (M2) activate Nrf2 in colon epithelial cells in vitro and in vivo. *Chem Res Toxicol.* **27(9)**: 1575-85.
15. Colla, G., Nardi, S., Cardarelli, M., Ertani, A., Lucini, L., Canaguier, R. and Rouphael, Y. 2015. Protein hydrolysates as biostimulants in horticulture. *Sci. Hortic.* **196**: 28–38.
16. Davies, K.M. 2007. Genetic modification of plant metabolism for human health benefits. *Mutat Res* **622**: 122-137.
17. Ding, Y., Mao, Y., Cen, Y., Hu, L., Su, Y., Ma, X., Long, L., Hu, H., Hao, C. and Luo, J. 2021. Small RNA sequencing reveals various microRNAs involved in piperine biosynthesis in black pepper (*Piper nigrum* L.). *BMC Genom.* **22**:1-12.
18. Dixon, R.A., Gang, D.R., Charlton, A.J., Fiehn, O., Kuiper, H.A., *et al.*, 2006. Applications of metabolomics in agriculture. *J Agric Food Chem.* **54**: 8984-8994.
19. Fernie, A.R. and Schauer, N. 2009. Metabolomics-assisted breeding: a viable option for crop improvement? *Trends Genet.* **25**: 39-48.
20. George, J.K., Shelvy, S., Fayad, A.M. *et al.* De novo transcriptome sequencing assisted identification of terpene synthases from black pepper (*Piper nigrum*) berry. Physiol Mol Biol Plants 27, 1153–1161 (2021). https://doi.org/10.1007/s12298-021-00986-4.
21. Gopalam, A. and Ravindran, P.N. 1987. Indexing of quality parametres in black pepper cultivars. Indian Spices, 22/23, 8–11.
22. Gorni, P.H., Brozulato, M.D.O., Lourenção, R., da, S. and Konrad, E.C.G. 2017. Increased biomass and salicylic acid elicitor activity in fennel (*Foeniculum vulgare* Miller). *Brazilian J. Of food technology*. e2016172.
23. Gupta, G., Rane, A.D., Vineet Keshore., Narkhede, S.S., Singh, G.V., Omkar Pawar and Raut, S.S. 2014. Study of variation in piperine content of *Piper longum* genotypes. *Medicinal Plants - International Journal of Phytomedicines and Related Industries*. 6(3): 209-212.

24. Gutierrez-Garcia, C., Ahmed, S.S., Ramalingam, S., Selvaraj, D., Srivastava, A., Paul, S. and Sharma, A. 2021. Identification of microRNAs from medicinal plant *Murraya koenigii* by high-throughput sequencing and their functional implications in secondary metabolite biosynthesis. *Plants*. 11: 46.

25. Hall, R., Beale, M., Fiehn, O., Hardy, N., Sumner, L., *et al.* 2002. Plant metabolomics: the missing link in functional genomics strategies. *Plant Cell*. **14**: 1437-1440.

26. Hasing, T., Tang, H., Brym, M., Khazi, F., Huang, T. and Chambers, A.H. 2020. A phased *Vanilla planifolia* genome enables genetic improvement of flavour and production. *Nature Food*. **1**: 811-819.

27. Hu, L., Xu, Z., Wang, M. *et al.* 2019. The chromosome-scale reference genome of black pepper provides insight into piperine biosynthesis. *Nat Commun*. **10**: 4702. https://doi.org/10.1038/s41467-019-12607-6.

28. Jiang, Y., Huang, M., Wisniewski, M., Li, H., Zhang, M., Tao, X., Liu, Y. and Zou, Y. 2018. Transcriptome analysis provides insights into gingerol biosynthesis in ginger (*Zingiber officinale*). *Plant Genome*. **11(3)**. doi: 10.3835/plantgenome2018.06.0034. PMID: 30512040.

29. Kita, T., Ima, S., Sawada, H., Kumagai, H. and Seto, H. 2008. The Biosynthetic Pathway of Curcuminoid in Turmeric (*Curcuma longa*) as Revealed by 13C-Labeled Precursors. *Biosci. Biotechnol.Biochem*. **72 (7)**: 1789–1798

30. Krauss, G.J. and Nies, D.H. 2015. (Eds.) *Ecological Biochemistry: Environmental and Interspecies Interactions*; Wiley-VCH Verlag GmbH and Co. KGaA: Weinheim, Germany.

31. Kunnumakkara, A.B., Bordoloi, D., Padmavathi, G., Monisha, J., Roy, N.K., Prasad, S., and Aggarwal, B.B. 2017. Curcumin, the golden nutraceutical: multitargeting for multiple chronic diseases. *British J Pharmacology*. **174(11)**:1325–1348. https://doi.org/10.1111/bph.13621

32. Kurian, A. 2012. Health benefits of herbs and spices. Handbook of Herbs and Spices 2nd ed. 72-88. [http://dx.doi.org/10.1533/9780857095688.72].

33. Lal, R. K. 2008. Stability and Genotypes × Environment Interactions in Fennel. *Journal of Herbs, Spices and Medicinal Plants*. **13(3)**:47-54.

34. Lu, M., Chen, C., Lan, Y., Xiao, J., Li, R., Huang, J., Huang, Q., Cao, Y. and Ho, C.T. 2020. Capsaicin-the major bioactive ingredient of chili peppers: bio-efficacy and delivery systems. *Food Funct*. **11(4)**: 2848-2860. doi: 10.1039/d0fo00351d. PMID: 32246759.

35. Matsuura, H. 2001. Saponins in garlic as modifiers of the risk of cardiovascular disease. *J Nutr*. **131**:1000

36. Newell-McGloughlin, M. 2008. Nutritionally improved agricultural crops. *Plant Physiol*. **147**: 939-953

37. Nicoletti, M. 2012. Nutraceuticals and botanicals: Overview and perspectives. *Int J Food Sci Nutr*. **63(Suppl.1)**: 2-6. [http://dx.doi.org/10.3109/09637486.2011.628012] [PMID: 22360273].

38. Pandey, M., Verma, R.K. and Saraf, S.A. 2010. Nutraceuticals: New era of medicine and health. *Asian J Pharm Clin Res.* **3**: 11-5.
39. Parthasarathy, V.A., Chempakam, B. and Zachariah, T.J. 2008. Chemistry of Spices. CABI International, London.
40. Prasad, S. and Tyagi, A.K. 2015. Ginger and its constituents: Role in prevention and treatment of gastrointestinal cancer. Gastroenterology research and practice, Article ID 142979. https://doi.org/10.1155/2015/142979.
41. Prasath, D., Kandiannan, K., Leela, N.K., Aarthi, S., Sasikumar, B. and Nirmal Babu, K. 2019. Turmeric: Botany and production practices. *Hort. Reviews.* **46**: 99-184, ISBN: 978-1-119-52106-8.
42. Rommens, C.M., Haring, M.A., Swords, K., Davies, H.V. and Belknap, W.R. 2007. The intragenic approach as a new extension to traditional plant breeding. *Trends Plant Sci.* **12**: 397-403.
43. Samira Shojaiefar., Aghafakhr Mirlohi., Mohammad, R., Sabzalian. and Hamid Yaghini. 2015. Seed yield and essential oil content of fennel influenced by genetic variation and genotype × year interaction. *Industrial Crops and Products.* **71**:97-105.
44. Sandeep, I.S., A. Kuanar, A. Akbar, B. Kar, S. Das, A. Mishra, P. Sial, P.K. Naik, and S. Nayak. 2016. Agroclimatic zone based metabolic profiling of turmeric (*Curcuma longa* L.) for phytochemical yield optimization. *Ind. Crops Prod.* **85**:229-240.
45. Saxena, A. and Cramer, C.S. 2013. Metabolomics: A potential tool for breeding nutraceutical vegetables. *Adv Crop Sci Tech.* **1**: 106. doi:10.4172/ 2329-8863.1000106.
46. Shukla, Y. and Singh, M. 2007. Cancer preventive properties of ginger: a brief review. *Food Chem Toxicol.* **45(5)**:683-90.
47. Singh, N., Srivastava, S. and Sharma, A. 2016. Identification and analysis of miRNAs and their targets in ginger using bioinformatics approach. *Gene.* **575**: 570-576.
48. Singh, S., Joshi, R.K. and Nayak, S. 2013. Identification of genotypes of turmeric through agroclimatic zone based evaluation of important drug yielding traits. *Ind Crops Prod.* **43**:165.
49. Singsaas, E.L. 2000. Terpenes and the thermotolerance of photosynthesis. *New Phytol.* **146**:1–2.
50. Sivaranjani, R. and Zachariah, T. 2022. Elicitors induced changes in essential oil constituents of turmeric (*Curcuma longa* L.) rhizome. *Journal of Horticultural Sciences.* **17(1)**: 237-244.
51. Smilkov, K., Ackova, D.G., Cvetkovski, A., Ruskovska, T., Vidovi, B. and Atalay, M. 2019. Piperine: Old Spice and New Nutraceutical? Current Pharmaceutical Design. **25**: 1729-1739.
52. Soltani Howyzeh, M., Sadat Noori, S.A., Shariati, V. M. and Amiripour, M. 2018. Comparative transcriptome analysis to identify putative genes involved in thymol biosynthesis pathway in medicinal plant *Trachyspermum ammi* L. *Sci. Rep.* **8**: 13405.

53. Sriram Sandeep, I.S., Ananya Kuanar., Abdul Akbar., Basudeba Kar., Suryasnata Das., Antaryami Mishra., Parshuram Sial., Pradeep Kumar Naik., Sanghamitra Nayak. and Sujata Mohanty. 2016. Agroclimatic zone based metabolic profiling of turmeric (*Curcuma Longa* L.) for phytochemical yield optimization. *Industrial Crops and Products*.85:229-240.

54. Taiz, L., Zeiger, E., Møller, I.M. and Murphy, A. 2018. *Fundamentals of Plant Physiology*; Oxford University Press: New York, NY, USA.

55. Vijaya Sree. N., Udayasri, P.V.V., Aswani kumar, Y., Ravi Babu, B., Phani kumar, Y. and Vijay Varma, M. 2010. Advancements in the Production of Secondary Metabolites. *Journal of Natural Products*. **3**: 112-123.

56. Wohlmuth, H., David, N., Leach., Mike, K., Smith., Stephen, P. and Myers. 2005. Gingerol Content of Diploid and Tetraploid Clones of Ginger (*Zingiber officinale* Roscoe). *Journal of Agricultural and Food Chemistry*. **53(14)**: 5772-5778. DOI: 10.1021/jf050435b.

57. Yao, D., Zhang, B., Zhu, J., Zhang, Q., Hu, Y., Wang, S., Wang, Y., Cao, H. and Xiao, J. 2019. Advances on application of fenugreek seeds as functional foods: Pharmacology, clinical application, products, patents, and market. *Critical Reviews in Food Science and Nutrition*. 1–11. doi:10.1080/10408398.2019.1635.

58. Zachariah, T.J. 2008. Ginger. In: Parthasarathy VA, Chempakam B, John Zachariah T(Eds.), Chemistry of Spices (pp.70–96). Pub; CABI, UK.

59. Zhu, S., Chen, X., Liu, X., Zhao, J. and Liu, T. 2019. Transcriptome-wide association study and eQTL analysis to assess the genetic basis of bulb-yield traits in garlic (*Allium sativum*). *BMC Genom*. **20**: 1-9.

Chapter 10

Papaya Breeding: Strategies and Opportunities

Jai Prakash[1]* and Preeti Singh[2]

[1]*Division of Fruits and Horticultural Technology, ICAR–Indian Agricultural Research Institute, New Delhi 110012*

[2]*Bihar Agricultural University, Sabour 813210, Bhagalpur, Bihar*

**e-mail: singhjai2001@rediffmail.com*

ABSTRACT

Papaya is an important tropical fruit crop cultivated widely and is extensively grown in tropics and is also extensively grown in the subtropical and frost-free regions of the world. Breeding strategies in papaya differs from other perennial fruit crops because of its short generation cycle that can be completed in a year. Papaya is propagated through seeds unlike majority of the fruit crops. Development of stable high yielding gynodioecious lines, lines with better fruit quality, better papain production, and resistance to biotic and abiotic stresses are the major breeding objectives. Sex determination in papaya plants is an important as the sex affects flower type, production, fruit shape and yield. Pyriform fruits occurring in the hermaphrodite plants are more preferred than the round fruits obtained from the female plants. Cultivated papayas are either dioecious (male and female plants) or gynodioecious (hermaphrodite and female plants. The genetic and molecular mechanism sex expression has been studied in details. Development of stress tolerant papaya genotypes using wild and related species are also important. In papaya applications omics based strategies are established to be important in developing productive and stress tolerate genotypes besides higher storability.

***Keywords**: Paraya, Sex forms, Quality, Disease resistance, Wild species.*

Papaya (*Carica papaya* L.), a crop of tropical origin is relished by the consumers due to its highly nutritious fruits and the medicinal properties possessed by different plant parts like immature fruits, latex, leaves and seeds. Papaya is among the few fruit crops which do not have long juvenile phase and comes to bearing in less than a year. So, it is also preferred by the growers owing to its short genration cycle, round the year production, high productivity and profitability. This fast-growing tree of tropics, is also extensively grown in the subtropical and frost free regions of the world. High phenotypic plasticity allows papaya plants to establish in diverse agro-ecosystems. India, Dominican Republic, Brazil, Mexico and Indonesia are the largest producers of papaya in the world (FAOSTAT, 6).

Papaya, an economically important fruit crop of *Caricaceae* family and is also closely related to Arabidopsis as they belong to the same order Brassicales (Bremerk *et al.,* 3). The 35 latex producing species of this family is divided into 6 generas, which are *Carica* (1 specie), *Horovitzia* (1 specie), *Cylicomorpha* (2 species), *Jarilla* (3 species), *Jacaratia* (7 species) and *Vasconcellea* (21 species) (Badillo, 2). Papaya is a diploid species with somatic chromosome number 18 and haploid genomic size of 372 Mb (Arumuganathan and Earle, 1). Owing to its small diploid genome size, fast growth, smaller tree size, large quantity of seed production (1000 in fruit from female plant and 800 in fruit from hermaphrodite plant), transformationability and nascent sex chromosome, papaya serves as a model plant for studying sex chromosome evolution and sex determination (Liu *et al.,* 10; Ming *et al.,* 12). Knowledge about genetics of papaya has been accelerated with the advancement in molecular markers, linkage and physical maps, comparative genomics studies, and the sequencing of genome (Tripathi *et al.,* 17). Papaya is trioecious in nature with sex forms like female, male and hermaphrodite. The morphology of papaya inflorescences and flower varies with environmental condition and sex of the plant that is controlled by an X/Y chromosome system. Female papaya plants are the stable ones, which bear pistillate flowers have five free petals and a rounded superior ovary that has five carpel, exhibits parietal placentation. The male plants bear staminate flowers, the stamens are ten in number and are arranged surrounding a rudimentary pistil or pistillode. The hermaphrodite ones, bears perfect flowers, also referred to as the elongate type, consists of five petals, five pairs of anthers, and a superior ovary. The flowers of male and hermaphrodite plants may undergo sex reversal due to some genetic causes and environmental fluctuations. Cultivated papayas are either dioecious (male and female plants) or gynodioecious (hermaphrodite and female plants). Sex determination for the three types of sex forms of papaya is genetically regulated by the pairing of sex chromosomes (Carvalho and Renner, 4), through a sex-linked region that behaves like an XY sex chromosome. Studies suggest that the Y chromosome contains a small specific region that controls expression of male (Y) or hermaphrodite (Y^h) types. All combinations among the Y and/or Y h chromosomes are lethal; therefore, the male (XY) and hermaphrodite (XY^h) types are heterozygous, whereas, the female is homozygous with two X chromosomes (XX) (Ming *et al.,* 14).

Genetic Diversity: A Pre-Requisite for Improvement

Papaya breeding is different from breeding of other fruit crops, as papaya has short generation cycle that can be completed in a year, so attaining homozygosity is not a very difficult task in this crop. Unlike other fruit crops, papaya is commercially propagated through seeds, so there is continuous need to maintain the hybrids and inbreds developed through selfing (gynodioecious varieties) and sib-mating (dioecious varieties). Papaya breeding objectives are more or less than same around the globe, which are as follows: to develop stable high yielding gynodioecious lines, lines with better fruit quality, better papain production, and resistance to biotic and abiotic stresses. Knowledge about the genetic variability existing within the crop species is essential to develop efficient crop improvement strategy, as genetic diversity enables the recombination between the genes (Moore, 15). Phenotypic variation in papaya germplasm ranges from moderate to high for the various morphological traits like tree habit, stem colour/pigmentation, petiole colour/pigmentation, leaf shape, leaf size, type and size of inflorescences and flowers, fruit colour, shape and size, and reactions to pests and diseases (IBPGR, 8). Molecular techniques have been utilized in papaya to assess the diversity at the genes level. Through these studies, it was clear that *C. papaya* diverged from *Vasconcellea* spp. early in its evolution (Ming *et al.*, 13). Initially, when the genus *Carica* and *Vasconcellea* were not separated and both belonged to the genus *Carica*, it was confused with a lot variability. However, restoration of *Vasconcellea* also known as highland or mountain papaya, with 21 species (Badillo, 2) and *Carica* with 1 specie is left with less variability. Moreover, it was identified that the closest relatives of papaya belong to the genera *Horovitzia* and *Jarilla* (d'Eeckenbrugge *et al.*, 5), whereas, *Vasconcellea* is closely related to *Jacaratia* (Carvalho and Renner, 4). In the caricaceae family, *Vasconcellea* is the one genus which shows potential and scope for improvement of the common papaya than the other closely related genus. Genetic variability in monotypic genus *Carica papaya* could be enhanced by introgressing desirable traits from those wild relatives. Transferring desirable genes from the species of related genera to papaya is difficult due to production of non-viable seeds. Viable intergeneric hybrids are obtained by sucrose treatment and embryo rescue technique in papaya.

Current Breeding Strategies Adopted in Papaya

Papaya breeding objectives revolve around the problems faced in papaya cultivation, which includes: a) exploration, conservation and characterization of genetic variability existing in papaya and related genera of caricaceae family, b) differentiation between the sex types occurring in papaya, c) development of genotypes with improved agronomic traits, d) breeding for cultivars resistant to biotic and abiotic stresses and e) delayed ripening. Conventional breeding strategies have resulted in many improved varieties in terms of yield and quality. But the time required in such improvement programme is around 12-14 years and also some outcomes like PRSV resistant variety could not be achieved from such programmes.

Sex determination in papaya plants is an important issue in papaya cultivation as the sex affects flower type, production, fruit shape and yield. Pyriform fruits occurring in the hermaphrodite plants are more preferred by the consumers than the round fruits obtained from the female plants. There are numerous factors like environment, hormones, and genetic and epigenetic background, that affects the occurrence of specific type of flowers in papaya plants. A number of biochemical and molecular markers have been identified related to the different sex types in papaya.

Two highland papaya species *i.e.*, *V.* × *heilbornii* (babaco) and *V. cundinamarcensis* (papayuela) are used as fresh fruit and its processed products is also famous in their area of production (d'Eeckenbrugge *et al.*, 5). Other than fruits, papain is the major industrial product obtained from immature fruits of common papaya. As reviewed by d'Eeckenbrugge *et al.* (5), some of the highland papaya species like *V. cundinamarcensis*, *V. × heilbornii* and *V. stipulate* has been identified with high papain activity when compared with common papaya. In *Vasconcellea* several species have been identified with resistance against several biotic and abiotic stresses.

Resistance breeding in papaya is another issue to be solved by understanding its genetics and genomics. Papayas susceptibility to numerous biotic and abiotic stress may be partially attributable to its narrow genetic base. A remarkable success in disease resistance has been achieved by development of transgenic cultivars "SunUp" and "Rainbow" with PRSV resistance through genetic engineering of papaya (Fitch *et al.*, 7). To develop cultivars resistant to pathogens, sources of resistance can be identified in related genera of Caricaceae. Species of *Vasconcellea* genera has been identified as a source of PRSV-P resistance (*V. cundinamarcensis, V. quercifolia, V. stipulata* and *V. cauliflora*), phytophthora resistance (*V. goudotiana*), *Fusarium oxysporum* resistance (*V. weberbaueri* and *V. monoica*), black spot resistance (*V. monoica, V. goudotiana,* and *V. cauliflora*) and *Mycosphaerella caricae* resistance (*V. goudotiana*). Under sub-tropical climatic condition, papaya is susceptible to low temperature and frost condition. Highland papaya like *V. cundinamarcensis* possesses low temperature tolerance and could be used as a source for imparting cold tolerance in papaya.

The papaya fruit is a fleshy berry and fruit growth follows a single sigmoid growth curve. Ripening is climacteric in papaya and the fruit has a very short shelf life. A number of ripening and fruit softening genes have been identified in papaya. This includes genes involved in cell wall synthesis and degradation, skin chlorophyll degradation, carotenoid biosysthesis, sugar accumulation, expansins and ethylene biosynthesis. All these makes papaya a valuable tool for the study of the complex regulatory networks in fruit ripening and may provide insights as to how ripening is controlled.

Future Opportunities

One of the main factor affecting the production and productivity of the papaya cultivars is the unstable gynodioecious forms due to the variation in gynoecium morphology and function. Although, physical mapping and sequencing of papayas sex chromosome has been completed (Ming *et al.*, 14), the identification of genes controlling the androecium and gynoecium morphogenesis could be done for better understanding of the sex determination in papaya and breeding of true hermaphrodite plants. The reason behind occurrence of variable gynoecium morphology in trioecious papaya due to genetic factors and environmental factors was studied (Zhou *et al.*, 18). Zhou *et al.* (18) were able to identify that proper auxin concentration and auxin homeostasis is vital for functional gynoecium morphogenesis in papaya flowers. Other factors affecting gynoecium morphology, whether at gene level or at the hormonal level can be identified in the future, so that the variations in gynoecium morphology can be managed. Papaya and its wild relative have same somatic chromosome number but differ on the molecular level. Genome sequencing of papaya was done in 2008 by a group of researchers from Hawaii in "Sunup" variety. It was the first reported genome sequencing performed in a fruit tree. Resequencing projects of wild relatives of papaya will greatly expedite our understanding about the genomes and would aid in prompt improvement programmes. For instance, *Vasconcellea cauliflora*, is resistant to PRSV, and *V. cundinamarcensis* could provide genes for cold tolerance and PRSV resistance (Manshardt and Wenslaff, 11). In this regard, Lin *et al.* (9) sequenced and assembled the chloroplast genome of *Vasconcellea pubescens* and also performed the comparative analysis of chloroplast genome of *V. pubescens* with common papaya. Another study conducted by Tineo *et al.* (16) analysed the complete plastidial genome of *Vasconcellea carvalhoae* from Peru and the phylogenetic analyses of *V. carvalhoae* support its sistership to *V. pubescens*.

REFERENCES

1. Arumuganathan, K. and Earle, E.D. 1991. Nuclear DNA content of some important plant species. *Plant Mol. Biol. Rep.* **9**: 208–218
2. Badillo, V. M. 2000. Carica L. vs. Vasconcella St. Hil. (Caricaceae) con la rehabilitación de esteúltimo. *Ernstia.* **10**: 74–79.
3. Bremerk, K., Chase, M.W. and Stevens, P.F. 1998. An ordinal classifi cation for the families of fl owering plants. *Ann. Mo. Bot. Gard.* **85**: 531–553.
4. Carvalho, F. A., and Renner, S. S. 2012. A dated phylogeny of the papaya family (Caricaceae) reveals the crop's closest relatives and the family's biogeographic history. *Mol. Phylogenet. Evol.* **65**: 46–53.
5. d'Eeckenbrugge, G. C., Drew, R., Kyndt, T. and Scheldeman, X. 2013. Vasconcellea for papaya improvement. *Genetics and genomics of papaya.* 47-79.
6. *FAOSTAT. Retrieved 30 September 2023.* Countries – Select All; Regions – World + (Total); Elements – Production Quantity; Items – Papayas; Year – 2020.
7. Fitch, M.M.M., Manshardt, R.M., Gonsalves, D., Slightom, J.L., Sanford, J.C. 1992. Virus Resistant Papaya Plants Derived from Tissues Bombarded with the Coat Protein Gene of Papaya Ringspot Virus. *Bio Technol.* **10**: 1466–1472.
8. IBPGR (International Board for Plant Genetic Resources). 1988. Descriptors for Papaya. Rome. ISBN 92-9043-129-6, pp. 31-33.
9. Lin, Z., Zhou, P., Ma, X., Deng, Y., Liao, Z., Li, R. and Ming, R. 2020. Comparative analysis of chloroplast genomes in Vasconcellea pubescens A. DC. and Carica papaya L. *Scientific Reports.* **10(1)**: 15799.
10. Liu, Z., Moore, P.H., Ma, H., Ackerman, C.M., Makandar, R., Yu, Q., Pearl, H.M., Kim, M.S., Charlton, J.W., Stiles, J.I., Zee, F.T., Paterson, A.H. and Ming, R. 2004. A primitive Y chromosome in papaya marks incipient sex chromosome evolution. *Nature.* **427**: 348–352.
11. Manshardt, R.M. and Wenslaff, T.F. 1989. Interspecific hybridization of papaya with other *Carica* species. *J. Am. Soc. Hortic. Sci.* **114**: 689–694.
12. Ming, R., Hou, S.B., Feng, Y., Yu, Q.Y., Dionne-Laporte. A., *et al.,* 2008. The draft genome of the transgenic tropical fruit tree papaya (*Carica papaya* Linnaeus). *Nature.* **452**: 991–996
13. Ming, R., Van Droogenbroeck, B., Moore, P.H., Zee, F.T., Kynd, T., Scheldeman, X., Sekioka, T., Gheysen, G. 2005. Molecular diversity of Carica papaya and related species. In: Sharma AK, Sharma A (eds) Plant genome: biodiversity and evolution, vol 1B: Phanerograms. Science Publishers, Enfi eld. pp 229–254.

14. Ming, R., Yu, Q., Moore, P.H. 2007. Sex determination in papaya. *Semin Cell DevBiol.* **18(3)**: 401–408.

15. Moore, P. H. 2014. "Phenotypic and genetic diversity of papaya," in Genetics and Genomics of Papaya (New York, NY: Springer). 35–45.

16. Tineo, D., Bustamante, D.E. and Calderon, M.S. 2022. Analysis of the complete plastidial genome of the newly highland papaya Vasconcellea carvalhoae (Caricaceae) from Peru. *Mitochondrial DNA Part B.* **7(10)**: 1882-1886.

17. Tripathi, S., Castro, L., Fermin, G. and Tennant, P. 2014. "Advances in papaya genomics," in Omics Technologies and Crop Improvement, ed N. Benkeblia (Boca Raton, FL: CRC Press), 219–252.

18. Zhou, P., Fatima, M., Ma, X., Liu, J. and Ming, R. 2019. Auxin regulation involved in gynoecium morphogenesis of papaya flowers. *Horti. Res.* **6**.

Chapter 11

Present Status and Future Opportunities for Exploring New Grape Rootstock in India

R.G. Somkuwar[1]*, Rutuja D. Nale[2] and Sharmistha Naik[1]

[1]*ICAR–National Research Centre for Grapes, Pune 412307, Maharashtra*

[2]*Mahatma Phule Krishi Vidyapeeth, Rahuri 413722, Maharashtra*

**e-mail: rgsgrapes@gmail.com*

ABSTRACT

Grape cultivation in India is unique as it is grown in almost all conditions from tropical to temperate agro-climatic zones. Earlier due to availability of healthy soil and sufficient irrigation water of good quality, grapevine growing on their own roots performed better. However, with the emergence of problems like soil salinity, poor quality water and the shortage of irrigation water, the yield and quality of grapes have started declining. Keeping in view the above problems, grape cultivation using rootstock was considered important. A significant portion of global viticulture is based primarily on grafting, where the rootstock is either a North American Vitis species or an interspecific Vitis hybrid and the scion is a cultivar of Vitis vinifera. Grape rootstocks have exhibited the ability to withstand biotic stress like phylloxera and abiotic stresses including drought, salinity and flooding. Tolerance of grapevines to salinity and moisture stress can be significantly improved by selecting rootstocks that produces deep tap root system and could restrict Cl and/or Na uptake or restricts their accumulation in scion leaves. In this changing environment, grapevine roots and rootstocks must adapt to water scarcity and salinity, affecting vineyard sustainability. Biotic factors like soil borne fungi are new challenges for vineyard sustainability. Nematode diversity and the emergence of virulent strains can impact rootstock performance and soil-born fungi can cause root damage, leading to various issues. These challenges pose significant concerns for viticulture today. Thus, choosing a rootstock is one of the important decisions, a grape grower should make while establishing new vineyard as it is a long-term investment. At present 90 percent rootstock used in vineyards is Dogridge, it is essential to have an alternative rootstock in view of changing climatic scenario. Rootstocks like 110-R, 99-R, and 1103-P are highly drought and salt-tolerant, with 110-R gaining popularity for its improved fruitfulness and moderate vigor, especially under moisture stress.

***Keywords**: Rootstock, Abiotic and biotic stress, New rootstocks.*

Grape cultivation in India is unique as it is grown in almost all conditions from tropical to temperate agro-climate. In India, around 72-74 per cent of grape is produced for fresh consumption and about 25 per cent for raisin making, while around 2 per cent collectively for juice and wine production. If we look 25 years back into the grape production scenario of the country, the area under the grape cultivation was about 45000 ha with the production of 1100 thousand tones. Presently, grapes are grown in over an area of 0.162 million ha with production of 3.445 million tons and productivity of 21.00 t/ha. The major grape growing states are Maharashtra (70.67 per cent), Karnataka (24.49 per cent), Tamil Nadu (1.43 per cent), Andhra Pradesh (1.34 per cent), Madhya Pradesh (1.02 per cent) and Mizoram (0.50 per cent) amounting to nearly 99 per cent of the total production. India ranks first in world for grape productivity and 7th for table grape export. During 2022-2023, India exported 0.267 million t of grapes worth Rs. 25434.2 million (APEDA, 1).

Earlier, because of availability of healthy soil and sufficient irrigation water of good quality, grapevine growing on their own roots were performing better. However, with the emergence of problems like soil salinity, bad quality water and the shortage of irrigation water, the yield and quality of grapes have started declining. Keeping in view the above problems, grape cultivation using rootstock was considered important. The initial reason for the wide adaption of rootstocks in many locations was, due to increasing occurrence of soil salinity, drought, and declining productivity of grape varieties (Somkuwar *et al.*, 32). Indian grape industry was historically based on the exclusive use of own rooted *Vitis vinifera,* which has no resistance to soil salinity and drought conditions. Therefore, rootstock resistant to soil salinity and drought were important considerations in Indian viticulture. The first rootstock was simply a selection from wild American species, some of which are still being used today. Initially, varieties of *V. vinifera* were utilized as rootstocks as they are moderately tolerant to soil salinity (contains higher concentration of total salt). Drought tolerance capacity varies among *Vitis* species. The two species *Vitis cinerea* and *Vitis champini* are considered to be tolerant to saline conditions. (Reynolds and Wardle, 23) listed seven major criteria for selecting rootstocks as nematode resistance, phylloxera resistance, adaptability to high soil pH, saline soils, low soil pH, adaptability to wet or poorly drained soils, and adaptability to drought. Rootstocks have not only potential for combating the soil problem but can also be a potential tool for manipulating the vine growth, yield, fruit quality and wine quality (Somkuwar *et al.*, 30).

A significant portion of global viticulture is based primarily on grafting, where the rootstock is either a North American *Vitis* species or an interspecific *Vitis* hybrid and the scion is a cultivar of *Vitis vinifera* (Weaver, 39). Grape rootstocks have exhibited the ability to withstand biotic stress like phylloxera and abiotic stresses including drought, salinity and flooding (Satisha *et al.*, 28). *V. riparia, V. berlandieri,* and *V. rupestris* species are resistant to phylloxera. Some rootstock cultivars such as Riparia Gloire, 1104-14 Mgt, SO4, K5BB (Kober 5BB), and St. George are resistant to phylloxera. Nematode resistant *Vitis* species include *V. champini, V. cinerea, and V. longii* (Mullins *et al.*, 18). *V. rupestris, V. berlandieri, V. aestivalis, V. champini* species and rootstock cultivars 'Barnes', 'Dog Ridge', 'B-45' were reported to be most resistant to pierce's diseases. *V. champini* and *V. vinifera* are tolerant to salinity (Leon *et al.*, 15). Tambe (35) reported that Dogridge, 1613C and Salt Creek were relatively salt tolerant under Indian condition. Rootstock such as St. George, 110R, 140Ru, 1103 P and 99R possess strong drought tolerant root structure compared with SO_4 and 5C.

Main Components of Root Architecture and Morphology

Different rootstocks have different habits of spread in the spoil. The root spread also depend upon the soil type and other cultural practices followed. Root penetration in soil is mostly horizontal, however, variations in the vertical and horizontal distribution of rooting depth may have a genotypic change. The root system development may significantly be influenced by the characteristics of the soil and vineyard management practices (Smart *et al.*, 29). The physical properties of the soil (structure and texture, soil strength, bulk density, etc.) and its chemical composition have strong effects on grapevine root development. Under good soil conditions, grapevines can have very deep roots, a wide depth distribution, and frequently can reach even up to 6 meters deep (Smart *et al.*, 29). The roots of 90-year-old vines of Bulgarian cultivar, had the deepest roots ever recorded in the literature, with depth ranging from 20 meters in *V. vinifera* own-rooted to 32 meters in *V. vinifera* grafted onto 41B rootstock (Pourtchev, 22). Carneiro and Menezes-Sequeira (2) found that rootstocks Castel 196-17 and 99 R had deeper roots compared to SO4, 140 Ru and 110 3P. (Perry and Lyda, 21) found that, four grape species *V. champinii* cv. Dogridge, *M. rotundifolia* cv. Noble, *V. vinifera* cv. Barbera, and *V. labrusca* cv. Concord have different root architectures (number, biomass, and depth of the root system). The longest root system was found in *V. champinii* cv. Dogridge, whereas *M. rotundifolia* cv. Noble showed the smallest root system.

While edaphic conditions tend to have a greater impact on the root system architecture, rootstock appears to have a greater influence on root density, as measured by root biomass or total number of roots (Somkuwar *et al.*, 34). It was found that different rootstocks possess varying total root and fine root counts and rootstocks have a significant impact on root distribution, number of fine roots, and root density under favorable soil condition. These characteristics may play a significant role in drought adaptation (Smart *et al.*, 29). The genetic influence of rootstocks on root density is probably of most significance, since fine roots are crucial for water intake and consequently for drought tolerance and recovery (Somkuwar *et al.*, 30).

Interactions Between Rootstocks and Scions

The way rootstocks interact with scions is still not fully understood. The fundamental change after grafting is the direct replacement of the root system. Different rootstock species may have varied root architecture, morphology, development, and distribution. The new roots will have a direct impact on water and mineral uptake, which will ultimately have an impact on the growth of the shoot system and modify vine physiology.

Root Systems

Grapevine roots perform a different function such as anchoring, storing reserves, absorbing, and transporting water and minerals, and supplying growth-promoting substances. Although individual roots may reach a depth of 600 cm or even more, most grapevine roots are found in the top 100 cm of soil (Somkuwar *et al.*, 30). In upper region, up to 10-60 cm fine roots are present which comprise most absorptive roots. Typically, the primary roots (6-100mm in diameter) are located 30-35 cm below the surface. The primary roots produce permanent roots (2-6 mm in diameter), which often develop more horizontally and are found in the upper 20–50 cm of soil (Mullins *et al.*, 18). Rootstock species may differ in their architecture and morphology, development, and dispersion of root systems, which are substantially determined by genetic characteristics. For instance, while the primary roots of *V. riparia* make wide angles and remain mostly superficial, those of *V. rupestris* form narrow angles and penetrate deeper. One of the distinctive features of grapevine is beginning of root growth is after budbreak. By absorbing water and nutrients, as well as the supply of hormones produced by the roots (such as cytokinins and gibberellins), this may have an impact on the growth of the shoots.

There has not been a lot of research on root distribution of grape rootstocks. Root studies are challenging because they cannot be easily seen in the soil or separated from it. Root dispersion was examined by Harmon and Snyder (11) on grafted vines that were on an average about 25 years old and growing in sandy loam soil with hardpan depths varying from 32 to 88 inches. The changes appeared to be loosely linked with the rootstock and to a lesser extent, with the scion. Williams and Smith (40) examined the root distribution of Cabernet Sauvignon grafted on the rootstocks Aramon Rupestris Ganzin No. 1 (A x R 1), St. George, and 5C, and observed significant variations in the number of root count among the rootstocks, with A x R 1 having the least number of roots and St. George having the highest number of roots per soil wall profile. The distribution of roots in the soil, irrespective of their size, was same for all rootstocks. The root distributions has considerable variation in rootstocks root size and total root numbers. In their investigation, St. George had a distinct root distribution pattern than 110R and A X R 1. Somkuwar *et al.* (30) reported significant variation in percentage of the dry matter among different part of Thompson Seedless grafted on different rootstock. Maximum percentage of dry matter was recorded in the canes followed by the primary arms, with Salt Creek showing the least and St. George most whereas, dry matter distribution was greatest in Dog Ridge and least in Salt Creek.

Challenges Related to Abiotic Stresses

The most significant abiotic constraints to viticulture have been recognized as water stress, increases soil salinity and temperature variation, which have a detrimental effect on vineyards. Climate change predictions indicate that these issues will become worse soon. Additionally, the demand on water supplies brought on by expanding industry, agriculture, and population has prompted to look for alternate irrigation water sources, which typically have high salt concentrations. Screening of rootstocks that adopt under salinity condition also need attention. Grapevine roots and rootstocks will therefore need to adapt to a future environment where water scarcity and salinity will limit their growth and function, limiting the effectiveness of the entire plant and, as a result, the sustainability of the vineyard.

Water Stress

Water stress is becoming a limiting factor for improving vineyard sustainability, nearly about two-thirds of the world's major viticulture regions have annual precipitation below 700 mm (Flexas *et al.,* 9) and it is predicted that soon, water availability will be a critical factor in areas where water is not a limiting factor (Santos *et al.,* 27). Modulating water availability with controlled deficit irrigation (RDI) enables a more effective use of water. However, rootstock to improve water use efficiency (WUE) plays a major role, as efficient, extensive, and deep root system provides access to a potential greater water supply to take profit of the rain water or irrigation, thus enhancing plant WUE (Medrano *et al.,* 17). Rootstocks can also lessen water stress by reducing scion vigour, due to leaf area thus leading to low transpiration rate (Clingeleffer *et al.,* 2). These characteristics, along with soil improvement to increase soil water storage capacity, are the two main ways to secure the environmental sustainability of vineyards. In recent decades, research on the physiology of grapevine roots under water shortage conditions has covered the effects of water stress from both the molecular and physiological point of view (Yildırım *et al.,* 41). It is difficult to identify quantitative trait loci (QTL) due to the intricacy of the characters associated with WUE and drought response. Rootstocks that require less water and are resistant to drought has been suggested as a measure of adaptation to climate change in viticulture (Sabir and Sahin, 26). Due to adverse soil and water qualities, the grapevines are grafted on different rootstocks, such as 'Dogridge', '110R', '140Ru', 'Salt Creek' etc. However, in India, 'Dogridge' and '110R' are popular among the farmers because of their ability to tolerate prevailing drought and salt conditions (Somkuwar *et al.,* 31). These selected rootstocks can change vineyard adaptation to drought situation through number of physiological mechanisms.

Ability to uptake and Transport Soil Water

Structure and growth of the grapevine root system are influenced by both environmental and genetic variables. Physical and chemical characteristics of the soil and different rootstock-scion combinations, together with rootstock genotypes, have an impact on the development of root system (Kocsis *et al.*, 14). Changes in root system density and distribution in response to water restriction may be influenced by rootstock genotypes and appears to be related to transcriptome controls that may encourage the transport of sugar and proteins, osmotic adjustment, or formation of suberin and wax (Yıldırım *et al.*, 41). Root hydraulic architecture affects the volume of soil primarily due to increased branching of the fibrous roots and an increased number of root tips accessing more water resources. The genotype primarily influences root density while root spread is mostly determined by soil characteristics (Keller, 13). For example, 1103 P (*V. rupestris* and *V. berlandieri*), the drought-tolerant rootstock produces a considerable amount of root biomass during the summer in the first few months after planting. In comparison to 110 3P, the less vigorous rootstock 101-14 (*V. rupestris* × *V. riparia*) forms roots more gradually and exhibits substantially less growth adaptability to soil moisture level or depth. The intrinsic capacity of roots to uptake water is also related to the expression of aquaporins, which is either genetically or environmentally determined, and whose contribution to the root hydraulic conductance ranges from 4 to 40 per cent (Lovisolo *et al.*, 16). This trait could be explored as a breeding target to improve rootstock response to water deficit.

Scion Water Use Regulation

Under conditions of adequate water supply and water stress, rootstocks play an essential role in regulating the rate of scion transpiration. The rootstock effect on vigour of scion cultivar accounts for a significant source of diversity in water consumption regulation, as low-moderate vigour conferring rootstocks exhibit lower transpiration rate (Clingeleffer *et al.*, 2). In contrast, the rootstock can modify the scion's stomatal sensitivity to water stress by shifting the level of stomatal closure towards lower (drought-resistant) or upper (drought-sensitive) leaf water potential (Ψ leaf), independently of the scion's near-isohydric or near-anisohydric behavior (Tramontini *et al.*, 36). Under drought condition, rootstocks have ability to control scion transpiration which involves a combination of hydraulic and hormonal root to shoot signaling.

Role of Rootstock on Carbon Balance

Under water stress, rootstocks have a significant impact on the scion's stomatal conductance and leaf photosynthesis. Corso *et al.* (6) described the activation of "primary and secondary mechanisms" of the detoxification of reactive oxygen species and the potential implications for drought tolerance by promoting higher water uptake capacity, active plant growth, and carbon assimilation. Significant amount of the carbohydrates fixed by leaf photosynthesis are used up by the respiratory activity of roots (Escalona *et al.*, 7). However, research on the effects of the rootstocks-scion interaction on the respiration costs of the roots are still lacking. Understanding the impact of water availability on vine root growth, activity, and distribution as well as the connection between root biomass, root respiration, and carbon allocation has been aided by several researchers (Romero *et al.*, 24). Numerous studies have illustrated the effect of rootstocks on scion vigour, particularly through changes in the xylem hydraulic function. High-vigor rootstocks retain more water uptake capacities and increase the resources in organs throughout the growing season. However, low vigour rootstocks showed a higher sink/source ratio, which might help vineyards be more resilient to climate change and make this adaptation compatible with grape quality in semi-arid conditions. Different rootstocks have different patterns of root development, longevity, and death, and these variations might have an impact on the overall carbon balance of the plant. Total plant carbon (C) storage contributed by the root system is approximately 9 to 26 per cent, and total C storage in the vineyard is around 5.7 to 7.2 t C/(ha/year) as per Brunori *et al.* (7). As a result, vineyard soils and root systems can favorably influence climate change mitigation measures by improving soil carbon storage (Funes *et al.*, 10).

Salinity

Salinity is becoming an increasing problem in viticulture due to the increase use of poor quality. It is estimated that about 6 per cent of the world's land and 20 per cent of irrigated land are affected by salinity and sodicity (FAO, 8) due to global change, precipitation pattern changes and increasing salinization in dry regions. Grapevines are considered as moderately sensitive to salinity and the damage is mainly caused due to chloride ions. The rootstock-scion combination, irrigation method, soil type, and climate all factors have direct impact on grapevine response to salinity. It is well known that there are differences in salt exclusion between grapevine species and cultivars where rootstocks are considered as one important means for improving grapevine salt tolerance. The salinity tolerance induced by rootstocks is attributed to the root system restricting the movement and/or avoiding absorption and accumulation of toxic ions from the saline soils (Somkuwar *et al.*, 33).

Earlier studies on the salt tolerance of grapevines indicated that high uptake and root-to-shoot movement of chloride lead to excessive accumulation in leaf tissues, affecting leaf function and damaging the leaves. Salt tolerance may be increased by roots excluding chloride from leaves rather than by chloride accumulation in leaves. Grapevine when grafted on salt tolerant rootstock (140 Ru), a much lower concentration was observed in xylem sap and shoot tissue than on a salt sensitive rootstock (K51-40). It is proved that chloride transportation takes place through symplastic pathway. However, the molecular mechanisms of chloride tolerance remain unclear. When grapevine rootstocks differing in salt tolerances were exposed to water stress and salinity stress, the expression of the vacuolar Na+/H+ antiporter, VvNHX1, in the leaves was investigated. VvNHX1 was up-regulated 7 days after salinity treatment in 110R (tolerant) but down- regulated in 1613C (less tolerant). Twenty-one days after, VvNHX1 expression decreased slightly in 110R and increased in 1613C. These differences correlated with a higher sodium content in the leaves of 1613C, suggesting this rootstock has a lower ability to exclude sodium ions from the shoot (Upadhyay *et al.*, 37).

Tolerance to salt in the field may be defined as the ability to maintain shoot growth and yield under high-salinity conditions. Under saline conditions, rootstock vigour significantly influences determining yield performance and hence salt tolerance of grapevines.

Challenges Related to Biotic Stresses

Rootstocks for grapevines are essential for the long-term sustainability of vineyards since they help to control pest-related issues. In most grape-growing areas where phylloxera is present, they have been essential component. During an extended period, the use of rootstocks for phylloxera management has been proved effective and well-founded, even though various reports have indicated that some places of the world have high infestation rates on the leaves. Nematodes have historically been viewed as relevant harmful agents for grapevines, which are typically managed in some parts of the world by utilizing tolerant or resistant rootstocks. However, nematode species in vineyard soils are varied, and the emergence of novel species or the development of virulent populations could negatively affect the performance of rootstocks. In India, the nematode problem in grape vineyard is not observed. Additionally, soil-borne fungus may cause damage to root system of grapevine rootstocks, leading to a variety of symptoms such as root necrosis, root mass decrease, and root rot. Nematodes and soil-borne fungus are two of the most problematic biotic elements in viticulture right now, and they both have significant effects on grapevine rootstocks.

Challenges Related to Vigor, Yield and Grape Composition Management

Global viticulture faces challenges in maintaining production, and rootstocks can play an important role in this area. Prior to selecting the rootstock for a new vineyard, it is important to consider the elements that restrict or significantly influence the performance of the vineyard. After eliminating the inappropriate rootstocks, it is required to select one from the potentially well-suited ones which is typically done with striving to meet yield and grape composition requirements. This section focuses on fruit quality attributes related to wine grapes, understanding that the requirements and effects of rootstock use on table grapes, raisins or grape juice may be different. Since the introduction of rootstocks towards the end of the 19th century, the effects of rootstocks on vine yield and wine grape composition have been extensively studied for all main cultivars under a wide variety of soil and climatic conditions.

In addition to the indirect processes linked to poorer yield and vigor that cause some rootstocks to produce berries with higher phenolic content, the rootstock could also exert some direct effects that are relatively unexplored. Rootstock stimulates plant vigour by improving physiological and metabolic activities of scions by increasing leaf exposure, water, and nitrogen availability at the time of its maturity (Somkuwar *et al.*, 31). Jogaiah *et al.* (12) reported variations in the concentration of bud phenolic compounds at budburst based on the rootstock in an experiment designed to investigate how rootstocks affected earliness in budburst. Somkuwar *et al.* (31) noted early bud sprout in vines grafted on 110 R due to higher enzyme activities. Changes and concentration of enzymatic activity can be used as biological indicators to identify the rootstock with early bud sprout. Wallis *et al.* (38) reported that rootstocks may have an impact on sap phenolic compounds. Depending on the rootstock, Nemeth *et al.* (19) reported variations in scion cane resveratrol. Particularly, rootstocks have been found to influence ABA regulation in vines when dealing with water stress, and this will influence the production of phenolic compounds, a process connected to ABA signaling in the berries (Rossdeutsch *et al.*, 25).

Main *Vitis* Species and Rootstock Families

Vitis Species

There are about 30 different wild *Vitis* species in North America. Only a few of these were utilized to breed rootstocks. The main characteristics of the most used species are presented in Table 1. There is also a significant intraspecific heterogeneity. As previously stated, *V. rupestris* and *V. riparia* were first found to be particularly suitable for use in grafting. They are phylloxera-resistant and have effective rooting and grafting abilities. However, none of them tolerate an excess of lime in the soil. Regarding the issue of vineyards grown on calcareous soils, *V. berlandieri* has been found to be the most phylloxera-resistant species and tolerant to limestone induced chlorosis. Early analysis of a large panel of accessions within this species variability and inter-varietal differences were not significant, especially when vines were grafted. This species was mostly used as hybrids with other species due to its high phylloxera resistance but difficult to rooting. It is the most recently ripening species in North America. *V. candicans* is a highly vigorous species which is characterized by interesting nematodes (*Meloidogynes*) tolerance, and some drought and salt tolerance. Through a few naturally occurring hybrids with *V. rupestris*, known as *V. champinii*, this species was mostly utilized in rootstock breeding. *Muscadinia rotundifolia* was more recently considered as a highly interesting species for rootstock improvement because of its high resistance to many pests and diseases (Bouquet, 2). However, the *Muscadinia* subgenus member *M. rotundifolia*, which has 2n=40 chromosomes, exhibits relatively poor roots and grafting abilities.

Table 1. Main characteristics of the most used *Vitis* and *Muscadinia* species used to breed rootstocks (adapted from Boursiquot, pers. commun.)

	Vitis rupestris	*Vitis riparia*	*Vitis berlandieri*	*Vitis vinifera*	*Vitis candicans*	*Muscadinia rotundifolia*
Earliness	Medium to late	Very early	Late	Variable	Late	Late
Conferred vigour	High to very high	Low to medium	High	Variable	High	Low
Resistance to limestone	Low excepted	Very low	Very high	Very high	Low	Very low
Resistance to drought	Medium	Low	High	Medium	High	Low
Resistance to salt	Medium	Almost none	Good chloride excluder	Fairly good	Fairly good	-
Rooting-grafting	Very good	Very good	Bad	Very good	Bad to medium	Bad
Resistance to phylloxera	High	High	Very high	None	Medium	Very high
Nematodes (Xiphinema and *Meloidogyne)	Variable	Variable	Sensitive	Variable	Good resistance	Good

Current Use of Existing Rootstocks

Information on mother vines and the propagation of grafted plants is more prevalent. Despite the large number of rootstocks are existing, only few of them are utilized throughout the world. The choice of proper rootstock is becoming increasingly difficult because of the availability of numerous new rootstocks (Somkuwar *et al.*, 33). In India, there are 16 registered rootstocks in the active germplasm site at ICAR-NRC Grape, Pune, promising ones are under evaluation for drought and salt tolerance. In important grape-growing regions of India, Dogridge rootstock is well-known for its resistance to salt and drought, and as a result, its compatibility with promising grape cultivars like Thompson Seedless and its mutants has been identified (Somkuwar *et al.*, 30) and it is therefore the most extensively used rootstock in India. As of late, Dogridge rootstock is facing a few problems, including uneven bud burst after pruning, decreased fruitfulness, and increased deadwood in the cordons, which have been related to its greater ability to induce vigor in scions when grown in dense black cotton soils. The systematic evaluation of rootstocks revealed rootstocks of V. berlandierii × V. rupestris group like 110-R, 99-R, and 1103-P etc. are most drought and salt tolerant. Among them, rootstock 110-R is becoming more popular due to its improved fruitfulness, moderate vigor, constrained uptake of chlorides, and increased water use effectiveness under moisture stress situations.

Grapevine Rootstock Breeding

Most cultivated grapevine plants are genetic chimeras with two distinct genotypes facing the aerial and soil environments, respectively, because of grafting. This chimerism provides the chance to independently breed root features without significantly changing the cultivar-specific fruit traits. In addition, choosing the right rootstocks can assist control particular scion phenotypes because of the interactions that exist between the two genotypes (Ollat *et al.*, 20,). In terms of the germplasm that can be used, rootstock improvement is much less restricted than grapevine scion selection, which is conditioned by the convenience of preserving traditional cultivars that winegrowers and consumers can associate to their characteristic grape and wine products. By using a variety of distant species of the Vitaceae family that are cross-fertile and graft-compatible with *V. vinifera* scions, traditional rootstock breeding can introduce genetic determinants of advantageous root traits through hybridization. Different grapevine wild relatives are naturally dispersed across a high diversity of soils and habitats, coexisting with many diseases. They provide a diverse source of genetic factors for adaptive resistance and tolerance features (Ollat *et al.*, 20). This genetic diversity has considerable potential for further exploitation. Despite the variety of soil environments that make up all of the vineyards in the world, less than ten rootstock genotypes, some of which were developed in the 19th century, are used to graft about 90 per cent of the current vineyards (Keller, 13). In fact, only four species *V. berlandieri*, *V. rupestris*, *V. riparia*, and *V. vinifera* represent nearly 90 per cent of the parentage of the most significant rootstocks, with only a few accessions in each case.

In addition to complex root performance phenotyping, characterizing interactions with scion genotypes during grafting is a crucial element of rootstock breeding. These interactions extend from grafting compatibility, scion vigor and hydraulic conductivity to fruit quality and production characteristics. The performance of rootstocks under various soils is another degree of complexity that must be considered. This depends not only on abiotic parameters like composition and structure but also on biotic elements like the soil microbiome that have not yet been fully characterized. The failure of traditional rootstock breeding is ultimately hampered by time-consuming phenotyping procedures, as is evident from the fact that just 20 new rootstocks have been successfully developed globally and added to the nursery supply in the 21st century. These time lapses can be reduced by using of high-throughput phenotyping approaches assisting root phenotyping. However, due to the accessibility of the roots and limitations on the amount of soil space, these procedures are not always simple to apply on grapevine rootstocks. For features when there are weak genetic-environmental interactions, phenotypic selection may not be the best option. Instead, genetic selection based on the identification of molecular polymorphisms linked to the desired phenotypes may be an option.

Way Forward

Viticulture faces major challenges from abiotic factors like water stress, soil salinity, and temperature variations and biotic factors like nematodes and soil born fungus. Salinity is a growing issue in viticulture due to poor water quality. Grapevines are moderately sensitive to salinity, primarily due to chloride ions. Rootstock-scion combinations, irrigation methods, soil types, and climate affect grapevine response to salinity. Tolerance of grapevines to salinity and moisture stress can be significantly improved by selecting rootstocks that produces deep tap root system and could restrict Cl and/or Na uptake or restricts their accumulation in scion leaves. The rootstock which produces good root architecture under limited soil moisture availability is one of the factors to be considered for its drought tolerant nature.

There is a need to focus on finding salt-tolerant rootstocks. In this changing environment, grapevine roots and rootstocks must adapt to water scarcity and salinity, affecting vineyard sustainability. Biotic factors like soil borne fungi are new challenges for vineyard sustainability. Nematode diversity and the emergence of virulent strains can impact rootstock performance and soil-born fungi can cause root damage, leading to various issues. These challenges pose significant concerns for viticulture today.

Rootstock improvement is done at very minimal rate. Apart from selection, little research has been done for improvement of rootstock through conventional and non-conventional breeding methods. Traditional rootstock breeding is slow due to complex phenotyping. High-throughput phenotyping and genetic selection based on molecular markers can expedite the process. Techniques like embryo rescue or protoplast fusion can further contribute towards rootstock improvement programs. Thus, choosing a rootstock is one of the important decisions, a grape grower should make while establishing new vineyard as it is a long-term investment. Rootstocks like 110-R, 99-R, and 1103-P are highly drought and salt-tolerant, with 110-R gaining popularity for its improved fruitfulness and moderate vigor, especially under moisture stress.

REFERENCES

1. APEDA, New Delhi. 2023. APEDA agri exchange database online (http://agriexchange.apeda.gov.in).
2. Bouquet, A. 1983. Contribution a l'Etude de l'Espece Muscadinia rotundifolia (Michx) Small et de Ses Hybrides avec Vitis vinifera L. Applications en Selection (Bordeaux, France: Universite de Bordeaux II).
3. Brunori, E., Farina, R. and Biasi, R. 2016. Sustainable viticulture: the carbon-sink function of the vineyard agro-ecosystem. *Agriculture, Ecosystems & Environment* **223**: 10–21.
4. Carneiro, L. C., and Menezes-Sequeira, L. 1989. Distribuição radical de porta-enxertos: contributo para o seu estudo. *Ciência e Técnica Vitivinícola*, **8**: 87-95.
5. Clingeleffer, P.R., Smith, B., Edwards, E., Collins, M., Morales, N. and Walker, R.R. 2011. Rootstocks, a tool to manipulate vine growth characteristics, fruit composition and wine quality attributes, water use efficiency and drought tolerance. Novello, V., Bovio, M. and Cavalletto, S., eds. *Proceedings of the 17th GiESCO*, **29:** 454
6. Corso, M., Vannozzi, A., Maza, E., Vitulo, N., Meggio, F., Pitacco, A., Telatin, A., D'Angelo, M., Feltrin, E., Simone Negri, A., Prinsi, B., Valle, G., Ramina, A., Bouzayen, M., Bonghi, C. and Lucchin, M. 2015. Comprehensive transcript profiling of two grapevine rootstock genotypes contrasting in drought susceptibility links the phenylpropanoid pathway to enhanced tolerance. *Journal of Experimental Botany* **66**: 5739–5752.
7. Escalona, J.M., Tomàs, M., Martorell, S., Medrano, H., Ribas Carbo, M. and Flexas, J. 2012. Carbon balance in grapevines under different soil water supply: importance of whole plant respiration. *Australian Journal of Grape and Wine Research* **18**: 308–318.

8. FAO, 2022. The World Map of Salt Affected Soil [WWW Document]. FOOD Agric. Organ, UNITED NATIONS. URL. https://www.fao.org/soils-portal/data-hub/soil-map s-and-databases/global-map-of-salt-affected-soils/en/

9. Flexas, J., Galmés, J., Gallé, A., Gulías, J., Pou, A., Ribas-Carbo, M., Tomàs, M. and Medrano, H. 2010. Improving water use efficiency in grapevines: potential physiological targets for biotechnological improvement. *Australian Journal of Grape and Wine Research* **16**: 106–121.

10. Funes, I., Savé, R., Rovira, P., Molowny-Horas, R., Alcañiz, J.M., Ascaso, E., Herms, I., Herrero, C., Boixadera, J. and Vayreda, J. 2019. Agricultural soil organic carbon stocks in the north-eastern Iberian Peninsula: drivers and spatial variability. *Science of the Total Environment* **668**: 283–294.

11. Harmon, F.N. and E. Snyder. 1934. Grape rootstock distribution study. Proc. Amer. Soc. Hort. Sci. **32**: 370-373.

12. Jogaiah, S., Maske, S.R. and Upadhyay, A. 2014. Rootstock induced changes in enzymes activity and biochemical constituents during budbreak in 'Thompson Seedless' grapevine. *Vitis* **53**: 57–64.

13. Keller, M. 2010. Managing grapevines to optimize fruit development in a challenging environment: a climate change primer for viticulturists. *Australian Journal of Grape and Wine Research* **16**: 56–69.

14. Kocsis, L., Tarczal, E. and Molnár Kocsisné, G. 2016. Grape rootstock-scion interaction on root system development. *Acta Horticulturae* **1136**: 27–32.

15. Leon, B., C.F. Ehlig, and R.A. Clark. 1969. Effects of grape rootstocks on chloride accumulation in leaves. *J. Amer. Soc. Hort. Sci.* **94**: 584-590.

16. Lovisolo, C., Lavoie-Lamoureux, A., Tramontini, S. and Ferrandino, A. 2016. Grapevine adaptations to water stress: new perspectives about soil/plant interactions. *Theoretical and Experimental Plant Physiology* **28**: 53–66.

17. Medrano, H., Tomás, M., Martorell, S., Escalona, J.-M., Pou, A., Fuentes, S., Flexas, J. and Bota, J. (2015) Improving water use efficiency of vineyards in semi-arid regions. A review. *Agronomy for Sustainable Development* **35**: 499–517.

18. Mullins, G.M., A. Bouquet, and L.E. Williams. 1992. Biology of the grapevines, Cambridge University Press, NY.

19. Nemeth, G., Molnár, Z., Podmaniczky, P., Sárdy, D.N., Kállay, M., Dunai, A. and Kocsis, L. 2017. Trans-resveratrol content in grape cane and root of different scion-rootstock combinations. Mitteilungen Klosterneuburg **67**: 256–264.

20. Ollat, N., Peccoux, A., Papura, D., Esmenjaud, D., Marguerit, E., Tandonnet, J.-P., Bordenave, L., Cookson, S.J., Barrieu, F., Rossdeutsch, L., Lecourt, J., Lauvergeat, V., Vivin, P., Bert, P.-F.and Delrot, S. 2016. Rootstocks as a component of adaptation to environment. Gerós, H., Chaves, M.M., Gil, H.M. and Delrot, S., eds. Grapevine in a changing environment (John Wiley: Chichester, England) pp. **68**: 108.

21. Perry, R.L. and S.D. Lyda. 1983. Root distribution of four Vitis cultivars. Plant and Soil **71**: 63-74.
22. Pourtchev, P. (2003) To the problem of depth of root system penetration of grapevine. *Soil Science, Agrochemistry and Ecology* **2**: 47-52
23. Reynolds, A. G. and D. A. Wardle. (2001) Rootstocks impact vine performance and fruit composition of grapes in British Columbia. *Hort Technol.* **11**: 419-427.
24. Romero, P., Dodd, I.C. and Martinez-Cutillas, A. 2012. Contrasting physiological effects of partial root zone drying in field-grown grapevine (Vitis vinifera L. cv. Monastrell) according to total soil water availability. *Journal of Experimental Botany* **63**: 4071–4083
25. Rossdeutsch, L., Edwards, E., Cookson, S.J., Barrieu, F., Gambetta, G.A., Delrot, S. and Ollat, N. 2016. ABA-mediated responses to water deficit separate grapevine genotypes by their genetic background. *BMC Plant Biology* **16**: 91.
26. Sabir, A. and Sahin, Z. (2018) The response of soilless grown 'Michele Palieri' (Vitis vinifera L.) grapevine cultivar to deficit irrigation under the effects of different rootstocks. Erwerbs-obstbau **60**: S21–S27.
27. Santos, J.A., Fraga, H., Malheiro, A.C., Moutinho-Pereira, J., Dinis, L.-T., Correia, C., Moriondo, M., Leolini, L., Dibari, C., Costafreda-Aumedes, S., Kartschall, T., Menz, C., Molitor, D., Junk, J., Beyer, M. and Schultz, H.R. 2020. A review of the potential climate change impacts and adaptation options for European viticulture. *Applied Sciences* **10**: 3092.
28. Satisha, J., S.D. Ramteke, and G.S. Karibasappa. 2007. Physiological and biochemical characterization of grape rootstocks. *South African Journal of Enology and Viticulture.* **28**(2): 163–168.
29. Smart, D.R., Schwass, E., Lakso, A. and Morano, L. 2006. Grape vine rooting patterns: a comprehensive analysis and a review. *American Journal of Enology and Viticulture* **57**: 89–104.
30. Somkuwar R G, Taware P B, Bondage D D and Navale S. 2012. Root length, root mass, and distribution of dry matter in different parts of 'Thompson Seedless' grapevine grafted on different rootstocks in heavy soil of Maharashtra. *Turkish Journal of Agriculture and Forestry* **36**: 543–552.
31. Somkuwar, R. G., Bhor, V. A., Ghule, V. S., Hakale, D., Shabeer, A., and Sharma, A. K. 2021. Rootstock affects stress relieving enzymatic activity during bud break in 'Red Globe'grapevine under semi-arid condition. *Vitis*, **60**(2): 93-99.
32. Somkuwar, R. G., Ghule, V. S., Deshmukh, N. A., and Sharma, A. K. 2023. Role of rootstocks in yield and quality of grapes (Vitis vinifera) under semi-arid tropics of India: a review. *Current Horticulture*, **11**(2): 9-16.
33. Somkuwar, R. G., Taware, P. B., Bhange, M. A., Sharma, J., and Khan, I. 2015. Influence of different rootstocks on growth, photosynthesis, biochemical composition, and nutrient contents in 'Fantasy Seedless' grapes. *International Journal of Fruit Science*, **15**(3): 251-266.

34. Somkuwar, R. G., Taware, P. B., Bondage, D. D., and Nawale, S. 2012. Influence of shoot density on leaf area, yield and quality of Tas-A-Ganesh grapes (*Vitis vinifera* L.) grafted on Dog Ridge rootstock. *IRJPS*, **3**: 94-99.
35. Tambe TB. Ph.D. (Agri.) Rootstock studies in grapes (*Vitis* spp.). Thesis submitted to MPKV, Rahuri (M.S.), 1999
36. Tramontini, S., Vitali, M., Centioni, L., Schubert, A. and Lovisolo, C. 2013. Rootstock control of scion response to water stress in grapevine. *Environmental and Experimental Botany* **93**: 20–26.
37. Upadhyay Anuradha, Upadhyay AK, Rita Bhirangi 2012. Expression of Na^+/H^+antiporter gene in response to water and salinity stress in salt tolerant and sensitive grape rootstocks. Biol Plant **56**: 762-766.
38. Wallis, C.M., Wallingford, A.K. and Chen, J. 2013. Grapevine root stock effects on scion sap phenolic levels, resistance to Xylella fastidiosa infection, and progression of Pierce's disease. *Frontiers in Plant Science* **4**: 502.
39. Weaver, R.J. 1976. Grape Growing. A Wiley Inter science Publication.
40. Williams, L.E. and R.J. Smith. 1991. The effect of rootstock on the partitioning of dry weight, nitrogen and potassium, and root distribution of Cabernet Sauvignon grapevines. *Amer. J. Enol. Viticult.* **42**:118-122.
41. Yildırım, K., Yagcı, A., Sucu, S. and Tunç, S. 2018. Responses of grapevine rootstocks to drought through altered root system architecture and root transcriptomic regulations. *Plant Physiology and Biochemistry* **127**: 256–268.

Chapter 12

Round the Year Production of Cole and Root Vegetable Crops

Pritam Kalia* and Shrawan Singh

Division of Vegetable Science, ICAR–Indian Agricultural Research Institute, New Delhi 110012

**e-mail: pritam.kalia@gmail.com*

ABSTRACT

Consumption of fruits and vegetables is of immense importance for good health and wellbeing. The Food and Agricultural Organization (FAO) prescribes intake of 400g fruits and vegetable per capita per day and on the same analogy, the Indian Council of Medical Research (ICMR) recommended 300 g/capita/day vegetables and 100 g/capita/day fruits for Indian population. Since consumers' preference is towards fresh vegetables, therefore regular supply of freshly produced vegetables remain a big challenge, particularly in case of seasonal vegetables and these vegetables are perishable in nature, therefore they have poor shelf-life. Further, some of these crops have high end market in India and also excellent export potential; therefore, tapping their economic and nutritional potential is essential to develop proper year round crop plans. With this scenario, the present paper highlights the efforts made in research and developments for year round production of Cole and root vegetables crops. The Cole vegetable crops include cauliflower, cabbage, broccoli, knol khol, Brussels sprouts and kale, while common root crops in India are radish, carrot, beetroot and turnip. The varietal development efforts in Cole vegetable crops led to breed a range of month specific varieties which can be grown from May to March months. Similarly, in root crops, the development of heat tolerant varieties in carrot (PusaVrishti) and radish (Pusa Chetki) have facilitated extending of their growing seasons. However, there are certain concerns for growing year round vegetables particularly Cole vegetables such as that of insect pest evolution, pesticide usage, resurgence of insect pests, pathogen survival and disease incidences, and excessive water requirement during summers.

Keywords: *Cole crops, Root crops, Production, Climate, Nutrition.*

Cole and root vegetables are important groups of vegetable crops as these are widely grown and popular in almost all the regions of the country. The word 'Cole' is abbreviated from `Caulis' which means stem. The term 'Cole' was used to refer the group of these plants originating from a single wild form referred as *Brassica Oleracea* var. *Sylvestris*. This group constitutes mainly cauliflower, cabbage, broccoli, knol khol, Brussels sprouts and kale. The Mediterranean Sea coast is considered the centre of origin of these crops from where they spread first in Europe and then to almost all the countries of the world ranging from temperate to tropical. Among the various Cole crops, cabbage and cauliflower are most widely grown on commercial scale and broccoli is picking in India. While crops like knol khol, Brussels sprouts and kale are grown on very small scale and their area is scattered. Root vegetables are distinguished by a conspicuous, fleshy subterranean structure that may be a root, tuberous root, or a hypocotyl with a taproot developing underneath it, depending on the crop. The Apiaceae (carrot), Brassicaceae (radish, turnip), and Chenopodiaceae (beet) families produce the most root vegetables. For prolonged availability of nutrients of these vegetables to consumers, their round the year production is important.

Cole Crops

Cole vegetables are important winter vegetables, however, to some extent, they are being cultivated in kharif and summer season also. They are widely recognized as valuable sources of dietary fibre, minerals such as calcium and magnesium, vitamin C, provitamin A carotenoids, and certain glucosinolates that may confer a chemoprotective effect in human body. These compounds are good for consumer health; however, their content remain in favourable range in fresh vegetables as compared to stored and/or processed vegetables. To ensure continuous year-round supply of these nutrient rich vegetables, it is better to develop technologies to ensure year-round availability of fresh vegetables to the consumers. The Cole vegetables and their growing seasons are given in (Table 1).

Root Crops

Carrot (*Daucus carota*), radish (*Raphanus sativus*), turnip (*Brassica rapa*) have both European and Asiatic type varieties. All cultivars of beet (*Beta vulgaris*), parsnip (*Pastinaca sativa*) and rutabaga (*Brassica napus* var. *napobrassica*) are European or temperate or biennial type and hence their seeds can only be produced in the hilly snowbound tracts. Temperate carrots, radish, turnip, beet, parsnip and rutabaga require exposure to low temperature for initiation of flower stalks. Root vegetables are low in calories and high in antioxidants. Each one contains a wide variety of vitamins and minerals. Some have nutrients in surprisingly high amounts. A cup of mashed turnips has 51 calories and 76 milligrams (mg) of calcium – as much calcium as half of a slice of cheddar cheese. Root vegetables also have a highwater content and relatively

Table 1. Economically important Cole vegetable crops

Crop	*Botanical Name*	*Edible Parts*	*Cultivation Scale**	*Main Growing Season in Plain and Hills*	*Other Seasons in Plains*	*Other Seasons in High Hills and Dry Temperate Zone*
Cauliflower	*B. oleracea* var. *botrytis* L.	Curd	Large**	September - February	June-October; December-April	March – September
Cabbage	*B. oleracea* var. *capitata* L.	Head	Large**	September - March	July-November	March – August
Broccoli	*B. oleracea* L. var. *italica* Plenck	Head	Medium**	September –February	-	March – July
Brussels sprouts	*B. oleracea* L. var. *gemmifera* DC.	Sprouts	Medium	October–February	-	-
Knol khol	*B. oleracea* L. var. *gongylodes* DC.	Knobs	Small	September – March	-	March – August
Collards	*B. oleracea* var. *viridis* L.	Leaves	Small	October–January	-	March – August
Kale	*B. oleracea* L. var. *acephala* DC	Leaves	Small	October–February	-	March – August
Marrow-stem kale	*B. oleracea* L. var. *medullosa* Thell.	Leaves	Small	September - January	-	March- August

*Based on cultivation area, **Based on varieties developed.

low carbohydrate content; one exception to this, however, is the parsnip, which contains on an average 12–13 per cent carbohydrate, comprising almost equal proportions of sugars and starch. Carrots are the most widely cultivated of root crops and are a particularly important source of β-carotene. Although carrots are poor sources of vitamin C, other root vegetables contain reasonable amounts. Root vegetables also contain significant quantities of magnesium, iron, calcium, and potassium. Root vegetables provide energy through their carbohydrate content and are naturally low in fat. They are a good source of fibre, ranging from 3 g fibre in 100 g of raw carrots and beets to 5 g in 100 g of parsnips. Cooking tends to lower the fibre content in beets and parsnips.

Carrot fibre is highly regarded and has become of interest to food processers due to the large quantities of carrot waste, known as pomace, created in the cut-and-peel and juice industries. The insoluble fibres, primarily cellulose and hemicelluloses, constitute the greatest portion of the total dietary fibre with small amounts of lignin. The soluble fibres include fermentable hemicelluloses and pectin. Raw root vegetables are also a good source of water ranging from 80 per cent in parsnips to 88 per cent in carrots and beets. Boiling the vegetables does not appreciably change the water content (< 2 per cent) when based on a 100 g serving. While cooking breaks down some nutrients, both carotenoids and anthocyanins of carrot are quite stable with light cooking. Root vegetables are also a good source of potassium providing ~ 350 mg in 100 g. Research has focused on the numerous health benefits of phytochemicals, including those that are found in the root vegetables. Carrots, in particular, have been studied for nutritive benefits and colour components and are widely considered a 'functional food.' The pigments in carrots serve an important role because they have been associated with reduced risks of atherosclerosis, cancer, and inflammation.

Beet greens by weight have even more micronutrients than beetroot, including vitamin A as β-carotene, vitamin C, vitamin K, calcium, iron, magnesium, potassium, copper, and manganese. Beet leaves also contain a substantial amount of lutein and zeaxanthin and, thus, have health benefits associated with carotenoids.

Technology for Year-Round Vegetable Production

The ultimate objective of a farmer is to earn higher profit from his piece of land per unit time which can be achieved through selling their produce out of the normal season when prices remain higher in the market. Climate resilient varieties of cauliflower, carrot, radish and turnip have been developed which can be utilized by integrating suitable agronomic management practices to harvest their potential during off-season. The following factors contribute in cultivation of year-round vegetables:

Breeding Varieties and Hybrids

Cauliflower

Cauliflower is fifth important vegetable crop which is also a leading Cole crop in India. It is a thermo-sensitive crop and Indian cauliflowers were evolved from Cornish type winter cauliflowers of England. Several local cultivars came up thereafter named after Indian *desi* months of maturity such as *Kunwari, Kataki, Agahani, Poosi etc.* These had tolerance to high temperature, black rot and downy mildew diseases. Cauliflower is grouped into four different broad groups, namely Extra-early or Early-I (temperature for curding: 25-30°C; maturity: September end – mid October), Early-II (20-25°C; mid-October – mid November), Mid-early (16-20 °C; mid November – mid December), Mid-late (12-16°C; mid December – mid January), late or snowball (10-16°C; February - March).On the basis of month of sowing, the available cauliflower varieties from public sector are: Pusa Meghna (May-June), Pusa Kartik Sankar and Kashi Aghani (June- July), Pusa Ashwini and Pusa Kartiki (June end to July end), Pusa Sharad (August), Pusa Paushja and Pusa Shukti (September- mid October), Pusa Snowball K-1, Kt-25, Pusa Snowball Hybrid-1, Pusa Cauliflower Hybrid-301 (October – November) and Pusa Snowball Cauliflower-2 (November) (Table 2). Some private sector hybrids are available which can be sown up to December end to harvest in the months of April- May.

Cabbage

In case of cabbage, Pusa Ageti was developed to grow cabbage for early harvest *i.e.* November month while predominant varieties are fit for main season cultivation (Table 3). Cabbage is the one among Cole crops where F_1 hybrids have become very popular. From public sector, first indigenous hybrid 'Pusa Cabbage Hybrid-1' (KGMR-1) was developed by utilizing self incompatibility system and recently, a CMS system based F_1 hybrid 'KTCBH-81'named as

Table 2. Varieties of cauliflower developed and released in India

Variety	*Source*	*Year*	*Maturity*	*Yield (t/ha)*	*Miscellaneous*
September-October maturity group (Sowing: June beginning; Transplanting: July 2nd fortnight; Temperature for curding: 20-30 °C)					
Pusa Kartiki	IARI, New Delhi	2015	2nd fortnight of October	24	Compact, white curds, field resistance to black rot
Pusa Ashwini	IARI, New Delhi	2015	1st fortnight of October	15	Compact, white curds
Sabour Agrim	BAU, Sabour	2013	2nd October	18	Compact white curds
Kashi Kunwari	IIVR, Varanasi	2005	October	21	White, compact, tolerant to high rainfall
Pusa Meghna	IARI, New Delhi	2004	1nd October	14	Compact, white curds
Pusa Kartik Sankar (F1)	IARI, New Delhi	2004	2nd October	16	Compact, white curd, field resistance to downy mildew
November-December maturity group (Sowing: July end; Transplanting: September beginning; Temperature for curding: 16-20°C)					
Kashi Agahani	IIVR, Varanasi	2008	December-January		White, compact curd, moderate resistant to downy mildew
Pusa Sharad	IARI, New Delhi	2004	November -December	24	White, compact curds
Pusa Hybrid 2	IARI, New Delhi	1994	November - December	23	White, compact
December maturity group (Sowing: August end; Transplanting: October end to mid-November; Temperature for curding: 12-16 °C)					
Pusa Shukti	IARI, New Delhi	2011	December-January	35	White, compact
Pusa Paushja	IARI, New Delhi	2008	December-January	30	White, compact
Arka Kanti	IIHR, Bangalore		October - November	10	Creamy white, compact
Late maturity group (Sowing: September end to mid-October; Transplanting: September beginning; Temperature for curding: 10-16 °C)					
Pusa Snowball-1	IARI RS, Katrain	1977	January – February	25-30	Very compact, white curds
Pusa Snowball K-1	IARI RS, Katrain	1983	February	30-35	Field resistant to black rot
Pusa Snowball Kt-25	IARI RS, Katrain	2002			Snow white curds, Very late
Pusa Snowball Hybrid-1	IARI RS, Katrain	2015	January - February	50-55	First CMS based hybrid in public sector

Central Pusa cabbage hybrid 81 has been identified for release through AICRP (VC). Another hybrid 'KTCBH-822' is in multi-location testing under AICRP (VC) trials.

Table 3. Important cabbage varieties in India

Variety	*Year*	*Source*	*Days for Maturity*	*Weight of Head (kg)*	*Yield per ha*	*Head Shape*	*Miscellaneous*
Golden Acre	1976	IARI RS, Katrain	60-65 days	1-1.5	25	Round, compact	Early, interior white excellent quality
Pusa Mukta	1985	IARI RS, Katrain	70-75 days	1.5-2.0	20-30t	Flattish-round, solid	Moderate resistant to black rot, Heads burst if harvesting delays
Pride of India		Dr. YS Paramar UH&F, Solan	70 - 75 days	1-1.5	20-28	Small round compact	-
Pusa Drum Head	1970	Japan; IARI RS, Katrain	90 days	3.0-4.0	50-54	Flat, less compact	Field resistant to black leg
Kinner Red		Dr. YS Paramar UH&F, Solan	70-80 days	0.5-.06		Oval, compact	Red cabbage
September		Germany; TNAU, Coimbatore	110 days	4.0-6.0	40-50	Oblong, very compact	For Nilgiri hills Long stalk
Pusa Ageti	2000	IARI, New Delhi	70-90 days	0.6-1.2	11-33	Flattish round	1st tropical variety, produce seeds in plains
Pusa Cabbage Hybrid-1	2012	IARI RS, Katrain	60-65	0.6-1.0	30-40	Round, compact	1st public sector hybrid in cabbage black rot resistant, better field staying capacity

Source: Kalia *et al.*, 4.

Broccoli

Broccoli has become in India in recent past for its taste and health benefiting properties. It is an exotic crop and its initial adaptation research work with exotic germplasm was carried out at IARI RS, Katrain and CSK Himachal Pradesh Krishi Vishvavidyalaya (CSK HPKV), Palamapur. A range of varieties were developed in broccoli which can be grown for harvesting in the months of November–January (Palam Samridhi, Pusa KTS-1) to February-March months (Palam Haritika, Palam Vichitra, Palam Kanchan). Presently, each of the maturity group has improved varieties which enable farming community across the country for year round cultivation to earn profits. The details of improved varieties are given in Table 4. Of late, CMS system has been transferred in to promising broccoli lines at IARI, New Delhi and Regional Station, Katrain for use in hybrid breeding. Also, new lines of green sprouting broccoli and purple heading broccoli are in advance stages of evaluation at IARI, New Delhi.

Table 4. Improved varieties of broccoli

Variety	*Year*	*Source*	*Days for Maturity*	*Head Weight (g)*	*Yield/ ha*	*Head Colour*	*Miscellaneous*
Palam Samridhi	1995	HPKV, *Palampur*	80-90	300-400	15-20	Green	Sprouting broccoli
Pusa Broccoli KTS-1		IARI RS, Katrain	85-95	350 -450	10-15	Reddish green	Sprouting broccoli
Palam Vichitra	2003	HPKV, *Palampur*	115-120	350-450	20-22	Purple	Heading broccoli
Palam Kanchan	2003	HPKV, *Palampur*	140-145	350-400	25-27	Yellowish green	Heading broccoli
Palam Haritika	2003	HPKV, *Palampur*	145-150	300-400	17-22	Green	Sprouting broccoli
Punjab Broccoli 1		PAU, Ludhiana	65-70 days	300-400	16-17	Bluish green	Sprouting broccoli

Source: Kalia *et al.*, 4.

In Knol khol, Brussels sprouts and Kale, the varieties recommended for normal winter season are Palam Tenderknob, Pusa Virat Hilds, White Vienna and Purple Vienna, Hilds Ideal and Central Kale Pusa 64(KTK64), respectively.

Root Crops

Among the root vegetable crops, which should form 1/3rd of the daily per capita/day requirement as per dietitians' recommendations, radish, carrot and turnip has both tropical and temperate versions. Tropical varieties grow well during August – October months, whereas European during October – February in plains of North India and March – May in hills. It was with the development of heat tolerant Pusa Chetki variety that it became possible to grow radish from early April – mid August, and other varieties which followed are Pusa Desi (mid-August to mid-October), Pusa Reshmi (mid-September to mid –November), Japanese White and Pusa Shweta (mid-October to mid-December), while European types Pusa Himani (mid-December to mid-February)and White Icicle and Rapid Red White Tipped (second fortnight of October to February end) are suitable for respective months. Thus, these varieties are now available which make a sequence for year round cultivation of radish. New antioxidant rich recent colourful releases of radish, namely Pusa Jamuni, Pusa Gulabi and Palam Hriday suitable for October sowing have further enriched year round radish cultivation sequence with diverse colours and nutrients.

In carrot, development of heat tolerant PusaVrishti variety has facilitated growing of red carrot during August – November in North India, while for main season (*i.e.* September – October) sowing Pusa Rudhira is a leading variety. Orange carrot varieties, namely PusaMeghali (tropical) suitable for October sowing and Nantes, PusaYamdagini and Pusa Nayanjyoti (temperate)can be sown from October – January months for harvest fromJanuary – April months followed by cold storage of late harvest for availability to consumers during summers. Heat tolerant tropical red carrot variety Pusa Vrishti can also be sown during March-April months for fresh harvest during May-June. In turnip, heat tolerant tropical variety Pusa Sweti is suitable for early season (August –November), whereas temperate types (Pusa Chandrima, Pusa Swarnima and Purple Top White Globe) are suitable for October-November sowing in North Indian plains. In beetroot, only few varieties (Detroit Dark Red and Crimson globe) are available which grows well in winter season and efforts are required to breed varieties for tropical climate to develop sequence for their year round cultivation.

Identification of Suitable Varieties for different Growing Season

Cauliflower

Based on temperature requirement for growth and curd development, different cauliflower varieties have been developed and now it has become possible to grow cauliflower for almost all the year-round in north Indian plain (Table 5).

Table 5. Improved Cauliflower varieties and hybrids for different seasons of the year

Maturity Group	*Varieties and Hybrids*	*Sowing Time*	*Availability Period*	*Av. Yield (q/ha)*
Early I	Pusa Meghna	End of May	End of August-October	100
	Pusa Kartik Sankar, Sabour Agrim			110
	Pusa Kartiki			120
	Pusa Ashwini, Pusa Deepali			120
Mid Early II	Pusa Sharad,	End of July-August	Nov.-Dec.	250
	Pusa Hybrid-2,			310
	Pusa Cauliflower Hybrid-3			390
Mid Late III	Pusa Paushja,	End of August-Sept.	Dec.-Jan.	325
	Pusa Shukti,			320
Snowball IV	Pusa Snowball K-1,	Sept.-Nov.	Jan.-March	350
	Pusa Snowball K-25			320
	Pusa Snowball Cauliflower Hybrid-1			380
	Pusa Cauliflower Hybrid 301			400

Carrot

Carrot was being grown only in one season from October to March during which cold environment was prevailing in north Indian plain. IARI has developed some climate resilient variety of carrot which has better tolerance to heat and high moisture and now carrot can be grown from July –August onwards and harvest of carrot in out of normal season can increase the profit margin of farmers and the availability of carrot can be ensured right from May - June to November – January (Table 6)

Table 6. Improved carrot varieties and hybrids for different seasons of the year

Cultivars	*Sowing Time*	*Availability*	*Yield (q/ha)*
Pusa Vrishti	July to October	October to November	180-200
Pusa Rudhira, Pusa Asita, Pusa Kulfi, Pusa Vasuda (F_1)	September to October	November to January	300-450
Pusa Yamdagini, Pusa Nayanjyoti(F_1)	September to March	November to April	270-370
Pusa Vrishti	March to April	May to June	150-160

Radish

Different cultivars of radish developed at IARI having specific growing requirements can be grown under north Indian plains for sowing right from August and continued for the entire year so as to make fresh radish available to the consumers (Table 7).

Table 7. Improved radish cultivars for growing in different seasons

Sl.No.	*Cultivars*	*Period of Sowing*	*Period of Harvesting*	*Av. Yield (q/ha)*
1.	Pusa Desi	Middle of August to Middle of October	Mid-September to Mid-December	175
2.	Pusa Mridula	First fortnight of September to Mid-November	2nd fortnight of October to 1st fortnight of January	180
3.	Japanese White Pusa Sweta Pusa Jamuni Pusa Gulabi	Mid-October to 2nd fortnight of December	Mid-December to 1st fortnight of March	250
4.	Pusa Himani	2nd fortnight of December to end of February	Mid-February to 3rd week of April	250
5.	Pusa Chetki	First week of April to mid-August	First fortnight of May to 2nd fortnight of September	170

Technologies for Ensuring Safe Vegetable Production

Incidence of insect-pests appears to be high in crops grown out of their natural seasons. In case of cauliflower, the incidence of insects is very high in summer cauliflower followed by early season and late snowball crop. Therefore, effective ways to crop control the insect pest is very much necessary before its commercial upscaling. Food quality and food safety are becoming important parameters for vegetable consumption as well as for export market potential. Vegetable crops are being attacked by many diseases and insect pests and different pesticides are generally used to protect the crops and sometimes it leads to very high dosages of harmful chemicals which are not very good for environment as well as human consumption. Some of the varieties and hybrids of different vegetable crops resistant or tolerant to diseases such as Pusa Drumhead (black leg) and Pusa Mukta (black rot) and in cauliflower (Pusa Snowball K-1, Pusa Snowball KT-25 for black rot) and Pusa Paushja, Pusa Shukti and Pusa Hybrid-2 (for downy mildew) (Table 6) are suitable to be included in cropping programme of areas where the diseases mentioned against them are problem. The use of these varieties/hybrids will greatly reduce cost of cultivation besides saving consumers from health hazards.

Monitoring of Pesticide Residues, Anti-nutrients and Dietary Elements

In vegetable crops, the microbial contaminants, chemical residues, heavy metals and anti-nutrients are major concerns for consumer health. The regular monitoring of produce from non-conventional periods should be a practice to ensure safety of produce for consumer health (Hanak *et al.,* 2). Since, most of the biochemical constituents are governed by complex genetics and influenced by environmental factor, therefore, their levels may go up or down depending upon the epigenetic factors. Thus, proper monitoring of consumer health related biochemical compounds needs to be done. In case of Cole crops, glucosinolates are major secondary metabolites which affect consumer acceptance and preference (Hopkins *et al.,* 3). These compounds also invite or distract insects.

Efficient Water Saving Technologies

Growing Cole crops and root crop during summer months or in Kharif season require frequent irrigations due to high rate of transportation and evaporation. Therefore, efficient and water saving irrigation methods are most essential for growing these crops out of season. Also, any unregulated/fluctuations in moisture levels in the root zone causes root hair development and bitterness in carrot, unpleasant taste of radish and turnip. In case of Cole crops, head splitting of cabbage, root rot of cauliflower, tuber cracking of knol khol are common problems related with imbalance in moisture regime. Drip irrigation, micro-sprinkler irrigation, sprinkler irrigation *etc.* are most efficient irrigation methods for vegetable crops (Nichols, 7; Kratky, 5).

Geographical Mapping for Production Sites

Since, India is a country with diverse agro-ecological regions, identification of suitable sites for production of crops during off-season in plains is a viable option. Such as hill region for growing Cole crops and root crops during summer season. However, fragile ecosystem of hill region should be considered with top priority. Further, these regions don't have water sources for regular irrigation. Therefore, such farming should be planned with effective soil and water conservation measures.

Proper Cool Chain for Transport of Produce

The productions in certain off-season pockets have utility when the produce taken off from the site and reach to the end users within time in desired quality. For this, effective cold storage and cold chain should be planned and created. This will help in balancing the price factor and also provide such vegetables to consumer year-round.

Exploring New Avenues of Uses

Since consumer prices remain in a specific range for maintaining the supply and intake by every consumer, So, sometimes, the growers get less returns for their produce due to availability of substitute vegetables and also due to high production and less demand. Therefore, new options such as processing and value addition also should be developed and explored for these crops to ensure proper price and utilize farm produce judiciously. Such products are sauerkraut from cabbage, kanji of black carrot, baby carrot, frozen cauliflower *etc.*

Novel Traits of Year-round Significance for Better Price and Consumer Health

Growing traditional crop varieties in off-season doesn't serve except some remunerative price to growers. Whereas varieties with added novel traits will have value addition for off-season varieties, such as orange gene in cauliflower, anthocyanin rich broccoli and cauliflower, high lycopene in radish and carrot, carotene rich carrot, anthocyanin rich carrot *etc*. These colourful varieties will add value to the crop and also nutritionally beneficial to consumers (Singh *et al.,* 10).

Protected Cultivation for Growing Year-Round Cole and Root Crops

Protected structures such as rain shelters and polyhouses for growing cauliflower and radish in high rainfall regions are an emerging option. Intense light during dry period in tropics increase temperature beyond the maximum limit of crop requirement, therefore, use of UV stabilised shade net (50 or 75 per cent) structures for cultivation of cauliflower during intense hot and dry months in tropics and sub-tropics is better option (Sanwal *et al.,* 8; Singh and Singh, 9, Spehia, 11). Production of Cole crops in net houses to avoid major insect pests attack and harvest pesticide free produce are good options for year-round production of Cole crops (Singh and Singh, 9; Unnithan, 12).

Use of Trench Systems in High Hills and Dry Temperate Snow Bound Areas for Producing Vegetables

Growing of Cole and root vegetables in high hills and dry temperate snow bound areas face problems due to heavy snow and extremely low temperature. In such situation, use of trenches covered with plastic/polycarbonate sheets favours growing these vegetables to meet the local demands (Angmo *et al.,* 1; Mishra *et al.,* 6).

Concerns Related to Year-Round Production of Cole and Root Vegetables

There are certain concerns for growing year round vegetables particularly Cole vegetables such as that of insect pest evolution, pesticide usage, resurgence of insect pests, pathogen survival and disease incidences, and excessive water requirement during summers *etc.* Since Cole and root vegetables are treasures of nutrient and form an integral part of our daily diet, therefore in order to ensure their round the year availability for nutrition security, there is need to develop nutrient rich, climate resilient biotic stress resistant, eco-friendly consumer sensitive varieties for different seasons for continuity, develop package of practices to harness full potential, and study nutrient stability over the seasons. In Brussels sprouts, knol khol and kale varieties are available for winter season cultivations only in plains, therefore efforts are required to generate germplasm for different temperature regimes to develop varieties/hybrids for different maturity periods. Additionally, since all the Cole vegetables are cool season crops, therefore they also find congenial environment for their cultivation during summer season in mid to high hills to dry temperate zone to ensure availability of these nutrient and anticancer rich vegetables to consumers during offseason.

REFERENCES

1. Angmo, S., Angmo, P., Dolkar, D., Norbu, T., Paljor, E., Kumar, B. and Stobdan, T. 2017. All year-round vegetable cultivation in trenches in cold arid trans-Himalayan Ladakh. Defence Life Science, *J*. ***2***: 54-58.
2. Hanak, E., Boutrif, E., Fabre, P., Pineiro, M. 2002. Food Safety Management in Developing Countries. Proceedings of the International Workshop, CIRAD-FAO, 11-13 December 2000, Montpellier, France, CIRAD-FAO. CIRAD CD-ROM, Montpellier, France.
3. Hopkins, R. J., van Dam, N.M. and van Loon, J. J. 2009. Role of glucosinolates in insect-plant relationships and multitrophic interactions. *Annual review of entomology*. **54**: 57-83.
4. Kalia, P., Singh, S., Prakash, C. and Dey, S.S. 2016. Cole vegetables enhancing volume of vegetable basket. *Indian Horticulture*. **61(6)**: 77-81.
5. Kratky, B.A. 2006. Plastic-covered rain shelters for vegetable production in the tropics. Proc. of the 33rd National Agricultural Plastics Congress. American Society for Plasticulture, Bellafonte, PA (published on a CD).
6. Mishra, G.P., Singh, N., Kumar, H. and Singh, S.B. 2010. Protected cultivation for food and nutritional security at Ladakh. *Defence Science Journal*. **60(2)**: 219.
7. Nichols, M. A. (2004). Irrigation system and cultural practices for crop production under control environment production system." In *International Symposium on Greenhouses, Environmental Controls and In-house Mechanization for Crop Production in the Tropics*. **710**: 71-78.
8. Sanwal, S.K., Patel, K.K. and Yadav, D.S. 2004. Vegetable production under protected conditions in NEH regions: problems ad prospects. ENVIS Bulletin: Himalayan Ecology. **12(2)**: 1-7.
9. Singh, S. and Singh, D. R. 2015. Standardization of technologies to ensure supply of high value vegetables in Nicobar Islands (LSRB-DRDO), CIARI, Port Blair.
10. Singh, S., Selvakumar, R., Mangal M., and Kalia P. 2020. Breeding and genomic investigations for quality and nutraceutical traits in vegetable crops-a review. *Indian Journal of Horticulture*. **77 (1)**: 1-40.
11. Spehia, R.S. 2015. Status and impact of protected cultivation in Himachal Pradesh, India. *Current science*. **108(12)**: 2254-2257.
12. Unnithan, L. 2020. https: //krishijagran.com/agripedia/cultivation-of-pesticide-free-cabbage-and-cauliflower-using-icar-iihr-neem-seed-powder-pellet-formulation.

Chapter 13

Improved Production Technologies in Vegetable Crops

S.N.S. Chaurasia*, Anant Bahadur, Swati Sharma, Jagdish Singh and T.K. Behera

ICAR–Indian Institute of Vegetable Research, Jakkhini-Shahanshahpur, Varanasi 221305, Uttar Pradesh

**e-mail: chaurasiaiivr@gmail.com*

ABSTRACT

Ever increasing population, urbanization, shrinking of land holding and mushrooming industrialization challenges vegetable production in next few decades. Vegetable growing area located densely more or less nearby the cities which is declining day-by-day. The traditional method of raising vegetable nursery, transplanting of seedlings without treatments, imbalance and excess use of chemical fertilizers without soil test, exposure to unfavorable soil and environmental conditions, thermal stress, water stress, salinity and pollutants etc. are the major causes for low productivity of vegetables. To overcome these problems various improved production technologies are being promoted worldwide for enhancing the vegetable production like improved nursery raising techniques, crop and season based method of seed sowing and planting, application of appropriate amount of organic manures and chemical fertilizer based on soil test, application of grafting technology, drip irrigation and fertigation, protected cultivation, soil-less and vertical farming, hydroponic technology, etc. have been practiced to get maximum production from limited space round the year. Organic farming practices offer better quality vegetables with improved the soil fertility. Protected cultivation has several advantages to produce vegetables of high quality with good yields, particularly during the off-season when prices are higher, thus using the land and other resources more efficiently. The expansion of grafting is likely due to its ability to provide tolerance to biotic stress, such as soil borne pathogens, and abiotic stresses, such as cold, salinity, drought, and heavy metal toxicity, besides enhancing yields and quality of the produce. Newer innovative grafting technologies like pomato, brimato, grafted cucurbits, etc. are the better options to realize higher productivity of vegetable crops with better quality.

Keywords: *Vegetables, Nursery, Soilless culture, Quality, Pest management.*

Our country is blessed with diverse agro-climates and distinct seasons, making it possible to grow a wide range of vegetable crops. The growth in vegetable sector is driven by consumer demand, and the need for farmers to enhance incomes through high value crops. In an estimate, 31 per cent (375 million) of Indians are vegetarian; therefore, vegetables play a significant role in nutrition of the diet of vegetarian, especially as sources of vitamins (A, C, B_1, B_6, B_9 and E), minerals, dietary fiber and phytochemicals. Leguminous vegetables are also a major source of protein for the vegetarian. Vegetables are important sources of many nutrients, including potassium, dietary fiber and folate (folic acid). Diets rich in potassium may help to maintain healthy blood pressure. Dietary fiber from vegetables, as part of an overall healthy diet, helps reduce blood cholesterol levels and may lower risk of heart disease. Folate (folic acid) helps the body form red blood cells. Vitamin 'A' keeps eyes and skin healthy and helps to protect against infections. Vitamin 'C' helps heal cuts and wounds and keeps teeth and gums healthy. The balanced diet contains adequate energy source, nutrients and vitamins, minerals, carbohydrates, fats, protein *etc.* Vegetable are the reliable source for many dietary factors. As vegetable contain many of the dietary factors like vitamins, minerals and amino acids; they are considered as protective supplementary food. They produce taste, increase appetite and produce fair amount of fibers. By adoption of modern production technologies, proper training of farmers, better marketing opportunities and developmental policies of Government culminated in significant increase in vegetable production and productivity in the country. Further, attempts have been made to critically analyze an overview of research developments in the field of vegetable research in India. Globally, India ranked second in vegetable production and contributed 15.8 and 14 per cent to global vegetable area and production, respectively. During 2021-22, India produced 204.84 million tons of vegetable from an areas of 11.35 million hectare area with an average productivity of 17.63 t/ha. India ranked first in production of okra in the world (73 per cent of world production) and second in brinjal (27.55 per cent), cabbage (13 per cent) and tomato (11 per cent). The major vegetable growing states in India are West Bengal (1.35 million ha), Uttar Pradesh (0.82 million ha), Bihar (0.83 million ha), Andhra Pradesh (0.65 million ha) and Gujarat (0.52 million ha) with production of 26.73, 17.70, 14.63, 11.85 and 9.38 million tons, respectively. There are several improved production technologies that can be used to enhanced vegetable productivity in India.

Hi-tech Nursery Production

Healthy nursery raising is the most crucial step in vegetable production, up to larger extent determines the productivity and profitability. A lot of problems are faced by growers in raising seedlings as majority of the seedlings are damaged due to biotic and abiotic stresses. Raising of vegetable seedlings under open field conditions is a very old practice and now proved uneconomical. Rains for several days, too low or high temperatures play a major role in growing healthy and disease-free seedlings. To reduce cost of production, early establishment in the field and round the year production, the seedlings are being raised in treys using growing media by automatic machine under controlled conditions. In this system of nursery raising each transplant grows in an individual cell so there is less competition among plants and greater uniformity. Less labour is required for mixing and sterilizing soil, filling of media, sowing of seeds and removal of transplants. Plug transplants establish better in the field because roots are not damaged in pulling.

a. Growing Media

Field soils are generally unsatisfactory for production of plants in containers because soils do not provide the aeration, drainage and water holding capacity required for proper growth and development of seedlings. They also need to be sterilization or fumigation to prevent the development of diseases and germination of weed seeds. Growing media for vegetable transplants raising contain a variety of soil-less ingredients such as peat moss, vermiculite, perlite, shredded coconut husks (coir), or composted materials plus starter nutrients and a wetting agent. Growing media are designed to achieve high porosity and water retention while providing adequate aeration. Pre-mixed media are commonly used in the greenhouse industry. Growers can obtain commercially available mixes with different types of biological fungicides added to the mix.

b. Media Treatment

In traditional method the nursery soil is treated by soil solarization method which took 30-40 days time in complete treatment, but with steam it needs 7-8 hrs only. All pathogens are killed by steam and only a few of the hardiest weed seeds will survive. The soil moisture content prior to steaming is important. Some moisture is necessary for the effective killing of microorganisms as well as the conductance of heat. Proper soil moisture for steaming is approximately the same for good planting conditions; soil squeezed in the hand should crumble easily. Prolonged steaming of soil at temperatures higher than 81°C can result in undesirable side effects, such as overkill of beneficial soil micro-flora and accumulation of ammonium and manganese.

c. Containers for Raising Transplants

The production of containerized vegetable transplants has changed in the past several years. Most container-grown vegetable transplants were produced in peat-based containers, but now the vast majority is grown in hardened plastic or polystyrene containers. Unlike polystyrene containers, plastic containers do not have the small pores and crevices that may harbor disease organisms and soak up residual chlorine. Numerous types (shapes, sizes and configurations) of containers are available. Select the type that the best fits in system. Plastic and polystyrene containers most often come in straight row arrangements. Polystyrene containers normally have inverted pyramid-shaped cells that taper towards the bottom. The number of cells in a container depends on the cell size. They may have from 12 to 338 cells/container. A polystyrene tray with a 4.5 cm cell size has 128 cells, while a tray with a 7.5 cm cell size has 72 cells. Generally, smaller cells are used for plants such as cabbage, broccoli, cauliflower, collard, kale and lettuce. Trays with 3 to 4.5 cm cells are well suited for producing transplants of these crops. These trays generally have from 200 to 338 cells/tray. Use larger cell sizes, 4.5 cm to 7.5 cm for production of tomato, pepper, watermelon, muskmelon, cucumber and squash transplants. These trays generally have from 72 to 200 cells/tray.

d. Seed Sowing in Containers

In most commercial facilities, trays are automatically filled with soilless mixtures, dibbled or pressed to a depth of about 1 cm, seeded, then covered with soilless mix or vermiculite. Many different types of commercial seeders are available for seeding plug trays. At commercial level of transplant production a automatic tray filler cum seeders is used to sow the seed in the trays. When selecting a seeder a grower should consider the ease of converting between different cell sizes, the flexibility of the seeders to handle raw or pelleted seed and seeding capacity in trays per hour. Proper sowing depth is more important for seeds with minimal food storage reserves (cabbage family) and those that need light for germination (lettuce, celery). Variable planting depth can result in problems such as uneven stands and "leggy" plants. Seed should be covered with fine vermiculite to ensure that the seed does not dry out. The capacity for filling the trays and seed sowing is 300-400 trays per hour as compare to 10-25 manually

e. Germination

High-quality seed is required for plug transplant production in order to produce acceptable plant counts and uniform transplants. Germination chamber is usually an insulated room in which temperature and relative humidity can be maintained at a precise level. Air circulation is important to ensure uniform temperature and humidity throughout the chamber. There should be a thermostat to maintain the temperature regime. If the temperature goes too high, the variability between the seeds is accentuated, resulting in both uneven germination and transplant development. Never allow the seedlings to elongate by leaving them in the germination. For best results, a controlled temperature room should be used for germination. Such rooms are insulated to allow for heating or cooling according to the specific germination needs of the seed. Warm temperatures and uniform moisture are needed to ensure successful germination and get the plants off to an even start. Many germination chamber systems are commercially available including custom built germination units. Many growers use bottom heat or root zone heating to provide warm, even temperatures.

f. Grafting

Quality seedlings with better productivity and tolerance to soil-borne diseases can be prepared through grafting in Solanaceous and cucurbitaceous vegetables. Most of the water melon, muskmelon, tomato and capsicum grown in protected condition in developed countries are the grafted seedlings. Now-a-days, besides tolerant to biotic stress, many grafted vegetables are grown to alleviate abiotic stresses such flooding, drought, low temperatures, salinity, *etc.* Many fancy vegetables, such as pomato and brimato have been produced in India through grafting technology.

Improved Nutrient Management Technologies

Vegetables are exhaustive crops with an exceedingly high turnover of plant nutrients. The nutrients required for optimum growth and development of vegetables should supply to the plants in right time and in a balanced quantity. The farmers of our country are applying only

NPK nutrients through chemical fertilizers in unbalanced manner. This leads poor yield of the vegetables, lower down the quality and also deteriorates soil. Neither the chemical fertilizers alone nor the organic sources exclusively able to sustain the soil fertility/productivity. Vegetable productivity may be increased up to 2-3 times more by adoption of appropriate nutrient management practices.

a. Organic Farming

Organic manures are the key characteristics including the long-term fertility of soil by maintaining organic matter levels, fastening soil biological intervention, nitrogen fixation as well as effective recycling of organic materials including crop residues, livestock wastes and weeds. Micro-organisms play an important role in various chemical transformations in soil and thus, influence the availability of major nutrient like N, P, K. and S to the plant. Use of chemical fertilizers, pesticides, irrigation and improved cultivars during 70s has resulted green revolution. But, due to over exploitation of these resources for last 4-5 decades has caused ill effects on all living things, soil, water, *etc.* The modern system of farming is felt unsustainable as evidenced by declining crop productivity, damage to environment and chemical contaminations in foods, fodder, water, *etc.* The negative effects of modern agricultural practices on the environment are manifested through soil erosion, water shortage, salination, soil contamination, genetic erosion, *etc.* A large quantity of pesticide is being used to control of insect-pests and diseases in vegetable crops like; brinjal, tomato, cauliflower, chilies, sweet pepper, cucurbits, *etc.* Vegetable growers are spraying pesticides at higher quantity and frequent intervals than recommended to control pests. Such contaminated vegetables carrying pesticides residues are causing several problems in human like; cancer, infertility, Parkinson, blindness, *etc.* Besides, these pesticides are also polluting soil and ground water. Herbicides and chemical fertilizers are the other harmful inputs which are frequently used for conventional vegetable production. The necessity of an alternative agriculture method is realized now, which can function in ecofriendly-system while sustaining the crop productivity. Organic farming is recognized at the best known alternative to the conventional agriculture where application of chemical fertilizers chemical insecticides and pesticides, chemical herbicides and other chemical inputs are restricted.

Use of biofertilizers is gaining momentum due to the increasing importance and maintenance of soil health, reducing environmental pollution, and the use of harmful chemicals in agriculture. Biofertilizers play a vital role in accelerating the microbial processes such as controlling soil-borne diseases, improving the soil health, and soil properties. Biofertilizers can provide an economically viable option to small farmers for realizing the ultimate goal of increasing crop productivity. These are low-cost, effective and renewable sources of plant nutrients to supplement chemical fertilizers. *Rhizobium, Azotobacter, Azospirillum,* phosphate solublizing micro-organism, vesicular arbuscular mycorrhiza (VAM), blue green algae, phyto-hormones, yeast, sea weeds, cyanobacteria, *etc.* are the important biofertilizers that can be used to enhance bio-vailability of plant nutrients.

b. Integrated Nutrient Management (INM)

INM I a complete and balanced methods of supplying nutrients to the plants by organic as well as inorganic means together to fulfill the nutrient requirements. INM is a system approach to nutrient management by tapping all the possible sources of organic, inorganic and biofertilizers in judicious and synchronous way to maintain soil fertility and crop productivity. For better utilization of resources and to produce good crops with less expenditure, INM is the best approach. All the possible sources of nutrients are applied based on economic consideration and supplemented with organic, inorganic and biofertilizers. Under INM system, the organic sources such as, FYM, compost, vermi-compost, green manures, plant residues/organic mulches, biofertilizers and other organic amendments are considered as per their availability and economics. Similarly, inorganic sources such as chemical fertilizers, micro-nutrients, growth promoters, *etc.* are applied to vegetables for realization of higher yields. The micronutrients like Zinc, Iron, Manganese, Copper, Molybdenum and Boron could be made available to the plants either by soil application or by foliar sprays. They play a major role in plant metabolic activities like chlorophyll syntheses, photosynthesis of enzymes and hormones, cell division and other physiological functions. A general INM package for vegetable comprised of soil application of 15-20 tones FYM or 10-12 tones vermi-compost/ha, besides application of recommended doses of NPK. As per soil and plant analysis data, micronutrients are also supplied either directly in soil or through foliar application. In absence of organic manures, green manure crops such as sun-hemp or *dhaincha* should be turn over to the soil at least once in 3 year period.

Improved Irrigation Method

The excess and under irrigation may reduce vegetable production and quality. Irrigation applied actual time of the crop results in effective irrigation and wastage of water, while delay in irrigation may results water stress and effect the crop yield. Water has to be applied in the right amounts at the right time in order to achieve the right crop result. Plant-water relations consist of a group of interrelated and interdependent processes. The internal water balance of plant depends on the relative rate of water absorption and water loss. Water deficit in plant causes: (i) decrease in stomatal opening (ii) reduction in transpiration and photosynthesis (iii) dehydration of protoplasm (iv) reduction in cell division and cell enlargement (v) increase in respiration rate a initial stage, (vi) hastening in maturity and (vii) accumulation of sugar particularly during later part of growth.

In vegetables, higher water and nutrient use efficiency (more yield with less water or nutrient) can be achieved by adopting drip irrigation and fertigation system. Studies conducted at ICAR-IIVR, Varanasi revealed that drip irrigation saves water in tune of about 35-62 per cent in tomato, 34 per cent in cucumber, 31-58 per cent in okra and 56 per cent in broccoli, besides 20-95 per cent increase in yield. Additional water saving in range of 15-35 per cent with improved water use efficiency can be achieved, if drip irrigation is combined with plastic mulching. In drip irrigation, 3-4 times higher water use efficiency have been achieved in different vegetable crops as compared to conventional surface irrigation.

Integrated Weed Management

The weeds are the uncultivated (volunteer) plants that are not desired by man with respect to place and time of occurrence. All weeds are unwanted plant but all unwanted plants may not be weeds. Finally, weeds can be defined as all that vegetation which are unwanted, self growing, harmful and persistent, interfere with agricultural operation reducing the fertility of the soil along with yield are said to be weeds. Weed management is utmost importance in the successful production of vegetable crops. Its control in vegetables especially important early in these as on when weed competition can substantially reduce vigour, uniformity and overall yield Weeds. Compete with the crops for water, soil, nutrients, light, space and crop yields are significantly reduced as reported by several studies. The vegetable as well as agricultural crops are infested with a variety of weeds. Out of over 800 weeds documented in India, 80 are considered as serious and nearly 200 as major weeds. They also harbour many insect-pests and microorganisms. On an average weed extract has two times more N and Ca and 25 per cent more potassium than the crop as observed in some studies. Reduction in economic yield of vegetables has been reported to be 60-82 per cent in potato, 25-30 per cent in peas, 70-80 per cent in carrot, 67 per cent in onion, 42- 71 per cent in tomato and 61 per cent in cauliflower. Farmers usually have a various methods for weed control and their choice of control measures is closely controlled by economics. Although, herbicides were introduced to horticultural crops in India in 1960, hand tools and animal drawn equipment still remain important methods of weed control in vegetable crops. Conventional methods of weed control have become an expensive input in the cultivation of vegetable crops. Owing to high cost and non- availability of labour at proper time therefore, no single method of weed control is adequate and cost effective. However, integrated weed management is a systematic approach to minimize weed impacts and optimize the land use by the different weed management practices. In spite of all these developments, vegetable production suffers to a great extent by weeds. The vegetable fields are usually infested by a wide spectrum of weeds. Weeds compete with vegetable crops for nutrients, moisture, light, space and other essential requirements. Therefore, yield is reduced in various proportions. The extent of yield reduction depends on the density and competitiveness of weed flora and availability of nutrients, moisture, space and light. The dry matter production of weed is negatively correlated with crop yield. The weeds are better adapted to adverse weather conditions, therefore they dominate over crops. The total annual losses of agricultural produce from various agencies are; weeds 45 per cent, insects 30 per cent, diseases 20 per cent and other pests 5 per cent. However, losses caused by weeds vary from crop to crop. The severity of losses is reported up to as much as 100 per cent. The yield losses in different vegetable crops are as follows: Potato-10-80 per cent. Carrtot-50-80 per cent, onion 50-85 per cent, tomato 20-90 per cent, chilli-20-80 per cent, okra-25-90 per cent, brinjal-15-35 per cent, cauliflower-36-60 per cent, cabbage-25-50 per cent, pea-20-45 per cent frenchbean 30-70 per cent, cowpea-20-60 per cent, root crop -30-90 per cent, cucurbits 45-50 per cent.

Protected Cultivation

Vegetable production in our county is significantly influenced by the seasonality and weather conditions. Extends of abundance and deficiency in production and availability cause considerable fluctuations in the prices and quality of vegetables. Seeking a balance between all season availability of high quality vegetables with minimum environment impact is major challenge for the modern technology of crop production. In open field cultivation, while the other factors could be taken care of, it is not possible to effect control on the microclimate around the plant. If any vegetables are chosen to grow outside of its region of natural adaptation, it must be provided all the necessary growth requirements in the new growing environment. In tropical region of the world where growing seasons are longer, growing vegetables indoor is not as population as growing them outdoor. On the contrary, the growing seasons in temperate regions are must shorter. In such conditions vegetables cannot be grown outside particularly in cold winter months. Therefore, they are grown indoor under artificial conditions which are called as controlled environment or protected cultivation. Protected cultivation practices can be defined as a cropping technique wherein the micro environment surrounding the plant body is controlled partially/fully as per plant need during their period of growth to maximize the yield and resource saving. In protected condition, vegetables can be successfully grown at any time of the year if they are provided with suitable environmental conditions that prevail in their normal growth conditions. Protected cultivation offers several advantages to produce vegetables of high quality and yields-particularly during the off-season when prices are higher, thus using the land and other resources more efficiently. This becomes relevant to vegetable growers in India who have land holding smaller than one hectare. They would be interested in a technology which helps them to produce more crops each year from their land, particularly during off-season when prices are higher. Other advantages of greenhouse cultivation include timely harvest as per market demand; higher yield due to better management, optimum use of land and increase in harvesting period and raising quality seedlings and plant propagation.

The main principle involved in protected agriculture are green house effect for heating cultivated space using sun rays and ventilation for cooling and air (CO_2) circulation. Cultivated area is isolated and covered with plastic film or glass which is transparent to incoming short waves infrared radiation from soil and plants. As a result protected space maintains higher temperature than the ambient. However the tapped heat is gradually dissipated through conduction, convection/ventilation and radiation. By incorporating heat sinks, reduced ventilation and application of radiation shields either rate of cooling is slow down or supplemented heat source placed inside. The term greenhouse means different things to different people. A greenhouse used to be a structure formed of glass, with a heating (or cooling) system that was used year-round, but especially in winter. A greenhouse is generally covered with a transparent material such as polythene or glass; now a day normally polycarbonate sheets are used because of their longevity (up to 12 year), durability, and transparency. The greenhouse covered with simple plastic sheet is termed as polyhouse. Depending upon the cladding material and its transparency, major fraction of sunlight is absorbed by vegetable crops and other objects. These objects in greenhouse in turn emit long wave thermal radiations for which cladding material has lower transparency. With the result, solar energy is trapped and it raises the temperature inside the greenhouse. This is popularly known as greenhouse effect. This rise in temperature in greenhouse is responsible for growing vegetable in cold climates. The greenhouse generally reflects back 43 per cent of the net solar radiation incident upon it allowing the transmittance of the photo synthetically active solar radiation in the range of 400 -700 nm wavelength. However in the summer season, due to the above-stated phenomenon, ventilation and cooling is required to maintain the temperature inside the structure well below 35°C. The sunlight admitted to the protected environment is absorbed by the crops, floor, and other objects. These objects in turn emit long wave thermal radiation in the infra red region for which the glazing material has lower transparency. As a result the solar energy remains trapped in the protected environment, thus raising its temperature.

Vertical Farming

With developing India, industries are also increasing and cultivable land is decreasing. This is very dangerous for farming, because in ancient time land is the compulsory source for farming, without land one can't do farming. But, nowadays this becomes just a myth because of "Vertical Farming". Vertical farming is the practice of producing food and medicine in vertically stacked layers, vertically inclined surfaces and/or integrated in other structures (such as in a skyscraper, used warehouse, or shipping container) by Modern methods of farming (Singh, *et al.*, 2020). With time changes, new technology also gets introduced for ease, better productivity and quality products. Without any doubt these technologies boost agriculture particularly vegetable cultivation. Most vertical farms use enclosed structures similar to greenhouse that stack vertically, either directly above each other or staggered for better natural light exposure. Saving of space is utmost important in vertical farming. Hydroponics method of crop raising in vertical farming is to cutoff the soil use. The use of aeroponics further reduces weight and water requirements. Most vertical farms are either hydroponic or aeoponic and do not have run off, which would make the potted plants heavier. Vertical farming typically uses the mix of natural lights and artificial lights. Artificial lights are often LED based and may be driven by a renewable power source such as solar power or wind turbines. Supporters of vertical farming praise the impact it can have now and in the future to increase food security and have a positive impact on human health. It could decreed forestation and pollution, and help urban areas be self-sufficient.

a. Hydroponic Systems

The word has its origin from the Greek words "*hydro*" meaning water and "*ponos*" together meaning working with water. The term hydroponics has been defined as the "cultivation of plants in water". It is a technique of growing plants without soil, where in, the nutrients for meeting the plant nutritional requirements are provided in solution form. Studies have demonstrated that may different aggregates or media can also support plant growth and, therefore, the definition of hydroponics has been broadened to "the cultivation of plants without soil" or "soilless culture". However, some prefer to use the term hydroponics for pure solution culture. Hydroponics is a method of growing plants without soil. Instead of having their roots supported and nourished by soil, the plant are supported by an inert growing medium like cocopeat and are fed via a nutrient-rich water solution and uses about 70 per cent less water than traditional farming. A study has shown that, compared to conventional farming, hydroponic farming could increase the yield per area of lettuce by around 11 times while requiring 13 times less water. Due to these advantages, hydroponics is the predominant growing system used in vertical farming

b. Aeroponics System

Aeroponic is the process of growing plants in an air or mist environment without the use of soil or an aggregate medium. The basic principle of aeroponic growing is to grow plants suspended in a closed or semi-closed environment by spraying the plant's dangling roots and lower stem with an atomized or sprayed, nutrient-rich water solution. The principles of Aeroponics are based on the possibility of cultivating vegetables whose roots are not inserted in a substratum (the case with hydroponics) or soil, but in containers filled with flowing plant nutrition. In these containers roots can find the best condition regarding oxygenation and moisture. These conditions allow for better plant nutrition assimilation in a more balanced way, with consequential faster development of the cultivated plants. Plant containers can be mounted on top of one another and because they are light and handy, they can be easily moved according to agricultural needs. The Aeroponic system is agriculture with a non-stop production cycle. In comparison with the traditional agriculture the most relevant advantages of aeroponics are limited water consumption, soil composition is not relevant because soil is never going to be used in the process, intensive production on a limited land surface area and has the highest output per square meter of land per year.

c. Aquaponics System

Aquaponics is a recirculating system that combines hydroponics (growing plants in water without soil) and aquaculture (fish farming) to create an efficient closed loop system. Aquaponics uses these two in a symbiotic combination in which plants are fed the aquatic animals' discharge or waste. In return, the vegetables clean the water that goes back to the fish. Along with the fish and their waste, microbes play an important role to the nutrition of the plants. These beneficial bacteria gather in the spaces between the roots of the plant and convert the fish waste and the solids into substances the plants can use to grow. The result is a perfect collaboration between aquaculture and gardening.

Precision Farming

Under or overuse of agri-inputs such as water, nutrients, pesticides, *etc.* may lead to loss in yield or economy. Timely and precise application of inputs enhanced the input use efficiency many folds. Furthermore, crop growth and health monitoring through remote sensing and imaging spectroscopy leads to timely control of pests and diseases. Precision farming has great potential to rationalize the input use and improve the economy of vegetable growers.

Precision farming in the form of site-specific management, offers a remedy to many of crop production concerns. The philosophy involves matching resource application and agronomic practices with soil properties and crop requirements as they vary across a site. Precision crop management strategies for improving resource use efficiency and reducing inputs and post-harvest losses are key issues to be taken care for sustained enhanced production system. Technological developments in the field of sensors, remote sensing at different platforms, geospatial sciences, big data analytics and ICTs have special advantage in terms of generating state-of-the art informatics -capturing the variability, vulnerability and dynamism of agricultural system for spatio-temporal monitoring crop, soil and micro-environment for precision agriculture. Remote sensing and specifically imaging spectroscopy have been found to very potential for soil and crop health monitoring for site specific nutrient management, discerning composite stresses (due to water, nutrient, pest and diseases) and stress levels for timely and precise application inputs. Low cost sensors and Wireless sensor network (WSN) or Internet of things (IoTs) and sensors on ground airborne (drone) and satellite platforms have been very useful to information retrieval. Development in big data analytics, computer vision helps in timely image/data analysis for information. Crop simulation models and decision support systems in GIS environment helps in making decision rules and e-agri advisory services to stake holders for efficient crop management. Variable rate technology using ground mobile and airborne (drone) platform helps in precise application of inputs

Although India has made considerable advance in agricultural research, but still the blanket recommendations of fertilizers for adoption over larger areas are in vogue. These blanket recommendations are no more useful to enhance productivity gains, which were witnessed between 1960's and 1980's. Now, to enhance growth rate in productivity, precision farming technology has to be developed. Precision farming is important because: (i) nutrient variability within a field can be very high affecting optimum fertilizer rates, (ii) yield potential and grain protein can also vary greatly even within one field, affecting fertilizer requirement, (iii) increasing fertilizer use efficiency will become more important with increasing fertilizer costs and environment concerns, (iv) irrigation at critical stages is very important and (v) pest and stress management at early stages helps the farmer to get maximum yield.

Grafting: An Innovative Technology

a. Alleviating Waterlogging Stress

Vegetables are frequently exposed to unfavourable soil and environmental conditions. These include thermal stress (high and low temperatures), water stress (drought and flooding), salinity and pollutants. An effective method of adapting plants to counteract environmental stresses is by grafting elite, commercial cultivars onto the selected vigorous/tolerant rootstocks. Grafting is now-a-days regarded as a rapid alternative tool to the relatively slow breeding methodology aimed at increasing environmental-stress tolerance of fruit vegetables. Grafting experiments was conducted at ICAR-IIVR during 2015-18 to alleviate waterlogging stress in tomato revealed that tomato cultivars grafted on brinjal rootstocks-IC-111056 and IC-354557 were able to survive 72-96 h of waterlogging stress, as they did not show any symptoms of wilting and leaf yellowing during or after exposure to waterlogging situations. Grafted plants had also higher CCI and chlorophyll fluorescence yields than non-grafted plants. On the other hand, self-grafted or ungrafted plants could not tolerate even 48 h of waterlogging stress.

b. Production of Grafted Pomato

AT ICAR-IIVR, Varanasi grafted pomatoes were successfully produced and demonstrated in field during 2019-20 and 2020-21 (Figure 1). In this study, maximum production was achieved when indeterminate hybrid tomato was grafted over potato rootstock. Experimental findings revealed that potato production ranged between 505 to 745 g while tomato production in range of 826-3090 g from single pomato plant. It was reported that the potato yields were reduced to 8.3-40 per cent and tomato by 12.4-52.3 per cent as compared to their sole crop production. In quality traits, such as Lycopene, β-carotene, acidity, ascorbic acid, reducing and non-reducing sugars and TSS no definite patterns have been observed and it varied considerably.

Figure 1. Grafted tomato on potato (Pomato).

c. Production of Grated Brimato

At ICAR-IIVR, Varanasi after successful field demonstration of grafted pomato (potato + tomato), dual grafting of brinjal and tomato (brimato) was demonstrated in field during 2020-21. In this study, brinjal hybrid-Kashi Sandesh and improved cultivar of tomato- Kashi Aman were successfully grafted onto brinjal rootstock- IC 111056. Experimental findings revealed that about 36.0 fruits with 2.383 kg yields were harvested in tomato/plant, while in brinjal 2.684 kg yields were obtained from 9.2 fruits/plant. Dual grafted brimato technology would be very useful for the urban and sub-urban areas, where limited spaces are available for accommodating vegetables in vertical garden or pot culture over terrace and compound. Research on commercial production of grafted brimato is continued at ICAR-IIVR, Varanasi.

Figure 2. Dual grafting of brinjal and tomato (Brimato)

d. Inter-Specific Grafting in Cucurbits

Cucurbits such as cucumber, muskmelon and watermelon can be grafted onto various inter-specific cucurbitaceous rootstocks such as sponge gourd, ash gourd, ridge gourd, inter-specific hybrids of pumpkin and melons, *etc.* have been used to enhance yields and tolerance to various soil borne diseases. Grafting study at ICAR-IIVR was carried out in cucumber, bitter gourd and muskmelon with using different inter-specific cucurbitaceous rootstocks. In cucumber, maximum number of fruits (23.33 and 24.67) and fruit yield (4.244 and 5.174 kg/plant) in Kashi Nutan and Kalyanpur Green was recorded with grafting on sponge gourd rootstock, respectively. These two cultivars produced 17.26 per cent and 115.13 per cent higher fruit yield over ungrafted control. Similarly, in muskmelon (cv. Kashi Madhu) 58.4 per cent higher fruit yields than the ungrafted plants were obtained when sponge gourd was used as rootstock.

REFERENCES

1. Anon., Second Advance Estimate of Horticulture Production released by the Ministry of Agriculture, July, 2021 report.
2. Bahadur Anant, Jangid, K.K., Singh A. K., Singh Umesh, Rai K.K., Singh M. K., Rai N., Singh P.M., Rai, A.B. and Singh, B. 2016. Tomato genotypes grafted on eggplant: Physiological and biochemical tolerance under waterlogged condition. *Veg. Sci.*, 43(2): 208-215.
3. Bahadur Anant, Rai, N., Kumar Rajesh, Tiwari, S. K., Singh, A. K., Rai, A. K., Singh Umesh, Patel, P. K., Tiwari Vivek, Rai, A. B., Singh Major and Singh, B. 2015. Grafting tomato on eggplant as a potential tool to improve waterlogging tolerance in hybrid tomato. *Veg. Sci.*, 42(2): 82-87.
4. Bahadur Anant, Singh J. and Singh K.P. 2004. Response of cabbage to organic manures and biofertilizers. *Indian J. Hort.* 61 (3): 178-179.
5. Bahadur Anant, Singh, J., Singh. K. P., Upadhyay, A. K. and Rai, M. 2006. Effect of organic amendments and biofertilizers on growth, yield and quality attributes of Chinese cabbage (*Brassica pekinensis* Olsson). *Indian J. Agric. Sci.* 76(10): 596-598.
6. Bahadur Anant, Singh, J., Upadhyay, A.K. and Singh, K.P. 2003. Effect of organic manures and biofertilizers on growth, yield and quality attributes of broccoli (*Brassica oleracea* L var. italica Plenck). *Veg. Sci.* 30 (2): 192-194.
7. Balraj Singh and Mahesh Kuamr, 2005, Effect of plant spacing and stem pruning on growth and yield of cherry tomato in greenhouse. *Haryana J. Hortic. Sci.,* 34(1-2): 179-180.
8. Bhatnagar, R. R., Vedprakash, R. C. and Srivastav, R. C., 1990, Production of vegetables in polyhouse greenhouse during winters in mid hills of Uttar Pradesh. *Prog. Hort.*, 22 : 97-100.
9. Chandra P., Sirohi, P.S., Behera T.K. and Singh, A.K. 2000. Cultivating vegetables in polyhouse. *Indian Horticulture.* 45: 17-25.
10. Chaurasia, S.N.S., Anant Bahadur, Yadava, R.B. and Krishna, H., 2021. Specialty Fertilizers for Vegetable Production, *Indian Journal of Fertilisers.* 17 (4): 388-394.
11. Chaurasia, S.N.S., De Nirmal, Singh K.P., Banerjee M.K. and Rai M. (2004) Nursery management in vegetables, *Technical Bulletin no. 15*, pub. IIVR, Varanasi. PP: 1-36.
12. Chaurasia, S.N.S., Singh, A.K., Singh, K.P., Rai A.K., Singh, C.P.N. and Rai, M. 2008. Effect of integrated nutrient management on yield and quality of cauliflower (*Brassica oleracea* var. botrytis L.) variety Pusa Snow Ball K-1. *Veg. Sci.* 35(2): 41-44.
13. Chaurasia, S.N.S., Singh, K.P. and Rai, M. 2005. Effect of foliar application of water soluble fertilizers on growth, yield and quality of tomato (*Lycopersicon esculentum* L). *Sri Lankan J. Agric. Sci.,* 42, 66-70.
14. Chaurasia, S.N.S., Singh, R. and Rai, M. 2009. Effect of integrated nutrient management and spacing on yield, quality and economics of broccoli (*Brassica oleracea* var. italica Plenck). *Veg. Sci.,* 36: 51-54.

15. K.J. Jeyabaskaran, P.S. Shirgure, Vikramaditya Pandey, A.K. Srivastava and S. Uma. 2021. Fertigation in Horticulture: A Guarantee to Economized Quality Production, *Indian Journal of Fertilisers*. 17 (4) : 364-383.

16. Kumar A. Integration of Vertical Farming and Hydroponics: A Recent Agricultural Trend to Feed the Indian Urban Population In 21st Century. 2019. *Acta Scientific Agriculture*. 3 (2): 54 -59.

17. M. Cicekli and N. T. Barlas. 2014. "Transformation of today greenhouses into high technology vertical farming systems for metropolitan regions,". *J. Environ. Prot. Ecol.,* 15(4): 1779–1785.

18. Nimje, P. M., Wajnan, O. D. and Shyam, M., 1990, Greenhouse technology for vegetable crop production. Proc. of XI Int. Cong. on the use of Plastics in Agric., pp. 83-90.

19. Sahoo, R.N., Tomar, R.K. and Arora, R.P. 2002. Precision farming: A prospective in 21st century. Proc. 2nd Int. Agronomy Congress on *Balancing food and environmental security – A continuing challenge*, IARI, New Delhi, Nov. 26-30, 2002: 1194-95.

20. Singh K.P. Bahadur A. and Chaurasia S.N.S. 2010. Protected Vegetable Cultivation. In; Agriculture Diversification; Problems and Perspectives (A.K. Sharma, S. Wahab and R. Srivastawa Eds.). IK International Publishing House, New Delhi. Pp. 153-175.

Chapter 14

Coconut Based Cropping Systems: Recent Developments in System Productivity and Carbon Sequestration

H.P. Maheswarappa[1]* and S. Sumitha[2]

[1]*University of Horticultural Sciences, Bagalkot 587104, Karnataka*

[2]*ICAR–AICRP on Palms, ICAR-CPCRI, Kasaragod 671124, Kerala*

**e-mail: maheshcpcri@gmail.com*

ABSTRACT

In the present scenario of fluctuation of coconut price and high production cost, increasing incidence of pests and diseases in addition to low and erratic rainfall, the pure crop of coconut is no more economical. India is a country where agriculture is the mainstay of our economy, whereas intervention can help a poor farmer earn much more than what he does by making the most optimum use of available limited natural resources. This era has seen a dynamic shift from sustenance production to commercial production. Coconut, being a perennial oil yielding crop with more than 50-60 years of economic life span, has a potential to act as a carbon sink. Carbon sequestration contribution will be even higher when the crops are grown as mixed cropping or in high density planting system, owing to high density of vegetation per unit area. Higher biomass production in plantation based ecosystems would enhance returns to soil through leaf litter, crop debris, root exudates etc. thereby enhancing soil carbon pools. Coconut-based cropping systems can improve the microclimatic condition by influencing air temperature, soil temperature, vapour pressure deficit and soil moisture content of plantations. Sustainability can be in built in the coconut system by introducing compatible crops as component crops and promoting positive interferences by judicious application of inputs especially fertilizers. All of these have a significant impact on modifying the reproductive and yield performance, soil and water conservation, nutrient status, and carbon sequestration potential of the systems and subsequently, a sustainable growth and productivity making coconut crops more resilient to climate change impacts.

Keywords: *Coconut, INM, Productivity, Intercropping, Carbon.*

India, the largest coconut producer in the world, accounts for 31 per cent of world production and produced 21,384 million nuts during the year 2019 from an area of 2.17 million ha with a mean productivity of 9,815 nuts per hectare (CDB, 11). Coconut farming and its allied activities provide sustenance security to millions of people in India. Hence, accomplishing and sustaining higher productivity and profitability from coconut- based farming is a preeminent goal in serving this farming community. Coconut -based farming system is a dominant production system in the agriculture sector (www.fao.org, 15). This has been described by various literatures which range from monoculture system to multiple intercropping systems. Coconut- based cropping system depicts the arrangement of multispecies utilizing the available space, horizontally and vertically, effectively on a sustainable manner. Multispecies cropping under coconut, particularly during the early growth stage (> 5 yr) and after the age of twenty-five years of plantation ensure maximum resource utilization and higher additional income per unit area of soil, water and light. Intercropping results in improvement of the soil properties and biological activities in the root region. Overall the soil environment is modified for the better crop growth and development. There are several forms in which coconut interspaces could be utilized. Broadly, when annual crops are used for planting, it is termed as intercropping. When perennial crops are incorporated in the system, it is called as mixed cropping. Multistoried cropping system is a form in which the annuals as well as perennials are grown as a system. High density multispecies cropping system (HDMSCS), comprising of annuals, biennials and perennials at a high population density, effectively utilizes the natural resources provided appropriate recycling of nutrients is followed (Rethinam and Sivaraman, 52).

Coconut Based Cropping System with Annual and Perennial Crops

A large variety of annual/biennial/perennial food, fruit, fodder, flower, vegetable, tuber, spice, cocoa and medicinal and aromatic crops can be grown as intercrops in coconut garden depending on the agro-climatic condition of the area. The crops selected for inter/mixed/under cropping should be shade loving or shade tolerant and offers minimum competition for light, water and nutrients by utilizing these resources from different layers of atmosphere and soil. The productivity of land is increased in the intercropping due to yield of intercrops in addition to coconut yield (Maheswarappa and Sumitha, 34). Various intercrop combination was described by many workers (Table 1) which provide additional income for coconut farmers. Data however, on the productivity of coconut-based farming systems are still deficient.

Intercropping Medicinal and Aromatic Crops in Coconut Garden

Promotion of conservation and cultivation enlarge the resource base of medicinal plants in the country which enable uninterrupted supply of genuine herbs to the industry. Medicinal and Aromatic Plants (MAPs) are looked upon not only as a source of health care products, but also as a source of income. As there is no scope for horizontal expansion of cultivated area, it is viable to grow MAPs as intercrops with the coconut growing region by devising suitable cropping schedules. The studies on intercropping of medicinal and aromatic plants in coconut garden conducted at different parts of India during 2006 to 2011 under All India Coordinated Research Project on Palms have indicated the suitability of lemon grass, garden rue, tulsi, kalmegh, arrow root and makoi for Karnataka (Basavaraju *et al.*, 5); *Alpinia galangal*, patchouli, lemongrass, *Aloe vera* and tulsi for Tamil Nadu (Thiruvarassan and Maheswarappa, 63), patchouli, palmarosa, mango ginger and citronella for Andhra Pradesh (Padma *et al.*, 47); shatavari, adulsa, arrow root, lemon grass and citronella for Maharashtra (Nagwekar *et al.*, 45); stevia, mango ginger, sarpaganda and patchouli for Chhattisgarh (AICRP on Palms AR., 2); sarpaganda, ashwaganda and arrow root for west Bengal (Ghosh *et al.*, 19); and sarpaganda, pipali, vedailota (*Paederia foetida*), citronella and patchouli for Assam (Nath *et al.*, 38). Noni is well suited intercrop in coconut garden (Maheswarppa *et al.*, 33). The yield of coconut was improved with the intercropping of medicinal and aromatic crops. Soil nutrient status and uptake of nutrients by coconut were improved with intercropping of medicinal and aromatic plants in coconut garden.

Table 1. Coconut based cropping system with annual and perennial crops

Sl.No.	*Coconut Based Cropping System with Annual and Perennial Crops*	*Reference*
	Coconut based cropping system (Model) in different agro climatic regions	
1	Assam	Nath *et al.*, 36; Nath and Deka, 37
2	Karnataka	Basavaraju *et al.*, 4
3	West Bengal	Ghosh and Bandopadhyay, 22
4	Maharashtra	Nagwekar *et al.*, 44
5	Tamil Nadu	Marimuthu, 35; Subramaniam *et al.*, 58
6	Andhra Pradesh	AICRP on Palms AR (1)
	Coconut + flower crops in different agro climatic regions	
1	Marigold and Gomphrena (Aliyarnagar,)	Rani *et al.*,53
2	Chrysanthemum and China aster (Arsikere)	Basavaraju *et al.*, 6
3	Lily and Heliconia (Ratnagiri)	Shinde *et al.*, 60
4	Tuberose and Gerbera (Kahikuchi)	Nath *et al.*,40
	Coconut + spices crops	
1	Ginger	Ghosh and Hore, 21
2	Black pepper	Maheswarappa *et al.*, 31; Subramanian *et al.*,59; Nath *et al.*, 39.
3	Cardamom	Korikanthimath *et al.*, 25
4	Tree spices	Parulekar *et al.*, 50
	Coconut + vegetables crops	
1	Elephant foot yam	Ghosh *et al.*, 20; Jaganathan *et al.*, 23; Gladis and Joseph, 17; Salam *et al.*, 55, Devi *et al.*, 14
2	Vegetables	Naveen kumar *et al.*,42; Desai *et al.*, 13

Sl.No.	*Coconut Based Cropping System with Annual and Perennial Crops*	*Reference*
	Coconut + medicinal and aromatic crops	
1	Medicinal plants	Bari and Rahim, 3; Basavaraju *et al.*, 5; Thiruvarassan and Maheswarappa, 63; Padma *et al.*, 47; Nagwekar *et al.*, 45; Ghosh *et al.*, 19; Nath *et al.*, 38
	Coconut +beverage crops	
1	Coconut + Cocoa	Bhalerao *et al.*,9; Phukon *et al.*, 49
	Coconut + fruit crops	
1	Coconut + Noni	Bhalerao *et al.*, 8; Siddappa *et al.*, 56; Maheswarppa *et al.*, 33; Khandekar *et al.*, 24
2	Coconut + fruits crops	Bhagya *et al.*, 7
	Coconut based cropping system with INM	
1	Organic and inorganic sources of manure	Subramanian *et al.*, 57; Maheswarappa *et al.*, 30; 32
	50 per cent of RDF (NPK) + 50 per cent N through organic recycling + VC	Farsanashamin and Anilkumar, 16
	75 per cent of recommended NPK + 25 per cent through organic recycling with VC	Naveen kumar *et al.*, 41
	100 per cent fully organic	Rani *et al.*, 54; Bhaleroa *et al.*, 10; Shinde *et al.*, 62
2	CBSC in Root (wilt) garden	Krishnakumar *et al.*, 26; Maheswarappa *et al.*, 29; 30
3	*HDMSCS*	Padma *et al.*, 46

Floriculture in Coconut Garden

Commercial flowers cultivation in India provides an opportunity for rural development owing to its higher returns per unit area and the new employment opportunities. Performance of commercial flower crops under coconut in different agro-climatic regions is one of the priority areas of research under AICRP on Palms to evaluate the impact of intercropping flower crops in coconut on productivity per unit area and economics of the system. Nearly 80 per cent of area under floricultural crops is concentrated in seven states comprising Tamil Nadu, Karnataka, Andhra Pradesh, West Bengal, Maharashtra, Haryana, Uttar Pradesh and Delhi. A major part of the area under flower cultivation is devoted to the production of Marigold, Jasmine, Rose, Chrysanthemum, Tuberose, *etc.*, with considerable increase in the area under cut flower cultivation. Therefore, coconut + flower crop sector can be identified as grey area which can be directly included in different farming systems as a viable diversification from the traditional field crops because of higher returns.

High Density Multispecies Cropping System (HDMSCS)

The cropping system studies carried out in different parts of the country through All India Coordinated Research Project on Palms (AICRP on Palms) indicated improvement in nut yield of coconut and productivity and income of the cropping system. The soil fertility and uptake of nutrients by coconut was also improved with intercropping. In the studies at Ambajipeta, Andhra Pradesh conducted during 1999 to 2003, the crop combination of coconut + cocoa + cinnamon + pepper + pineapple + banana + elephant foot yam + colocasia + turmeric was found highly productive and remunerative. The cropping system studies conducted during 2004 to 2008 have also identified suitable intercrops with coconut like banana, drumstick, french bean, ladies finger and redgram for Karnataka and turmeric, ginger, banana, tapioca and pine apple for Maharashtra. Similarly, the crop combinations of coconut + black pepper + bottle gourd + cowpea was found suitable for Chhattisgarh; coconut + banana + tuberose and coconut + bitter gourd + bottle gourd for Orissa; coconut + black pepper + pineapple for West Bengal and coconut + black pepper + turmeric/ginger for Assam. The high-density multispecies cropping system (HDMSCS) is the growing of number of compatible crops in a unit area to meet the diverse needs of a farmer such as food, fuel, timber, fodder and cash. This system aims at maximizing production per unit of land area and is ideally suited for smaller holdings. The sustainability of production is well addressed in this system through efficient utilization of natural resources and biomass recycling. The productivity of land increases in the high-density multispecies cropping system due to crop diversification and intensification. The results of the studies on HDMSCS conducted during 2008 to 2013 at different Centres of AICRP on Palms located in different parts of the country have indicated improvement in coconut yield and productivity of the land in the high-density multispecies cropping systems (Table 2) are highly productive and remunerative than monocrop of coconut. Growing of intercrops in high density multispecies cropping system improves the available nutrient status of soil due to addition and recycling of organic matter and the manures and fertilizers applied to intercrops.

Table 2. Coconut based cropping system comprising of following crops was recommended for different regions

Karnataka	Coconut + Cocoa + Lime + Drumstick
Andhra Pradesh	Coconut + Cocoa + Banana + Pine apple + Tomato
Tamil Nadu	Coconut + Cocoa + Banana + Pine apple + DrumstickCoconut + Black pepper + Cocoa + Banana + Elephant foot yam
Maharashtra	Coconut + Nutmeg + Cinnamon + Banana + Pine apple
Chhattisgarh	Coconut + Guava + Cinnamon + Banana + Colocasia + Mango ginger- Bottle gourd + Cowpea- Elephant foot yam
Odisha	Coconut+ Guava + Banana + Pine apple
Assam	Coconut + Turmeric + Pine apple + Lemon + Banana + Elephant foot yam

Integrated Nutrient Management under Coconut Based Cropping System

The most important challenges in soil health management are the availability of the organic manure *in situ*. The high cost of external organic manure input considerably increases the cost of production of the produce which hinders the interest of the farmer to apply organic manure as per the recommended soil health management practices to improve the soil health. In this context, CBCS which involves cultivating compatible crops in the interspaces of coconut offers a plenty of biomass available *in situ* to meet the organic matter requirement to maintain the soil health in a very economically profitable manner. Hence, this system offers a very practical solution to the challenges of the soil health management.

Application of 50 per cent of RDF (NPK) + 50 per cent N through organic recycling with vermicompost + vermiwash + biofertilizer and *in situ* green manuring recorded higher system productivity followed by 75 per cent of recommended NPK + 25 per cent through organic recycling with vermicompost or fully organic - recommended from Aliyarnagar, Ambajipeta, Arsikere, Bhubaneswar, Jagdalpur, Kahikuchi, Mondouri, Navsari, Ratnagiri, Sabour and Veppankulam centres for adoption by farmers. The results of the studies at different Centres of AICRP on Palms showed that the available N, P_2O_5 and K_2O were higher in the cropping system compared to monocrop of coconut. Thus, the productivity of coconut gardens can be improved and sustainability of production and income can be achieved by growing various compatible crops in coconut garden (Table 3). The effective and efficient utilization of soil space and solar radiation can be ensured in the cropping.

Shinde *et al.* (61) studied the INM Practise on coconut nut yield (5-year average) was higher yield under T_1 (147.2 nuts palm^{-1} year^{-1}), followed by T_2 (138.4 nuts palm^{-1} year^{-1}) and T_3 (123.6 nuts palm^{-1} year^{-1}),whereas, monocrop had recorded a significantly lower number of nuts (97.2 nuts palm^{-1} year^{-1}). Rani *et al.* (54) observed that the treatment comprising of 75 per cent of recommended dose of NPK + 25 per cent of N through organic recycling with vermicompost (T_1) recorded significantly higher nut yield(18082 nuts/ha), followed by application of 50 per cent of recommended NPK + 50 per cent through organics (T_2) and it was on par with fully organic treatment (T_3) treatments. Coconut based cropping system recorded the highest mean nut yield of 16978 nuts/ha/year and it was 17.5 per cent higher compared to monocrop of coconut. Bhalerao *et al.* (10) noticed significant difference among the INM treatments in respect of nut yield it was highest with the treatment T_2 (palms are treated with 50 per cent RDF + 50 per cent organic manures (146 nuts palm^{-1}), followed by T_3 (136 nuts palm^{-1}) and significantly lower (117 nuts palm^{-1}) in monocrop. Enhancement of coconut yield due to combined application of inorganic fertilizer and organic manure has been reported by previous workers in coconut (Farsanashamin and Anilkumar, 16; Maheswarappa *et al.*, 32; and Krishnakumar and Maheswarappa, 27).

Carbon Sequestration in CBSC

Climate change will affect coconut through higher temperatures, elevated CO_2 concentration, precipitation changes, increased weeds, pest, and disease pressure, and increased vulnerability of organic carbon pools. But, yields are projected to go up in Kerala, Tamil Nadu, Karnataka and Maharashtra (Das *et al.*,12; Malhotra *et al.*, 28). Apart from this, coconut plantations especially under coconut based cropping system are a prominent system to sequester the carbon in the above ground standing biomass as well as in the belowground. Carbon sequestration in the soil organic carbon pool is appreciable in the coconut based cropping system. Increased sequestration of C in agricultural soils has a potential role to reduce the climate change caused by the rising atmospheric carbon level. Evaluating the best crop management practices that can enhance soil organic carbon (SOC) sequestration will improve the amount of carbon fixed as SOC. Moreover, increased SOC improves soil structure, fertility, and water-holding capacity and thereby enhances and sustains the soil health and crop productivity. In the integrated nutrient

Table 3. Coconut and intercrops yield per hectare basis at different centres (Mean 2016-17 and 2018-19)

Treatments	*Arsikere*		*Veppankulam*		*Mondouri*		*Navsari*		*Ratnagiri*	
	Yield/ ha	*Net Returns (Rs/ha)*	*Yield/ ha*	*Net Returns (Rs/ha)*	*Yield/ ha*	*Net Returns (Rs/ha)*	*Yield/ ha*	*Net Returns (Rs/ha)*	*Yield/ ha*	*Net Returns (Rs/ha)*
T- 1: (75 per cent of RDF + 25 per cent organic)	9000	125573	20160	277280	18408	387454	21120	556744	14690	641540
T- 2: (50 per cent of RDF + 50 per cent organic)	9710	134039	21960	350690	18089	276788	24728	719356	13965	551335
T- 3: (100 per cent organic)	9490	139816	21060	337698	18196	265192	23320	611723	13705	529265
T- 4 (Monocrop) (RDF)	9270	85510	18900	96300	18231	98705	19888	85716	13430	201527

management experiment, after organic recycling, builtup of soil carbon in coconut basin was observed (Palaniswami *et al.*, 48).

Ranasinghe and Thimothias, (51) have shown that coconut monocrop plantations under different growth conditions have the potential to sequester carbon between 0.4 -1.9 Mg C ha-1 month-1, depending on the agro-climatic and soil conditions. In the same plantation, C stocks of palms varied between 17 and 25 Mg C ha-1 depending on the growth condition. Coconut stem was found to be the main C storage organ which stored about 56-70 per cent of the total C stock of palms. Investigation was conducted at Horticulture Research Station, Arsikere, Karnataka, during *kharif* and summer seasons of 2012-14 to study the performance of different cropping sequences with vegetable crops and integrated nutrient management practices on soil carbon sequestration and above ground carbon sequestration under coconut based cropping system (CBCS). The study revealed that carbon sequestration by coconut palm (above ground) was the highest under CBCS when compared to monocropping. Soil carbon sequestration recorded at 0-15 cm depth was significantly the highest under S4 treatment (Organic alone - 5 tonnes FYM+50 per cent N by vermicompost+50 per cent N by CCP+vermiwash spray + *Azotobacter*) (10.30 Mg C/ha, and 10.74 Mg C/ha during 2012-13 and 2013-14) and M_2S_4 interactions (green manure-cucumber and organic alone) (11.56 Mg C/ha) when compared to other treatments (Naveen Kumar, 43). Bhagya *et al.* (7) studied the effect of cropping system on above and below ground carbon sequestration in a 50-year-old plantation intercropped with seven-year-old fruit crops. Among the different cropping systems, coconut (*Cocos nucifera*) + jamun (*Syzygium cumini*) system sequestered the highest above ground carbon (60.93 t/ha) followed by coconut + mango (*Mangifera indica*) system with 56.45 C t/ha, coconut + garcinia (*Garcinia indica*) sequestered 53.02 C t/ha, whereas, coconut alone had sequestered 51.14 C t/ha. The below ground soil carbon stock in the rhizosphere of 0-60 cm depth was the highest in coconut + mango (82.47 t/ha) system followed by coconut + jamun (79.13 t/ha), coconut + garcinia (78.69 t/ha) and it was the lowest in coconut monocrop (47.06 C t/ha). The total carbon sequestration by coconut + jamun (140.06 t/ha) followed by coconut + mango systems (138.91 t/ha), coconut + garcinia (131.72 t/ha), whereas it was only 98.2 C t/ha under coconut monocrop. Ghavale *et al.* (18) studied the carbon sequestration and productivity potential of twelve coconut hybrids and three varieties. The above ground standing biomass and carbon stock recorded was the highest in the variety East Coast Tall (312 kg/plant and 27.32 t/ha, respectively) followed by hybrid WCT x GBGD (308.69 kg/plant and27.01 t/ha) and the lowest was in hybrid MYD x ECT (138.71 kg/plant and 12.14 t/ha, respectively). The highest soil carbon stock 39.12 t/ha and 37.16 t/ha at 0-30 and 31-60 cm depth was recorded in the rhizosphere of hybrid ECT x MYD and the lowest soil carbon stock (35.52 t/ha and 34.71 t/ha) was observed in hybrid PHOT x GBGD. Productivity and carbon sequestration potential as influenced by integrated nutrient management (INM) practices in coconut based cropping system was studied by Shinde *et al.*, 61. Treatment T_1 sequestered the highest amount of above ground (coconut + nutmeg) carbon stock (31.1 t ha^{-1}) followed by T_2 (30.3 t ha^{-1}) and T_3 (27.3 t ha^{-1}). In contrast, the treatment T_4 - coconut monocrop alone had sequestered 25.6 t ha^{-1}. The below-ground soil carbon stock in the rhizosphere of different crops was the highest in the nutrient management under T_3 followed by T_2 and T_1, whereas, it was the lowest in coconut monocrop.

REFERENCES

1. Annual Report for 2008-09, pp. 1-102. All India Coordinated Research Project on Palms, CPCRI, Kasaragod, Kerala.
2. Annual Report for 2011-12, pp. 1-105. All India Coordinated Research Project on Palms, CPCRI, Kasaragod, Kerala.
3. Bari, M. S. and Rahim, M. A. 2012. Economic evaluation and yield performance of some medicinal plants in coconut based multi-storeyed agroforestry systems. *Agriculturists*. **10 (1)**: 71-80.
4. Basavaraju, T. B., Hanumanthappa, M., Nagaraja, K and Boraiah, B. 2008. Coconut based cropping systems for *maidan* tract of Karnataka. *J. Plantation Crops,* **36(3)**: 290–295.
5. Basavaraju, T. B., Nanjappa, H. V., Umesha, K., Vasundhara, M and Arulraj, S. 2011. Intercropping of medicinal and aromatic plants in coconut gardens. *J. Plantation Crops,* **39(2)**: 299–304.
6. Basavaraju, T. B., Prashanth, M and Maheswarappa, H. P. 2018. Performance of flower crops as intercrops in coconut garden in southern dry region of Karnataka. *J. Plantation Crops*. **46 (1)**: 52–6.
7. Bhagya, H.P., Maheswarappa, H.P., Surekha and Ravi Bhat. 2017. Carbon sequestration potential in coconut based cropping system. *Indian J. Hortic*, **74(1)**: 1-5.
8. Bhalerao, P. P., Maheshwarappa, H. P., and Patil, S. J. 2016. Evaluation of Noni (*Morinda citrofolia*) as a mixed crop in coconut garden under South Gujarat condition. *Current Horticulture*, **4(1)**: 52-54.
9. Bhalerao, P. P., Maheswarappa, H. P., Sumitha, S. and Elain Apshara, S. 2018. Performance of cocoa clones as intercrop in coconut gardens under south Gujarat condition. *Int. J. Innov. Hortic,* **7(2)**: 120-122.
10. Bhalerao, P. P., Maheswarappa, H. P. and Sumitha, S. 2021. Effect of integrated nutrient management in coconut (*Cocos nucifera L.*) based cropping systems in South Gujarat condition. *Current Horticulture*, **9(2)**: 52-55.
11. CDB. 2020. Coconut Statistics 2019-20. Coconut Development Board, India http: //www. coconutboard.gov.in/presentation/statistics.
12. Das Manish., Vanita Jain and Malhotra, S.K. 2016. Impact of climate change on medicinal and aromatic plants. *Indian J. Agric. Sci.* **86(11)**: 1375–82.
13. Desai, N., Mamatha, B., and Patil, C. 2017. Assessment of Yield and Economics of Vegetables as Intercropping System in Coconut Garden for Additional Returns. *Int. J. Plant Sci.* **12(2)**: 164-168.
14. Devi, L. G., Gladis, R. and Joseph, B. 2011. Performance studies of various crops in coconut based cropping system compared to sole cropping. *J. Progressive Agric*, **2(1)**: 47-50.
15. https: //www.fao.org/home/en

16. Farsanashamin, P and Anilkumar, A. S. 2016. Sustainable soil fertility management in coconut based multi-storeyed cropping system. *J. Plantation Crops,* **44(1)**: 1–7.

17. Gladis, R., and Joseph, B. 2015. Multiple Cropping Systems Involving Cassava and Elephant Foot Yam in Coconut Gardens. *J. Root crops,* **39(2)**: 242-246.

18. Ghavale, S. L., Shinde, V. V., Wankhede, S. M., Maheswarappa, H. P. and Haldankar, P. M. 2020. Carbon sequestration and productivity potential of coconut (*Cocos nucifera* L.) hybrids and varieties under coastal eco-system of Maharashtra. *Curr. J. Appl. Sci. Technol.* **39(22)**: 30-37.

19. Ghosh, D.K., Bandopadhyay, A., Maji, M.K. and Mahapatra. S. 2007. Studies on the performance of medicinal plants under coconut plantation in West Bengal. *Indian Coconut Journal,* **38**(8): 15-18.

20. Ghosh, D. K., Hore, J. K., Bandopadhyay, A., and Maji, M. K. 2008. Effect of spacing and seed corm size of elephant foot yam on economics of a coconut based cropping system. *J. Crop weed,* **4(1):** 15-19.

21. Ghosh, D. K. and Hore, J. K. 2011. Economics of a coconut-based inter-cropping system as influenced by spacing and seed rhizome size of ginger. *Indian J. Hortic,* **68(4):** 449-452.

22. Ghosh, D.K. and Bandopadhyay, A. 2011. Productivity and profitability of coconut based cropping systems with fruits and black pepper in West Bengal. *J. crop weed.* **7(2):** 134-137.

23. Jaganathan, D., Suja, G. and Byju, G. 2021. Enhancing farm income by coconut based cropping systems involving climate resilient tuber crops. *Indian Coconut Journal,* 17.

24. Khandekar, R. G., Nagwekar, D. D., Sawant, V. S., Gurav, S. S. and Haldankar, P. M. 2014. Performance of *Morinda citrifolia* as mixed crop in coconut under Konkan region of Maharashtra. *Int. J. Res. Emerg. Sci. Technol.* **1(4):** 63-6.

25. Korikanthimath, V. S., Rajendra, H., Ravindra, M., and Hosmani, M. M. (2000). Coconut based cardamom multistoreyed cropping system compatibility and yield pattern. *Crop Research* (*Hisar*), **20(2)**: 343-349.

26. Krishnakumar, V., Maheswarappa, H. P., Jayasekhar, S. and Shanavas, M. 2011. Economic evaluation of high-density multi-species cropping system in root (wilt) affected coconut (*Cocos nucifera* L.) area in Kerala. *J. Plantation Crops,* **39(1)**: 125–130.

27. Krishnakumar, V and Maheswarappa, H. P. 2010. Integrated nutrient management for root (wilt) diseased coconut (*Cocos nucifera* L.) palms. *Indian J. Agric. Sci.* **80(5)**: 394–398.

28. Malhotra, S. K., Maheswarappa, H. P., Selvamani, V., and Chowdappa, P. 2017. Diagnosis and management of soil fertility constraints in coconut (*Cocos nucifera L*): A review. *Indian J. Agric. Sci.* **87 (6)**: 711–26.

29. Maheswarappa, H. P., Anithakumari, P. and Sairam, C. V. 2003. High density multispecies cropping system for root (wilt) affected coconut gardens- Its impact on productivity and economic viability. *J. Plantation Crops*, **31**: 23–27.

30. Maheswarappa, H. P., Anithakumari, P., Kamalakshiamma, P. G. and Shanavas, M. 2005. Influence of integrated nutrient management and high density multispecies cropping system

on soil properties, plant nutrition and yield in root (wilt) affected coconut palms. *CORD* **21(2)**: 18–9.

31. Maheswarappa, H. P., Krishnakumar, V., Srinivasa Reddy, D.V., Dhanapal, R. and John Zachariah, T. 2012. Performance of different varieties/hybrids of black pepper (*Piper nigrum* L.) as mixed crop in coconut garden. *J. Plantation Crops*,**40(2)**: 82-87.
32. Maheswarappa, H. P., Dhanapal, R., Subramanian, P. and Palaniswami, C. 2013. Evaluation of coconut- based high density multi-species cropping system under organic and integrated nutrient management. *J. Plantation Crops*, **41(2)**: 130–135.
33. Maheswarappa, H. P., Subramanian, P and Geetha Kumari. A. 2017. Influence of organic nutrient management practices for noni (*Morinda citrifolia*) when grown as an intercrop in coconut garden. *Indian J. Agric. Sci.* **87 (9)**: 1155–7
34. Maheswarappa, H. P. and Sumitha, S. 2018. Doubling farmers' income through palm based cropping under different agro climatic regions of India. *In*: XXI Biennial National Symposium of Indian Society of Agronomy "Doubling Farmers' Income Through Agronomic Interventions Under Changing Scenario". 24–26 October, 2018 at MPUAT, Udaipur, Rajasthan. pp 45-47.
35. Marimuthu, R. 2005. Multispecies cropping system in coconut garden. *Madras Agric. J*, **92(7-9)**: 404-406.
36. Nath, J. C., Saud, B. K., Chowdhury, D., Deka, K.K. and Sarma, U. J. 2008. Coconut based high density multispecies cropping system in Assam. *J. Plantation Crops*, **36(2)**: 98–102.
37. Nath, J. C., and Deka, K. K. 2010. Second generation coconut based high density multispecies cropping system in Assam. *Improving productivity and profitability in coconut farming*, 167.
38. Nath, J. C., Deka, K. K., Saud, B. K. and Maheswarappa, H. P. 2015. Intercropping of medicinal and aromatic crops in adult coconut garden under Brahmaputra valley region of Assam. *J. Plantation Crops*, **43(1)**: 17–22.
39. Nath. J.C**.**, Sumitha. S. and Maheswarappa. H.P. 2018. Black pepper, promising intercrop in coconut gardens of Assam. *Indian Coconut Journal.* **61 (8)** : 27- 28.
40. Nath, J. C., Deka, K. K., Maheswarappa, H. P. and Sumitha S. 2019. System productivity enhancement in coconut (*Cocos nucifera* L.) garden by intercropping with flower crops in Assam. *Indian J. Agric. Sci.* **89 (11)**: 1842–5.
41. Naveen Kumar, K. S., Maheswarappa, H. P. and Basavaraju, T. B. 2016. Effect of integrated nutrient management practices on growth and yield of vegetable crops grown as intercrops in coconut garden. *Indian J. Agric. Sci.* **86(10)**: 1361–5.
42. Naveen Kumar, K. S., Maheswarappa, H. P. and Basavaraju, T. B. 2017. Productivity and economic benefits of coconut based vegetable cropping systems under central dry zone of Karnataka. *J. Plantation Crops*, **45(1)**: 49–56.
43. Naveen Kumar, K. S. 2015. 'Performance of coconut based cropping system with vegetable crops under integrated nutrient management in central dry zone of Karnataka'. Ph D thesis, UHS, Bagalakot.

44. Nagwekar, D. D., Desai, V. S., Sawant, V. S., Haldankar, P. M., Arulraj, S., and Jadhav, B. B. 2010. Intercropping with fruits and annual spices in coconut under Konkan condition of Maharashtra. *Improving productivity and profitability in coconut farming*, 174.

45. Nagwekar, D.D., Sawant, V.S., Haldankar, P.M., Jadhav, B.B., Arulraj, S.and Maheshwarappa, H.P. 2013. Performance of medicinal and aromatic plants as intercrops in coconut plantations in Konkan region of Maharashtra. *J. Plantation Crops,* **41(3)**: 384-388.

46. Padma, E., Ramanandam, G., Kumar, K. R., Rao, N. C., and Maheswarappa, H. P. 2016. Performance of coconut-based high-density multispecies cropping system under East Godavari Dist. of Andhra Pradesh. *Int. J. Innov. Hortic,* **5 (1):** 29-33.

47. Padma, E., Ramanandam, G., Dorajee Rao, A.V.D., Kalpana, M. and Maheswarappa, H.P.2018. Performance of Medicinal and Aromatic Crops as intercrops in coconut garden under East Coast of Andhra Pradesh. *Ind. J. Pure App. Biosci.* **6(2):** 421-426.

48. Palaniswami, C., Thomas, G. V., Dhanapal, R., Subramanian, P., and Maheswarappa, H. P. 2010. Nutrient management in coconut based high density multispecies cropping system. *Coconut based cropping/farming systems. CPCRI (ICAR), Kasaragod, Kerala*, 80-96.

49. Phukon, R. M., Nath, J. C., Sumitha, S., Maheswarappa, H. P., and Apshara, E. 2021. Early evaluation of cocoa (*Theobroma cacao L.*) clones as intercrop in coconut gardens under red river region of Assam State, India. *J. Plantation Crops,* **49(2):** 77–82.

50. Parulekar, Y. R., Khandekar, R. G., Haldankar, P. M., Babu, N., Cheriyan, H., Pethe, U. B. and Rathod, R. R. 2020. Integrated multi-storeyed cropping of tree Spices in coconut plantations for sustainability under aberrant coastal Climate. *J. Adv. Agric. Technol.* **6(1)**: 105-113.

51. Ranasinghe C.S. and Thimothias K.S.H. 2012. Estimation of carbon sequestration potential in coconut plantations under different growth conditions. *J. Natl. Sci. Found.* **40(1)**: 77-93.

52. Rethinam, P. and Sivaraman, K. 2009. Poly culture in coconut for humid tropics and coastal plains. National Resources India Foundation, New Delhi, pp.: 1 -392

53. Rani, S., Rajakumar, D., Shoba, N. and Maheswarappa, H. P. 2018. Productivity and economic advantages of flower crops in coconut based intercropping system. *Indian J Hortic.* **75(2)**: 279–82.

54. Rani, S., Maheswarappa, H.P. and Sudhalakshmi. C. 2021.Impact of temperature and rainfall on production and productivity of coconut. *Indian J Hortic.* **78(3)**: 287-291

55. Salam, P. K., B. Singh, D. S. Ram and R. K. Patel.2017. Effect of spacing and size of planting material on elephant-foot yam grown as intercrop in coconut garden. *J. Root crops,* **42(2)**: 62-65.

56. Siddappa, R., Basavaraju, T.B., Maheswarappa, H.P. and Manjunath Hubballi.2018. Performance of noni (*Morinda citrifolia*) in coconut garden as a mixed crop in southern dry zone of Karnataka. *J. Pharmacogn. Phytochem.*157-160.

57. Subramanian, P., Srinivasa Reddy, D. V., Palaniswami, C., Gopalasundaram, P., and Upadhyay, A. K. 2005. Studies on nutrient export and extent of nutrient recycling in coconut based high density multispecies cropping system. *CORD,* **21(1)**: 20-27.

58. Subramaniam, P. S., Mohandas, S., Marimuthu, R., Gurusamy, A., Natarajan, C., and Arulraj, S. 2010. Multispecies cropping system in coconut for East Coast region of Tamil Nadu. *Improving productivity and profitability in coconut farming*, 162.

59. Subramanian, P., Maheswarappa, H. P., Zachariah, T. J., Surekha, R., Selvamani, V., and Bhat, R. 2016. Performance of black pepper in coconut based high density multi-species cropping system under different nutrient managements. *J. Plantation Crops*, **44(2):** 90-95

60. Shinde, V.V., Ghavale, S.L. and Maheswarappa. H.P. 2017. Intercropping flower crops a best option in coconut based cropping system of Maharashtra. *Bioinfolet*, **14 (2) :** 142 - 145

61. Shinde, V. V., Maheswarappa, H. P., Ghavale, S. L., Sumitha, S., Wankhede, S. M. and Haldankar, P. M. 2020. Productivity and carbon sequestration potential of coconut-based cropping system as influenced by integrated nutrient management practices**.** *J. Plantation Crops*, **48**(2): 103–11.

62. Shinde, V.V., Sumitha, S. and Maheswarappa. H.P. 2021. Soil fertility properties, leaf nutrient status and yield of coconut and intercrops as influenced by integrated nutrient management (INM) and coconut based cropping system in coastal plain of western India. *Bangladesh J. Bot.* (Accepted)

63. Thiruvarassan, S. and Maheswarappa. H.P. 2014. Performance of medicinal and aromatic plants as intercrops in coconut garden. *J. Plantation Crops,* **42(2)***:* 238-240.

Chapter 15

Floriculture: Road Map for Future

K.V. Prasad

ICAR–Directorate of Floricultural Research, Pune, Maharashtra

e-mail: director.dfr@icar.gov.in

ABSTRACT

The main stay of Indian floriculture is growing of traditional flowers in open field conditions which is mostly in the hands of small and marginal farmers. From 3.40 lakh hectares of land, India produces about 19 lakh MT of loose flowers and 8.90 lakh MT of cut flowers per annum contributing significant revenue to the exchequer. The domestic trade from loose and cut flowers is estimated to be around Rs.19000cr which is poised to grow to Rs 55000cr by 2025. The export of floricultural produce is hovering around Rs 500cr during last 5 years. The domestic trade is mostly unorganised while the export trade is better organised.

India has a sizeable domestic trade of foliage and flowering ornamental plants, landscaping plants, lawn grass, trees, palms, dry flowers, and value-added products. The trade in these commodities is a silent contributor to the economy and is not captured in the overall trade. India also exports some of the above commodities albeit in low volumes. India is the sixth largest exporter of essential oils which is not accounted for under floriculture sector although majority are extracted from flower crops. Considerable trade is on the anvil in nutraceutical pigments derived from flowers both in domestic and export markets through contract farming of marigold.

Owing to the favourable agroclimatic conditions, adequate natural resources in terms of land, water, diverse soils and other inputs India has enormous potential to become a leader in floriculture. In spite of these strengths, India imports some of the exotic floricultural produce from different countries to a tune of Rs.110-120cr per annum.

However to attain the leadership position there is need to develop forward and backward linkages for domestic and export markets, value chain for each commodity and link the same to the One District One Crop (ODOC), encourage start-ups and FPOs.

A national policy on floriculture needs to be put in place to address various issues like support for R&D, import substitution, subsidies for domestic and export commodities, custom hiring, crop insurance, post-harvest infrastructure including cold chain, integrated flower auction centres dedicated to flowers, skilling the stakeholders and capturing the data of those commodities that are not crystallised at present. This calls for integrating the initiatives with the national programmes/priorities/Missions like Per Drop More Crop (PMSKY), Crop insurance (PMFBY), StartUp India, Green India Mission (GIM), National Biodiversity Mission, National Mission on Waste to Wealth and Skill India *etc.*,

This paper dwells in detail on steps already taken and those that needs to be taken to create a roadmap for a vibrant floriculture sector.

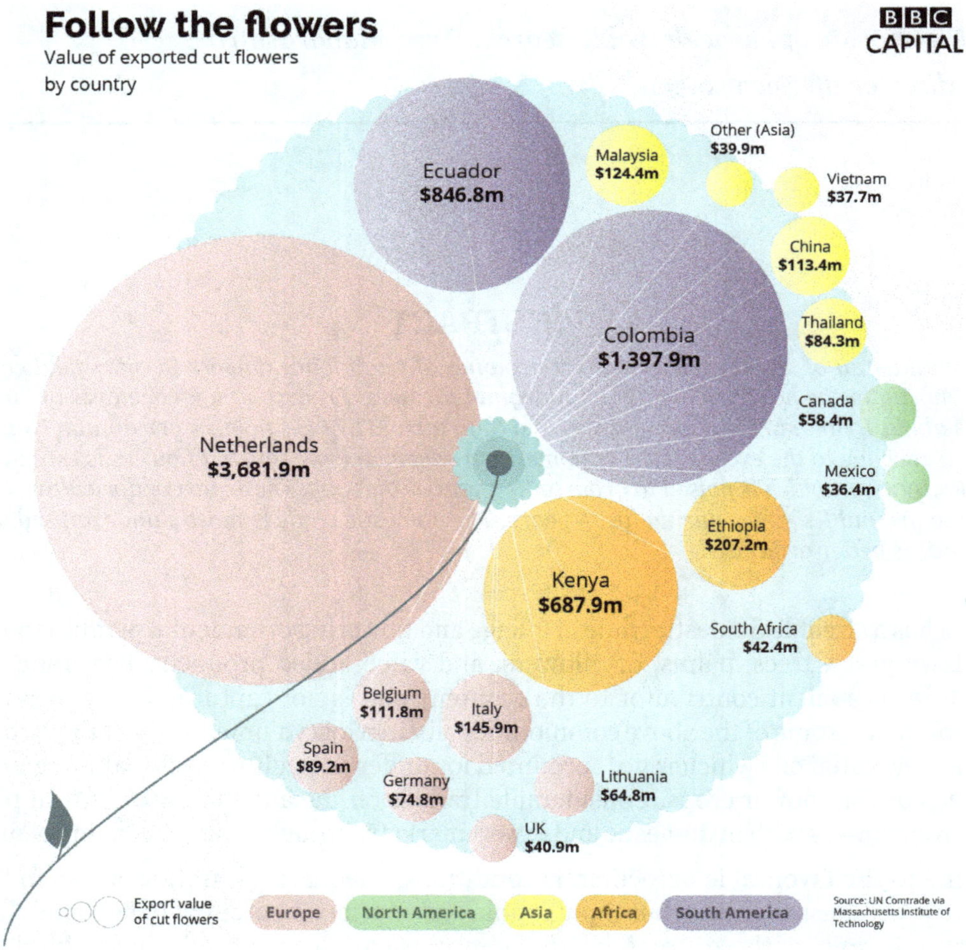

Figure 1. Major players in the global export of flowers (2020).

Global Trade

Floriculture Industry is part of Lifestyle Horticulture industry that is worth US $ 300 Billion. It is 352 most traded commodity contributing 0.05 per cent of total global trade. It is valued at UD$8.45 billion (2020).

The Netherlands is the leading producer, importer and exporter (Figure 1) of floricultural produce in the world (US$3681.9 m) followed by Columbia (US$1397.9m), Ecuador (US$846.8 m), Kenya (US$ 687.9m) and Ethiopia (US$ 207.2 m). The top 5 producers contribute nearly 85 per cent of the overall trade (Figure 2).

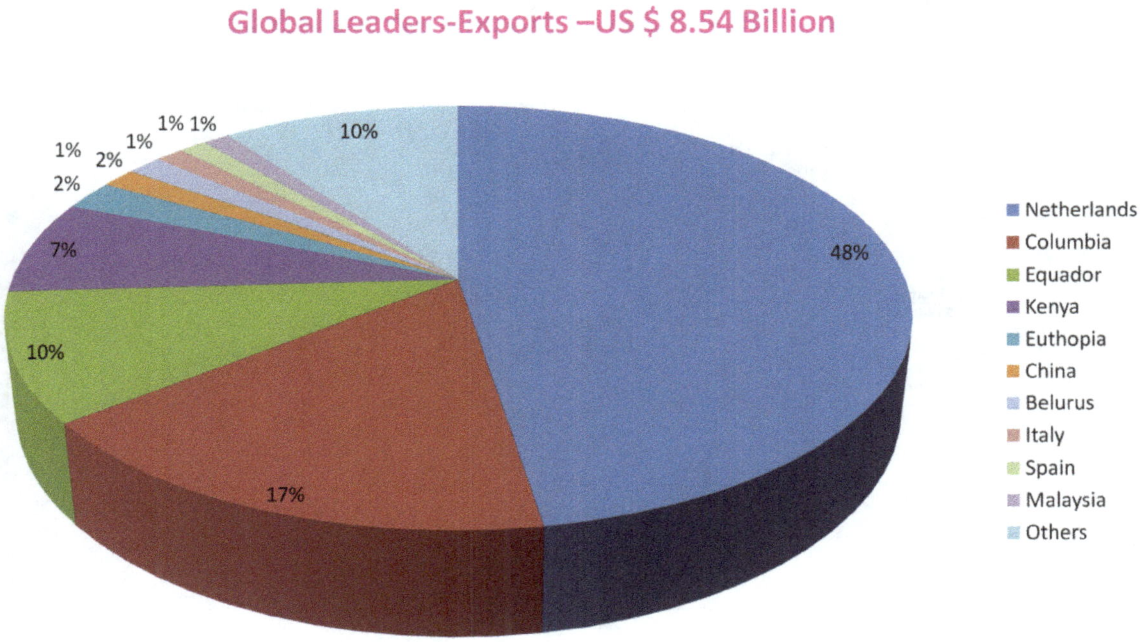

Figure 2. Share of different countries in global exports.

The United States of America is the largest importer of floricultural produce (US$ 1735.7m) followed by Germany (US$ 1204.1m), the Netherlands (US$ 920.9m), United Kingdom (US$ 973m) and Russia (US$445.5m). Together the top five countries consume 74 per cent of the global imports (Figure 4).

Indian Scenario

Floriculture is in Indian Culture and consumption of floricultural produce is increasing which is driven by the young middle-class population (65 per cent population that is less than 35 years of age). The consumption which was restricted to the maga cities and metros earlier is now percolating to the Tier II and Tier III cities which are more than 50 in number with 10 lakh population. The booming economy (US$ 3.75 Trillion) is the main driving force with dispensable double income groups. Floricultural sector is poised to grow further as India aims to attain US$

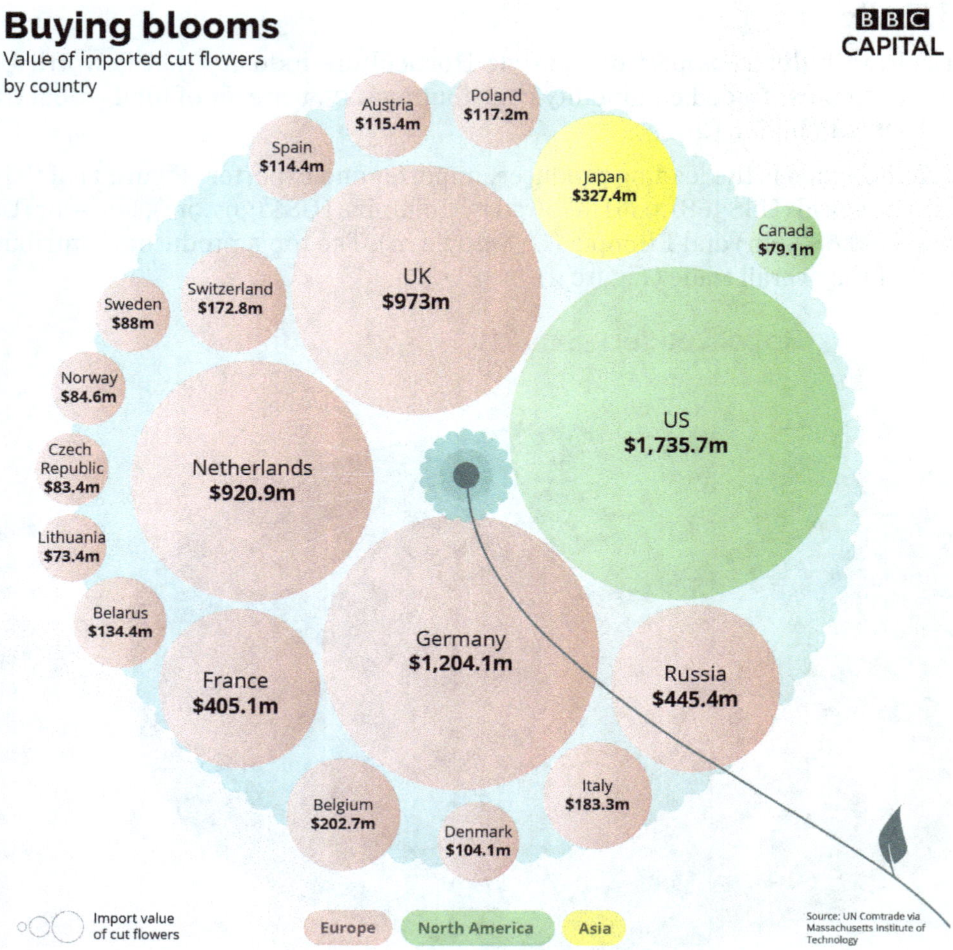

Figure 3. Major importers of floricultural produce.

5.00 Trillion economy soon. Proactive policies of the Government of India through central sector schemes like Mission for Integrated Development of Horticulture (MIDH) and State specific interventions paved the way for the growth of the floricultural sector that contributed Rs 25000 crores to the exchequer from domestic trade and Rs 772 crores through exports.

Till the advent of protected cultivation in greenhouses during the early 90's, Indian farmers followed traditional practices of growing loose flowers like rose, jasmine, marigold, chrysanthemum, tuberose, crossandra *etc.*, under open field conditions. Booming Indian economy, liberalized seed policy and ever-growing habit of 'saying it with flowers' paved the way for state-of-the-art greenhouse technology during early 90's. Liberalization from all the fronts has not only facilitated import of latest technology, but also encouraged and enhanced import of internationally acclaimed planting material of different varieties. The Government

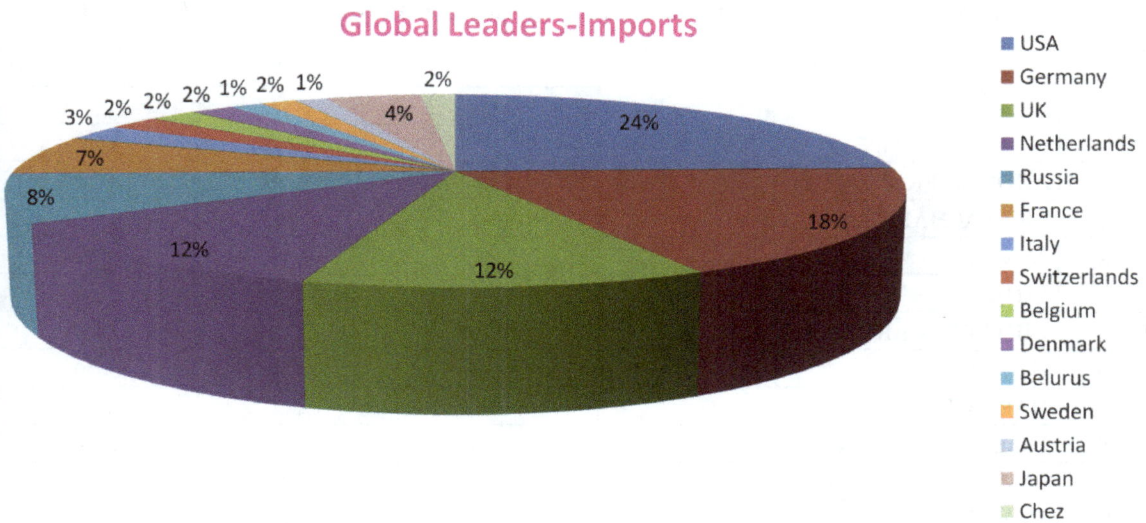

Figure 4. Leading importers of floricultural produce.

of India made various provisions and introduced a variety of incentives helping to establish 100 per cent Export–Oriented–Units (EOU's) resulting in a vibrant floriculture industry in India. Indian floriculture industry, which was hitherto individual driven, became corporate driven during 90's. Open filed cultivation of traditional flowers still remains the backbone of Indian floriculture with 95 per cent of area out of 282380 ha of land producing 2301000 MT of loose flowers (Table 1). Cut flowers are grown in the remaining 5 per cent of the area producing 762000 MT of different cut flowers (NHB 2019-20). Andhra Pradesh, Tamil Nadu, Karnataka, Gujarat, West Bengal, Madhya Pradesh are the major loose flower producing states whereas West Bengal, Karnataka, Odisha, Assam are the major cut flower producing states (Table 2).

Table 1. Area and production of flowers during the last two decades

Year	*Area (ha)*	*Production (MT)*	*Year*	*Area (ha)*	*Production (MT)*
2001-02	106000	535000	2002-03	70000	735000
2003-04	101000	580000	2004-05	118000	659000
2005-06	129000	654000	2006-07	144000	880000
2007-08	166000	868000	2008-09	167000	987000
2009-10	183000	1021000	2010-11	191000	1031000
2011-12	254000	1652000	2012-13	233000	1729000
2013-14	255000	2297000	2014-15	248000	1658000
2015-16	278000	1526000	2016-17	306000	1699000
2017-18	342000	1760000	2018-19	339386	1991381
2019-20	305000	3063000	2020-21	282380	2813470

Table 2. Area and production of loose and cut flowers in different states (2020-21)

Sl.No.	State/UT	TOTAL		
		A	P	
			Loose	Cut
1	ANDHRA PRADESH	14.58	113.26	17.55
2	ARUNACHAL PRADESH	0.00	0.00	0.00
3	ASSAM	5.31	35.58	57.80
4	BIHAR	1.21	10.95	0.19
5	CHHATTISGARH	12.23	57.20	115.37
6	GUJARAT	20.57	194.83	0.00
7	HARYANA	1.61	17.12	0.77
8	HIMACHAL PRADESH	0.381	1.547	11.242
9	JAMMU and KASHMIR	0.10	0.15	0.67
10	JHARKHAND	1.13	2.14	3.63
11	KARNATAKA	46.66	417.55	19.06
12	KERALA	0.35	0.63	0.12
13	MADHYA PRADESH	37.65	426.62	0.00
14	MAHARASHTRA	11.93	73.37	41.73
15	MANIPUR	0.07	0.01	0.17
16	MEGHALAYA	12.47	0.00	0.35
17	MIZORAM	0.08	0.00	0.80
18	NAGALAND	0.04	0.04	0.24
19	ODISHA	5.64	31.72	33.69
20	PUNJAB	2.20	13.78	0.00
21	RAJASTHAN	3.67	8.42	0.00
22	SIKKIM	0.24	16.50	0.09
23	TAMIL NADU	45.61	590.32	2.49
24	TELANGANA	4.06	58.57	7.90
25	TRIPURA	0.00	0.00	0.00
26	UTTAR PRADESH	22.76	49.66	69.78
27	UTTARAKHAND	1.61	3.02	11.36
28	WEST BENGAL	30.04	74.95	219.58
29	OTHERS	0.21	0.88	0.04
	TOTAL	282.38	2198.83	614.64

Owing to the wide range of agroclimatic conditions a number of loose and cut flowers are grown across the country (Tables 3 and 4) which can be identified as dedicated clusters.

Table 3. Crop Wise Loose Flower Producing States

Major Loose Flowers	*Major Producing States*
Rose	Karnataka, Tamil Nadu, Andhra Pradesh, Uttar Pradesh, Rajasthan, West Bengal, Bihar, Jharkhand, Telengana
Chrysanthemum	Tamil Nadu, Karnataka, Andhra Pradesh, Kerala, Odisha, Maharashtra, Telangana
Marigold	Uttar Pradesh, Haryana, Punjab, Karnataka, West Bengal, Maharashtra, Jammu and Kashmir, Tamil Nadu,
Tuberose	Tamil Nadu, Karnataka, Andhra Pradesh, Uttar Pradesh, Maharashtra
Jasmine	Tamil Nadu, Andhra Pradesh, Karnataka, Kerala, Maharashtra
Crossandra	Tamil Nadu, Andhra Pradesh, Karnataka, Kerala
Barleria	Tamil Nadu, Andhra Pradesh, Karnataka, Kerala
Spider Lily (Hemarocallis)	Maharashtra, Gujarat, Chattisgarh, Jharkhand

Table 4. Crop-wise major cut flower growing regions

Major Cut Flowers	*Major Production Centers*
Rose	Bangalore, Pune, Nasik, Dehradhun, Kolkata, Hosur
Carnation	Sangli, Satara, Bangalore, Ooty, Kulu, Shimla, Solan, Kolkata, Hosur
Chrysanthemum	Bangalore, Deharadun, Pune, Kolkata, Hosur
Gerbera	Sangli, Satara, Pune, Bangalore, Solan, Hosur
Orchids	Trivendrum, Chennai, Mangalore, Guwahati, Gangtok, Shillong
Anthurium	Coorg, Kerala, Bangalore, Mangalore, Aizol, Dhimapur, Shillong
Lilium	Ooty, Bhimtal, Kulu, Solan, Palampur,
Heliconia and Bird of Paradise	Bangalore, Trissur, Pune, Hosur
Limonium and Lisianthus	Ooty, Kodaikanal, Hosur

Floriculture Exports from India

The transformation of India floriculture from pushcart transportation to charted flight transportation is phenomenal. Floricultural exports from India comprises of fresh cut flowers (to Europe, Japan, Australia, Middle East and USA) Loose flowers (for expatriate Indians in the Gulf, South Asian Countries) cut foliage (to Europe) Dry flowers (To USA, Europe, Japan, Australia, Far East and Russia) Potted Plants (Limited to very few countries). Out of these components dry flowers contribute a major share to the total export.

The floriculture exports which stood at Rs. 63 crores during 1996-97 increased almost 9 times to Rs. 575 crores during 2020-21 (Table 5).

Table 5. Export trends during the last three years

Country	*2020-21*			*2021-22*			*2022-23*		
	Qty in MT	*Rs. Lacs*	*US$ Mill*	*Qty in MT*	*Rs. Lacs*	*US$ Mill*	*Qty in MT*	*Rs. Lacs*	*US$ Mill*
U S A	3,139.19	15,895.51	21.51	3,520.05	22,231.02	29.83	2,631.82	18,078.94	25.54
Netherland	1,603.87	10,930.30	14.74	2,206.37	14,710.77	19.73	1,552.80	12,844.60	18.15
U Arab Emts	1,659.88	3,443.22	4.67	3,074.79	3,659.65	4.9	3,973.82	4,733.00	6.69
U K	860.91	3,312.10	4.51	917.69	3,962.30	5.33	635.34	3,519.69	4.97
Germany	1,054.69	3,213.11	4.32	1,208.63	5,071.42	6.81	710.17	3,089.33	4.36
Malaysia	516.38	963.55	1.31	882.13	1,832.28	2.45	1,223.63	3,049.30	4.31
Canada	485.65	2,301.48	3.1	768.59	3,356.77	4.51	696.5	2,859.30	4.04
Italy	235.31	1,767.47	2.4	261.18	2,136.76	2.86	207.57	1,930.31	2.73
Singapore	1,418.93	1,119.58	1.52	1,932.30	1,558.16	2.09	2,009.12	1,894.29	2.68
France	262.64	769.49	1.04	302.59	1,578.39	2.12	246.82	1,727.53	2.44
Others	4457.87	13,882.64	18.74	8522.85	17,043.96	22.84	7136.82	17,054.22	21.21
Total	**15,695.32**	**57,598.45**	**77.86**	**23,597.17**	**77,141.48**	**103.47**	**21,024.41**	**70,780.51**	**88.35**

The export basket (Figure 5) comprises of dry flowers (71 per cent), fresh cut flowers (18 per cent) live plants (9 per cent), fresh bulbs(1 per cent) and foliage (1 per cent).

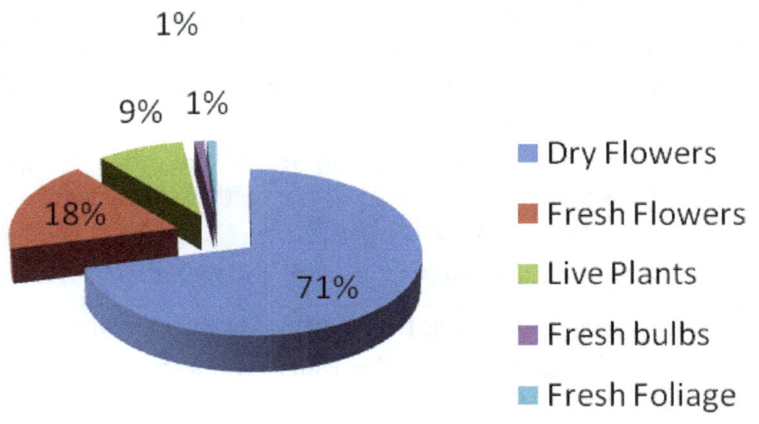

Figure 5. Composition of Indian floriculture exports.

Import of Floricultural Produce

India also imports (Table 6) sizable amount of floricultural produce (Rs 160. Cr during 2021-22) mostly temperate orchids, tulips and other bulbous flowers from the Netherlands, tropical orchids from Thailand, Cala Lily from New Zealand, protea from South Africa *etc.*, Scope exists to minimize the imports by developing dedicated cluster for the exotic flowers.

Table 6. Import of floricultural produce (2020-21)

Countries	*Qty*	*Rs. Crore*
ITALY	899.26	40.18
NETHERLAND	1118.96	35.53
CHINA P RP	1056.41	26.04
U S A	37.59	13.52
THAILAND	213.4	11.9
U K	10.21	7.33
SPAIN	153.7	4.13
IRAN	3.69	3.63
FRANCE	4.52	1.88
KENYA	8.26	1.64
Top 10 Total	3506	145.78
Other Countries	452.57	14.6
Total	**3958.57**	**160.38**
Per cent Share of Top 10 Countries	**88.56**	**91.25**

Challenges

Although floriculture is practiced across the country the data captured on Area and Production is sketchy with many missing values each year when we analyze the data captured by NHB. The data mostly pertains to loose flowers and cut flowers leaving aside many silent contributors like ornamental cut foliage, plant nursery, dry flowers, essential oils, natural colours and dyes, nutraceutical pigments and tissue culture plants.

The Floriculture sector is highly unorganized and the markets are mostly confined to roadsides in rural as well as urban areas. The post harvest losses therefore range from 10-30 per cent at various stages of the supply chain.

To address the challenges in different verticals the following measures are suggested.

<table>
<tr><td rowspan="3">Loose Flowers/ Traditional Flowers Align with Floriculture Missions planned by Govt of Odisha and UP</td><td>Data Rationalisation</td><td>❑ Authentic Data capture (Area, Production)
❑ Authentic Value Estimates (Many silent contributors like nursery industry)
❑ Authentic estimation of Post harvest losses (30 per cent)
❑ Dynamic data on Market arrivals and consumption patterns</td></tr>
<tr><td>Cluster Development</td><td>❑ Declare floricultural activity as an essential services
❑ Bring 100 per cent area under micro-irrigation PMKSY
❑ To promote mulching for floricultural crops: reduce manual weeding
❑ Incentives to small scale industry to make tools and gadgets for floriculture sector
❑ Promotion of solar powered AC push carts for retail sale of flowers (Green India)
❑ Promotion of foldable crates for packaging
❑ SHGs for custom hiring of equipment and tools
❑ Dedicated cold store for smaller markets
❑ Bring under the ambit of PMFBY</td></tr>
<tr><td>Development of Value Chain</td><td>Establish Integrated Markets (Flori- Mall) that contain
❑ Cold chain
❑ Pilot plant for essential oils extraction
❑ Pilot plant for nutraceutical pigment extraction
❑ Pilot plant for natural flower dyes extraction
❑ Pilot plant for gulkand making (roses)
❑ Pilot plant for Dry flower making and Solar Dryers for Dry flower units
❑ Pilot plant for Herbal teas
❑ Pilot plant for handmade paper
❑ Pilot plant for agarbatti and floral gulal
❑ Pilot plant for Floral Meal for poultry
❑ Vermicompost units</td></tr>
</table>

		❑ Women SHG's for making value added products (mala/veni) for domestic and export markets on the lines of Value Chain Model for jasmine developed under NAIP by TNAU, Coimbatore
Cut Flowers	Cluster Development	❑ Subsidy to continue for erection of polyhouse, replacement of cladding materials, hail proof nets, *etc.* along with input support ❑ Identification and promotion of certified nurseries for QPM ❑ Promotion for use of alternate energy source (Green India) ❑ Less dependence on Water Soluble Fertilizers (WSF) (One India one Fertilizer) ❑ Indigenous varieties for cut flower (Research) ❑ Develop Multi-tier greenhouse ❑ Intensify rain water harvesting, solar farming ❑ Promotion of crop specific SHGs ❑ Introduction of new technologies and technology demonstration on retractable greenhouses at ICAR- institutes ❑ Bring under the ambit of PMFBY
	Export Promotion	❑ Develop India GAP ❑ IPR protected varieties
	Development of Value Chain	❑ Greenhouse clusters with florimalls (essential oil extraction, pigment and dye extraction, gulkand making, petal embedded handmade paper, dry flower, vermi-compost)
Pot Plants (Nurseries)	Cluster Development	❑ Accreditation of nurseries ❑ Development of unified standard for planting materials ❑ Support for infrastructure, input cost and linkages with flori-malls ❑ Identify Nursery activity and pot plant rentals as agriculture activity ❑ Develop diagnostic labs at ICAR institutes to address phyto-sanitary issues ❑ Indexing of mother plant/planting materials ❑ Standards for domestic markets ❑ Align with StartUp India programme ❑ Intensive training and capacity building ❑ Sensitization of nurserymen on IPR, bio-security issues, emerging threats ❑ Establishment of Garden Centres (One stop Solution for all Garden Needs)
	Export promotion	❑ IPR protected material ❑ Support for soilless media ❑ Develop Export standards for all categories ❑ Develop Good Management Practices

B. Bedding Plants		❑ Accreditation of nursery ❑ Indexing of mother plant/planting materials ❑ Development of diagnostic tools
Florists and Floral Decoration		❑ Import restrictions on plastic flowers ❑ Support to florist: cool box, air conditioner, subsidized floral preservatives, tools and gadgets, *etc.*
Turf and Landscaping		❑ Import of new lawn grasses ❑ Development of standards for different components of landscaping ❑ Carbon trading ❑ Support for certification programme ❑ Rationalization of water use and use of STP water ❑ To bring the landscape services to minimum tax regime
Tissue culture	Cluster Development	❑ Support for establishment of accredited diagnostic labs. ❑ PPP mode for production of disease free plants ❑ Accreditation of Tissue culture labs ❑ Dedicated cold chain ❑ Support for air cargo ❑ Protocols for sea transportation to nearby counties
Essential oils	Cluster Development	❑ Cluster development for essential oil bearing crops ❑ Support to SHGs for oil extraction units, ❑ Modernatization of essential oil extraction units ❑ Mobile oil extraction units on custom hiring ❑ Tools and gadgets for efficient harvesting of flowers ❑ Development of cold chain for volatile ❑ Align Agri Startups
Nutraceutical pigments	Cluster Development	❑ Development of cluster in selected pockets - pigment bearing crops ❑ Mobile extraction units on custom hiring ❑ Tools and gadgets for efficient harvesting of raw materials ❑ Support to cold chain ❑ Contract farming for native floras ❑ Create and sustain SHGs – cultivation, extraction ❑ Creation of required infrastructure
Flower Seed Production	Cluster Development	❑ Support for creation and strengthening of seed production cluster ❑ Integrating apiculture with floriculture ❑ Development of SHGs ❑ Support for creation of required infrastructure- processing and packaging

Way Forward

Over the years the Indian farmers have perfected the art and science of growing flowers. The growth of floriculture sector from 40000 ha during 1990 to 2.83 lakh ha during 2021-22 producing 2.31 million tonnes stands a testimony. The demand for flowers is mostly during festivals and weddings resulting in gluts often during the absence of events. Due to drastic fall in prices the farmers tend to discard the flowers as waste. During the Covid Pandemic floriculture sector faced stupendous distress as the demand/consumption was near to zero. To address such challenges there is a need to develop value chain for all the flower crops with dedicated integrated markets illustrated in Figure 6.

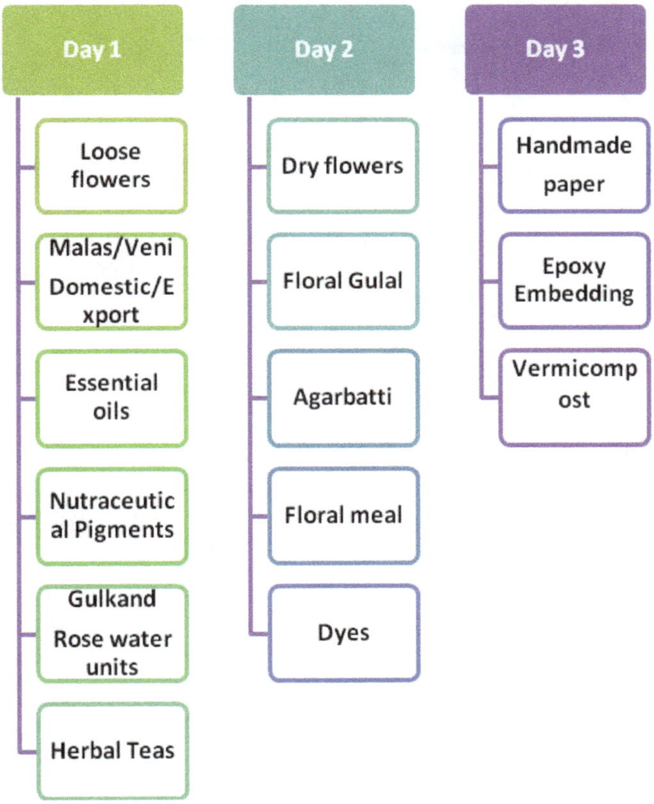

Figure 6. Model system for value addition in integrated markets.

There is need to develop dedicated markets in identified clusters with integrated markets with facilities for value addition. Depending on the arrival and the demand for a given day excess arrivals can be utilised for essential oil extraction, extraction of pigments, Gulkand and rose water making besides all fresh flower products on day one. The remaining quantities can be utilised for making dry flowers, floral gulal, agarbatti, floral meal and foral dyes on second day of arrival. Unsold flowers on day three can be converted to handmade paper, epoxy embedding and finally vermin-compost. Through such system it is possible to convert a major portion of the production to wealth with zero wastage.

Chapter 16

Advances in Protected Cultivation of Horticultural Crops

Balraj Singh

Sri Karan Narendra Agriculture University, Jobner 303329, Rajasthan

e-mail: drbsingh2000@gmail.com

ABSTRACT

Due to the increasing demand of the high quality fresh horticultural produce by world markets has led production of more and more horticultural crops under protected conditions. Recently, there has been demand, acceptance and interest by consumers in alternative crops species having unique visual appearance, tastes as well as source of biochemical compounds for increasing their immunity across the world. To meet out this demand of high quality produce area under protected cultivation across the world has increased many folds in the last two decades and is presently estimated more than 5.0 million hectares. As a result of significant research work carried out in the area of protected cultivation several advancements have been made and several innovative technologies have been developed in India and other countries in different areas viz.; greenhouse structures, cladding material, energy saving, automation, greenhouse environment control, precision input control technology, soilless cultivation, fertigation, vertical farming, development of new varieties/hybrids, grafting technology, identification of new crops, biological management of biotic stresses, pollination management, abiotic stress management, use of bio stimulants, mechanization and use of robotics, nursery production systems, on farm value addition and marketing of the produce. In India varieties and hybrids mainly in vegetables viz.; Standard Tomato, Parthenocarpic Cucumber, Cherry Tomato, Muskmelon, Summer Squash, Bitter gourd have been developed and released for protected cultivation. Some of these important crops and varieties are, Cherry tomato (Pusa cherry tomato-1), standard tomato (Pusa Rakshit), Parthenocarpic cucumber (Pusa Seedless Cucumber-6), Melon (Pusa Sunheri) from Indian Agricultural Research

Institute, New Delhi. Similarly other institutions like GBPUA and T, Pant Nagar, KAU, Kerala, IIHR Banglore, IIVR, Varanasi have also developed varieties and hybrids for tomato and parthenocarpic cucumber in last few years for protected cultivation under Indian conditions. New design of greenhouses have been developed at CPCT, IARI in which the side ventilated has been increased from 3.0 to 5.0 meters and ventilations through roof ventilators has been increased from 1.0 to 1.8 meter. Low pressure drip irrigation systems has been designed and installed for small protected structures having small area (up to 1000-2000 Sq. meters). Good Agricultural Practices (GAP) for protected cultivation of Tomato; Capsicum and Partheonocarpic Cucumber crops have been standardized for their use during greenhouses cultivation. Battery operated opening and closing of the shade net inside the greenhouses have been developed for its operation during the non availability period of electricity during day time for summer seasons. For an semi-arid and arid regions, greenhouse designs have been developed where sprinklers are installed on the roof gutters for their use during peak summer period for cooling down the inside temperature of the greenhouses. Self driven trolleys for greenhouses operators like trellising, training and pruning have been developed, which can be easily operated and used by a single worker. Indigenous bumble bees have been collected and being reared for pollination of greenhouses tomato crops, on the other hand stingless bees and carpenter bees have been found effective pollinators for small protected structures for crops like Cucurbits, Sweet pepper and Chillies. A tractor operated soil heating machine (up to soil depth of 30 cm) has been developed for controlling soil borne problems. Production of greenhouse horticultural crops is a very management intensive method, and therefore there is always need for adopting new and innovative technologies and advanced cultivation practices for successful production, maximum returns and sustainability of the technology under Indian conditions by greenhouse growers for meeting out the increasing demand of high quality fresh and safe produce.

***Keywords**: Protected cultivation, Vegetable crops, Fruit crops, Micro-irrigation, Fertigation.*

INTRODUCTION

The basic purpose of greenhouse designs in different regions of the world is to efficiently use of available natural resources such as solar radiations and temperature, skilfully while not depending upon external energy sources. In past traditional greenhouses were not achieving such goals and so research efforts have been made for developing advancements in these main critical aspects *viz.* Improving solar radiation transmission in winter in colder regions, Reducing excessive day time temperature in tropical, semi arid and arid regions, Increasing night time temperature during winters in that using external energy in colder regions.

Although natural ventilation is the most essential and economical tool that usually the growers are using for maintaining the greenhouse climate (*i.e.* temperature, humidity and CO_2) within the best ranges for the crop in different regions of the world for different seasons. During the last two decades research has widened our knowledge that how to optimise natural ventilation designs of greenhouses for different regions. Recent research has strongly improvised the important effect of wind direction in relation to the greenhouse vents on multi span greenhouses. When comparing wind ward with leeward ventilation, research indicates that for different types of greenhouses, when flap roof vents are used, windward ventilation provides higher rates of ventilation than leeward ventilation. In most of the studies in last decade showed higher ventilation rates when vents are perpendicular to dominant wind than when it is parallel.

On the other hand greenhouse height of the roof slope showed that increasing the roof slope by 25-30° increases the ventilation rate by almost a factor of two and also improves cooling of the cropped area inside the greenhouse. Evaporative cooling systems (fogging, pad and fan *etc.*) consume large amounts of good quality water which is a very scarce resource in many regions of the world, so their use is only justified in high value crops or nurseries. Active cooling by using cold water (chillers) is too expensive and it has very limited use across the world. Use of geothermal energy for cooling and heating of the greenhouses during peak winters and peak summer months is cost effective technology is different parts of the world. New designs of Roof retractable greenhouses have been made for tropical, semi arid and arid regions for removal of the excessive temperature accumulated inside the structures. Similarly, new designs of insect proof net houses have been made which are stronger to bear the high wind velocity during cyclones. In these net houses more steel cables are used instead the use of GI pipes and grouting is done at least 2-3 meters in the soil. In arid and semi arid reason use of sprinklers the gutters have been found highly efficient for reduction of thermal heat transmission through roof area during the peak summer periods in India. For arid and semi-arid regions new design of insect proof net house have been designed in which aluminium coated shading net (Aluminates) is provided 50-60 cm above the screen roof in rolling mode. During peak summers are aluminates can be spread over the roof, whereas during the winter months it can be closed for better transmission of the optimum sun light inside the net house for better crop growth and development. At Centre for Protected Cultivation Technology, Indian Agricultural Research Institute, New Delhi a new design of greenhouse has been installed in which side ventilation has been increased up to 5.0 meters; whereas ventilation area through roof ventilations have been increased up to 1.8 meters which is use full for arid and semi-arid regions for vegetable cultivation.

Cladding Material

Majority of the greenhouses across the world are using plastic film covers, as they are cheaper than the rigid materials and technically adaptable to all kinds of protected structures. Presently commercial greenhouses use plastic films with additives that stabilize them against Ultra Violate and chemicals like sulphur and chloride, which can extend their use to maximum of four growing seasons. Mostly multilayer, UV stabilized, anti dust, anti drip and that filters UV by limiting the infestation by pests without affecting the activity of pollinators inside the structures. For greenhouse growers durability of the plastic films is a major issue, as the price of replacement is quite high. Presently for covering the other low cost temporary protected structures like plastic low tunnels non-woven materials are also being used looking to the cost and environmental pollution. Similarly for mulching under open fields or under greenhouses, biodegradable (or compostable) plastic material is being recommended for use against the use of synthetic plastic films in different thickness (for 8 to 50 micron). New technologies or advancements in the area of covering materials, light emitting diode (use of LED) lighting, alternation nutrient resources and sensors that are expected to contribute to more digital, automatic and advanced greenhouse production system.

Crops, Varieties and Production Technologies

Vegetable Crops

Predominantly Parthenocarpic Cucumber, Sweet pepper, Tomato, Cherry tomato, water melon, Muskmelon, Summer squash (Zuccuni), Egg plants of different size, shapes and colours are being grown under different protected structures, mainly under greenhouses in different regions and countries of the world from colder to hot humid regions. Beyond this lettuce beans, few herbs like thyme, oregano, basils *etc.* are also cultivated under protected environment in different Countries as per the market demands. But India only few vegetables like Sweet pepper, parthenocarpic cucumber, tomato and cherry tomato crops are being cultivated different area and states mainly under greenhouses. Some other crops like summer squash (zuccuni), muskmelon and water melon are cultivated normally under low cost temporary structures like plastic low tunnels, walk in tunnels *etc.*, mainly in states like Punjab, Haryana and Rajasthan for their off season production and marketing for fetching high price of off season produce form the markets. But now new vegetables crops like have been introduced for protected cultivation *viz.*; okra (mainly under insect proof net houses), egg plants (under plastic low tunnels and insect proof net houses), bitter gourd and long melons (under plastic low tunnels and greenhouses), leafy vegetable crops like Pak choi, Coriander, Beet Leaf, Methi, Mint *etc.* during peak summer months under shade net houses along with some herbs like Basils, Thyme, Oregano *etc.* more vegetable crops needs to be suggested for their protected cultivation on the basis of their techno-economic viability for different regions and seasons looking to their demand and availability of varieties and seeds.

During the previous decade varieties and seeds of the commonly given vegetables like parthenocarpic cucumber, sweet pepper, tomato and cherry tomato were being supplied and sold by limited MNC'S through their import, but now varieties and hybrids have been developed by Indian public sector Research institutes like Indian Agricultural Research Institutes, New Delhi, Indian Institute of Horticultural Research, Bangalore, Kerala Agricultural University, Kerala, GBPUA and T, Pantnagar and others and research work is still going on for development of vegetables and hybrids in other vegetables for protected cultivation.

Fruit Crops

Among the fruit crops only one or two crops like Strawberry and Papaya was being cultivated mainly under plastic low tunnels by using raised beds equipped under drip irrigation and plastic mulching and nutrients but new fruit crops like Banana, Blueberry, Raspberry and Blackberry have been introduced for protected cultivation and mainly under greenhouses in India. More fruit crop like dragon fruit, pitaya and grapes *etc.* are required to be introduced looking their market demand and problems encountered under their open field cultivation. Gynodioecious varieties in papaya have been developed by some public sector institutions in India, which can be well suited for its protected and open field cultivation.

Ornamental Crops

Mainly cut flower purpose ornamental crops like cut flower roses, Carnation, Gerbera, Lilium, Orchids and Chrysanthemum *etc.* are presently being cultivated under greenhouses across the world mainly their domestic and expert demands. But in India only gerbera, carnation, chrysanthemum and in some areas cut flower roses are being grown for protected cultivation in different regions of the country. But recently some new floricultural crops like Anthurium, Orchids, Lilium and Gladiolus have been introduced for protected cultivation looking to their increasing demand during off seasons. Few varieties in ornamental cut flowers crops like chrysanthemum, Orchids, Anthurium have been developed for their protected cultivation. There is need to introduce bedding plants production on sizeable scale under protected conditions in different parts of the country looking their increasing demand in last few years.

Production Technology

Grafting in Vegetable Seedlings

Application of grafting technique in vegetable seedlings produced plants or samplings which are more vigorous and it is a complete non-chemical answerer to all soil borne problems created by fungus, bacteria, nematodes and for problematic soils (acidic or saline soils). This technology has been applied on commercial scale in vegetable belongs to Solanaceae and Cucurbitaceous family, but in India this technique has been standardized at IARI, New Delhi in world bank funded NAIP project and now it is being applied by few private nurseries and seed company in case of water melon, muskmelon and tomato only. In case of tomato, capsicum and brinjal, wild brinjal Solanum torvum is being used for grafting and some more root stock in brinjal has been identified which are compatible and resistant or tolerant to various soil borne problems. In some countries grafting has been greatly expanded to include various types of stress tolerance, increasing the yield and duration of the crops and also for enhancing quality of the produce. Nowadays rootstock breeding programmes in vegetables have been initiated for achieving all desirable traits in the root stocks and this is being used as an alternative breeding in horticultural crops for getting compatible rootstocks for improving nutrient uptake and utilization efficiency. Commercial application of this technology in vegetable seedlings across the world can drastically reduce the use of hazardous chemicals like Methyl bromide or others being used as fumigants.

Pollination Management and Fruit Setting

Success of fruit setting in vegetable under protected environment is linked to pollination management in India is a major challenge under different protected cultivation structures. Farmers are growing selected vegetables like tomato, cherry tomato, capsicum, cucurbits, brinjal, chilli *etc.* Pollination management in these vegetable crop under different protected structures *viz.;* semi climate control green houses, naturally ventilated green houses, insect proof net-houses *etc.*, is a very important area of research and development. Presently, various methods of pollination management in vegetable crops have been developed and standardized in several developed countries like Israel, Spain, Morocco, Italy *etc.* by utilizing bumble bees and *Osmia* sp. Limited work has been done in India by different institutes, but now the AICRP on Honey bees and Pollinators has advanced the research work on domiciliation and utilization of indigenous bumble bees, stingless bees and carpenter bees for pollination of crops grown under protected cultivation. Various methods that are used for pollination of different crops in protected structures is as follows:

a. **Use of electric vibrators:** Vibrators are used for buzz pollination of tomato and other solanaceous flowers to release the pollen from anther column. This technique is more labour intensive, time consuming and have certain operational limitations.

b. **Use of air blowers:** Air blowers are used in pollination of tomato crop in protected structures. The use of air blower requires precision, high speed of air blower damages the flowers thus results in poor fruit set and reduction in yield. If the air blower speed is slow pollen will not be released on to the stigma and cause poor fruit setting. This limitation with the use of air blowers hinders large scale adoption of air blowers for pollination of crops under protected cultivation.

c. **Use of bumble bee:** Bumble bees are excellent pollinators of tomato and solanaecous crops due to their limited flight range and temperature tolerance. Bumble bees are collecting the pollen and nectar rewards from the flowers use vibrate the thoracic and wing muscles to release the pollen from anther columns. This process is called as 'sonication'. Bumble bees are extensively used for pollination of solanaceous crops in several countries. In India, AICRP (HB and P) center at Solan, Himachal Pradesh has developed technology for laboratory mass rearing of indigenous bumble bee, *Bombus haemorrhoidalis* and the results of pollination trials with this bumble bee species under net house conditions found promising with increase in the qualitative and quantitative characters of the fruits.

d. **Use of honeybees and other pollinators under protected conditions:** Various honey bee spp. *Apis melifera, A. cerena* and *A. florea* were deployed for pollination of crops under protected conditions. Honey bees are not effective pollinators under protected cultivation as they have long foraging range compare to small protected structures that are most common in India. High mortality rate of bees was observed when colonies were introduced small protected structures due to darting of bees to nets and protective structures. Trials with introduction of bee colonies with two openings *i.e.* one inside green house and other to outside showed non-preference of bees to forage inside the green house. Further, the vent created in the structure created space for entry and establishment of harmful insects like white fly, aphid, thrips in protected structures. In some cases dogs and squirrel gain access inside the structures through the openings and usually destroyed the crop. Indigenous bumble bees have been collected from Himalayan regions by Solan centre of AICRP on honey bees and pollinators and being reared for pollination of greenhouses tomato crops, on the other hand stingless bees and carpenter bees have been found effective pollinators for small protected structures for crops like Cucurbits, Sweet pepper and Chillies.

Irrigation and Fertigation

Mostly pressurized drip irrigation method is used under protected conditions for production of different horticultural crop and water soluble fertilizers are being used across the regions and countries for application of nutrients this drip system which is called fertigation. In several countries automated drip fertigation systems are used, where the cost of the controller is usually very high. At the same time the cost of water soluble fertilizers is also high in comparison to traditional fertilizers used under open fields. In recent past irrigation and fertigation schedules for different horticultural crops for types of soils and regions have been standardized by various organizations and research institutions, which can be used in economical way. Similarly very simple and low cost irrigation controllers have been developed by research institutions in India which can be used by common greenhouse growers in different area of the country. Recently there has been an increasing demand for biochar application, a carbon rich material made from biomass, in protected horticultural production systems. POP for organic vegetable cultivation under protected conditions have been standardized for different crop in some countries.

Among recent innovations, a blended strategy continuing modeling and monitoring approaches is giving importance. Crop monitoring tools are increasing used for getting fast indications of the actual nutrients status of the crop. Nutritional imbalances can be presented by accurate monitoring of the crop marginal stapes.

Soilless Cultivation and Vertical Farming

Soilless cultivation usually refers to cultivation of plants in soilless media is which plant root grow either is giving media bags, Containers or directly is nutrient solution, known as "Hydroponic system" may be achieve or passive both forms. In recent decades, adoption of nutrient solutions for optimization of crop nutrition has fallowed and increasing trend in different regions and countries. Soilless media like perlite, rock wool and coco peat in bags or containers or in slabs is a growing trend in vegetable cultivation; whereas leafy vegetables and herbs are grown in hydroponic system (NFT). Recently 'vertical farming systems' in tower shapes with aeroponic methods have been started in some crops. This system provides 10 times more number of plants per unit area, water and fertilizer savings and year round clean production but lot of energy request in there. In recent years strawberry and leafy vegetables and herbs are being grown in substrates culture especially in 'A' shape or "multi story" systems. Aquaponics is being used in some regions mainly in Latin America where different vegetables are grown on small piece of land or back yard of the house.

Integrated Pest Management under Protected Conditions

Pest and diseases under protected conditions are changing as fast as the new production methods are adopted, therefore, within few years we are required to have new control measures for pests and diseases under Indian conditions for which IPM strategy can be more effective for control of pests and diseases. Most innovative technique for management of these problems under greenhouses or other protected structures can be divided in to two systems, one as preventive and second as curative. Several of these technique, which are alternatives to chemical control have only partial effect on management of pests and diseases problems, but if applied in combined way resulted in to effective control of these problems.

Preventive

Legislation for insect of plastic mulch from infected areas, healthy seeds, resistant varieties, use of insect proof nets, on roof ventilation areas or to provide physical barriers between invading pests and the crops. Mesh size of the Insect proof net is very important factor and it should not be less than 40 meshes. Use of insecticides impregnated insect proof nets for lowering during the population of sucking pests like Aphids and White flies *etc.* use of colored plastics as physical barriers pest to modify the behaviour of the insects by altering their vision use of colored insect proof nets for effective reduction in the population of sucking pests under greenhouses and other protected structures. Solarization for decreasing the amount of diseases inoculums like *Pythium spp* and *Phytophora spp,* whereas the use combination of biofumigation and solarisation gave effective results.

Curative Measures

Use of Chromo tropic traps for effective monitoring of pests under protected conditions use of pheromone traps. Mass trapping method by using semiochemicals is used for Lepidoptera and Thrips by providing environmental friendly control method use of resistant root stocks for grafting and resistant varieties in various crops. Use of tomato Mi-1 gene for resistant against nematodes and other sucking pests, proper fertigation of the crop, use of banker plants in to a crop is a preventive and long term significant method for aphid control.

Biological Plant Protection

Use of biological control agents under protected environment had shown to be a economically viable alternative to use of synthetic pesticides. Use of biological control agents has increased in several regions across the world mostly in vegetables and also in ornamental crops up to some effect. Beside biological control agents like *Coniothyrium minitans* and *Thrichoderma spp.* can be used for management of soil borne pathogens being used as Agricultural practice in greenhouses production system. For effective control of various devashing pests in greenhouses parasitoids and predators have been found highly effective in different regions and different crops. In many cases of pest management such as the biological control *Lyriomyza spp,* parasitoids are best and often only form of biological control is available.

Mechanization, Automation and Use of Robotics

Greenhouses production is a highly intensive activity which requires a high amount of labour, mostly manual. The need of mechanization, automation and use of robotics is among other reasons, due to high labour cost which is a more immediate problem in industrial and developed countries than in developing or under-developed countries.

Generally two moments are the most decisive and expensive for a successful crop: the start cultivation of and the harvest. In greenhouses the crop maintenance is most intensive practice. Whereas for the start of cultivation different innovative technologies, such as *e.g.*, the automatic sowing, transplanting, cutting, grafting and spacing machines have been introduced so far, the process of crop maintenance and harvest are, apart from some developments, as yet not fully automated. During the first phase of the production cycle, the position, shape, size and other characteristics of the plant are relatively well defined and uniform, thus, basic industrial automation, based on mechanical solutions, using few sensors and limited computing power can perform sufficiently. Crop maintenance and harvesting are much more difficult to automate, because of the position, shape, size and colour of produce very widely and these high-tech mechanization, robotic or mechatronic systems are required. The situation is more complex for the most important greenhouse cultivated vegetables such as tomatoes, cucumbers and bell peppers, where multi-harvesting processes during the long production cycle are needed.

Some examples are the current commercial robot developments for cutting roses, harvesting strawberries and cucumbers, intelligent automatic guided vehicles that can change speed automatically adjusted to the human workload and the new development in pepper harvesting machines. Success of such machines will not only depend on the development in mechanization, automation and use of robotics, but also on combination of efforts in other adjacent areas such as the plant breeding, greenhouse structures, maintenance and energy saving measures, plant protection and staff training and skill development. Self driven trolleys for greenhouses operators like trellising, training and pruning have been developed, which can be easily operated and used by a single worker. A tractor operated soil heating machine (up to soil depth of 30 cm) has been developed for controlling soil borne problems. Low cost controllers for automation the irrigation and fertigation units for protected cultivation have been developed by different Indian Research Institutes for the use of small and marginal greenhouse growers in different parts of the country.

Conclusion

Protected cultivation has shown yield enhancement up to 4-5 times in several vegetable crops, which is a potential technology to achieve vertical growth in horticulture, but would need investment and region specific technology upgradation and use of advance technology developed within the contrary or abroad. The future of protected cultivation technology in India is very vast but highly depending upon the availability of the technologies and recommendation of the technology implementation. It is believed that protected cultivation technology has to play a significant role under varied agro climatic conditions of the country as a means for sustainable crop diversification, intensification and for vertical growth of productivity of horticultural crops leading to optimization of water and fertilizer use efficiency in an environment of water scarcity in addition to better control of product quality and safety, in line with the market demands, standards and regulations. In the near future, the first and most important requirement for the use of protected cultivation technology is for its large scale use for raising disease and virus free healthy planting material in all kinds of horticultural crops and secondly to use the technology for hybrid seed production of vegetables and thirdly production of fresh safe food for better economic viability in the country and mostly in cluster approach. Further, for sustainability of the technology it is utmost important to use the innovative and advanced technologies looking to their techno-economic feasibility and then develop a large skilled manpower in form of rural youths in two sets, one set for designing, fabrication/installation and thereafter maintenance of the protected structures and the other set for entire crop production management system and smart marketing of the high quality produce to various markets within the country or through export to various destinations.

Chapter 17

Cosmic Farming: An Effective Tool to Mitigate Problems Associated with Sustainable Agriculture and Environment

R.K. Pathak[1]* and Neeru Dubey[2]

[1]*Fivefold Path Mission, Former-Director, ICAR–CISH, Lucknow, Uttar Pradesh*

[2]*Ernst and Young LLA, New Delhi*

e-mail: pathakramkripal@gmail.com

ABSTRACT

Soil fertility and environmental quality are critical and crucial for survival of humanity at the planet earth. Agrochemicals based farming over 5-6 decades has badly affected the soil biology and environmental ecology in many regions of the world. Ultimately the three basic elements of Nature i.e., soil; water and air have been polluted beyond imagination. It is well established that any amounts of agrochemicals pumped in cannot restore soil fertility as well as sustainable agriculture production. It is only possible by mediation everlasting sources of energy. Cosmic farming, being promoted by institute is based on systematic and synergistic mediating of infinite source of energy by few simple techniques. Since human body is a mini replica of cosmos, hence food produced and consumed in consonance of nature's gesture without use of agrochemicals will be Sattvic, full of nutrition and therapeutic values. Horticultural crops are well suited for cosmic production and consumption. The same technique is equally effective for all crops and in each ecological situation. Looking the current plight of small and marginal farmers and pathetic situation of indigenous cow, both can be addressed with assertive promotion of cosmic farming even in remote villages with enormous implications.

Kewwords: *Agnihotra, Aura energy, Bioenhancer, Cowpathy, Resonance point.*

INTRODUCTION

On tracing history of ancient agriculture when agrochemicals were not used/invented, there are number of evidences that country has been practicing cosmic farming for centuries. There used to be record quality production of large number of crops. India was leader in production and export of many spices, silk fabrics, cotton and many others. It is pertinent to mention that people respected the five sources of the energy *i.e., Pancha Maha Bhutas* (earth, water, air, sun and ether (Aakash). In turn, these provided their energy and India was at the helm of supremacy in every sphere of life. Integration of Biodynamic and Homa organic farming are capable of enhancing the rhizosphere and biosphere simultaneously, thereby assuring quality production in all crops and in all situations under current situations.

Food produced by enhancing soil biology and environmental ecology will be *Sattvik* (genuine), full of nutrition, therapeutic value, taste and quality. With over two decades of intensive research and field visits on organic farming activities, it is confirmed that cow- based formulations can pave a way for sustainable production and nutritional security through horticultural crops, thereby resolving the problem associated with mal and under nutrition with small and marginal farmers and pregnant woman facing with resource constraints. This can also ensure breathable air, drinkable water and eatable foods particularly to small and marginal farmers.

Some Facts of Conventional Agriculture

- Conventional Agriculture (CA) is based on fossil fuel and imported sources of energy which are dwindling and becoming expensive to farmers, governments and to the society.
- Predomination of monoculture of rice: wheat, indiscriminate use of fertilizers, agrochemicals, more water and heavy machineries.
- This has disturbed the soil biology, environmental ecology, thereby polluted the soil, water and environment.
- Pollution of the basic elements is major threat to humanity.
- Broken social fabrics and climatic disasters in one or another region of world is routine.
- This calls for revamping back to ancient system which could resolve the current crises of humanity.

In fact, Mahatma Gandhi long back cautioned that "Agriculture can be sustainable and self-reliant only if farmers use locally available resources and inputs *i.e.*; farm waste, cow dung and other biomass for preserving the soil as a living material". In consonance of this statement, let us examine the history of agriculture in India.

In order to resolve these challenges, policy makers and few institutions including NGOs had been trying to conceive alternative technology which can address issues of production and simultaneously address issues of soil, water and particularly deteriorating environmental pollution. As alternative of chemical led agriculture, two systems *i.e.,* Conservation Agriculture (CA) and Jaivik Agriculture (JA) are being promoted for over two to three decades with promising response, but there are number of apprehensions in mind of farmers, policy makers and to the society.

Looking the apathy of small and marginal farmers, facing resource constraints, we were compelled to strive for a cheap, effective and affordable source of energy for sustainability in agriculture. While continuously thinking and tracing history of agriculture, it came to our mind that mediation cosmic energy could provide an alternative and resolve the current crises at planet earth. Hence farming based on mediation cosmic energy is now been promoted as "Cosmic farming, can provide a way for sustainable agriculture.

Cosmic Farming

"Cosmic farming pertains with systematic and synergistic mediation of energy through five-ultimate sources *i.e.*, soil, water, air, fire and ether in agriculture". In fact, no chemical fertilizer or even organic manure is needed as food to the plants. Cosmic energy is the fundamental source of plant growth and responsible for all the mysteries of the nature. It can be mediated through primitive microorganisms in soil and also by the living cells of the plants. Therefore, plants have capacity to produce all food elements needed for their growth, without feeding them specific elements. Soil humus and plant cells even can trans mutate one element in to another through biologically active soil as per need of the plant (Pathak, 16).

Some Facts of Cosmic Energy

After working over two decades, we are of the view that for sustainable agriculture/ horticulture, one has to change the mind set and conceive a farming system approach based on ancient wisdom, which can enhance the rhizosphere and biosphere simultaneously for sustainable production.

- Enormous quantum of cosmic energies through solar, lunar, planets, earth are freely available;
- Ancestors respected *Panch Mahabhutas* and these provided free energy for wellbeing of mankind;
- Somehow, these free sources of energy have been ignored in recent century.
- Energy from cosmos, planets, moon is still in prime forms;
- Cosmic energy is available all time, while solar in day hours;
- Sun is giant power house-Infinite, non-polluting source and freely available;
- If solar energy is not mediated, it is of no use;
- Mediation of one percent of solar energy can meet world's energy need;
- Soil microbes, cow and Agnihotra have capacity to mediate cosmic energy;
- Natural resources *i.e.*, soil, water, air, bio diversity of flora and fauna are available as grace of Cosmos.

With little support and skill up gradation, every farmer can be trained and tuned to mediate these energies for the wellbeing of humanity. The proposed hypothesis will be accepted, if it is capable of addressing the following issues (Pathak, 16).

Issues to be Addressed with Cosmic Farming

- Continuous improvement in physical, chemical and biological properties of soil
- Improvement in soil organic carbon (SOC), content.
- Input generation at the farm/cosmic village.
- Par excellence produce quality (*Sattvic*), full of nutrition and therapeutic values.
- Cost effective and ecofriendly technology.
- Well suited for small and marginal farmers of the country.
- Helpful maintenance of natural resources to hand over to the coming generations in improved conditions or as we inherited.

Cosmic Energy: Source or Cause of Production

The Sun is source of all energy directly or indirectly available on the earth. Surprisingly the power of photosynthesis to convert the solar energy into life force rests only with the green leaves. For growth of any crop, sun light, water, air (oxygen and carbon dioxide oxide), are essential elements. Fertile soils as supporting medium are needed.

On introspection, it is evident that nature has made elaborate arrangement to mediate cosmic energy, with little efforts (Pathak, 16, www.cosmicfarmingindiacom.).

Nature's Gesture to Mediate Cosmic Energy

- Soil microbial consortia, many soil dwellers including earthworms have capacity to mediate cosmic energy;
- Nodules in leguminous and few other plants (BNF) have specific bacteria capable of mediation free N available in the biosphere;
- Specific bacteria solubilizers for all elements to be utilized by the plants;
- Millions of lips beneath foliage are busy in devouring carbon dioxide and exhaling oxygen essential for survival of human being and other living one;
- Chlorophyll in foliage mediates solar energy in process of photosynthesis, the basic step in preparation of food needed for survival of everyone.
- Without plants we cannot breathe, nor drink and not eat.
- Cow with hump which is pyramid shape, belly is mini replica of cosmos and horns have capacity to mediate cosmic energy.

The main responsibility on all of us; particularly the researchers in agriculture is that natural resources are used in such a precise way so these are handed over to coming generation, in improved forms or as inherited. Through this, one can be instrumental in assuring breathable air, drinkable water, eatable food and fertile soil for future generations (Pathak, 16).

After closely observing impact of different organic farming systems being implemented by few organizations in different parts of the country with positive response, it come to understanding that looking the environmental crises, probably ancient technology which was based on mediation energy from the five-ultimate source of energy *i.e.*, the Earth, Water, Fire, Air and Space could be viable alternative. In fact, ancient people respected and worshiped natural elements to get the required energy for agriculture without a gram of agrochemicals. Role of basic elements and few simple affordable techniques have been dealt.

Soil

In Vedas, soil has been respected as "Mother" and human beings as her sons. Even as on today earth is recognized as Annapurna –meaning thereby that all elements needed by plants can be obtained from soil. The soil conducts many functions which help stabilize climate. If there is healthy soil, with plants flourishing on it, then it can sequence large amount of carbon-if the soil is left bare and continuously tilled, it leads to soil desertification. Keeping the soil healthy benefits those who farm it. Healthier the soil with more plants and organic matter has more carbon. Living soil is one of basis of cosmic farming. Life, in the basic form of microbes, has been companion of earth since its inception. It has close bond between the environment of the earth and the microscopic organisms thriving upon it. Within soil procreating in high concentrations, the bacteria and fungi ensure soil fertility by recycling the elements in the chemical laboratory that constitutes in their bodies, making them available to plants. As nitrogen is converted into nitrate, phosphorus is solubilized into phosphate, Sulphur to sulphate, chlorine to chloride, boron to borate and so on through elements (Ghosh, 4).

It is generally admitted that chemicals and salts introduced into the soil through agrochemicals, turn the soil sterile: soil organisms perish; earthworms and many other soil dwellers disappear; the soil becomes dead. On this substratum crops, are grown. In fact, insect pest occurrences are symptoms of a failing crop, are indication of reduction of aura energy of rhizosphere not the cause (Tomkins and Bird, 24). These can be easily managed with humus and optimum plant nutrition.

The soil food web is essential community of organisms that live in the soil in close harmony. Every farm has its own soil web. Healthy soils contain massive population of bacteria, fungi, protozoa, nematodes, soil arthropods and earthworms. A teaspoon of fertile soil may contain between 100 million to one billion microbial lives. Fungi and bacteria in soil have considerably more nitrogen in their bodies than other organisms. Nutrient cycling happens when other sets of soil organisms are present to consume the nutrient rich bacteria and fungi and release nutrients to plants in available forms. These rapid interactions and countless exchanges of nutrients between soil organisms occur in the soil where highest concentrations of organisms exist because root exudates provide food for bacteria and fungi (Ghosh, 4). It is close, intelligent and intense symbiosis between soil and plants. Plants provide nutrients to soil, which are utilized by soil bio-life in return and it supply of all elements which are utilized by plants for their satisfactory performance. A strong connection exists between a fertile soil and healthy crops, healthy animals and last but not least, healthy human beings must be made known far and wide.

One of the main reasons of insect pest occurrence at the farm is poor soil health. Poor soil aeration always encourages disease in crops. Insects and fungi are not the real cause of plant disease but only attack unsuitable varieties or crops imperfectly grown. In fertile land; soil and the plant come into gear in two ways simultaneously. In establishing and maintaining these contacts humus is essential. Certain soil fungi directly connect the humus in the soil with the roots of the crop. These fungus tissues may contain as much as 10 per cent of nitrogen in the form of protein, which is digested in the active roots and carried by the transpiration current to the seat of carbon assimilation in the green leaves (Tomkins and Bird, 23).

The other soil organisms are also involved in nutrient recycling. Nitrogen fixing bacteria convert atmospheric nitrogen into useable form as they colonize in the roots of legumes. Mycorrhizal fungi colonize root systems of other crops such as vegetables like onion, maize, sorghum *etc.* In doing so, these specialized fungi recycle nutrients by secreting enzymes that solubilize calcium phosphate and pump the phosphorus directly to the plants, thus making an otherwise unavailable element now available to plants. Mycorrhizae association with crops play beneficial role for crops by aiding in disease suppression and moisture absorption.

Air

Air is important element to plants and human being. It is evident that plants obtain around 94 per cent element (carbon and oxygen 44 per cent each, and 6 per cent of hydrogen) from the biosphere. Soil fauna is not only important for water circulation but is equally important for oxygen which is necessary in the soil because plants can assimilate only oxidized forms of most of the elements. For instance, nitrogen is not assailable by plants as ammonia but as nitrate, Sulphur as sulphate, and phosphorus as phosphates. These transformations are done by specific soil microbes (Ghosh, 4; Ram and Pathak, 21).

Poorly aerated soils lead to the development of anaerobic microbes; these transform nitrates back into ammonia, sulphates into sulphur and phosphates back to phosphorus. With this depletion, there is no possibility for plants to grow and the use of fertilizers becomes obligatory.

Water

Water is next to air and one of the important elements needed by plants and human being. But due to mismanagement, almost 50 per cent of rainfalls runs of carrying the life sustaining soil with it. It is true that rainfall comes pouring down within a few rainy days in the monsoon season. These downpours act as a source of destruction rather than anything else. Since ancient time, people had regarded water as a divinity, a being with great strength and vigour, to be revered. Owing to indiscriminate use of agrochemicals and many other factors, no water is safe for drinking and even it is adversely impacting farming in many ways.

Shree Paranjape promoter of Homa Therapy stated in an article written in 2002 regarding water quality. He emphasized that water resources on earth surface are finding difficult to absorb energy from sun (due to pollution). This will result in depletion of marine life and imbalance in nature what is being experienced now a days' in most of the regions. In continuation of this, he suggested that if Homa Farming if practiced on massive scale, the atmosphere is healed and the water resources get purified, leading to better absorption of sun's rays' (Paranjape, 12). Activities like massive plantation of selected trees, plants, arresting of rain water in field, water bodies through appropriate structure by involving large number of people, promoting *Homa*

Jaivik Krishi and regular placement of Agnihotra ash in water bodies and production and use of bio-enhancers are effective tool to enhance soil humus important ions in quality production.

Techniques for Enhancement of Soil Fertility

Mulching

Practice of mulching is integral component in cosmic farming. Mulches keep the soil cool and prevent water loss through evaporation. It provides optimal situations by management of soil temperature, moisture and dark conditions, friable soil for root development, enhancement of earthworm and termite activities thereby soil fertility. This in turn will enhance other soil life for plants. The mulch ultimately decomposes and enriches the soil. Once the soil gets better aeration through these processes, giving them better access to subsoil moisture, and additionally, capillary action is improved.

I advocate for with locally available biomass, if possible, a combination of monocot and dicot is ideal proportion. In fruit plants even drooped foliage and local weeds can be used in mulching.

> *Things to ponder-We change our clothing with change in season, but plants have to face natural vagaries. Proper mulching provides cushion to root systems thereby they continue their activities, hence play important role in quality production (Pathak and Ram, 20).*

Earthworm

Earthworms are known as pulse of the soil. They play major role in maintenance of soil fertility and quality production. Earthworm is the chief operator of nature's microbe's factory. Local earthworms and uncountable microorganisms in the soil are the chief source of plants nutrition. They eat gravels, limestone, organic biomass, move upwards and downwards making capillaries in the soil. This encourages better infiltration of water and helpful in upward movement of nutrients in available forms. Earthworms' excrement contains 5 times more of N, 7 times more of P, 11 times more K and 2 times more Mg and Ca than normal soil. By these activities they are capable of minimizing run off water, better aeration of roots, hence vigorous plants. Earthworms also play important role in minimizing root diseases. In conducive situations earthworm can add more than 100t/ha fertile (cocoon) every year (Ismail, 7; Tomkins and Bird, 24).

Cow

The ancient Indian knowledge system embodies a wealth of wisdom and experience gained over millennia direct observations and transmitted most often orally over generations. Many citations in Vedas establish the significance of Cow in day-to-day life and sustenance of well-being of humanity. Cow is a highly revered animal in Indian culture and we are duty bound to protect it to reap benefits (Rig-Veda*1.164.27, Rig-Veda 1.164.40). Even, the other ancient scriptures of *Sanatam dharma* like *Upanishads, Puranas, Mahabharata, and Manu smriti* extol the virtues of cow. There is a reference in *Bhagavad-Gita* that for agriculture, rearing of cows and honest exchange of merchandise – these constitute the natural duty of human Vaisya (Ch-18 -44).

Maharshi *Dhavantari* offered a wonder medicine '*Panchgavaya*' (a combination of cow urine, milk, dung, ghee and curd). Indian cow breeds are unique and distinct species, both in their appearance and characteristics. Cow is the backbone of Indian culture rural economy, and sustains human life; represent cattle wealth and biodiversity (Dhama *et al.,* 3).

Cow is special creature on planet earth. It is interesting to record indigenous breeds of cow (*Bos indicus*) with hump and horns have capacity to mediate cosmic energy. That lives on left over material by human consumption and convert these into valuable products for agriculture and human health. Dung which permeates through her intestine is enriched with microbial consortia, while urine and butter milk are used for preparation of composts, bio- enhancers and bio-pesticides.

The hump of cow supplements length of spine (Nene, 10; Pathak, 17; Pathak and Ram, 20).

Things to ponder-It is interesting to record that in general leftover materials from human consumption such as *bhusa* (wheat straw), paddy straw, sugarcane leaves, grasses, cakes, *chuni*, choker *etc.* are fed to cow; but in return cows provide three direct products *i.e.,* dung, urine and milk. All these are of much use in agriculture and human health. It is praiseworthy to note that even after consumption of chemically loaded materials; all cow products are free from any contaminations. This conversion is not possible by any industry or laboratory (Pathak, 19).

Cow Aura Energy

Aura Energy is almost a new concept in cosmic farming, but there is immense possibility of use of cow- based products in cosmic farming. Cow products have maximum "Aura energy' light emitting from animate objects (Murthy, 8). Cow products- based formulations are used in various forms to address nutrient requirement, insect pest management. In fact, nutrient deficiency, pest and disease incidence occurs in plants, when the Aura energy of soil and plants are depleted. Such plants show negative Aura energy which can be quantified with Universal Scanner (Table 1).

Table 1. Aura energy of cow products

Cow Product	*Aura Energy (m)*	*Impact*
Milk	12-13m'	Has been used for seed treatment and management of viral infection during ancient time in India. It has unique blending of 101substances, 19 amino acids. Microbes like *Lacto bacillus* produces organic acids which promote crop growth and suppress pathogens.
Cow curd	6.5-7m'	An important component in Panchgavaya and good for human health. Fermented butter milk is used for seed treatment and for foliar application for the management viral diseases.
Cow ghee	14m'	Richest source of energy, helps in quick combustion of dung, by burning it cleanses air pollution. It is a power full vehicle for capturing solar energies and its transmission in larger area through Agnihotra fumes.
Cow dung	6m'	Rich source of beneficial microbes, used for seed coating and preparation of tree paste, composts and bio enhancers.
Cow urine	8-9 m'	It is disinfectant, prophylactic, antibacterial, antifungal, antiviral, contains amino acids, cytokinin's, which play role in immunity enhancement and soil fertility.

Somehow, utility of cow has been ignored on the name of advancement and easy access to agrochemicals in agriculture. Looking the current crises in agriculture and human health, there is need for introspection of current agricultural practices and again reverting back to ancient agricultural practices based on mediation cosmic energy from cow- based formulations depicted in Figure 1.

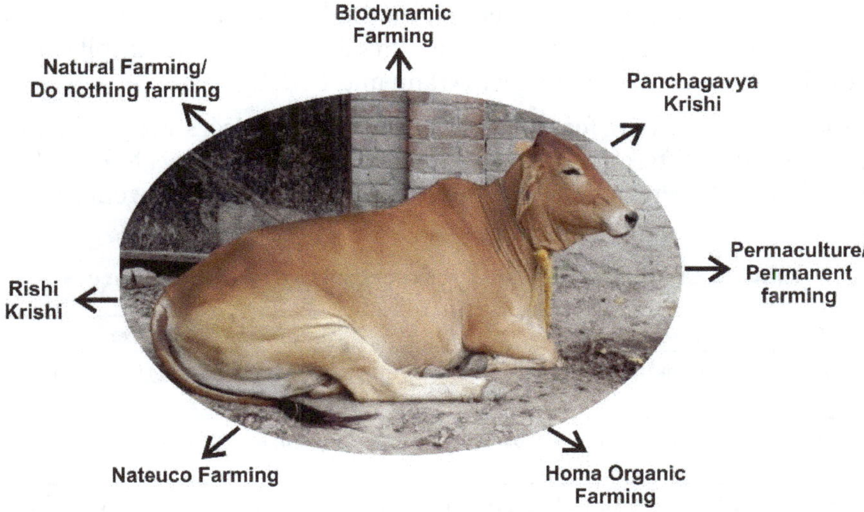

Figure 1. Cow based organic farming systems.

Cow Based Formulations

After visit of many organic growers, we could notice that in various organic farming systems growers are using few cow products on various names with appreciable response. On tracing history, *Vrikshayurveda*, composed between 6,000 BC and 1,000 A D which is a testimony for importance of cow in agriculture. Vedic literature has outlined a systematized agricultural practice that recommended use of *Panchgavaya* (mixture of the five products of cow in a specific ratio) and *Kunapajala* (decomposed products of fish or animals in liquid form), to enhance the biological properties of soil plants' health (Nene, 11). In consonance of this, it is pertinent to mention that in all organic farming systems, prevalent in the country are using specific cow-based formulation with specific names. All these formulations are potent source of beneficial microbial consortium. Looking the impact of these products, generic name "Bio-enhancer" was given, (Pathak, 19).

Bio-enhancer

Basically, bio-enhancers are prepared by fermenting cow products over specific duration. For improving quality few other ingredients are also used in few bioenhancers. Specific bioenhancers are available for all crop activities. Bio-enhancers have impact to improve soil fertility, crop productivity and quality, insect, pest and disease management. It is pertinent to record that these are potent source of all macro, micro nutrients, trace elements, PGPR activities, immunity enhancer and sustain abiotic and biotic stresses (Pathak, 17). These can be easily prepared by every farmer as per need. The main bio-enhancers which are commonly used in organic farming are Beejamrita, Jeevamrita, Panchgavaya, Amritpani, Vermiwash *etc.* (Ram and Pathak, 22). Assertive promotion of bio-enhancers will have many implications in cosmic farming.

After visualizing the impact on few crops, methods of preparation, nutrient status and elucidation of microbial profiling for most of bio-enhancers have been compiled by ICAR-CISH, Lucknow, India in a book on Bio- enhancers (Ram and Pathak, 21).

It is praiseworthy that cow formulation based bioenhancers are available for all crop activities, and are used in different organic farming systems. It is advocated that farmers should prepare and use it for different activities without depending upon any agencies as summarized in Table 2.

Table 2. Crop activities and perspective use of bio-enhancers

Sl.No.	*Crop Activity*	*Suitable bio-enhancer*
1.	Seed/plant parts treatment	Amritpani/Beejamrita/CPP (cow pat pit)/cow urine + dung powder + Agnihotra ash *etc.*
2.	Enhancing decomposition of biomass	Jeevamrita/CPP/Panchgavaya/Agnihotra ash water, Kunapjal, *etc.*
3.	Enhancing nutritive value of compost	CPP/Agnihotra ash enriched water, Biosol/Vermi wash/Kunapjal, Panchgavaya *etc.*
4.	Enhancing soil fertility	Amritpani/Jeevamrita/CPP/Biosol/Kunapajala/Panchgavaya *etc.*
5.	Improvement in crop vigour	Panchagavya/Kunapjal/Biosol/Vermi wash *etc.*
6.	Enhancing biotic and abiotic stress	CPP/Biosol/Kunapajal, Panchgavaya, Vermi wash *etc.*
7.	Insect pest management	Kunapajala/vermi wash/biosol/DasparniPanchgavaya, Biodynamic liquid pesticides.
8.	Seed/grain storage	Agnihotra ash/Panchgavaya

Note: These are field base experiences, for its acceptance, there is need of initiation of systematic research its validation at few select institutions for its assertive promotion.

Historical Perspectives of Homa Therapy

There is one verse in Bhagavad Gita that, having created mankind along with Yajna, at the beginning of creation, the creator, Brahma, said to them, "You shall prosper by this; may this yield the enjoyments you seek (ch-3vs-10). In Valmiki Ramayana, there is reference that Bhagavan Ram was daily performing Agnihotra in his life. Even on the day of crown ceremony both Mata Sita and Ram performed Agnihotra. Ancient people have in depth knowledge of great culture of the country. Since human being were fully aware that directly and indirectly because of routine activities are daily polluting environment. Hence to counter act this, Yajna was mandatory routine in every household. This has been elaboratory emphasized in the Vedas, ancient most body of knowledge known to human being.

There are number of evidences that till Yajna was performed regularly, country was at whelm of affairs and has supremacy in every aspect of life including education and agriculture for number of centauries. On introspection of the current activities promoted by untiring efforts of Vasant Parajnje and many others, there are number of evidences that assertive promotion of Homa Organic Farming can provide an viable alternative in the current circumstances.

Homa Organic Farming

Krishna ordained "All beings are evolved from food; production of food is dependent on rain; rain ensures from sacrifice, and sacrifice is rooted in prescribed action. Know, that prescribed action has its origin in the Vedas, and the Vedas proceed from the Indestructible (God); hence the all -pervading Infinite is always present in sacrifice. and the Vedas proceed from sacrifice (Bhagwat Gita, ch-3, 14-15).

Homa is a Sanskrit word used as synonymously with *Yajna.* It is the technical term from the Vedic science of bioenergy denoting the process of removing the toxic conditions of the atmosphere through agency of fire. The most basic Homa (*Yajna*) is called Agnihotra and is tuned to the biorhythm of nature *i.e.*, sunrise/sunset. Regular performance of Agnihotra establishes the healing energies necessary for a healthy environment. Agnihotra fire ceremony is a technique to heal the Earth. Environment: purify air, land, and water resources using healing fires. These healing fires have profound healing effects on all life forms including agriculture. This ancient technology continued for thousands of years. It is based on the principle that 'Heal the atmosphere and healed atmosphere will heal you. The basic principle in Homa Therapy.

> *Things to ponder- In fact, conventional farming has polluted the three-ultimate sources of energy, i.e., soil, water and air, thereby these are not able to provide their energy. But regular performance of Agnihotra and use of ash obtained has capacity to combat with environmental pollution and enhance Aura energy of the five sources of Panch Mahabhotas (soil, water, air and space) thereby remarkable impact- Pathak.*

It is pertinent to mention that environmental pollution is one of the major problem, currently humanity is facing. Most of the international efforts could not provide any viable solution. As alternative, this calls to see ancient history of farming. It is pertinent to mention that our ancestors were fully aware that with most of human activities, environment is polluted, which need to be counteracted intensively.

Agni means- fire and Hotra means -purification. It is antidote for pollution control ancient science of healing. Somehow, due to few problems mass adaptation has been ignored. It is well accepted that environmental pollution has become one of the major problems for survival of humanity at the planet earth. Most of international efforts have not given expected impacts.

On deep pondering of history, it is apparent that nature has its own way and play major role for resolving any crises faced by the humanity. Param Sadguru Shree Gajanan, who is said to be incarnation for New Age to save Planet Earth, and known as Son of Man, took birth as "Kalki Avatar.

With his dedicated efforts, standardized cheap, effective technique from Aakalkot, Shivapuri, Solapur, Maharashtra, India with long experience to combat environmental pollution problem.

After getting convinced its impact for humanity, he entrusted its further dissemination to a select group of people to the universe. He standardized simple technique which is based on locally available inputs and can be performed by everyone, now has spread in the universe as the Homa therapy.

Agnihotra is a gift to humanity from ancient- most Vedic Sciences of bio energy, medicine, agriculture and climate engineering. A simple technique which can be practiced by every citizen of the world has been standardized for addressing pollution. Paranjape, 13; 12) took it to entire world as on today it is practiced for human health and agriculture in more than 100 countries.

It is praiseworthy that with intensive efforts 'What has been done and what can be done' has been reviewed by Berk, 2.

It is the basic fire in HOMA THRAPY. It is the process of purification of the atmosphere through the agency of fire, prepared in copper pyramid tuned to the biorhythm of sunrise/sunset. By practice of Agnihotra, one starts noticing that the tension of mind disappears and begins to experience peace. The mind is reshaped so nicely, so delicately, so effortlessly by sitting in Agnihotra atmosphere.

Extension activities on Homa Therapy, outside India was initiated by Shree Vasant V. Paranjape, one of his devoted disciples, since 1977. As a result, now millions of people in different parts of the world are practicing Agnihotra daily and harvesting spiritual and material happiness.

The major strength of Homa Therapy is Agnihotra. It is science of Pyramidology, burning of organic substances (cow dung patties, cow ghee with unbroken rice), bio rhythm of nature *i.e.*, sunrise and sunset, electromagnetic effect of chanting specific Mantras.

Nature of Agnihotra- Tremendous amounts of energy are gathered around the Agnihotra copper pyramid just at Agnihotra time. A magnetic type field is created, one which neutralizes negative energies and reinforces positive energies. Therefore, a positive pattern is created by one who regularly performs Agnihotra, merely by his/her performance.

When Agnihotra is performed, Agnihotra smoke gathers practices of harmful radiation from the atmosphere and on very subtle level neutralizes their radioactive effect. Nothing is destroyed, merely changed from one form to another form.

In Agnihotra fire there is not just energy from the fire, but subtle energies are generated or thrust into the atmosphere by fire. Observance of discipline is prerequisite *i.e.*, quality of materials (cow dung cakes, cow ghee, brown rice), sonic power of chanting specific mantras at right time of sunrise and sunset of the place are essential. Much healing energy emanates from the Agnihotra pyramid.

Effect of Agnihotra

Agnihotra renews the brain cells. It revitalizes the skin and its purifiers the blood. It provides the holistic approach to life. It has the ability to neutralize pathogenic bacteria in the air, soil and water.

One sits near Agnihotra fire and breathes in the smoke, which goes quickly into bloodstream and lungs has excellent effect on circulatory system and even more so if Agnihotra ash is ingested. The smoke has a good effect on the brain and nervous system.

If plants are kept in Homa atmosphere where vibration of Agnihotra pyramid fire are maintained, one can actually see growth of plants and its vigour. Plants receive nutrition from Agnihotra atmosphere, become happy and grow well. Just as Agnihotra pyramid fire gives nourishment to plant; it provides the same for human life and animals.

The sun brings and takes the energy, which makes all conditions conductive to an anti-pollution change. It calms the world. The pyramid act as generator and the fire as turbine. The cow dung, ghee and rice then interact to form a composition which is thrust, surrounds, neutralizers. This is how the Agnihotra fire physically heals the atmosphere.

Thousands of people on all continents belonging to different races, languages, religions and spiritual groups who practice Agnihotra have remarked that simply by performing daily HOMA, (*i.e.*, Agnihotra sunrise/sunset) they feel like a protective film surrounds them.

Materials Required in Agnihotra

It is matter of great concern that regular performance of Agnihotra is possible by every family in the world. Materials required for Agnihotra can be arranged anywhere in the world with little efforts.

Pyramid: Pyramid is made from copper is universally acknowledged for its excellent conduction of electricity and heat. The word 'pyramid' means fire in the middle. Pyramid has special power to attract cosmic energy.

This ancient –word meaning for pyramid is closely connected with the inexplicable energies emanating from the central shape. The pyramid shape is widely experienced to generate and store a special energy field, which possess bacteriostatic properties. The inverted pyramid shape *allows* controlled generation and multidimensional dissipation of energy. It acts as a generator of unusual energy fields and spreads in its surrounding areas wherever Resonance Point (RP) has been established. The Agnihotra has an obvious link with fire, heat, electro-magnetic force and because of its inverted pyramid shape have connection with the cosmic energy fields. The dimensions of the pyramid are: 14.50 cm x 14.50 cm at the top, 5.25 x 5.25 cm at the bottom and 6.25 cm in height with three steps.

Rice: Rice is widely grown in all part of the world. Brown *i.e.*, unpolished, unbroken rice (*Akshata*) is used for performance of Agnihotra. Highly polished rice loses nutritional value and hence brown rice is used. If rice is broken, the subtle energy structure is disturbed and it is not fit for Agnihotra healing fire. If organically produced rice is available, it is to be preferred.

Dried Cow Dung cakes - Cow dung is treated as medicine in all ancient culture whether they may be Indians of North or South Americans, Scandinavians, East or West Europeans, Africans and Asians. Cow's dung contains Menthol, Phenol, Formalin, Phosphoric acid, Potash, Ammonia, Indole, and Nitrogen. Make a pan cake like patties and dry them in sun. Agnihotra fire is to be prepared from this dried cow dung.

Cow Ghee: The Russian scientist Sorovich claimed that burning of 10 gram of cow ghee generates one tone of oxygen. Cow ghee comprises of glycerol and number of saturated and unsaturated fatty acids such as Steric acid, Oleic acid, Linolenic acid, Palmitic acid, Myristic acid and Laric acid. When ghee is burnt along with other materials in prescribed disciplines, it becomes much more energetic. Traditionally, homemade ghee has been found ideal for performance of Agnihotra. This can last without refrigeration for long time. Ghee is a very special medicinal substance. It helps in rapid combustion of the materials and keeps the fire alight. On combustion and oxidation, these form hydrocarbons, aldehydes and formaldehyde responsible for fragrance and purification of the area by minimizing pathogenic bacteria.

Agnihotra Timings

The time component of Agnihotra is one of the most essential aspects to be observed. Light radiated by sun at sunrise and sunset is termed as diffused light in the sky which has great ecological significance. Since sun at the rise or set becomes near to horizon in transverse/oblique position, solar rays have to cover a longer distance to reach to the earth and in this course solar rays (electromagnetic waves) of visible spectrum of lower wave lengths *i.e.*, violet, indigo, and blue and probably ultraviolet, X-rays and Y-rays are scattered and lost in the atmosphere. The other visible rays of higher wavelengths *i.e.*, yellow, orange, red and infra-red of invisible spectrum reach to the earth and the sky looks yellowish red at the time of sunrise and sunset. One can say that solar ray reaching to the earth at both times contains greater proportion of infra-red which is lethal but more penetrating and greater energetic in nature.

How to Prepare Agnihotra Fire

- ❑ Place a flat piece of dried cow dung cake at the bottom of the copper pyramid.
- ❑ Arrange pieces which have been coated with ghee in the pyramid in such a manner as that allow free air to pass.
- ❑ Apply a little ghee on the small piece of cake and light it and insert in the mid of pyramid.
- ❑ Soon all the dung in the pyramid will catch fire.
- ❑ One can use a hand fan to blow the air and help the flame. However, do not blow through the mouth to avoid bacteria from the mouth getting into the fire.
- ❑ Do not use any mineral oil or similar material to start the fire. At sunrise and sunset, a good fame should be ready in the pyramid.

Agnihotra Process

- ❑ Take a few grains of rice in a copper dish or in left palm and apply a few drops of ghee.
- ❑ Exactly at sunrise utter the first mantra and after the word SWAHA add a few grains of rice by right hand (as little as one can hold in the pinch of your fingers will be sufficient) in the fire.
- ❑ Utter the second Mantra and after the word *SWAHA* add a few grains of rice by right hand in the fire.

Mantras

There are vibrations that exist everywhere. It is only vibrations when someone tries to understand deep into it. Where there is vibration there is also sound. When someone performs Mantras, the sounds activate special vibrations that will create specific atmosphere for its impact.

Then the desired results are realized. These vibrations exist for everything, so anything can be activated, controlled or changed by Mantras. When one utters the Mantra with a pure mind into the pyramid at Agnihotra time, the ash gets energized from cosmic energy retains that energy and healing properties of the ash become more powerful.

Agnihotra Mantras

At Sunrise

Soraya Swaha, Soraya Idam Na Mama

(Add the first portion of rice mixed with ghee into the fire)

Prajapataye Swaha, Prajapataye Idam Na Mama

(Add the second portion of rice mixed with ghee into the fire)

This completes morning Agnihotra.

At Sunset

Agnaye Swathe, Aganye Idam Na Mama

(Add the first portion of rice mixed with ghee into the fire)

Prajapataye Swaha, Prajapataye Idam Na Mama

This completes evening Agnihotra

(a' is pronounced like a in 'father')

For timings and other information following websites will be helpful

www.homatherapyindia.com; www.agnihotra.com.au; www.cosmicfarmingindia.com.

One should try to observe exact times for performance of Agnihotra. After Agnihotra, try to spare as many minutes as one can for meditation. One should sit at least till the fire extinguish. It creates pollution free healing atmosphere.

Just before the next Agnihotra collect the ash and keep it an earthen or glass container. It can be used for plants or making folk medicines.

For Agnihotra materials-Please contact on following mail

Shree Rishikesh Paranjape, Dhule, Maharashtra

E-homatherapy@gmail.com, 09225135175

Caution- If exact time is missed, and materials are not pure, then anticipated results can be assured.

Resonance point- A special configuration of Homa Pyramids is installed to activate a Resonance point on a Homa Farm. It requires in all 10 Pyramids which are charged with special Mantras, once in the beginning of Homa Therapy. Volunteers of Fivefold Path Mission are authorized for installation of RP and providing initial training. They will provide all logistic support with little expenditure for their travel and stay at the site. With establishment of Resonance Point, up to 80 hectares area can be managed by same human efforts with exciting results. The Resonance Point, act as Energy Center connecting farm with biosphere, elements and the cosmos. Agnihotra and Homa tuned to biorhythms of sun and moon cycles and the positions of constellations in the Zodiac. These performances heal the atmosphere and the healed atmosphere heals the plant, animal and human life. In one way, it can be claimed as nanotechnology. The resultant Agnihotra ash which is used one kg per acre or 5-25 gram per plant and ash enriched water all has remarkable impact in soil health and crop production.

Since every component of environment is very much polluted, anticipated results may not be attained in full, as claimed earlier with Agnihotra. Hence, to accelerate its impact, few other Homas in conjunction with Agnihotra are being performed with encouraging impact.

Om Trayambakam Homa

This needs to be integrated with Agnihotra for their synergistic impact. Normally Homa farmers, perform 1- 4 hours of Trayambakam homa daily and for 12 to 24 hours particularly on New Moon and Full Moon and node days. As customary in beginning and at the end of Trayambakam Homa, *Vyahruti* is to be performed.

Vyahruti Mantra

Bhooh swáhá, agnaye idam na mama

(Offer a drop of ghee after swáhá)

Bhuwah swáhá, wáyawe idam na mama

(Offer a drop of ghee after Swáhá)

Bhoor bhuwah swah swáhá, prajápataye idam na mama

(Offer a drop of ghee after Swáhá)

Some people prefer to perform *Vyahruti* Homa at start of crop sowing, harvesting and even threshing. *Vyahruti* can be performed at any time, except sunrise and sunset Agnihotra.

There are two basic energy systems in the physical world: Heat and Sound. In performing Agnihotra, these two energies *i.e.*, the heat of the fire and sound of chanting specific Mantras, are combined to achieve the desired, physical, physiological and spiritual benefits for humanity (**Narang, 9**).

Note- Imagine with four drops of ghee and 4 offerings, one can convey gratitude to mother earth, air, sun and ultimately *Prajapati* creature of this universe.

Agnihotra fire nourishes the Earth-soil to plants, air to water, all of humankind, animal and plant kingdoms. Agnihotra fires along with several hours of Om Trayambakam Homa are performed daily, with more hours of Om Trayambakam Homa on full moon and new moon days are effective tool to purify every component, the soil, water and nearby environment. Agnihotra ash is power full tool to cosmic farmer.

Things to ponder- Homa therapy is unique technology, capable of simultaneously encouraging all production factors i.e., soil biology (soil bio life, enhancing N-fixing, P-solubilizing microbes, discouraging pathogenic organisms, water quality, availability, improving the environmental ecology of biosphere including predators, parasites, birds and cattle productivity- Paranjpe, 15).

Agnihotra ash is full of subtle energy and therapeutic values. It is used in powder form or enriched water and a potent bio enhancer, known as Biosol. Agnihotra ash is used for seed storage, treatment of seed/seedling, enhance soil fertility, water quality, availability, and manage pest and disease in plants.

Use of Agnihotra Ash

- ❑ Mix Agnihotra ash with grains and seeds before their storage. It keeps the attack of pests at zero level. The grain can be used either for human consumption or as seed.
- ❑ One kg Agnihotra ash is spread over one-acre land before crop sowing/planting. The ash not only helps to maintain soil fertility but also energies the soil to capture cosmic energies that aid in plant growth.
- ❑ Agnihotra ash increases the moisture retaining capacity of the soil. It renders the nutrients and elements already existing in the soluble and easily available to plant in great extent.
- ❑ Agnihotra ash has been observed to be rich source of microbes *i.e.*; P- solubilizers and N-fixtures hence are effective to release soluble phosphates from the soil and make them bio available.
- ❑ Equally effective to enhance availability of N from atmosphere,
- ❑ Thus, it plays important role in the growth of plants.
- ❑ For seed treatment place seeds fully covered in Cow urine for 20-30 minutes, drench urine and carefully cover it with dung powder and Agnihotra ash before sowing. Seeds with hard seed coat can be kept for 1-2 hours.
- ❑ Five grams of Agnihotra ash can placed in hills before transplanting vegetable/flowers and herbal seedlings.
- ❑ In fruit plant, incorporate 25-50 g Agnihotra ash before planting.
- ❑ Fine powder of Agnihotra ash can dusted on foliage, fruits wherever insect infection is problem.
- ❑ Agnihotra ash can is mixed with organic mulch all-round the tree trunk. It will help in quick decomposition of mulches and provide desired nutrients required by the plants.
- ❑ In perennial fruits mulching can be done with its own dropped foliage. It is advisable that some legumes are sown in the orchard where some solar radiation is available, mix with dropped foliage and make a thick mulch leaving tree trunk and drench it with Agnihotra ash water/Biosol.
- ❑ An excellent plant food can be made by making a solution of Agnihotra ash+ cow urine and water. For this, incorporate one Kg Agnihotra ash + 5 liters cow urine in 200 liters water. Incubate the mixture for three days and stir it thrice a day. After filtration Agnihotra enriched water can be used as foliar spray or drenching on mulches.
- ❑ One liter of this solution can be sprayed over a compost heap of 5X1X 1M. The heap should be covered with gunny bag/sugarcane trash/coconut fronts or paddy straw.

- This will be helpful in quick decomposition of partially or undecomposed materials thereby improve nutritive value of compost.
- Agnihotra ash is placed in a gunny bag and kept at main water source used for irrigation of field or orchard.
- One kg Agnihotra ash is placed at weekly interval in wells/ponds used for irrigation of field. This will be helpful in improving in water quality and also availability.
- A thick tree paste can be prepared by incorporating cow dung+ cow urine+ Agnihotra ash and clay.
- In fruits like mango, cashew, litchi, guava, aonla, sapota *etc.* the paste should be smeared on the trunk, if possible, on main branches and cut ends.
- Twice pasting *i.e.*, once immediately after fruit harvest and the other in spring – February-March will help in management of borers, gummosis and will help in better vascular activities.
- Use of Agnihotra ash in vermi composting- In process earthworms should be put in the compost material at the end of the afternoon and moistened with Agnihotra ash solution after sunset.

A very potential bioenhancers *i.e.*; Biosol is prepared by using Agnihotra ash and few other organic products in Homa atmosphere. Brief description of biosol and its use in agriculture.

Biosol

It is one of the most potent bio-enhancer developed by Gloria and named as "Gloria Biosol" from Peru, South America (Weir, 25). It is prepared after series of operations and processes in Agnihotra environment. It leads to biodegradation of organic substances *i.e.*, cow dung, cow urine, vermi compost, Agnihotra ash along with a Shree Yantra in a biodigester. It takes four weeks in preparation. It is a powerful bio food and bio fertilizer for plants with rich in nutrients and rich source of beneficial microbial populations.

For preparation of Biosol a rigid bio digester of 200-1000 liters capacity is used. In these two outlets one for release of air at top and the other at bottom are made to get the Biosol. At base one Shree Yantra facing upward is placed. A mixture of cow dung, vermi compost, cow urine and Agnihotra ash mixed in water is placed and properly sealed. At weekly interval, upper valve is opened to release the gas. In this way in four weeks Biosol is ready. If properly stored can be used for six months. Normally it is diluted with water (1:10-15) times, sprayed on plants at monthly intervals. Its soil application also has been found effective (Pathak, 19).

Biosol rich source of macro and micro nutrients and other growth promoting factors and have better response as compared to other bio enhancers with many benefits. But its soil application has also been found effective (YouTube Link: https:www.youtube.com/chanel/UCL1W-cEC5OcE_49KFXeOgiw, Facebook Link: HTTPs://www.facebook.com/proghort.journal.3).

Use of Biosol along with Homa ashes, especially Agnihotra ash provides a promising supplement at a very low cost, affordable technology, even by the marginal farmers.

Effect of Homa Organic Farming on Plants

In Homa farming, there are two components *i.e.*, Homa atmosphere and use of Agnihotra ash, both are synergistically interwoven to give desired results. If plants are kept, where vibration of pyramid fire are maintained, one can watch impact of subtle energy on plant growth and its performance. When Homa is done, its smoke goes to leaves, act as catalyst to enhance chlorophyll, thereby enhanced photosynthesis, which results in better yield of crops.

Biosol rich source of macro and micro nutrients and other growth promoting factors and have better response as compared to other bio enhancers with many benefits. But its soil application has also been found effective (YouTube Link: https:www.youtube.com/chanel/UCL1W-cEC5OcE_49KFXeOgiw, Facebook Link: HTTPs://www.facebook.com/proghort.journal.3).

Impact of Biosol on Crops

- ❑ Foliar and soil application of biosol have been observed to have better impact than alone.
- ❑ Soil application of Biosol increased root nodules maximally as compared to other Homa treatments and control.
- ❑ Rust incidence and insect attack were significantly low (10-30 per cent) with foliar application of Biosol.
- ❑ Soil applications of Biosol have been observed to improve soil biological properties including soil enzymes such as dehydrogenase and phosphatases.
- ❑ Invariably significant impact on crop yield, quality parameters, macro and micro nutrient contents of produce.

Use of Biosol along with Homa ashes, especially Agnihotra ash provides a promising supplement at very low cost, affordable technology, even by the marginal farmers.

There are number of experiences with Homa Therapy and use of Agnihotra ash that improvement in water quality with respect to its biological properties, its availability and even mitigation of drought can be addressed by massive performance of Homa (Berk, 2). In one study, on effect of application of Homa ash for enhancing soil fertility revealed interesting observations. Addition of one per cent ash resulted in increase in activities of nitrogen fixers and phosphate solubilizing microbes, while reduction in pathogenic fungal flora. Ratio of positive to negative micro-organisms was 100:0; soil contained only effective micro-organisms after the treatment with Agnihotra ash (Berde *et al.*, 1).

In fact, systematic research of Homa Organic farming has not been done at any institution but with efforts of members of Fivefold Path Mission few individuals and institutions got interesting results which on various crops as summarized, Pathak, 18. Since promising response of homa therapy and use of biosol on large number of crops are available, just to give a glimpse of its impact two select horticultural crops wherein number of repeated agrochemicals, particularly pesticides are used *i.e.*, mango and cabbage have been included in which, I was directly associated.

Use of Biosol along with Homa ashes, especially Agnihotra ash provides a promising supplement at a very low cost, affordable technology, even by the marginal farmers.

Homa Production of Mango-India

Lucknow in Uttar Pradesh is main mango producing region known for quality production. Infestation of mango hopper, mealy bug and powdery mildew are the serious constraints. Due to repeated spray of hazardous pesticides, there is harmful impact on activity of pollinating agents. Even if pests are managed by pesticide application, it has adverse impact on pollination and to environment. As a result, unfertilized fruit lets in large number can be seen per bunch initially but all these drops within two weeks. This results in physiological problem of "clustering" locally known as *Jhumka*. Experience of homa farm have shown that, in homa atmosphere and with use of Biosol, these problems has been resolved and record production of high-quality fruit can be viewed from Y-tube-https://www.youtube.com/results?search_query=homa+farmig+in+mango

Agnihotra and plants: The ghee which is used in performance of Agnihotra is thrust into the atmosphere and attaches itself to the molecular structure of the soil, allowing the soil to retain more moisture. Thus, plants grown in Agnihotra atmosphere are better able to withstand droughts. Agnihotra causes a change in the cellular structure of plant, which sends more nutrients to the fruits and less to the leaves, stem and roots. Performance of Agnihotra in the garden yields fruits and vegetables are superior.

Note- It is pertinent to note that with Homa atmosphere and use of Agnihotra, is unique technology, wherein all production factors *i.e.*, soil with respect to its structure, organic carbon, pH, availability of elements, soil bio life including earthworm activity, water quality, availability, cows, honey bees, environment, predators, all gets boost, which are essential component of production (Pathak, 18)

In preliminary study at UAS, Dharwad, India, Biosol has been observed to be rich source of beneficial bacteria, fungi, actinomycetes, N-fixers, P-solubilizers. Impact of Biosol has been reported in cabbage and tomato. Enhanced availability of nitrogen phosphorus, potash, Cu, Zn, Mn, Fe, dehydrogenase and phosphatase in soil has been reported. Similarly, enhancement in yield and quality with respect to ascorbic acid, lycopene, total free phenols, total sugars have been observed. Besides these attributes, reduction in leaf spot and fruit borer in tomato and in cabbage black rot, black spot, head borer, diamond black moth *etc.* have been reported.

Benefits

- Rejuvenation of all kind of soils *i.e.*, acidic, alkaline, barren *etc.*
- Prevention, control eradication of plagues and diseases in all crops with short and long vegetative cycles.
- Cereals like rice, corn, wheat, millets, sugarcane, cotton *etc.*
- Vegetables like tomato, onion, cabbage, cucumber, cauliflower, beans, potato *etc.*
- Fruits like banana, mango, orange, lemon, papaya, pineapple, apple, pomegranate, guava *etc.*

Figure 2. View of preparation of Biosol and its impact in cabbage production Maheswar, Khargoan, MP, India.

- ❑ Nuts like peanut, walnut, cashew nut, coconut *etc.*
- ❑ Coffee, cocoa, cotton, *etc.*
- ❑ Forest trees and Pastures.
- ❑ Crops become superior in quantity, taste, texture, color and dieses resistance.
- ❑ Homa Therapy controls and eradicates weeds.
- ❑ Homa Technology is cheap and no agrochemicals are required.
- ❑ Homa improves the health of cattle and no vaccinations are required.
- ❑ Homa Therapy improves quality and quantity of milk.
- ❑ Earthworms and honeybees are beneficially affected by Homa and they produce more.
- ❑ It is necessary to promote Homa/Cosmic seed village to assure quality seeds to cosmic farmers.

Homa Therapy in Cattle Health

There is some information that regular performance of Agnihotra and administration ash to cattle improves their health, production and even there is claim for large number of females in Homa atmosphere. We request for its verification and validation.

Summary

Integration of biodynamic farming and Homa Organic Farming, high quality production can be obtained in any crop in any ecological situations. Through these efforts one can provide solution to convey strong message to the farming community *i.e.,* small and marginal farmers in remote areas who can keep few humped cows and use their products as per crop need without depending on any external inputs. By assertive promotion of this, the plight of small and marginal farmers across the country, pathetic situation of cows can pave a way for sustainable agriculture. Cosmic production and consumption will be instrumental in fulfilling the dreams of honourable Prime Minister "*Atman Nirbhar Bharat*". Food produced with energy derived by enhancing soil biology and environmental ecology will be "*Sattvic*" rich in nutrition, with better quality and higher therapeutic values, while those produced with energy derived by killing millions of soil microbial consortia in soil and biosphere will definitely be "*Tamsik*" poor in taste and nutrition and cause human health ailments.

Conclusion

"What technology has done and persists is doing nothing short of raping the land the air we breathe, of all nutrients, destroying plant life with pesticides, insecticides, polluting human beings with chemical fertilizers and food additives. All this is taking its toll now. Man cannot survive this era without mass attempt of counteracting the destruction. Assertive promotion of Homa Therapy is an answer to resolve current crises being faced by humanity. Methods of experimental science may soon prove to be useful to show that Homa Therapy may be the one single practice which can take care of the body and mind. Involvement of every citizen of world for tree plantation and practice of Homa Therapy is the need of hours and calls for unconditional support of every citizen at planet. Current Corona Pandemic has given a challenge and also an open opportunity to strive our ancient technology by initiating systematic research and validation. This is high time to save humanity at the planet earth.

Our moto should be- Plant-Plant-Plant: Agnihotra-Agnihotra- Agnihotra

REFERENCES

1. Berde, C., Kulkarni, A., Potphode, A., Gaikwad, A., Gaikwad, S. 2017. Application of Agnihotra ash for enhancing soil fertility. *International Journal of Sciences, Engineering and Technology Research*. **7:** 2546-2551.
2. Berk, U. 2020. Impact of Homa Organic Farming in mitigating Soil, Water and other Environmental Crises. *International Journal of Plant and Environment*. **6:** 56-67.
3. Dhama, K., Rathore, R., Chauhan, R. S., Tomar, S. 2005. Panchgavaya (Cowpathy): An overview. *International Journal of Cow Science*. **1(1)**: 1-5.
4. Ghosh, S. 2002. Principles of Organic Farming in Tropics. In: Alvares *et al.* (eds) The Organic Farming Reader Other India Press. Goa. 53-68.
5. https: //www.youtube.com/results?search_query=homa+farmig+in+mango
6. https: //www.youtube.com/results?search_query=hortlak-43+bird+ulrich+brush+johnson

7. Ismail, S. 2002. The contribution of soil fauna-especially earthworms to soil fertility. In: The Organic Farming Reader. Alvares *et al.* (eds.) Other India Press. Goa. pp 177-163.

8. Murthy, M. 2005. Vedic Aura Energy - A Remedial Measure for Plant Diseases. In: Proceeding of National Conference on Bridging Gap between Ancient and Modern Technologies to Increase Agricultural Productivity. Asian Agri-History Foundation of Udaipur, Rajasthan, pp129.

9. Narang, I. 2004. The Science of Agnihotra, Published by Maharishi Dayanand Charitable Trust, Dayanand Vihar, Delhi.

10. Nene, Y. L. 2003. Cow and Agriculture, Paper presented in National Seminar on Cow in Agriculture and human health. Agri History Society, Udaipur, Rajasthan, pp 55.

11. Nene, Y. L. 2018. The concept and Formulation of Kunapajal, the World's Oldest Fermented Liquid Organic Manure, Asian Agri-History. **22**: 6-13.

12. Paranjape, V.V. 1977. Light towards divine path. Booklet, Published in U.S.A. by The Agnihotra Press, INC. Post office Box 13, p. 31.

13. Paranjape, V.V. 1989. Homa Therapy Our Last Chance, Fivefold Path Mission, Inc, Madison, U.S.A.

14. Paranjape, V. V. 1989. Purified with Fire. In: Secrets of the Soil (Tompkins-Bird Ed.), Rupa and Co. New Delhi, pp 243-254.

15. Paranjape, V. V. 2004. Homa Therapy Our Last Chance, Fivefold Path Publication, Dhule, Maharashtra, India.

16. Pathak, R. K. 2013. Homa Jaivik Krishi: A Ray of Hope for Sustainable Production in Fruit Crops. Lead talk delivered in Silver Jubilee Celebration of Aspee College of Horticulture and Forestry, at Navsari Agricultural University, Navsari, Gujarat, pp: 192-201.

17. Pathak, R.K. 2019. Cosmic Farming: A Ray of Hope for Sustainable Horticulture, Lead Talk delivered in International Seminar on Modern Agriculture Approaches in 21st Century, Organized by Samarg Vikas Welfare Society (SVWS), Lucknow pp: 1-7.

18. Pathak, R. K. 2019. Cosmic farming: A ray of hope for sustainable horticulture and nutritional security. In: Progressive Horticulture Conclave on Futuristic Technologies in Horticulture, pp1-14.

19. Pathak, R.K. 2021. Cosmic Farming in Horticultural Crops, Mrs. West Ville, Publishers, New Delhi.

20. Pathak, R.K. and Ram, R. A. 2014. Homa Jaivik Krishi: A Ray of Hope for Sustainable Horticulture, In Proceeding of Horticulture for Economic Prosperity and Nutritional Security in 21st, Centaury, Westville Publishing House, Ed. Hazarika, T.K. and Nautiyal, B.P. pp: 39-46.

21. Ram, R. A. and Pathak, R.K. 2017. Bio-enhancers. LAP Lambert Academy Publishing, Germany, ISBN 978-3-330-33128.

22. Ram, R.A. and Pathak, R.K. 2019. Bio-enhancers for soil, plant health and insect pest management in organic production of horticultural crops. *Journal of Eco-Friendly Agriculture* **14 (2)**: 28-33.

23. Tomkins, P. and Bird, C. 2004. Secrets of Soil, Rupa Publications India limited Ansari Road, New Delhi-110002. India.

24. Tomkins, B. and Bird, C. 2018. The Secret Life of Plants, Harper and Row, New York.

25. Weir, M. L. 2009. Preparation and application of Gloria Biosol. Bringing Homa Organic Farming into mainstream of Indian Agriculture System. Fivefold Path Publication, Dhule, Maharashtra. pp 95-96.

Chapter 18

Soil and Water Conservation Techniques for Improving Horticultural Production through Exploitation of Degraded Lands in India

M. Madhu, A.C. Rathore* and Charan Singh

ICAR–Indian Institute of Soil and Water Conservation, 218 Kaulagarh Road, Dehradun 248195, Uttarakhand

**e-mail: rathoreac@gmail.com*

ABSTRACT

In India, average productivity of fruits and vegetables is 13.83 t ha-1 which is much lower than average productivity of USA (28.17 tha^{-1}), Spain (28.0) and Turkey (24.60 tha^{-1}). The average productivity of fruits and vegetables in the country can be enhanced from 13.83 to 27.0 t ha^{-1} yr^{-1} by adopting soil and water conservation practices for utilization of waste lands/degraded lands. It has been proved that such a system can reduce soil erosion from 42 to 1.5 tonne ha^{-1} and can save soil and essential plant nutrients. Water harvesting can be harnessed both for irrigation to fruit plants at critical stages as well as for other purposes like pisciculture. Fruit trees with vegetables as a part of Agri-horticultural systems provide subsidiary sources of income through live stock rearing and have great scope for the degraded lands or waste lands of India.

***Keywords**: Soil, Water, Erosion, Conservation, Horticultural crops.*

Land, a non-renewable natural resource, is the central resource for all production systems, which has suffered from different types of degradations over the years leads to land degradation which is caused by biotic and abiotic pressures. An ever-increasing population places enormous demands on land resources. This is particularly acute in India, which has only 2.4 per cent of the world's geographical area but supports over 16 per cent of the world's population. It has 0.5 per cent of the world's grazing area but has over 18 per cent of world's cattle population. These pressures have led to drastic changes in the proportion of land utilised for agricultural activities, urbanisation and industrial development. Approximately 36.70 per cent (120.72 million ha area) of total geographical area in India is under various kinds of land degradation due to water, wind, salt, *etc.* (Table 1) spread all over country (NAAS, 9). Hilly regions are more vulnerable to soil erosion than the other states of India due to undulated topography, medium to very steep slopes and poor soil depth. It has been reported that 5334 million tonnes of productive soil is lost every year, out of which major portion is lost from about 80 million hectares of cultivated land in India.

Table 1. Status of degraded lands in India

Degradation Type	*Arable Land (m ha)*	*Open Forest (m ha)*
Water erosion (>10 tonnes/ha/yr)	73.27	9.30
Wind erosion	12.40	-
Sub Total	85.67	9.30
Chemical degradation		
Exclusively salt-affected soils	5.44	-
Salt-affected and water eroded soils	1.20	0.10
Exclusively acidic soils (pH<5.5)	5.09	-
Acidic (pH<5.5) and water eroded soils	5.72	7.13
Sub-Total	17.45	7.23
Physical degradation		
Mining and industrial waste	0.19	-
Water logging (permanent surface inundation)	0.88	-
Sub-Total	1.07	-
Total	104.19	16.53
Grand Total (Arable land and open forest)	120.72	

Erosion Status in different States of India

The erosion status of different states in India is presented in Table 2. The overall percentage distribution of total geographical area (TGA) under erosion classes 1 (very low), 2 (low), 3 (moderate), 4 (severe) and 5 (very severe) is 25.72, 21.37, 27.96, 10.69 and 14.31 per cent, respectively. On an average basis, the total soil erosion from the terrestrial land surface of India is estimated at 4.87 Pg yr^{-1} (or billion Mg yr^{-1}). Of the total loss, the maximum area (24.96 per cent) occurred in the moderate erosion category (10–20 Mg ha^{-1} yr^{-1}), whereas the minimum area (9.54 per cent) was in the severe erosion category (20–- 40 Mg ha^{-1} yr^{-1}). About 12.77 per cent of the total geographic area in India is affected by extremely severe erosion (>40 Mg ha^{-1} yr^{-1}). In the extremely severe category, Madhya Pradesh has the largest area (5.45 Mha) >

Uttar Pradesh (5.10 Mha) > Andhra Pradesh (4.94 Mha) Chhattisgarh (3.85 Mha) > Arunachal Pradesh (3.55 Mha) > Maharashtra (3.04 Mha).

Table 2. Areas (Mha) under different erosion class and states of India

States	*Total Geographical Area (Mha)*	*Erosion Class*					
		Very Low (>5 Mg ha^{-1} yr^{-1})	*Low (5-10 Mg ha^{-1} yr^{-1})*	*Moderate (10-20 Mg ha^{-1} yr^{-1})*	*Severe (20-40 Mg ha^{-1} yr^{-1})*	*Very Severe (>40 Mg ha^{-1} yr^{-1})*	*Total Soil Loss (Tg yr^{-1})*
Andhra P.	26.61	0.12	8.17	12.01	1.32	4.99	531 ± 104
Arunachal P.	7.43	0.76	0.98	0.93	1.20	3.56	237 ± 74
Assam	6.72	0.01	1.94	1.62	1.04	2.13	176 ± 41
Bihar	9.01	6.09	1.73	0.88	0.27	0.05	51 ± 5
Chhattisgarh	13.36	3.70	1.52	1.60	2.68	3.86	318 ± 77
Delhi	0.09	0.01	0.06	0.02	0.00	0.00	0.97 ±.15
Gujarat	15.16	11.84	2.05	0.69	0.48	0.10	74 ± 9
Haryana	4.33	3.34	0.77	0.16	0.05	0.02	18 ± 3
H.P.	3.32	0.39	0.24	1.21	0.62	0.87	82 ± 17
J&K	6.74	4.00	0.26	0.51	0.41	1.56	109 ± 31
Jharkhand	7.85	1.47	1.33	2.06	1.56	1.43	162 ± 27
Karnataka	18.30	0.00	2.96	11.37	2.01	1.96	351 ± 67
Kerala	3.78	0.00	3.21	0.49	0.09	0.00	34 ± 10
M.P.	30.38	6.84	5.44	7.06	5.59	5.46	604 ± 103
Maharashtra	30.01	14.00	5.95	4.69	2.33	3.04	372 ± 46
Manipur	2.18	0.00	1.00	0.59	0.59	0.00	34 ± 7
Meghalaya	2.24	0.33	0.52	0.40	0.22	0.77	56 ± 15
Nagaland	1.65	0.15	0.13	0.15	0.10	1.14	63 ± 24
Orissa	15.11	7.26	3.03	2.69	1.27	0.86	162 ± 11
Punjab	4.94	4.29	0.34	0.17	0.08	0.07	21 ± 3
Rajasthan	31.62	5.48	8.17	10.92	5.17	1.89	488 ± 63
Sikkim	0.59	0.33	0.01	0.02	0.06	0.19	12 ± 4
Tamil Nadu	11.97	0.00	5.73	5.22	0.52	0.50	162 ± 30
Tripura	1.03	0.00	0.57	0.22	0.07	0.18	18 ± 3
U.P.	23.24	0.00	2.44	13.33	2.38	5.10	544 ± 113
Uttarakhand	4.38	0.68	0.62	0.75	0.61	1.71	122 ± 35
W. Bengal	8.20	3.55	2.87	1.41	0.31	0.06	64 ± 8
Total	**290.24**	**74.64**	**62.02**	**81.15**	**31.02**	**41.53**	**4,876.50**
(per cent TGA)		**(25.72)**	**(21.37)**	**(27.96)**	**(10.69)**	**(14.31)**	

The areas in the low erosion category (<5 Mg ha–1 yr–1) followed the order: Maharashtra (14.0 Mha) > Gujarat (11.84 Mha) Orissa (7.26 Mha) > Madhya Pradesh (6.83 Mha) and Bihar (6.09 Mha). The remaining states had areas ranging between 0.001 Mha in Tamil Nadu to 5.47 Mha in Rajasthan under the low erosion category (Mandal *et al.*, 2020).

Fruit Species Suitable for Degraded Lands

The fruit crops have assumed an important role in the diversification of agriculture so as to optimize the use of varied soils and ecological situations. Fruit cultivation has shown its promise as one of the most remunerative diversification of land use in hill and mountain ecosystem as it provides congenial and varied agro-climatic conditions varying from subtropical to temperate ones. Fruit plantation is also ecological recommended as an ideal land use for steep undulating slopes of the hills, otherwise unsuitable for cereal crops. Rainfed horticulture has assumed greater attention due to better economic returns from relatively poorer lands. Fruit trees provide a better substitute in areas where cropping may not be possible due to limitation of soil, slope, non-availability of irrigation *etc*. The degraded lands in rainfed areas generally earmarked for fruit cultivation lack fertile soil and moisture required for establishment and effective growth of plants. Hence, soil and moisture conservation is a precondition for survival of plantations. As majority of rainfed farmers are marginal or small with low risk bearing capacity, the long gestation period of fruit plants is a stumbling block. To overcome this, research efforts have been made to introduce sandwich agriculture or intercropping (millets and legume) into the system (Das *et al.*, 2). Similarly, vegetable/spice crops like melons, pumpkins, cucurbita, gourds, tomato, brinjal, chilli, cauliflower, cabbage, colocasia, turmeric, zinger, coriander, spinach, methi, *etc*. can be grown on such lands during different season without manuipulation in soil surface. Fruit species which have deep tap root system, sunken stomata, high density of stomata in the down side of leaf, thorny growth, less water requirement for completion of biological/ physiological activities, are some important characteristics of trees make more suitable to degraded lands (Table 3).

Table 3. Some of the important fruit species suitable for degraded lands

Fruit Species	*Scientific Name*	*Suitable Cultivars*	*Spacing (m)*	*Pit Size (m^3)*
Bael	*Aegle marmelos*	NB5, NB9, CISH B1-2	8-10	1.0
Aonla	*Emblica officinalis*	NA7, NA10, Chakaiya	6-8	1.0
Ber	*Zizyphus mauritiana*	Umran, Apple ber, karaka	6-8	1.0
Guava	*Psidium guajava*	L-49, Sweta, A.safeda	4-6	0.42
Jackfruit	*Artocarpus heterophyllus*	Ceylon jack	8-10	1.0
Mango	*Mangifera indica*	Dashehari, D51, Mallika	8-10	1.0
Pomegranate	*Punica granatum*	Bhaga, Mridula, Ruby, Jyoti,	4-6	0.216
Mandarin	*Citrus reticulata*	Kinnow	4-6	0.216
Lemon	*Citrus limon*	Eureka round and oval, Pant lemon, Pusa Abhinav	4-6	0.216
Kagzi lime	*Citrus aurantifolia*	Pramalini, Pusa Uday, NRCL-7-8	3-6	0.216

Fruit Species	*Scientific Name*	*Suitable Cultivars*	*Spacing (m)*	*Pit Size (m^3)*
Pumelo	*Citrus maxima*	NRC Pumelo 5-6	6-8	1.0
Grape fruit	*Citrus paradisi*	Flame Seedless, Ruby,	6-8	1.0
Kumquat	*Citrus japanica*		3-6	0.216
Peach	*Prunus persica*	Pratap, Prabhat, Sarbati	6-8	1.0

Fruits and vegetables are considered as protective food since they are rich sources of vitamins and minerals which not only help in conserving soil but also provide nutritional security. Total fruit and vegetable production of India is 282.4 million tonnes against target production (318.0 million tonnes) which is 36.0 million tonnes short of target set by planning commission of India in 2021. Average fruit and vegetable productivity of India is 13.83 tha^{-1} which is much lower than USA (28.17), Spain (28.0) and Turkey (24.60 tha^{-1}) globally (NHB, 10). This gap in production as well as poor productivity of fruits and vegetables of our country is primarily due to lack of soil and moisture conservation practices and loss of top fertile soils every year in pre and post monsoon season. The present production is insufficient to meet the fruit and vegetable requirement due to low per capita availability against recommended quantity of 120 and 300 gm day-1, respectively because most of the production (80 per cent) comes from rainfed areas (NRAA, 11). This compels cultivation of fruits and vegetables on degraded lands or waste lands with using soil and water conservation practices besides high yielding cultivars in India.

Soil and moisture conservation techniques/conservation horticulture is a resource saving production system, which conserve, improve and make more efficient use of natural resources through integrated management of available soil, water and biological resources combined with external inputs. It also contributes to environmental conservation as well as to enhanced and sustained production. This system utilizes soils and water for the production of crops with aim to maintain soil productivity by crop residues on soil surface which minimizes damage to the root environment. It provides an optimum environment to the root zone to maximum possible depth. It ensures that water entered into the soil goes down to ground and retained for subsequent crop production. It favours beneficial biological activity in the soil to maintain and replenish soil health which contributes to soil organic matter and humus. It also absorbs, retain, chelate and slow release of plant nutrients. Conservation horticulture also includes conservation of genetic resources, developing cultivars; fruit species produces more with less water, water saving technologies, cropping system adjustment, architectural manipulation of canopy and root, integrated nutrient management, *etc.*

Soil and water conservation practices or conservation horticulture practices/technologies like micro-irrigation and fertigation save water and fertilizer by 40-50 per cent and increase fruit yield by 50-75 per cent and reduce soil loss up to 20 per cent from fruit based land use systems. Manuring and mulching both conserve moisture by 15-20 per cent and increase fruit yield by 20-25 per cent. Canopy architecture saves moisture by reducing evaporation losses up to 10 per cent. Thus, conservation horticulture practices/technologies can enhance Indian fruit production to the tune of 10-15 per cent which is essential for nutritional security in the country. This paper review the soil and water conservation technologies practiced in India and its scope for enhancing horticultural production through utilization of degraded lands.

Limitation of Degraded Lands

Horticulture sector needs to improve horticultural productivity of degraded lands to nourish the Indian mass using soil and water conservation practices. It is well known fact that the horticulture sector lack appropriate technologies, infrastructure support, marketing, extension network, organized planning and execution of research/education and diversity in farming system for the degraded lands. Because of decrease in productive land, reduction in the forest and undulated topography resulted into soil fertility depletion caused by soil erosion of unprecedented proportion. The following limitations of degraded lands are as follows.

1. Presence of high per cent of sand, gravels, boulders, riverbed materials
2. Poor soil depth and poor water holding capacity
3. Low in carbon contents
4. Severe soil erosion and degradation of land.
5. Undulating topography and land features and small fragmented holdings limit mechanized farming.
6. Climatic factors *i.e.* Temperatures aberrations, erratic rains, hail during flowering and fruiting, insufficient chilling in temperate fruits *etc.*
7. Lack of hardy fruit/vegetable cultivars *i.e.* less productive cultivars, alternate bearing, non-availability of elite planting material.
8. Rainfed production systems and poor irrigation infrastructure.
9. Adoption of conventional farming and lack of suitable technologies.
10. Lack of appropriate training to growers, lack of adequate and timely supply of inputs.
11. Lack of proper transport system.
12. Lack of skill for postharvest technologies

Soil and Water Conservation Measures/Techniques

The success of a properly planned alternate land use system, such as fruit cultivation, largely depends on the soil and water resource conservation programme (Hudons, 3). As the topography of most degraded lands does not permit for adequate water infiltration, it is essential to have some measures to arrest the water and save the thin soil layer from displacement. In the strict sense, such obstructions to the natural flow of runoff water are termed as the SWC interventions. By this the soil loss is reduced. In addition to this, the SWC interventions in the upper region of a topo-sequence help in recharging the underground reservoir (Sharma and Mishra, 16). Above all, SWC interventions are the first and the foremost in any land use planning from point of view of resource conservation (soil, water and vegetation) and later resource generation. Experience has shown that in extremely degraded lands, natural conservation processes become very slow. In such situations, SWC interventions improve soil moisture status and promote ecological succession; thus providing better vegetative biomass for resource conservation as well as ensuring production on sustained basis.

The SWC measures are broadly classified into two categories (i) Vegetative or Biological, and (ii) Mechanical or Engineering or Structural. The combination of biological and mechanical measures is called bio-engineering measures. In situations where biological measures only are insufficient to check erosion to a desired level due to high velocity of runoff or discharge, mechanical measures are needed. The mechanical measures not only check soil erosion but also conserve moisture for crop growth. Basically these measures constitute a series of mechanical barriers constructed across the slope to reduce or break the length of slope only and/or both the length and degree of slope to dissipate the energy of flowing water. The important principles to be kept in view while planning structural measures are:

1. Increasing the time of concentration of runoff and thereby allowing more of it to be absorbed and held the soil.
2. Intercepting a long slope into several short ones so as to maintain less than a critical velocity for the runoff water.
3. Protection against damage due to excess runoff.

The selection and the suitability of engineering control measures for different locations depend on soil depth, rainfall, land slope and crops to be raised. It is unrealistic to have all standard recommendations/specific procedures for SWC interventions for adoption in different soil and climatic environments. The philosophy behind the introduction of any system lies on the consideration of better soil and moisture conservation practices. After carefully analyzing the following relevant factors, and giving due respect to the wisdom of the local people involved in the activities of greening the degraded land, the SWC interventions should get priority to achieve success in the alternate land use programme. Water conservation is inseparable from soil conservation. There are two approaches, *viz.*; mechanical and vegetative which have been used in soil and water conservation to intensify horticulture production. Mechanical soil and water conservation measures are required for controlling soil erosion, retaining maximum rainfall within the slope and safe disposal of excess run off from the top to the foot hills. These structures are used in case of extreme soil degradation, where other approaches are not possible or less effective. The vegetative barrier such as cover crops, residue mulches and grass contour barriers apart from conserving soil and water also offer additional benefits such as fodder and fuel wood. Various technologies adopted for conservation of resources in the fruit and vegetable crops have been discussed as under:

Micro Site Improvement

The micro-site improvement is a form of hydrological intervention for improving point water storage particularly in degraded soil with limited water holding capacity. In degraded soil, perennial vegetation in terms of tree components can be raised successfully through restructuring of the soil at the point where the tree is planted. The restructuring of soil through soil ameliorants can be formed as micro-site improvement. This is nothing but the replacement of native soil with a synthesized mixture of high water holding material. The micro-site improvement ensures betters moisture regime around the plant ecosystem by holding more moisture and the same time improving aeration and drainage characteristics. The soil for this purpose can be amended in many ways through ameliorants (organic/inorganic) in different proportions. Also, some sort of mulching either vegetative mulch (straw, grass *etc.*) or stone mulch should be spread in the tree basin to reduce evaporation losses. The improved micro-sites act as catch pits for arresting runoff as it increases the amount of rain water storage in soil in the root zone and also allows better recharge of ground water when implemented on a large scale. By this technique a satisfactory survival rate of seedlings can be achieved and vigorous growth of plant is ensured. The amendment of sand/coarse soil with organic and inorganic materials combined with farm yard manure (FYM) offer tremendous scope in improving the soil moisture retention (water holding capacity) characteristics of the micro-site and vertical drainage. Some water absorbing polymers (hydrophilic compounds) are nowadays available for use in the micro-site like : neopeat, soil rite, water works *etc.* It is always desirable to have bigger size pits for this purpose, provided hard rocks are not encountered at shallow depths. However, cost becomes a limiting factor at times, but it could be reduced by using mechanical pitting augers. Similar, experiences have been recorded with bentonite by Muller and Engelthaler, (8) in Argentina. Other clay minerals available for the purpose are perlite and vermiculite; these materials are being used in nurseries only. In most situations, tank silt is a locally available cheap and practical material. Use of tank silt is nothing but a practice of "soil harvesting and recycling" for benefit of both plant as well as reservoir. In a laboratory study, a mixture consisting of tank silt (33 per cent), alfisol (42 per cent), poultry manure (15 per cent), and paddy husk (10 per cent) improved the soil moisture content by 59 per cent on first day and 237 per cent on the 30th day over the soil (alfisol) alone (Katyal and Osman, 4). This indicates a prolonged availability of soil moisture by mixing indigenously available cheap materials with native soil. In a shallow alfisol, the addition of bentonite in the pit mixture may be preferred in low dose (Upto 10 per cent) for hardy plants (Mishra and Osman, 6). However, all the conventional and non-conventional amendments should be tested in the desired environment before final adoption.

Application of 75 per cent soil with 25 per cent gravels, 60kg FYM yielded 45.0 kg tree^{-1} (9.00 tha^{-1}) which was slightly lower than peach tree planted in the pit mixture with 100 per cent soil without gravels (50 kg tree^{-1} or 10.2 tha^{-1}) than other combinations along with cultivation of black gram – toria in the interspaces of peach plants. Peach based agri-horticulture system revealed that the practice is economically viable with a B:C ratio of 3.50, calculated for a period of 20 years life of peach trees. Using the crop combination of Black gram – Toria for the 20 years, the payback period was calculated to be 5 years along with net return of Rs. 65,560 ha^{-1}. Toria cultivation is more profitable than urad, due to the absence of weeding operations during rabi. Yields of 3.55 and 3.58 q/ha were obtained for urad and toria, respectively. Black gram residue addition also helps in improvement of site soil conditions (Rathore *et al.,* 13).

Soil Working and Pitting Techniques for Planting of Fruit Trees

The soil working techniques depend upon the climate and edaphic factors (Table 4). The soil depth and proportion of gravel or boulders at the site, decide the depth and size of pits/ trenches. The common soil working techniques are as under (Khanna, 5).

Pits

Ordinary Pit is an ordinary pit is recommended for clayey and saline/alkaline soils in all rainfall classes. It is a pit dug up 30 cm deep with 30 cm base and filled back with earth. It has a high crest and a deep part of the water storage ring, away from the highest point of the mound.

Table 4. Type of pits/trenches under broad rainfall groups

Type	*Rainfall Distributor*	*Rainfall Amount*		*Rainy Days*		*Suitable Conditions*
		Summer (cm)	*Winter (cm)*	*Summer (No.)*	*Winter (No.)*	
Pits : Ordinary	SH and WL	25-75	Up to 38	15-50	Up to 15	Clayey, saline and sandy soils
		SL and WH	Up to 25	25-75	Up to 15	15-30
		SL and WL	Up to 25	Up to 38	Up to 5	Up to 15
		SH and WH	25-75	25-75	15-20	10-30
Saucer			As above			Loamy Soils. Sloppy land
Ring			As above			Dry areas
Ridge-ditch:						
Vertical Sides	SH and WL	50-75	Up to 38	30-40	Up to 15	Loamy soils. Most parts of dry zone
Small	SH and WH					
Large	SH and WL	25-75	25-75	15-50	10-30	Deep Soil
Shallow	SH and WH	38-50	Up to 38	30-40	Up to 15	III – distributed rainfall
	SH and WL	25-50	25-50	15-50	15-30	Top soil washed away
	SH and WH		As above			Gravels or murrum exposed
Sides sloped	SH and WL	35-75	Up to 38	20-50	Up to 28	Shallow surface soil sub-soil gritty, but fine clay with crushing.
						More rainy days.
Slanting sides	SH and WL	25-38	Up to 38	15-30	Up to 15	Soils friable
Side half slanting		Up to 25	Up to 38	Up to 5	Up to 15	Rainfall low
		Up to 25	25-75	Up to 15	15-30	III – distribution
Sides double slanting	SH and WH	50-75	Up to 38	30-50	Up to 15	Loamy clay soils
	SH and WH	25-75	30-50	30-50	10-20	Good well distribution rainfall
Shallow trench	SH and WH	38-75	Up to 38	20-50	Up to 15	Alkaline/Saline deflocculated, clay soils, scanty rainfall
	SH and WH	25-75	25-75	15-50	10-30	
Deep trench	SH and WL	25-38	Up to 38	15-30	Up to 15	High salts, scanty rainfall
	SL and WL	Up to 25	Up to 38	Up to 5		

SH = Summer high; SL = Summer low; WH = Winter high; WL = Winter low.

This helps in minimising water logging and concentration of salts near the seedlings. Saucer Pit is suitable for all loamy kinds of soils in all rainfall classes of the dry zone. The crescent trench is necessary only on sloping lands. The pit is made in the centre of 1 m radius prepared by giving the soil a slope from periphery towards the centre. Ring Pit is suited for sandy soils in dry areas. The crest is not required to be very high because deep planting will be necessary to place the roots about 30 cm below the surface where a zone of moisture accumulation usually occurs in case of very sandy soils. The ring trench will allow water to infiltrate near the planting site and thus improve its moisture status.

Ditches

This is suited for moist part of the dry zone with large number of rainy days and for loamy soils. Large Ridge-Ditch is suitable for low rainfall and deep soil conditions. This traps a large amount of moisture in the worked up soil if the soil is sandy and the rainfall is rather low but the number of rainy days is not small. It is also suitable for low and ill-distributed rainfall for non-sandy soils. Slanting Sides is suitable for clayey profiles where water remains standing for a relatively longer period and there is greater seepage of moisture in a vertical direction and roots lie in dry zone, the following techniques may be suitable.

Trenches

For peninsular region where clay and murrum form impermeable crust, which on wetting and drying bind granular murrum in hard clod, shelved trenches are recommended. In shelved type of trench, planting is done at ground level or little above it. Shelved trenched may be shallow or deep. Shallow Shelved Trench, in this technique water will be retained mainly in deeper part of the trench. Deep Filled Shelved Trench is suitable for area with high rainfall which is received in fewer rainy days as heavy or torrential showers. It is recommended for sandy and gravely soils and also for saline and alkaline soils. Double Trench is suited for stony and detrital slope with small amount of soil in the interspaces of mall and large pieces of stones where water holding capacity is very low. It impounds water in the filled portion. Sowing or planting is done at the top of the ridge. Trench Ridge is suitable for salt-affected soils and provides essential mechanism forefficient leaching of salts from the seed bed and impounding of water not in direct contact with the worked up soil and thus minimizes damage from water-logging. The shallow trench ridge is suitable for alkaline, saline and deflocculated clay soils in areas receiving rainfall over a large number of rainy days. A deep trench ridge is suited to soil with a high salt content in scanty rainfall received in a few violent storms. It is spacious enough to impound water and has a high ridge to store moisture and allow leaching.

Drilling

Auger holes are made either manually or power driven which economizes the cost of soil working. The soil from the holes is excavated through long-armed spatula. Planting in auger hole is, however, suited for the tree species, which have more vertical growth of the tap roots. Soil working should be completed well in advance to allow the excavated soil to weather and the refilled pits get a chance to settle with two or three showers.

Trenches conserve more soil, moisture, nutrients and enhance growth and productivity of the trees as compared to normal planting method. Experiment was carried out on planting methods for plum fruits in trenches and holes during1995-2002 at Bulgaria. Results showed that maximum canopy volume (10.85 m^3) and fruit yield (20.84 kg $tree^{-1}$) were recorded in plum fruit planted in trenches as compared to canopy volume (9.74 m^3) and fruit yield (17.86 kg tree^{-1}) in hole method of planting over 7 years of plantation (Table 5). The plum trees planted in trenches had grown vigorously and bore fruit profusely. Water erosion risk was brought down to the minimum by combining the trench planting method with sod on the inter row alleys.

Table 5. Planting of Plum fruit (cv Stanley) in trenches- A promising method for establishing orchard in the mountain regions

Years	*Canopy Volume (m^3)*		*Fruit Yield (kg tree^{-1})*	
	Trench	*Hole*	*Trench*	*Hole*
1995	-	-	-	-
1996	2.16	2.35	-	-
1997	4.93	4.64	1.2	2.0
1998	9.57	8.10	5.7	6.0
1999	14.11	12.70	15.0	14.0
2000	15.00	13.57	28.0	22.0
2001	15.20	15.25	40.0	35.0
2002	15.00	11.55	56.0	46.0
Total	75.97	68.16	145.9	125.0
Average	10.85	9.74	20.84	17.86

Source: Dinkova *et al.* (2006).

In the coast of Granada, an intensive irrigated agriculture based on subtropical crops has been established. These trees have been planted in terraces of highly sloped areas. In this fragile Mediterranean agro-ecosystem, the removal of native spontaneous vegetation cover and substitution by orchards, increase the susceptibility to soil degradation and eventually brings up the destruction of these structures by rainfall events. To study this net change, we monitored the soil loss and runoff over a two-year period with a mature mango (*Mangifera indica* L.) orchard. The studied treatments were bare soil (BS) and natural spontaneous vegetation (NSV), each twice replicated. The erosion plots were 4 m x 4 m in area and were located in the orchard terraces (65° slope). The average annual soil loss by erosion for BS and NSV was 2.5 and 0.3 Mg ha-1 yr^{-1}, and for runoff 34.1 and 6.8 mm yr-1, respectively. Therefore, soil erosion and runoff from BS plot were 8 and 5 times higher than in NSV, showing the importance of plant covers in the taluses of terraces in reducing this impact. Thus, the removal of plant cover from the taluses under these conditions, represent a high risk of slump and collapse, causing serious environmental and economic problems for farmers of subtropical crops.

Water Harvesting in Poly-lined Water Harvesting Pond/Tank

This is one of the most important components in utilizing the stored water in a part of the catchment area as well as downstream land (Mishra and Sharma, 7). This technique has wide scope in the areas having moderate to high runoff producing potential. In the initial stages of greening the degraded lands as a part of alternate land use, the availability of stored water is very important from the point of view of off season/supplemental or life saving irrigation. Particularly for raising tree components, arrangements for irrigation should be thought of in the dry periods, at least in the first two years. Hence, provisions must be made to store some water within the easy reach of the farmers, for conveniently using the harvested water, when required. Major storage structures need to be located in community areas or Government lands.

Assessment of runoff and storage capacity and information on soil, land, geology are pre-requisites in designing such structures. As far as possible, local materials should be used for making hydraulic structures for the purpose of impounding and storing water. Depending on the value of harvested water in water scarcity areas, lining of storage structures may also be thought of. The drainage lines (water courses) should be properly aligned and maintained in order to feed the storage structures. Where the runoff amount is more and it is required to store water at number of points; a net work of small water pools may be created along the waterways with offset feeder channels. From the hillock areas water can also be collected by the drain running at the base of the hillock and it is diverted to water harvesting bodies through gullies or water ways.

The harvested water from springs/streams is stored in ponds/tanks before application to the crops. To prevent seepage losses, lining of the tanks/ponds is to be resorted to. Conventionally cement lining was being done but since it is very costly an alternate low cost lining by low density polyethylene (LDPE film) (800 to 1000 gauge thickness) is preferred. Geomembrane and Silpaulin sheet (200 GSM) have also been used for lining of the small ponds (Sharda and Juyal, 15). Polylined ponds cost about half of the cement lined ones even after taking into account their respective life periods. Polylined tanks/ponds (20-50 cum storage capacity) are particularly suitable for hilly areas of Himalayan region where cement lined tanks are prone to failure by earthquake and unequal settlement. A series of poly lined ponds may be constructed along a stream in the hill areas. The low-cost water harvesting poly-ponds can be integrated with gravity fed micro-irrigation system (drip) in the hilly areas for irrigation of high value horticultural crops. The naturally available gravity head in the terraced lands in hills provide an ample opportunity in this regard without resorting to use of external pumping device (Patra *et al.*, 12).

Micro Water Harvesting Structures

Creation of different size of micro catchments (V, D, L shapes with polythenes) in dry land areas is necessary for *in situ* water harvesting. They can store water in the root zone of plants for their survival. On hill slopes, soil and water conservation measures may be followed with horticulture land use system. Analysis revealed (Tables 6) that V shaped micro catchment with polythene lining at bottom produced maximum fruit yield of 23.03 (kg tree^{-1}) followed by V shaped micro catchment where as moisture conservation was maximum (17.5 per cent) under diagonal catchment followed by L shaped (15.4 per cent). Study conducted at Dehradun revealed that maximum fruit yield (12.4 kg tree^{-1}) of lemon was recorded in the treatment of V shaped micro catchment with grass mulch followed by V shaped treatment over other treatments as shown in Table 6.

The peach plants were planted at 6m apart in square system in the circular trench (1m diameter x 0.30m depth) and rectangular trenches (1m x 0.30m x 0.30m) on degraded lands (Table 7). The maximum fruit yield (137.0 kg tree-1or 36.0 tha-1) of peach planted in circular trench with cluster bean was harvested from T_{12} treatment among all other treatments. The benefit cost ratio for all twelve treatments was calculated using 8 per cent discount rate assuming

Table 6. Moisture conservation in different micro catchments with ber plantation

Year	*Total Rainfall (mm)*	*Micro Catchment Shapes*					
		L	*L+P*	*V*	*V+P*	*D*	*C*
Runoff (per cent of rainfall)							
1987	257	53	58	61	66	43	2
Moisture (per cent)		6.4	5.0	5.5	4.9	7.5	4.5
1990	601	66	75	70	78	56	2
Moisture (per cent)		15.4	12.0	12.5	11.9	17.5	11.5

Source: CSWCRTI Annual report 1990-91.

economic life of peach (16 years) on degraded lands. The BC ratio varies between 3.82 (T_{10}) to 5.24 (T_{12}) among all treatment with IRR ranging from 71-147 per cent due to very high jump in income flow. The Peach tree planted in circular trench with cluster bean was most effective in term of producing lowest run off of 3.40 per cent and sediment loss of 0.68 tha-1 which reduced soil loss by 76.22 per cent and run off by 84.78 as compared to control.

Table 7. Performance of peach under various intercrops and water harvesting structures on degraded lands at Chandigarrh

Treatments		*Soil Loss ($t\ ha^{-1}$)*	*Runoff (per cent)*	*Fruit Yield ($kg\ tree^{-1}$)*	*B:C Ratio*
Pure peach	T_1	2.12	10.32	127.8	5.07
Peach + Trench	T_2	1.82	4.98	123.3	4.86
Peach + Circular trench	T_3	1.80	4.17	120.5	4.63
Peach + Sorghum	T_4	1.92	4.63	122.4	4.76
Peach + Pearl millet	T_5	1.20	4.96	120.6	4.58
Peach + Cluster bean	T_6	0.85	4.18	125.3	4.92
Peach + Sorghum + Trench	T_7	1.68	4.37	123.2	4.70
Peach + Pearl millet + Trench	T_8	1.87	4.54	122.5	4.64
Peach + Cluster bean + Trench	T_9	0.88	3.96	116.8	4.16
Peach + Sorghum + Circular trench	T_{10}	1.17	4.12	113.5	3.83
Peach + Pearl millet + Circular trench	T_{11}	1.32	4.27	125.5	4.79
Peach + Cluster bean + Circular trench	T_{12}	0.68	3.49	137.0	5.24

Canopy and Root Modifications

Canopy and root modification has become a priority for increasing fruit yield and quality under rainfed conditions. Canopy alteration has been a common practice in temperate fruits but it is lacking for subtropical and tropical fruits. In high density plantation of fruit spp. or for meadow orcharding, canopies are manipulated to optimize the quality production under high density planting of guava. It is important to start shaping the tree immediately after planting for keeping desirable plant height. Results (Tables 8 and 9) showed that maximum canopy spread (3.48 m) recorded under 6X6 m spacing whereas plant height (3.09 m) and moisture conservation (8.60 per cent) was observed in the plants managed with 8 no. of shoots under sunhemp mulch and 6X3 m spacing. The iron drums were used as root trainers to allow roots

Table 8. Plant height (m), canopy spread (m^2) and soil moisture (cm) as influenced by crop rotation, plant density, canopy management and mulching in guava plants planted on boulder riverbed lands

Shoots Bush^{-1}	*Sun Hemp Manuring - Toria*						*Fallow -Toria*					
	6X6m			*6X3m*			*6X6m*			*6X3m*		
	SM	*WM*	*C*	*SM*	*WM*	*C*	*SM*	*WM*	*C*	*SM*	*WM*	*C*
Plant height												
2	2.59	2.43	2.28	2.75	2.60	2.45	2.50	2.35	2.12	2.65	2.60	2.20
4	2.72	2.54	2.30	2.85	2.75	2.51	2.61	2.49	2.17	2.78	2.64	2.23
6	2.76	2.61	2.43	2.93	2.79	2.60	2.68	2.52	2.28	2.85	2.71	2.37
8	2.95	2.72	2.55	3.09	2.91	2.71	2.85	2.65	2.35	2.95	2.80	2.44
Canopy spread												
2	3.05	2.88	2.65	2.71	2.73	2.20	2.24	2.16	1.99	2.16	2.09	1.90
4	3.16	3.06	2.72	3.08	3.01	2.36	2.32	2.26	2.09	2.20	2.18	1.99
6	3.24	3.11	2.83	3.16	3.13	2.56	2.66	2.37	2.19	2.28	2.22	2.06
8	3.48	3.16	2.95	3.25	3.20	2.41	2.82	2.45	2.30	2.32	2.23	2.19
Soil moisture												
2	4.70	3.65	2.81	5.83	4.70	3.12	3.13	3.08	2.00	4.26	3.13	2.32
4	5.58	4.10	3.51	6.76	5.68	3.66	4.93	4.05	2.43	6.11	4.03	3.06
6	6.28	5.75	4.30	7.20	6.70	5.25	5.57	5.03	3.12	6.54	4.97	3.65
8	7.07	6.61	5.80	8.60	6.82	5.67	6.34	5.88	4.64	7.85	5.49	4.82

SM= Sunhemp mulch, WM=Weed mulch and C= Control

Table 9. Performance of ber under various root trainers and water harvesting structures.

Treatment	*Tree Volume (m^3)*	*Plant Height (m)*	*Fruit Yield (kg tree^{-1})*	*Soil Loss*
Control	11.2	3.0	13.61	VII
V shape	14.3	3.8	18.29	V
V+Polythene	10.4	2.8	23.03	III
L shape	11.1	3.0	16.09	IV
L+Polythene	8.5	3.1	12.26	I
L+Irrigation	15.1	3.2	14.09	VI
Diagonal	8.0	2.8	17.0	II

Source: CSWCRTI Annual report 1990-91.

in deeper zone vertically. During the initial growing phase, these iron drums of different shape and size guided plants root to go deeper for mining moisture and nutrients from deeper zones which is otherwise remains unutilized. Similarly, roots of ber plant were trained in root trainers with micro water harvesting structures (V shape + polythene) produced maximum fruit yield that all other treatments under arid conditions at Agra, Uttar Pradesh.

Drip Irrigation/Micro Irrigation

The drip irrigation can be defined as slow application of water on, above, or beneath the soil surface near or into the plant root zone at the rate equivalent to consumptive use by surface or sub-surface emitters. It is one of the advanced micro irrigation methods becoming popular in our country. However, the concept of micro irrigation is not new in India as traditionally pitchers, porous cups and perforated bamboos were used for irrigation in various parts of the country. Drip system basically consists of main line, sub-mains, laterals, valves, drippers or emitters, riser valves, vacuum breakers, pressure gauges, water meters, filters, fertilizer tanks or venturi pumps, flush valves and pressure regulators. From the main, feeder lines run across the field and laterals are inserted on the same. Low-density polyethylene pipes are laid along the plant or tree rows with outlets or drippers inserted at appropriate intervals. These are designed to supply water at the desired rates (1 to 10 litres/hr) directly to the soil. Low pressures ranging from 0.35 to 2 kg/crn2 are sufficient to operate the system. Usually, 3.75 to 7.00 cm diameter pipes are used for mains and 2.5 to 3.75 cm for sub-mains. For laterals, pipes of various diameters ranging from 6 to 18 mm with a pressure rating of 4 atmospheres are generally used (Table 10).

Table 10. Effect of irrigation and mulching on water use in fruit production of Kinnow mandarin

Treatments	*Irrigation Applied*			*Rainfall (m^3ha^{-1})*	*Percolation Losses ($m^3\,ha^{-1}$)*	*Water Use ($m^3\,ha^{-1}$)*	*Fruit Yield ($t\,ha^{-1}$)*
Irrigation Based on Net CPE	*Numbers ($Plant^{-1}$)*	*Amount (m^3)*	*(ha^{-1})*				
80 mm	3.7	1.58	632.0	564.0	108.0	995.0	35.56
160 mm	1.7	1.20	480.0	564.0	120.0	828.0	31.27
240 mm	1.0	0.85	340.0	564.0	56.0	704.0	28.36
Control	-	-	-	564.0	28.0	608.0	21.48
Mulching							
Sal		0.83	332.0	564.0	92.0	724.0	30.12
Lantana		0.91 3	64.0	564.0	76.0	776.0	28.80
Control		0.98	392.0	564.0	68.0	820.0	26.53

Source: CSWCRTI publication, 2011.

The performance of these systems on horticultural and other row crops has been very encouraging. It has been observed that there is about 50 to 70 per cent water saving and 25 to 30 per cent saving of fertilizers and pesticides when they are applied through the system. For irrigation purposes with advanced technologies, the system can be controlled centrally; hence, also reduces the labour requirements. Water is delivered to growing crops with the help of micro irrigation systems directly to the active root zone, thus minimising losses of water, which ultimately helps in improving productivity and quality of farm produce. It ensures higher yield with savings in time and labour. Data presented in Table 10 revealed that irrigation applied at 80 mm net CPE produced maximum fruit yield of 35.56 t ha^{-1} with use of 995 $m^3 ha^{-1}$ water followed by irrigation applied at 160 mm net CPE and minimum values were observed in the treatment which received 608 m^3 ha^{-1} water and produced 21.48 t ha^{-1}. Table 8 indicate that drip irrigation increased fruit yield by 50 per cent in sweet lime, 98 per cent in pomegranate and 75 per cent in papaya (Table 11). Significant yield response coupled with enhanced quality of produce is possible through micro irrigation systems for increasing productivity.

Table 11. Drip irrigation increases fruit yields (tha^{-1})

Fruits	*Conventional*	*Drip Irrigation*	*Increase in Fruit Yield (per cent)*
Sweet lime	100.0	150.0	50.0
Pomegranate	55.0	109.0	98.0
Papaya	13.4	23.5	75.0

Source: Report of task force on micro irrigation, Angle Research.

Fertigation Techniques

Fertilizers are delivered to growing crops with the help of micro irrigation systems. Fertigation provides nitrogen, phosphorus, potash as well as the essential trace elements directly to the active root zone, thus minimising losses of expensive nutrients, which ultimately helps in improving productivity and quality of farm produce. Fertigation ensures higher and quality yield along with savings in time and labour, which makes fertigation economically viable. As evident from Table 12-13, fertilizer applied in the ratio of 19:19:19 and 18:40:0 in grape increased fruit yield by 29 and 37 per cent and saved water by 20 and 25 per cent, respectively with C: B ratio of 1:3. Fertigation is ideally suited for high-tech horticultural production systems since it involves not only the efficient use of the two most precious inputs *i.e.*, water and nutrients but also ensures their simultaneous availability to plants (Table 12). Though micro irrigation

Table 12. Research studies carried out at various research centres on different aspect of fertigation

Sl. No.	*Fruit Crop*	*Fertigation*		*Fertilizer Saved (per cent)*	*Yield Increased (per cent)*	*Location*	*C:B Ratios*
1	Orange	Drip	6:12:6	28	50	PKV, Akola	1: 2.8
2	Grapes	Drip	19:19:19 13:0:46	25	29	Jain R&D farm, Jalgaon	1: 3.0
3	Grapes	Drip	18:44:0 16:8:24	20	37	UAS, Bangalore	1:3.2
4	Sapota	Drip	19:19:19	20	74	UAS, Bangalore	1:4.0

Source: Summer school, 2003.

has found widespread use in horticultural crop production in India, fertigation is confined to few high value crops. Significant yield response coupled with enhanced quality of produce is possible through hi-tech productivity using micro irrigation and fertigation.

Table 13. Effect of moisture conservation treatments on lemon (eureka round) fruit yield

Treatment	*Tree ht (m)*	*Fruit Yield*		*Gross Return ($Rs\ ha^{-1}$)*
		No.	*kg tree^{-1}*	
T1: V-shaped micro catchment with 5 per cent inward slope and tree basin method.	3.4	108	10.8	4300
T2: V-shaped micro catchment with 5 per cent inward slope and vertical + surface grass mulched in basin.	3.4	56	5.6	2300
T3: V-shaped micro catchment with 5 per cent inward slope and runon surface grass mulched.	3.5	124	12.4	5100
T4: V-shaped micro catchment with 5 per cent inward slope and runon surface sprayed with 10 per cent Na_2CO_3 solution.	3.3	55	5.5	2200
T5: V-shaped micro catchment with 5 per cent inward slope (slope, control)	3.5	76	7.6	3000
T6: Flat control (farmer practice)	3.1	63	6.3	2600

Source: CSWCRTI Annual report 1990-91.

Fertilizers are delivered to growing crops with the help of micro irrigation systems. Fertigation provides nitrogen, phosphorus, potash as well as the essential trace elements directly to the active root zone, thus minimising losses of expensive nutrients, which ultimately helps in improving productivity and quality of farm produce. Fertigation ensures higher and quality yield along with savings in time and labour, which makes fertigation economically viable. Fertigation is ideally suited for high-tech horticultural production systems since it involves not only the efficient use of the two most precious inputs *i.e.*, water and nutrients but also ensures their simultaneous availability to plants. Though micro irrigation has found widespread use in horticultural crop production in India, fertigation is confined to few high value crops. Significant yield response coupled with enhanced quality of produce is possible through hi-tech productivity using micro irrigation and fertigation. As evident from Figure 1, eight year young mango plants irrigated at 50 kPa soil moisture tension with 100 micron black plastic mulch and fertigation with 100 per cent of recommended dose was found better for water saving by 25 per cent and increased fruit yield 67 per cent over to conventional method of irrigation, fertilizer application and mulch (Singh, 17).

In situ Moisture Conservation through Mulching and Organic Manure

Mulching and organic manuring are practices of covering the soil surface and enriching soil profile with nutrients around the plants to make conditions more conducive for plant growth. Use of dry leaves, straw, hay, stones *etc.*, as a mulching material has been prevalent for ages.

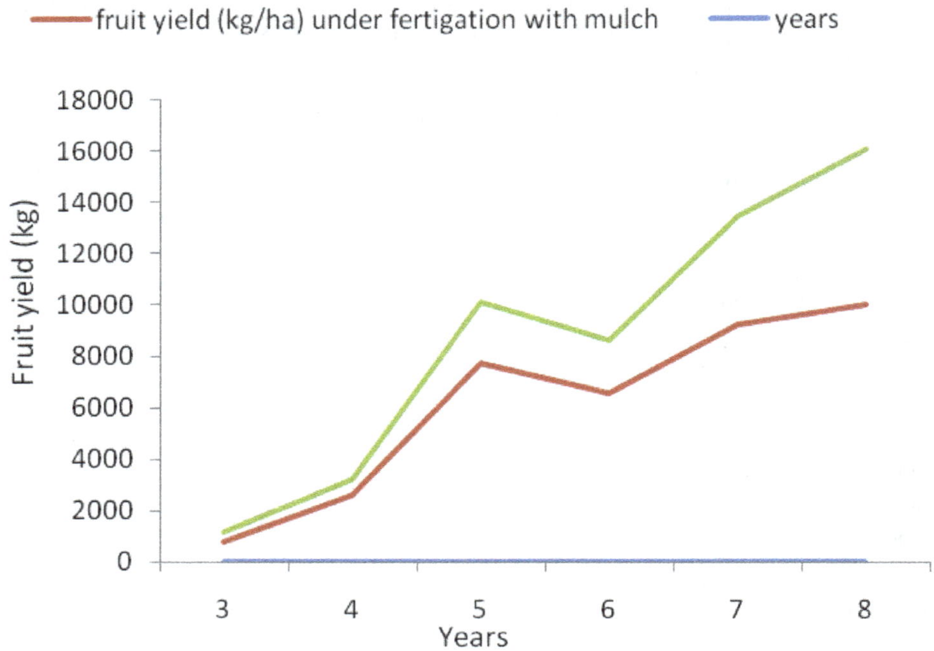

Figure 1. Fruit yield (kg tree^{-1}) under fertigation with mulch.

Results in Table 14 indicate that organic mulch (weed mulch, leave mulch *etc.*) improves fruit yield by 24-35 per cent by conserving moisture in the soil profile by 12-19 per cent in fruit plants of mango, guava, mandarin and apple. Data in Tables 15-16 shows that sunhemp mulch applied at the rate of 10 kg per tree in mango with 50 per cent of each of vermi-compost and NPK fertilizers conserved more moisture (10.2 per cent) as compared to other treatments, where as maximum fruit yield of mango (19.17 kg tree^{-1}) was observed in the treatment which received poultry manure + NPK (50 per cent each) than other treatments.

Table 14. Research trials on effect of various types of mulches on different fruit plants

Sl.No.	*Fruit Crop*	*Moisture Conserved over Control (per cent)*	*Yield Increase (per cent)*	*C:B Ratio*	*Location*
1	Mango	18	35	1: 2.5	CSWCRTI, Dehradun
2	Guava	19	24	1:3	
3	Mandarin	15	27	1: 3	
4	Apple	12	27	1:3.5	Solan

A study was conducted with different types of organic and synthetic mulches for moisture conservation in established apple orchard at YSPUHF Solan. Data revealed that apple tree with dry grass mulch recorded maximum fruit yield (73.4 kg tree^{-1}), net return (Rs. 1243.3 tree^{-1}), moderate soil moisture (18.3 per cent) and soil temperature (20.0 °C) followed by black polythene. The minimum values of fruit yield (56.0 kg tree^{-1}), net return (Rs. 476.0 tree^{-1}) and soil moisture (16.8 per cent) were observed in clean cultivation (Tables 15 and 16).

Table 15. Moisture conservation (per cent) and fruit yield (kg tree^{-1}) of mango as influenced by organic manures combined with fertilizers and mulches

Treatments	*Moisture Conservation (per cent)*			*Fruit Yield (kg tree^{-1})*		
	Sunhemp Mulch	*Grass Mulch*	*Mean*	*Sunhemp Mulch*	*Grass Mulch*	*Mean*
T1 (FYM)	7.10	6.84	6.97	12.67	11.25	11.96
T2 (VC)	9.67	8.76	9.22	13.82	11.84	12.83
T3 (PM)	7.15	5.45	6.30	12.83	12.78	12.81
T4 (NPK)	5.98	4.69	5.34	12.32	11.99	12.16
T5 (FYM+50 per cent NPK)	9.98	7.88	8.93	15.63	14.88	15.26
T6 (VC+50 per cent NPK)	10.01	8.61	11.8	22.19	19.19	20.69
T7 (PM+50 per cent NPK)	8.92	7.42	8.17	20.16	18.17	19.17

Source: ICAR-IISWC Publication, 2018

Table 16. Effect of moisture conservation techniques on soil moisture, soil temperature, fruit yield and net return of apple at YSPUHF Solan

Mulch Treatment	*Soil Moisture (per cent)*	*Soils Temp. (°C)*	*Fruit Yield (kg tree^{-1})*	*Net Return (Rs. tree^{-1})*
Pine needle	18.0	20.3	62.3	733.6
Dry grass	18.3	20.0	73.4	1243.3
Oak leaves	17.6	20.5	57.9	615.1
Black polythene	19.2	23.0	68.1	1065.8
Clean cultivation (control)	16.8	21.0	56.0	476.0

Source: Sharda (14).

Modified Plant Ideotype with Integrated Nutrient Management

Application of Liquid fertilizers 6.0 g $N_{19}P_{19}K_{19}$ @ 6.0 g liter^{-1} at fortnightly interval plus 4.5 g $N_0P_{52}K_{34}$ liter^{-1} at monthly interval with 10.0 tha^{-1} farm yard manures and modified plant ideotype improved fruit productivity of tomato in the riverbed lands. The net return among various treatments showed an increasing trend from T_1 to T_5 treatments. The highest net return (10.75 thousand USD ha^{-1}) was recorded with T5 treatment followed by T_4 (8.77), T_3 (7.73), T_2 (4.82), and minimum with T_1 treatment (3.58 thousand USD ha^{-1}). The benefit-cost ratio (BCR) of different treatments of pruning, trellising, liquid fertilizer application was calculated and shown in Fig 2. The highest BCR (3.08) was recorded in T_5 treatment followed by T_4 (2.73), T_3 (2.58), T_2 (2.15), and minimum with T_1 treatment (1.91). The BCR indicated that all the treatments were beneficial to the growers as the return was higher than investment, but farmer practice (T_1 and T_2 treatments) realized low BCR than improved practices (T_3, T_4, and T_5). From the economic point of view, it is apparent from the above results that the treatment T5 was more profitable compared to other treatments. Some authors also observed higher BCR (3.7) under different nutrient levels, pruning and training system of tomato. The net return ranged from 7.95 to 11.06 thousand USD obtained from the combination of technologies in tomato compared to traditional farming practices (3.58 – 4.82 thousand USD). The marketable produce harvested was more in advanced treatment (T_3–T_5) over farmers' practice due to balanced nutrient management, aeration in the plant canopy, better photosynthesis by the plant foliage, and its distribution to all developing fruits.

Soil Erosion in Fruit-based Land Use

Soil loss in India is the prime concern as many states are affected by soil erosion of varying magnitude. The area affected is 11.22 per cent, 6.46 per cent, 9.92 per cent, 7.14 per cent and 3.99 per cent from the moderate (10-15 t/ha/year), moderate severe (15-20 t/ha/year), severe (20-40 t/ha/year), very severe (40-80 t/ha/year) and extra severe (>80 t/ha/year) erosion categories, respectively (Anonymous, 1). Soil formation takes almost 600 -1000 years to build 2.5 cm of topsoil but it is eroded in no time. It has been reported that 5334 million tonnes of productive soil is lost every year out of which major portion is lost from about 80 million hectare of cultivated land in India. Fruit base land uses can reduce the soil loss upto permissible limit *i.e.* 15-20 t ha^{-1}. Besides minimizing soil loss, it also provides nutritional security. Data presented in Table 17 revealed that soil erosion was recorded 20-22 t ha^{-1} under mango-based land use in Philippines. This soil loss is within the permissible limit of country.

Table 17. Average soil erosion rate (t ha^{-1} yr^{-1}) under various land uses in Philippines

Sl.No.	*Various land uses*	*ENRAP*	*DAVID*
1	Grasslands/Pasturelands	267.80	-
2	Upland agriculture	112.80	-
3	Open grasslands	79.60	210.72
4	Fruit crops (Mango, Jack, *etc.*)	22.10	20.30
5	Trees, grasses, shrubs	12.50	-
6	Secondary forest	3.00	6.95
7	Paddy rice irrigated	2.30	0.45
8	Gmelina, coffee	1.00	-
9	Primary forest	-	507.99
10		-	2.22

Source: Francisco, 1994.

Fruit Based Agri-horticulture System for Stabilization of Degraded Lands

There is large area under degraded lands in Doon valley which can be restored by cultivation of fruit crops along with leguminous intercropping. Results (Table 18) revealed that out of five intercrops, which were grown in association with mango and litchi, Cowpea-toria performed well. Fruit yield of mango and litchi was recorded maximum (54.13 kg $tree^{-1}$ and 34.9 kg tree-1, respectively) in cowpea-toria intercropping treatment. Likewise results showed that yields of aonla was maximum (7082 kg ha^{-1}) in the treatment of maize + live weed mulch with minimum soil loss (4.8 tha^{-1}) as compared to other treatments (Table 19).

Table 18. Evaluation of mango-litchi based agri-horti system on degraded lands in Doon Valley (Average 1995-2010)

Cropping Systems	*Tree Height (cm)*	*Crown Spread (m)*	*Crown Height (m)*	*Trunk/Stock dia. (cm)*	*Fruit Yield (kg/tree)*
MANGO					
Cowpea-Toria	3.11	3.45	2.59	14.3	54.13
Okra-Toria	3.08	3.57	2.65	14.1	48.25
Sesame-Toria	3.24	3.55	2.74	15.0	45.05
Blackgram-Toria	3.07	3.63	2.66	14.4	51.09
Redgram	3.17	3.47	2.63	14.20	52.50
Control	3.03	3.35	2.51	135	44.60
LITCHI					
Cowpea-Toria	2.33	2.75	1.92	7.9	34.9
Okra-Toria	2.18	2.57	1.83	7.8	32.6
Sesame-Toria	2.18	2.68	1.87	7.60	31.9
Blackgram-Toria	2.03	2.44	1.67	7.80	34.0
Redgram	2.56	2.73	2.03	8.30	33.1
Control	2.04	2.25	1.57	7.10	30.1

The economic evaluation of first phase of mango based AHM indicated that all the models were economically viable and profitable on degraded lands as they were having positive NPV values as well as BCR > 1.7 (Tablse 18a, 18b). Further, both phase, they were having significantly higher NPV and BCR, and lesser PBP than the sole mango, which indicated that the intercropping in mango inter tree spaces provide higher economic benefits from the system as mango has sparse foliage, which permits intercropping and these intercrops provided additional returns and reduced the PBP. Besides additional income, the intercrops also improved soil fertility, soil moisture, moisture retention capacity and maintained soil microbial population on degraded lands.

Table 18a. Economic evaluation of agri-horticultural models on degraded land (1-10 years)

Economic Parameter (ha^{-1})	*Sole Mango*	*Mango + Cowpea – Toria*	*Mango + Urd – Toria*	*Mango + Sesame – Toria*	*Mango + Okra – Toria*	*Mango + Pigeon Pea*
Net present value (USD)	1,792	5,068	3,208	2,126	2,814	2,701
B:C ratio	1.82	2.67	1.98	1.70	1.86	1.93
Payback period (years)	10	5	6	8	7	8

Table 18b. Economics of different agri-horticultural models on degraded land (11-15 years)

Economic Parameter (ha^{-1})	*Sole Mango*	*Mango + Turmeric*	*Mango + Colocasia*
Net present value (USD*)	9,664	12,229	11,818
B:C ratio	3.91	2.85	2.76
Payback period (years)	1	1	1

Source: Agroforestry Systems, 2013.

Table 18c. Economics of litchi based agri-horticultural models under degraded lands (USD ha^{-1})

Economic Parameters	*Sole Litchi*	*Litchi + Cowpea-Toria/ Turmeric*	*Litchi + Sesame – Toria*	*Litchi + Urd – Toria*	*Litchi + Okra – Toria/ Colocasia*	*Litchi + Pigeon Pea*
Net present value (USD)	19,872	23,983	20,698	21,900	22,775	21,013
B:C Ratio (BCR)	3.5	4.70	5.00	5.80	4.60	5.50
Payback period (Years)	7.0	5.0	4.0	4.0	5.0	4.0

Source: Current Science 2014.

Table 19. Effect of various soil and moisture conservation treatments on runoff (per cent), soil loss (per cent), grain yield of intercrops under varying spacing of *Aonla* based agri-horti system on 2 per cent slope in Doon Valley

Treatment	*Runoff (per cent)*	*Soil Loss (per cent)*	*Yield (kg ha^{-1})*			
			Maize	*Wheat*	*Mustard*	*Aonla*
Spacing 6 X 6 m						
T_1- sole maize	20.70	10.80	3366	2082	372	2828.17
T_2- maize + cowpea	19.50	7.90	3221	2203	447	4038.66
T_3- maize + live weed mulch	18.50	5.40	3235	2289	482	5257.46
T_4- maize + *insitu* sunhemp mulch	17.50	4.80	3237	2511	608	7082.89
Mean	19.10	7.20	3265	2271	478	4803.18
Spacing 8 X 8 m						
T_1- sole maize	21.10	10.30	3408	2246	396	2219.88
T_2- maize + cowpea	19.60	8.60	3277	2398	473	2790.84
T_3- maize + live weed mulch	18.60	5.90	3264	2586	545	3235.44
T_4- maize + *insitu* sunhemp	18.30	5.10	3265	2700	522	4580.16
Mean	19.50	7.50	3279	2486	484	3207.36

Source: IJSC, 2015.

All the intercrops grown in association with litchi tree performed well and indicated that the practice was economically viable and profitable in respect of NPV, BCR and PBP (Table 18c). Litchi + (Cowpea – Toria)/Turmeric land use realized maximum net present value (23,983 USD) with 4.6 cost benefit ratio. On the whole, intercropping with litchi is successful up to 15 years with suitable crops. Litchi is a slow growing tree, which transmits the light through the canopy for intercropping and these intercrops provided additional returns and contribute to reduce the PBP. Besides additional income, the intercrops also improved soil health on degraded lands.

Grassed Waterway

These channels are laid out preferably on natural drainage lines in the watershed. As far as possible, natural courses should be used without much disturbances for draining out the excess water. At appropriate locations, basin should be created with land use for temporary detention of runoff water. Results (Tables 20-21) revealed that maximum soil was deposited (360 cum) in segment III with lemon + grass treatment followed by II (250 cum) with lemon + finger millet-lentil sequence and minimum soil deposition (90 cum) in I segment where lemon + rice bean-Gobhi sarson treatment was adopted. Grass grown in bed has deposited more soil than agricultural crops in association with lemon. The fruit yield of lemon in different segments varied from 47.5 to 50 q ha^{-1} after four years of planting. The advantages of treating grassed water way are conservation of soil and moisture along with production of fruit and inter-crops from the area which has been otherwise unutilized and unproductive.

Table 20. Hydrological behaviour of grassed waterway

Segment #	*Length (m)*	*Slope (per cent)*	*Rainfall and No. of Events (mm)*	*Runoff in 22 Events (mm)*	*Silt Deposited (cum)*
I	300	0.9	1134 in 62 events	249	90
II	240	1.1		334	250
III	180	1.3		498	360

Source: IJSC, 2020.

Table 21. Productivity (q ha^{-1}) of lemon with *kharif* and *rabi* season crops during 2005-2009

Segment #	*Crops*	*2005-2006*	*2007-2008*	*2008-2009*	*Average*
I	Rice bean	2.0	1.8	2.5	2.1
	Gobhi Sarson	3.0	2.9	3.1	3.0
	Lemon	20.2	40.2	47.5	35.9
II	Finger millet	4.2	4.6	7.6	5.5
	Lentil	3.7	3.5	3.7	36.5
	Lemon	22.8	42.2	50.0	38.3
III	Grass	80.5	140.8	180.3	133.9
	Lemon	19.5	38.4	48.2	35.4

Source: IJSC, 2020.

Fruit Productivity under Saline Irrigation Water

Irrigation of saline water to the mango plants grown in coastal region our country is common practice. 12 year old mango plants planted in high density orcharding (3 m apart) on southern aspect slope were selected for the study during 1996-1999. Four levels of saline water and two rootstocks were tested on mango cultivar Osteen. Data presented in Table 18 revealed that maximum fruit yield of mango (13.7 and 12.6 kg tree^{-1}) was harvested from Osteen cv grafted on gomera 1 and gomera 3 rootstocks, respectively. Inverse relationship was noticed with fruit yield and salinity levels. As salinity levels increased, fruit yield was decreased. Maximum fruit yield was weighed under minimum salinity level (1.0 dSm^{-1}) and least fruit yield with maximum salinity level (2.50 dSm^{-1}). Fruit yield of mango was increased increasing age of mango trees from 10.0 -12.0 and 8.5 – 10.7 Kg tree^{-1} with gomera 1 and gomera 3 rootstocks, respectively during 1996-1999 (Table 22).

Table 22. Effect of electrical conductivity of irrigation water and rootstocks on fruit yield of mango (cv Osteen) under high density orcharding.

Treatments	*Gomera 1 Rootstock*		*Gomera 3 Rootstock*	
Salinity Levels (dSm^{-1})	*Kg Tree^{-1}*	*No. of Fruits*	*Kg Tree^{-1}*	*No. of Fruits*
1.00	13.7	22.4	12.6	20.6
1.50	10.7	18.9	9.8	16.5
2.00	9.2	16.0	8.1	14.0
2.50	8.7	15.5	7.2	13.3
Years				
1996	10.0	18.0	8.5	15.2
1997	9.0	16.0	8.4	15.2
1998	11.3	18.2	10.1	16.0
1999	12.0	20.5	10.7	18.9

REFERENCES

1. Anonymous. 2008. Harmonization of degraded lands/wastelands datasets of India – A GIS based approach. NBSS and LUP, Nagpur (Personal Communication).

2. Das, S.K., Sharma, S., Sharma, K.L., Saharan, N., Nimbole. and Reddy, Y.V.R. 1993. Landuse option on a semi-arid. Alfisol. *American, J. Alternative Agric.* **18(I)**: 34-39.

3. Hudons, W. 1987. Soil and Water Conservation in Semi-Arid Areas. FAO Soils Bulletin 0.57, FAO, Rome, 172 p.

4. Katyal, J.C. and Osman, M. 1994. Dryland Horticulture. In J.L. Chadda (Ed.) Horticulture Research and Education in India. Indian Council of Agricultural Research, New Delhi.

5. Khanna, L.S. 1977. Principles and practices of silviculture. Khanna Bandhu, Dehradun, pp. 469.

6. Mishra, P.K. and Osman, M. 1995. Pit amendment for higher water retention in degraded alfisols. Proc. All India Seminar on Controlled Irrigation Systems for Horticultural Orchard Crops of Arid Zones. The Institution of Engineers (India), Bhubaneswar, p. 55-59.
7. Mishra, P.K. and Sharma, S. 1992. Watershed approach in wasteland development. Interaction session on Watershed Development with Focus on Wastelands Development, Regional Centre for Wastelands Development, Report o. **22**: 39-44.
8. Muller, J. and Engelthaler, Z.A. 1985. Bentonite application to sandy soils of Argentina. Report of U 100 Czechoslovakia Joint Programme for International Cooperation in the field of ceramic, building material and on-metallic Minerals based Industries, Pilsen, Czechoslovakia.
9. National Academy of Agricultural Sciences. 2010. Degraded and Wastelands of India- Status and Spatial Distribution. New Delhi: 158
10. NHB. 2017. Horticultural Statistics at a Glance, Indian Horticulture Database, National Horticultural Board, Ministry of Agriculture and Farmers Welfare, GOI (http: // nhb.gov.in)
11. NRAA. 2011. Report on challenges of food security and its management, National Rainfed Area Authority NASC Complex DPS Marg New Delhi India
12. Patra, S., Singh, D., Ojasvi, P. R. and Rathore. 2018. Integration of Drip Irrigation and Water Harvesting System for Vegetable Production - A Case Study. Popular Kheti. **6(1)**: 22-24
13. Rathore, A.C. Mehta, H. and Sharma, N. K. 2018. Performance based land use system in gravelly riverbed lands. *Field Forester*. 3(1): 167-171.
14. Sharda, V. N. 2006. Development of horticulture. NATP, Hill and Mountain Agro-Ecosystem, Final Report, Saraswati Press, Dehradun: P 62-63.
15. Sharda, V.N. and Juyal, G.P. 2013. Water Resource Development in the Indian Himalayan Region: Potential and Strategies. In: Water Management in the Hill Regions. -Evidence from Field Studies (Eds. Palanisami, K., Sharda, Y.N. and Singh, D.Y.). Bloomsbury Publishing India Pvt. Ltd., New Delhi.
16. Sharma, S. and Mishra, P.K. 1994. Strategy for ground water monitoring and management in small watersheds. International Workshop on Groundwater Monitoring and Recharge in Semi-arid Areas. National Geophysical Research Institute, Hyderabad, SI-163 to 174.
17. Singh, A. K. and Singh, Man 2003. Use of modern tools for increasing on farm water use efficiency. Summer school (July, 16- 5, August,) at WTC, IARI, New Delhi.

Chapter 19

Horticulture Crops: An Opportunity for Livelihood Security and Economic Development under Degraded Land

O.P. Aishwath

ICAR–National Research Centre on Seed Spices, Tabiji, Ajmer 305206, Rajasthan

e-mail: aishwath9@gmail.com

ABSTRACT

Land resource, in recent years, are under tremendous pressure with highly competing and conflicting demands of rising population. Not only the demands on food, fodder and fuel have largely upsurge but also claims on land for settlement, urban growth, industrial expansion, roads, railways and other usage are on progressive rise. Land use can create diverse cultural landscapes of outstanding aesthetic, economic and ecological value, but it may equally result in land degradation. Accelerated land degradation limiting crop production over extensive areas has become a subject of serious global concern in recent years. A complex paradoxical situation is confronted today in which more production is required from the scare natural base to meet the multiple demands of the fast exploding population for food, fuel, fibre, fertilizer, timber and other products on one hand, while on the other hand vast areas are either going out of cultivation or show a declining productivity due to degradation of various kinds at an unacceptable rate. Land degradation refers to a decrease in soil productivity including its present as well as potential capabilities for general crops. Ecologically none of the land or soil is degraded but it is a shifting paradigms for one use to another or the succession of one plant or animal species to another or shifting of land or nutrients from a particular location to other one. Horticultural crops are versatile in nature, as these are annuals, perennials, trees, bushes, herbs, shrubs, creepers, climbers etc. resultant bears wide adaptability. Some of the crops have inherent capability for absorbing noxious and greenhouse gasses, mobilization and utilization of nutrients, arresting of nutrients to check the runoff. Thus, reclaiming degraded soils can also increase terrestrial C pools, which could off-set some of the anthropogenic emissions. For example- Physically degraded land; Khirni, Custard apple, Drumstick, Sponge gourd, betel nut, orchids, cashew, aonla, bajradanti, Indian bdellium, kalihari, musli, tulsi etc. are plastic for water eroded land and soil; ber, ker, kherjri, kachri, tumba, Periwinkle, neem, bael, asgandh, isabgol, beet, vetiver etc. for wind eroded land and soils of various rainfall areas; apricot, karonda, phalsa, aloe, cowpea, amaranths, portulaca, ajwain, cumin, fenugreek, kalmegh, senna etc. for nutritionally eroded soil; Annonas, cashew, pomegranate, fennel, kaner, gonkhru, senna etc. for compacted and rocky land; Chironji, custer bean, jojoba, fig, sweet potato, henbane, lemon, colocasia, patchouli, lemon grass etc. for various rainfall situations of undulated lands. For chemically degraded land and soil- Date palm, sugar beet, karonda, mahua, phalsa, turnip, bael, cumin, dill etc. for saline soil; Aegle mamelos, Date palm, sugar beet, karonda, mahua, phalsa, turnip, bael, fennel etc. for saline sodic soils; Date palm, jamun, olive, amaranthus, Indian spinach, guggal, isabgol, dill, lemongrass, vetiver etc. for alkali and sodic soils; Carambola, pineapple, drumstic, sweet potato, carnation, cinchona etc. for acidic soil; Rauvolfia, Rosa, Salvia, Rhus etc. for metal contaminated soil. Besides the efficient utilization of degraded land and soils through horticultural crops, these could be remedies for those problems and improve environment quality, resulted economic development is inevitable.

***Keywords**: Livelihood, Horticultural crops, Degraded land, Economic development.*

Land is the most important natural resource for terrestrial ecosystem which comprises soil, climate, relief, water and vegetation. It is a non-renewable resource at a human time-scale and therefore, adverse effects of degradative processes on land quality are irreversible. Land degradation refers to a temporary or permanent decline in the productive capacity of the land or its potential for environmental management. Largest area is under water erosion followed by chemical degradation and wind erosion (Table 1). There are some phenomenon differences in 'Land' and 'Soil' degradation as reported. Land degradation includes soil degradation, landscape alteration, water deterioration, nutrient depletion, loss due to biodiversity, climate change, erosion by wind and water, reduced vegetation, pollution, drought, compaction by animal or machinery, sedimentation *etc.* However, soil degradation includes physical degradation (crustings, compaction, runoff and erosion, endangered or extinct soil, un-optimal soil temperature, inhibited aeration and desertification), chemical degradation (acidification, salinization, decline in CEC, nutrient depletion, elemental imbalance, leaching and pollution/ contamination), biological degradation (loss of soil biodiversity, soil-borne pathogens, decline in soil organic matter, emissions of greenhouse gasses and loss of soil C sink capacity) and ecological degradation (disruption in nutrient cycling, perturbations of the hydrological cycle, decline in net biome productivity, loss of nutrient and carbon, decline in use efficiency of inputs, inhibited denaturing of pollutants and shrinking down of biodiversity (Lal, 34). Wastelands include degraded forests, overgrazed pastures, drought-struck pas-tures, shifting cultivation land, eroded valleys, hilly slopes, waterlogged/marshy lands, saline/alkali soils, diara and barren land *etc.* It may be defined as the 'Land which is lying unproductive or which is not being utilized to its potential or produces less than 20 per cent of economic potential' (Chaudhari *et al.*, 22). National Remote Sensing Agency (NRSA, 36) carried out a district-wise mapping of wastelands on 1:50,000 scale using satellite data placed at 63.85 mha.

Table 1. Harmonized area statistics of degraded of India

Degradation Type	*Arable Land (mha)*	*Open Forest (<40 per cent canopy) (mha)*	*Data Source*
Water erosion (>10 tonnes/ha/yr)	73.27	9.30	Soil loss map of India-CSWCRTI
Wind erosion (Aeolian)	12.40	-	Wind Erosion Map of India-CAZRI
Sub total	**85.67**	**9.30**	
Chemical degradation			
Exclusively salt-affected soils	5.44	-	Salt-Affected Soil Map of India, CSSRI, NBSS and LUP, NRSA and others
Salt-affected and water eroded soils	1.20	0.10	
Exclusively acidic soils (pH <5.5)	5.09	-	Acid Soil Map of India-NBSS and LUP
Acidic pH <5.5) and water eroded soils	5.72	7.13	
Sub-total	**17.45**	**7.23**	
Physical degradation			
Mining and industrial waste	0.19	-	Wasteland Map of India-NRSA
Water logging (permanent surface inundation)	0.88		
Sub-total	**1.07**		
Total	**104.19**	**16.53**	
Grand Total (Arable land open forest)	**120.72**		Maji *et al.*, 35

On-site impacts of land degradation on productivity are easily masked due to use of additional inputs and adoption of improved technology but in long run the negative effects leads to desertification. Therefore, sustainable means for utilization and rehabilitation of degraded lands needs to be explored. Hence, Horti-remediation and utilization of degraded land is a multifaceted option for revolutionizing agriculture in a sustainable way. Ecologically none of the land or soil is degraded but it is a shifting paradigms for one use to another or the succession of one plant or animal species to another or shifting of land or nutrients from a particular place to other one. Potentially, India accounts for nearly 8 per cent biodiversity of the world despite having only 2.4 per cent of the total land area. Horticultural crops includes various type of vegetation like annuals, perennials, trees, herbs, shrubs, bushes, creepers, climbers, grasses and medicinal plant includes both lower and higher plants with wide adaptability. The potential for their performance under stressed environment can be further harvested by using trait specific varieties. Besides the efficient utilization of degraded land, these crops could have remedies for the reclamation and rehabilitation of degraded land/soil to bring the sustainability in agriculture and environment quality too by reducing emission (Aishwath and Lal, 1). These crops are not only hardy for most marginal Eco-geographic conditions where other crops have productivity below 20 per cent of their biological potential but also rich sources of food. These could provide good income to the peasants of these areas by their consistent employment generation potential from the fruits, fuel-wood and industrial raw materials. Fruits and vegetables are rich in the much needed vitamins and minerals to fight the prevalent malnutrition. A number of them also provide protein and energy rich food too. The plantation crops, medicinal and aromatic plants, spices and condiments and ornamental plants form a strong base for the industry; these provide valuable products for export and import substitution (Eswaran *et al.*, 27; Aishwath, 5).

Horticultural Crops for Physically Degraded Land and Soils

Water Eroded Lands

In most of the arid and semiarid regions, physiographic conditions of the land are undulated caused serious problem of water erosion even with small precipitation. Slope of the land not only encourages the runoff of water but also takes away the precious layer of soil resulting lowering of soil depth. It has been estimated that about 5,333 million tons of soil is lost through erosion every year owing to agricultural activities alone in India (Aishwath *et al.*, 14).

To combat against these problems of erosion, there is vast potential of horticultural crops. Some fruits trees and bushes, creeper and climber type vegetables and ornamental cover crops are potential to combat against erosion and adaptive to these conditions given in Table 2. Aromatic grasses on sloppy lands have wide potential for the extraction of oil. In *Cymbopogon species*, *Cymbopogan martinii* produced oil yield of 68.3 kg ha^{-1} year^{-1} followed by *C. winterianus*, 41.3kg ha^{-1} year^{-1}. The cultivation of other aromatic plants like *Ocimum kilimandscharicum* and *Anethum sowa* could also be recommended in degraded habitat orchards as inter-crop. At many a places, *Cymbopogons* are utilizing for dual purpose to control the gully erosion and oil extraction in Uttar Pradesh, Madhya Pradesh, Rajasthan and Gujarat (Verma and Chinnamani, 53). Similarly, *Eucalyptus globules* and *Eucalyptus robusta* utilized in humid regions for dual purpose to control the soil erosion and stabilization of gully erosion as well as essential oil extraction (Ramamohan Rao, 42). Lavender (*Lavandula officinale*) and rose (*Rosa domascena*) could also be grown in sloppy and erodes lands or terraces in hill areas (Shiva *et al.*, 48).

Table 2. Horticultural crops found suitable for eroded land

Sl.No.	*Common Name*	*Botanical Name*	*Suits in Erosion Type and their Control*
Fruits crops			
1	Almond	*Prunus dulcis*	Rill and gully erosion
2	Aonla	*Phyllanthus emblica*	Rill and gully erosion
3	Apricot	*P. armeniaca*	Rill and gully erosion
4	Apple	*Malus pumila*	Sheet and rill erosion
5	Avocado	*Persea americana*	Sheet, rill and gully erosion
6	Ber	*Ziziphus mauritiana*	Sheet, rill and gully erosion
7	Bread Fruit	*Artocarpus incisa*	Sheet, rill and gully erosion
8	Khirni	*Manilkara hexandra*	Gully erosion
9	Mahua	*Bassia latifolia*	Sheet, rill and gully erosion

Sl.No.	*Common Name*	*Botanical Name*	*Suits in Erosion Type and their Control*
10	Mulberry	*Morus alba*	Rill and gully erosion
11	Sharifa	*Annona squamosa*	Rill and gully erosion
12	Tamarind	*Tamarindus indica*	Gully and ravine
	Vegetables crops		
1	Ash gourd	*Benincasa hispida*	Sheet and rill erosion
2	Cowpea	*Vigna unguiculata*	Sheet and rill erosion
3	Drumstick	*Moringa oleifera*	Sheet, rill and gully erosion
4	*Kachra*	*Cucumis callosus*	Sheet and rill erosion
5	Pointed gourd	*Trichosanthes dioica*	Sheet and rill erosion
6	Sponge gourd	*Luffa cylindrica*	Sheet and rill erosion
7	*Teasel gourd*	*Cucumis dipsaceus*	Sheet and rill erosion
	Plantation and other crops		
1	Betel nut	*Areca catechu*	Sheet, rill and gully erosion
2	Cashew	*Anacardium occidentale*	Erosion in coastal belt
3	Cocoa	*Theobroma cacao*	Erosion in coastal belt
4	Coffee	*Coffea arabica*	Sheet, rill and gully erosion
5	Orchids	*Cymbidium sps*	Sheet and rill erosion
6	Palmyrah palm	*Borassus flabellifer*	Bund lands
7	Tea	*Camellia sinensis*	Sheet, rill and gully erosion
	Medicinal and aromatic plants		
1	Aonla	*Emblica officinalis*	Gully erosion
2	Bael	*Aegle marmelos (bael)*	Gully erosion
3	Bajradanti	*Barleria prionit*	Sheet and rill erosion
4	*Chitrak*	*Plumbago sps*	Sheet and rill erosion
5	Dill	*Anethum sowa*	Sheet and rill erosion
6	Eucalyptus	*Eucalyptus sps*	Gully erosion
7	Giloy	*Tinospora cordifolia*	Sheet, rill and gully erosion
8	Indian bdellium	*Commiphora wightii (Guggal)*	Landslide, rill and gully erosion
9	*Kalihari*	*Glories superba*	Sheet and rill erosion
10	Ketki	*Agave sps*	Landslide, rill and gully erosion
11	Lavender	*Lavandula officinale*	Sheet and rill erosion
12	Musli	*Chlorophytum sps*	Sheet, rill and gully erosion
13	Rosa grass	*Cymbopogan sps*	Sheet, rill and gully erosion
14	Rose	*Rosa domascena*	Sheet and rill erosion
15	Tulsi	*Ocimum sps*	Sheet and rill erosion
16	Velvet bean	*Mucuna sps*	Sheet and rill erosion

Wind Eroded and Sandy Wastelands

Wind erosion takes place where soil texture is coarse, moisture and organic matter is low leads to low binding capacity of sand particles. The suitable horticultural plants for arid, semi-arid and coastal sandy wastelands have been identified. Such plant types must have characteristics of binding the sand, should be capable of growing under low fertility and soil moisture and atmospheric stress conditions (Verma and Chinnamani, 53).

Arid Regions

The prominent species suitable for arid environment are jharber (*Z. nummularia*), boradi (*Z. mauritiana var.rotundifolia*), ber (*Z. mauritiana*) among fruits, ker (*Capparis decidua*), gonda/ labhera (*Cordial myxa*), kherjri (*Prosopis cineraria*), kamat (*Acacia sengal*), phog (*Calligonum polygonoides*), drumstick (*Moringa oleifera*), cluster been (*Cyampsis tetragonoloba*), cowpea (*Vigne unguiculata*), water melon, long melon, kachri (*Cucumis melo*), phoot (*Cucumis melo var. momordica*), arya (*C. melo*) buffalo gourd, *Ameranthus* and *Aloe barbendensis* among vegetables, tumba (*Citrullus colocynthis*), agave, *Balanites aegyptica, Euphorbia antisyphilitica, Periwinkle, Datura mental, Lawsonia inermis* among medicinal plants and *Tecomella undulate* (rohida), *Azadirachta indica* (neem) *Albizzia lebbak* (siris), *Ailanthus excelsa, Eucalyptus camaldulensis, Parkinsonia aculeata, Tamarix articulata* among ornamental plant species (Aishwath *et al.*, 18; Aishwath *et al.*, 15 and 16)

Semi-arid Regions

Under the semi-arid conditions, the choices of suitable species gets enlarge. Thus, fruits like guava, custard apple, mango, pomegranate, bael, fig, tamarind, khirni (*Mimusops hexandra*), Jamun (*Syzygium cuminii* and *S. Jambos*), palmyrah palm and lemons, vegetables like kundru (*Coccinia cordifolia*), Dolichos bean, French bean round melon, chillies, brinjal, okra, bitter gourd, cumin, sweet potato and potato, medicinal and aromatic plants like *Cassia angustifolia, Withania somnifera, Ocimum sanctum/basilicum*), *Vetiver zizanioides,* liquorice (*Glyeyrrhiza glabra*), *Cymbopogon martini*, henbane, *Plantago ovata*, mahua (*Madhuca lutifolia*), *Terminallia bellerica, T. chebula* and commercial flowers like marigold, *Jasminum, Murraya exotica, nerim odorum, Tabernaemontana coronaria, Hibiscus, etc.*, can also be grown successfully (Singh, 50; Aishwath *et al.*, 18).

Coastal Sands

In the coastal sand, the suitable fruits are date palm, coconut, ber, pomegranate, *etc.* Several vegetables like brinjal, chillies, turnip, beet, water melon, pumpkin, long melon, onion, garlic, *Dolichos*, bean, French bean, cluster bean, and drumstick, medicinal plants like senna, periwinkle, vetiver, liquorice and asgand and commercial flowers like marigold, Jasminum, *Nerim, Murraya exotica, etc.* would be successful (Chadha and Pareek, 21).

Horticultural Crops for Nutritionally Eroded Soils

Most of Indian soils (arid and semi-arid areas) are generally poor in nutrients due to low organic matter, low clay content, high nutrient erodibility with soil and water, narrow soil column and low water retention of soil. In these regions vegetation stand is poor and high temperature causes fast mineralization of soil organic matter and release of nutrients in soil and losses through water runoff. The major portion of soil is composed of sand (>85 per cent) resulting lower retention power of nutrient, high infiltration rate (9-16 cm hr^{-1}) and poor moisture storage capacity. Lower organic matter content also causes the poor store house of nutrients. Lower fertility and moisture stress which offer a limited choice for increasing crop production under varying rainfall situation. At most of the places these soils are having the problem of salinity and sodicity. Rains and flood irrigation cause volatilization of ammonical nitrogen with high soil pH. Nutritional imbalance in soil also causes the imbalance in the plants. For example, excessive absorption of Na^+ ions leads to a decrease in the absorption of Ca^{++}, Mg^{++} and K^+ ions; high bicarbonates in soil solution cause 'Lime Induced Chlorosis' and high soil pH might accentuate deficiencies of micronutrient cations (Aishwath and Anwer 6). In most of the cases alkali soil contains high amount of soluble and exchangeable Ca^{++} and very frequently $CaCO_3$. These high concentrations of free calcium and calcium carbonate ions react with phosphate ions and forms the tri-calcium phosphate, which is insoluble and beyond the reach to the plants.

Increase in salinity/alkalinity can also adversely affect the soil biology, mineralization of organic matter and transformation of applied fertilizer nutrients. It is well known fact that grasses can squeeze the soil by extracting the nutrients with their complex root system. Here, vast potential of aromatic grasses could be utilized with limited nutrient and water (Aishwath *et al.*, 18). Some important horticulture crop/plants have lower requirement of nutrient can be grown as diversified crop (Table 3). The root colonization with VAM fungi have special structures known as vesicles and arbuscules. The arbuscules help in the transfer of nutrients from fungus to the root system and vesicles, which are 'saclike' structures store P as phospholipids. VAM have also been associated with increased plant growth and with enhanced accumulation of plant nutrients, mainly P, Zn, Cu, and S through greater soil exploration by mycorrhizal hyphae. It has also been reported that VAM simulate plant growth by physiological effects other than by enhancement of nutrient uptake or by reducing the severity of diseases caused by soil borne pathogens (Dehne, 26). Hence the untamed potential of horticultural crops with VAM can be harvested in these poor fertile soils. Recently, possibilities have been explored for the cultivation of rose scented geranium (*Palargonium sps*) in semi-arid conditions by multiple cuttings/ harvesting (Aishwath and Anwer, 7; Aishwath and Tarafdar, 12; Prakasha Rao, 41).

Table 3. Horticulture crops for nutritionally eroded (poor fertile) soils

Sl.No.	*Common Name*	*Botanical Name*	*Climatic conditions*
		Fruits	
1	Apricot	*Prunus armeniaca*	Temperate regions
2	Ber	*Ziziphus mauritiana*	Arid and Semi-arid regions
3	Cactus pear	*Opuntia ficus*	Arid regions
4	Karonda	*Carissa carandus*	Arid and Semi-arid regions
5	Marula nut	*Sclerocarya birrea*	Arid regions
6	Phalsa	*Gewia subinaequalis*	Arid regions
7	Wood apple	*Feronia liminia*	Semi-arid regions
		Vegetables	
1	Aloe	*Aloe babdensis*	Arid regions
2	Amaranath (chaulai)	*Amaranath spp.*	Semi-arid and tropical areas

Sl.No.	*Common Name*	*Botanical Name*	*Climatic conditions*
3	Cassava	*Manihot esculenta*	Semi-arid and tropical areas
4	Cluster bean	*Cyamopsis tetragonolobas*	Arid and semi-arid regions
5	Cowpea	*Vigna unguiculata*	Semi-arid regions
6	Curry leaf	*Murraya koenigii*	Semi-arid and tropical areas
7	Drumstick	*Moringa oleifera*	Semi-arid and tropical areas
8	Ker	*Capparis decidua*	Arid regions
9	Khejri	*Prosopis cineraria*	Arid and Semi-arid regions
10	Teasel gourd	*Cucumis dipsaceus*	Semi-arid regions
Ornamental			
1	Amaranths	*Amaranths sps*	Arid and Semi-arid regions
2	Cornflower	*Centaurea cyanus*	Semi-arid regions
3	Garden pea	*Pisum sativum*	Semi-arid regions
4	Heliotropium	*Heliotropium indicum*	Arid and Semi-arid regions
5	Portulaca	*Portulaca gilliesii*	Arid and Semi-arid regions
Spices			
1	Ajwain	*Tracyspermum ammi*	Arid and semi-arid regions
2	Cinnamom	*Cinnamomum verum*	Tropical regions
3	Coriander	*Coriandrum sativum*	Semi-arid regions
4	Cumin	*Cuminum cyaminum*	Arid and semi-arid regions
5	Dill	*Anethum graveolens and sowa*	Semi-arid and tropical areas
6	Fenugreek	*Trigonella foenumgraecum*	Semi-arid and tropical areas
Medicinal and aromatic plants			
1	Artimisia	*Artemisia pallens*	Temperate regions
2	Ashwagandha	*Withania Somnifera*	Semi-arid and tropical areas
3	Basil	*Ocimum basilicum*	Arid and Semi-arid regions
4	Kalihari	*Gloriosa superba*	Semi-arid regions
5	Kalmegh	*Andrographis paniculata*	Semi-arid regions
6	Khas	*Vetiveria zizanioides*	Semi-arid and tropical areas
7	Lavendula	*Lavendula officinalis*	Temperate regions
8	Lemon grass	*Cymbopogon sps*	Arid and Semi-arid regions
9	Mint	*Mentha sps*	Semi-arid and tropical areas
10	Musk	*Abelmoschus moschatus*	Semi-arid and tropical areas
11	Musli	*Chlorophytum sps.*	Semi-arid regions
12	Progostemon	*Pogostemon cablin*	Semi-arid and tropical areas
13	Pilu	*Salvadora sps*	Arid and Semi-arid regions
14	Senna	*Cassia angustifolia*	Arid, semi-arid tropics
15	Vanila	*Vanilla planifolia*	Semi-arid and tropical areas

Horticultural Crops for Compacted and Rocky Soil

Soil compaction is a process where soil particles are pressed together so that the pore space between particles is reduced. Reduced pore space limits water movement and storage and also limits air space. A compacted soil is tight with higher bulk density. Compacted soils present barriers to root growth and can limit yield. Healthy roots require pore spaces in the soil for growth, water uptake and gas exchange. Root growth of most species is highly sensitive to compaction and becomes very slow or completely ceases when penetrometer resistances begin to exceed 2.0 MPa. There is no threshold to absolutely determine that a soil is compacted or not. But, it is suggested that plant growth will not be impaired if soil bulk density is not higher than 1.3 g/cm^3. In developed country, the most common cause of soil compaction on farms is vehicle traffic, particularly the use of heavy equipment with poor weight distribution on wet soils. Compaction is a common problem in vegetable production systems because farm operations often must be conducted within narrow time windows that do not allow for adequate soil drying before entering the field. While in India, it may cause by poor organic matter and sodicity in arid and semiarid regions. The most important fruit and plantation crops for compacted and rocky soils are Annonas (*Annona equamosa*), date palm (*Phoenix dactylifera*), cashew (*Anacardium occidentale*) fig (*Ficus carira* var. *Hortinesis*), kronda (*Carissa carandus*) pomegranate (*Punica granatum*) and tamarind (*Tamarindus indica*); curry leaf (*Murraya koenigii*), *khejari* (*Prosopis cineraria*), radish (*Raphanus sativus*) and sponge gourd (*Lufa aculangula*) among vegetables; bougainvillea (*Bougainvillea campanulata*) and kaner (*Thevetia peruviana*) among ornamental crops; in spices medicinal and aromatic plants- fennel (*Foeniculum vulgare*), *chhota gonkhru* (*Tribulus statis*) *bada gonkharu* (*Pedalium murex*), khas (*Vetiveria zizanioides*) are most suitable. Some of the crops require compaction (*Gonkharu*, tumba and Khas) for better growth, hence along the road sides are good opportunity to utilize unused space. However, other medicinal plant senna (*Cassia angustifolia*) could withstand with the soil bulk density of about 2.0 gcm^3. In Ohio (USA) radish is used as bio tillage under conservation tillage. The turgor pressure generated by radish leads to open the soil and after senescence earth worms take it up and rest of the matter decomposed by soil microorganisms leaves open hole for seeding next crop (Aishwath and Lal, 8). Possibilities have been explored for the cultivation of rose scented geranium (*Palargonium sps*) in semi-arid conditions by multiple cuttings/harvesting by Kothari *et al.* (33).

Crops for Plateau Undulating Land Relief

This situation could be sub-divided further as low, medium and high rainfall areas so as to manage them accordingly by selecting suitable plant species.

Low Rainfall Areas

In these areas, lower piedmonts and pediment plains commonly occur for which the suitable fruit plants are bordi, ber, ker, *Opuntia tuna*, chironji (*Buchanania lanzan*) and pomegranate; the vegetables are drumstick, *Acacia sengal*, custer bean, watermelon, *etc.* and the medicinal plants are guggal (*Commifera wightii*), jojoba (*Simmondsia chinensis*) senna, neem, *Balanites aegyptiaca,* periwinkle, *etc.* (Aishwath *et al*, 14).

Medium Rainfall Areas

Under this situation variety of plants can be chosen includes mango, custard apple, chironji (*Buchanania lanzan*), pomegranate, guava, aonla, fig, cashewnut, olive, walnut, almond, apricot, chestnut, peach, plums, cherries, *etc.* among fruits; cowpea, brinjal, okra, ridge gourd, round gourd, bitter gourd, bottle gourd,onion, tomato, sweet potato, potato, *etc.*, among vegetable crops; *Ocimum species, henbane, Terminallia bellerica, T. arjuna, T. chebula, etc.* among the medicinal plants.

High Rainfall Areas

These region lies in tropical climate and the plant species grown successfully includes mango, jamun, pineapple, lemon, mandarin, sweet orange, apple, kokum among the fruits; brinjal, tomato, cauliflower, parwal (*Trichosanthes dioica*), ash gourd, cabbage, ginger, turmeric, colocasia, *etc.* among vegetable crops; and palmarosa, Opium poppy, patchouli, (*Pogostemon pactouli*), lemon grass, java, citronella, (*Cymbopogon winterianus*), *etc.* among the aromatic and medicinal plants.

Horticulture Crops for Chemically Degraded Land and Soils

Salt Affected Soil (Salinity Alkalinity and Sodicity)

Agricultural losses caused by salinity are difficult to assess but estimated to be substantial and expected to increase with time. Secondary salinization of agricultural lands is particularly widespread in arid and semiarid environments where crop production requires irrigation schemes. Globally least 20 per cent of all irrigated lands are salt-affected, with some estimates being as high as 50 per cent (Pitman and Läuchli, 40). However, in India salt affected soils occupy about 6.73 million ha area out of which 3.77 million ha is alkali/sodic in India. Due to poor physical properties, excessive exchangeable sodium and high pH, most of these lands support a very poor vegetation cover. There are two important approaches for utilizing saline alkali or calcareous environment. The first one is to modify the environment so as to suit the plant or modifying the plant or substitute the plant species so as to suit the environment. First approach has been used more extensively because it enables the plants to respond better not only to water but also to other production inputs which is not far sustainable. However, second option is cost effective and most sustainable (Aishwath and Nibauria, 9).

Fruits like *Aegle mamelos* (beal), *Emblica officinalis* (Gooseberry/Aonla); vegetables like *Aloe bardabensis*, *Capparis deciduas* (kair/kurrel) and medicinal and aromatic plants like *Azadirachta indica* (neem), *Cassia senna* (senna), *Catharanthus reseus* (periwinkle, sadabahar), *Euphorbia antisyphalitica*, *Glycyrrhiza glabra* (Liquroce/mulhatti), *Jatropha curcas* (purging nut/jamlaghota), *Lepidium sativum*, *Matricaria chamomilla*, *Ocimum sanctum* (Tulsi), *Ricinus communis* (Castor/arindi), *Salvadora persica* (jaal/piluva) and *Simmondsia chninsis* (Jojaba/ hohoba) could withstand under high saline soil and water conditions. Some other fruit and vegetable crops (Tables 4 and 5), medicinal (Table 6) and aromatic plants (Table 7) are listed with their tolerance level. Some important fruits, vegetables, medicinal and aromatic plant tolerant to alkalinity are listed in Table 8 and seed spices in Table 9. (Rao *et al.*, 43; Singh *et al.*,

49; Patra and Singh 38; Aishwath and Pal, 11; Aishwath, 3; Dagar, 23; Aishwath, 2; Aishwath, 4; Aishwath, 5)

Table 4. Fruit and vegetable crops for saline/sodic soil and their tolerance level

Sl.No.	*Categories*	*Name of Crops*
1	High tolerance ESP 40-50 ECe 9-12 dSm^{-1}	Ber and date palm, sugar beet (its varieties Ramonskaaya 06, Maribo Resistapoly),
2	Medium tolerance ESP 30-40 ECe 9-12 dSm^{-1}	Aonla, custard apple, jamun, karonda, khirni, lasora, mahua, phalsa, and tamarind; vegetables- garlic, onion and turnip,
3	Tolerance ESP 20-30 ECe 9-12 dsm^{-1}	Bael, citrus, guava and mango; seed spices-coriander and fennel
4	Susceptible (ESP <20, ECe < $6dsm^{-1}$)	Banana, mango, papaya, pineapple, and ficus species; vegetables- pea and cowpea

Source: Pathak, 37

Table 5. Vegetable crops and their tolerance level to salinity with 50 per cent yield loss

Crops	*Botanicals Names*	*Tolerance*	*C 50*
Asparagus	*Asparagus officinali*	4.1	14.3
Broccoli	*Brassica oleracea*	2.8	8.2
Cabbage	*Brassica oleracea*	1.8	6.8
Carrot	*Daucus carota*	1.2	4.5
Chinese cabbage	*Brassica rapa*, subspecies pekinensis and chinensis	3.2	8.2
Lettuce	*Lactuca sativa*	1.3	5.2
Onion	*Allium cepa*	1.1	4.1
Potato	*Solanum tuberosum*	1.7	6.5
Radish	*Raphanus sativus*	1.3	5.3
Spinach	*Spinacia oleracea*	2.0	8.6
Sweet potato	*Ipomoea batatas*	3.8	11
Turnip	*Brassica rapa* subsp. rapa	0.9	6.5

Table 6. Medicinal plants tolerant to salinity

Species	*Adaptability Class*	*Salinity Limit (dsm^{-1})*
Abutilon indicum (L.)	G	16
Acacia jacquemontii Benth	G	12
Alhagi maurorum and *A. pseudoalhgi*	FH	18
Alove vera (*A. barbadensis* Mill)	FH	18
Balanites aegyptiaca (L.)	G	14
Barleria prionnitis L.	FH	24 (pH 9.6)

Species	*Adaptability Class*	*Salinity Limit (dsm^{-1})*
Calligonum polygonoides L.	G	30
Calotropis procera	G	30
Capparis deciduas	FH	24-100
Cassica auriculata L.	G	22
Clerodendrum phlomidis L.	G	14
Cordia gharaf (Forsk)	G	18
Colopospermum mopane Kirk	G	16 (pH9.6)
Crotalaria burhia	H	42
Dichrostachys cinerea (L.)	G	16 (pH9.4)
Haloxylon recurvum	H	28
H. salicornicum	FH,G	24
Indicofera oblogifolia	G	20
Leptadania pyrotechnica	G	-
Lycium barbarum L.	G	24
Maytenus emarrginatus	G	14 (pH10)
Mimosa hamata Willd	G	20
Opuntia elatior Mill	FH	18 (pH 9.2)
Plantago ovata	G	10.0
Salsola baryososa	H	51
Salvadora oleoides	FH	48 (pH 9.7)
S.persica L.	FH	60
Suaeda fruticosa (L.) Forsk	H	72
Tamarix aphylla (L.)	FH	60
T. dioica Roxb.	H	50
T. eriocoides Rottl.	H	48
T. indica Willd	H	42
Withania somnifera	G	4.0
Ziziphus nummularia	G	20 (pH9.7)

*G = Glycophyte; FH = Faculative halophyte; H = True halophyte

Source: Dagar and Singh 24.

Table 7. Aromatic plants and their tolerance level to soil salinity

Sl.No.	*Name*	*Botanical Name*	*Tolerance Level (dSm^{-1})*
1	Dill	*Anethum graveolens*	5.0
2	Java citronella	*Cymbopogon winterianus*	5.0
3	Lemongrass	*Cymbopogon citrates*	11.0
4	German Chamomile	*Matricaria chamimilla*	12.0 (up to 5.5 beneficial)

Sl.No.	*Name*	*Botanical Name*	*Tolerance Level (dSm^{-1})*
5	Palmarosa	*Cymbopogan martinii* var Motia	12.0
6	Jamarosa	*Cymbopagon fiexuosus,*	13.0
7	Vetiver	*Vetiveria zizanioides*	12.0
8	Ashwagandha	*Withania somnifera*	4.0

Source: Farooqi, and Sreeramu, 28; Dagar *et al.*, 25; Patra *et al.*, 39; Aishwath and Pal, 10.

Table 8. Tolerance limit of horticulture crops to soil alkalinity

Sl.No.	*Name*	*Botanical Name*	*Tolerance Level (pH)*
		Fruits	
1	Date palm	*Phoenix dactylifera*	8.5
2	Guava	*Psidium guajava (ECe 8-9)*	8.2
3	Jamun	*Syzygium cumini*	10.0
4	Kronda	*Carissa carandus*	8.0
5	Limen	*Citrus indica*	8.5
6	Mango	*Mangifera indica*	8.5
7	Olive	*Olea europaca*	8.0
		Vegetables	
1	Amaranthus	*Amaranthus sps*	8.5
2	Beet root	*Beta vulgaris*	8.5
3	Chilli	*Capsicum annum*	8.0
4	Carrot	*Docus carota*	8.1
5	Cow pea	*Vigna unguiculata*	8.0
6	Cluster bean	*Cyamopsis tetragonolobas*	8.0
7	Indian spinach	*Beta vulgaris*	8.5
8	Lablab bean	*Dolicos lablab*	8.5
9	Radish	*Raphanus sativas*	8.3
10	Tomato	*Lycopersicon esculentum*	8.5
		Medicinal plants	
1	Guggal	*Commiphora wightii*	8.5
2	Henbane	*Hyocyamus niger*	8.5
3	Isabgol	*Plantago ovata*	12.0
4	Khasi kateri	*Solanum viarum*	8.0
5	Senna	*Cssia angustifolia*	8.5
		Aromatic plants	
1	Amrette	*Abelmoschus moschata*	8.6
2	Chamomile	*Matricaria chamimilla*	9.5
3	Dill	*Anethum graveolens*	8.8

Sl.No.	Name	Botanical Name	Tolerance Level (pH)
4	Lavender	*Lavandula species*	8.4
5	Lemongrass	*Cymbopogon sps*	9-10
6	Mint	*Mentha arvensis*	8.5
7	Palmarosa	*Cymbopogan martini* var Motia	10.0
8	Rose	*Rosa sps*	8.0
9	Vetiver	*Vetiveria zizanioides*	9.0

Source: Sharma *et al.,* 47; Aishwath and Anwer, 7; Gautam and Nayak, 30.

Table 9. Salinity/alkalinity tolerance limit in seed spices

Sl.No.	Name of Crops	EC dSm^{-1} (soil/water)	Remarks
1	Coriander	4.0	Sensitive to higher pH in calcareous soil.
2	Cumin	13.5	Sensitive to higher pH in calcareous soil.
3	Fennel	8.0-10.0	Tolerance improves with advancing age of crop. Sensitive to very high pH in calcareous soil.
4	Fenugreek	2.0-4.0	Seedling stage is more tolerant than the later.
5	Dill	5.0-12	Depends upon the soil texture and salinity type. It is also tolerant to high pH (9.8)
6	Celery	9.0-11.0	Slightly sensitive to sodicity.
7	Ajwain	6.0-8.0	slightly sensitive to high pH in calcareous soil. Medium tolerant to sodicity.
8	Anise	6.0-8.0	This tolerance is for germination.
9	Nigella	2.0-7.5	Depend upon the types of salinity. Somewhat sensitive to high pH in calcareous soil but not in sodic.

Source: Aishwath *et al.,* 17 and 13.

Horticultural Crops for Acidic Soils

The extent of area undergoing chemical degradation through acidification (only strongly acidic pH < 4.5 and moderately acidic pH 4.5-5.5 soils) was estimated at 16.03 M ha. However, through further harmonization of the available data on different possible classes with different stake holders, the extent of this chemically degraded area has been worked out as 10.81 M ha of arable land (exclusively acidic soils <5.5 pH) - 5.09 M ha; acidic < 5.5 pH) and water eroded soils 5.72 M ha and 7.13 M ha of acid soils under open forest according to the Acid soil map of India. In addition to fruit crops, most vegetables grow best in acidic to near neutral soil pH and some of them could tolerate more soil acidity. Various range of tolerance of these crops is listed here. Many of the seed spices comes up well in acid soils, however coriander and anise are prime for both yield and quality. Coriander could give about double yield in these conditions than the traditional growing areas with proper agronomic practices. Therefore, problematic/ degraded land somehow opportunities for the growers, if crops could be diversified based on soil suitability or the right plant at right location. Some important horticultural crops are listed in Table 10 suitable for acidic soils (Julie, 32).

Table 10. Horticultural crops for acidic soils

Sl.No.	*Crop Name*	*Tolerable pH Limit*
	Fruits	
1	Apple (*Pyrus malus*)	5.0-6.5
2	Avocado (*Perrea Americana*)	4.5 – 5.5
3	Blueberry (*Vaccinium sps.*)	4.5-5.0
4	Carambola (*Averrhoa carambola*)	5.2 – 6.2
5	Durian (*Durio zibethinus*)	5.0-6.5
6	Guava (*Psidium guajava*)	4.5 – 8.2
7	Indian cherry (*Malpighia puniciflolia*)	6.0-6.5
8	Litchi (*Litchi chineusis*)	5.5 -6.5
9	Pineapple (*Ananas comosu/sativus*)	4.5 – 6.0
10	Passionfruit (*Passiflora edulis*)	5.5 - 7.5
	Vegetables	
1	Brinjal (*Solanum melongena*)	5.5- 6.0
2	Carrot (*Docus carota*)	4.3-8.1
3	Cow pea (*Vigna unguiculata*)	5.5-6.5
4	Cucumber (*Cucumis sativus*)	5.5-7.0
5	Drumstic (*Moringa oleifera*)	5.5-7.5
6	Melon (*Cucumis melo*)	5.5-6.5
7	Sweet potato (*Ipomoea batatas*)	5.5-6.0
8	Yam (*Dioscorea spp.*)	5.0-7.0
	Ornamental, spices, medicinal and aromatic plants	
1	Anthurium (*Anthurium sps*)	5.0-6.0
2	Carnation (*Dianthus caryophyllus*)	5.5-7.0
3	Chrysanthemum (*Chrysanthemum sps*)	6.0-7.0
4	Cinchona (*Cinchona sps*)	4.6-6.5
5	*Coleus* (*Plectranthus scutellarioides*)	5.5-7.0
6	Coptis (*Coptis aspleniifolia*)	5.5-6.8
7	Coriander (*Coriandum sativum*)	5.0-7.5
8	Rose (*Rosa sps*)	6.0-7.5
9	Liquorice (*Glycyrrhiza glabra*)	5.5-7.5

Horticultural Crops for Metal Contaminated/Polluted Soils

Heavy-metal contamination is one of the world's major environmental problems, posing significant risks to agro-ecosystems. Increasing emissions of heavy metals are dangerous because they may get into the food chain with risks for human health (Angelone and Bini, 20). For the re-cultivation of degraded soils and the reclamation of industrial sites, stress tolerant plants are required. Biotechnological efforts are underway to improve plant stress tolerance and the ability to extract pollutants from the soil with the aim of using plants for soil cleanup (Salt *et al.,* 44). Conventional technologies for metal-contaminated soil remediation have often been expensive and disruptive. Novel, environmentally friendly and inexpensive solutions are phyto-remediation of heavy metal and hydrocarbon contaminated soils. An unprecedented and rapid change in environmental conditions heavy metal contamination leads to further adverse effect on common field crops. However, some of the seed spices and aromatic plant can be taken successfully without affecting the quality of essential oil of the crops. Since these metals do not interfere directly with cellular oxygen metabolism, the question arises as to the reasons of the observed oxidative stress. To understand the mechanism for exposure to heavy metals also provoked pronounced responses of antioxidative systems, but the direction of the response was dependent on the plant species and tissue analysed. However, some common reaction patterns can be found. In most cases, exposure to heavy metals (Cd) initially resulted in a severe depletion of GSH (Glutathione) in *Rauvolfia serpentina* (Grill *et al.,* 31) and in pine (Schützendübel *et al.,* 45).

Spices and aromatic crops can provide economic return and metal-free final product and essential oil in metal contaminated soil. The study in Bulgaria using coriander, sage, dill, basil, hyssop, lemon balm, and chamomile grown at various distances from the smelter revealed that aromatic crops may not have significant phytoremediation potential, but growth of these crops in metal contaminated agricultural soils is a feasible alternative. Herbage and essential oil yields of basil, chamomile, dill, and sage were reduced when they were grown closer to the contaminants. Metal removal takes place from the site with the harvestable plant parts was as high as 180 g ha^{-1} for Cd, 660 g ha^{-1} for Pb, 180 g ha^{-1} for Cu, 350 g ha^{-1} for Mn, and 205 g ha^{-1} for Zn. It also reported that high concentrations of heavy metals in soil or growth medium did not result in metal transfer into the essential oil. Of the tested metals, only Cu at high concentrations may reduce oil content (Valtcho *et al.,* 52). No detectable amount of Cd, Cu, or Pb was found in the oils of plant species like peppermint, basil and dill. These can be grown in soils enriched with Cd, Pb, and Cu medium without risk for metal transfer into the oils, and without significant alteration of essential oil composition that may impair marketability (Valtcho *et al.,* 51). Zheljazkov and Jekov (54) investigated by Furnace Atomic Absorption (FAA) and Inductively Plasma (ICP) techniques. The content of Cd, Pb, Cu, Mn, Zn and other trace elements determined in essential oils and plant extracts from the genera *Rosa*, *Lavandula*, *Mentha*, *Salvia*, *Ocimum*, *Foeniculum*, *Coriandrum*, *Anethum*, *Hyssopus* and *Rhus*. On the other hand the heavy metal content in oils from different regions were compared to heavy metal content in some oils obtained from species grown on heavily polluted soils and under severe atmospheric pollution. They found that the concentration of the most hazardous heavy metals in all of the tested oils and plant extracts was very low, near the detection limits of the used apparatuses. Therefore, most of the essential oil and medicinal crops could be grown successfully on heavy metal polluted soils and under atmospheric pollution as substitutes for some other edible crops.

Horticultural Crops for Flood and Water-Logged Areas

Water logging in any area may be caused when inflow into the ground water reservoir exceeds the outflow from it resulting in rise in water table. There are some other causes such as low lying areas near to the river bank and river basins *etc*. There has been no accurate estimate of the area afflicted in India under water logging. There has been no accurate estimate of the area afflicted in India under water logging. Framji (29) reported that nearly 3.5 million hectares area became waterlogged with the rise in water table in the root zone. Recently, Sehgal *et al.* (46) reported about 20 million hectares area is under flooding causes water logging. According to CWC, 1997 the area under flood and water logging is about 36 Mha. Recently, it has been reported about 20 million hectares area is under flooding causes water logging. Horticulture crops thrive well under water logging conditions are listed in Table 13. Besides the nutritive value of rice, it is used against pimple, boils, headache, epilepsy, Rheumatism, paralysis, skin disease, skin infection and cough is also a water logged medicinal crop. Some of the varieties used as medicinal recommended for submergence conditions are - *Alcha, Baissor, Gathuwanor, Karhani, Kalimoonch, Maharaji, Bhajari, Dhanwar, Orissa Mehar, Saraiphol, Karnataka Kari bhatta, Karikagga, Atikaya, Mullarya, Kerala Nivara, Erumakkari, Himachal Pradesh Katheri* (Amudha *et al.*, 19). Most of the seed spices are susceptible for the water logging (except anise). Some of the vegetable like agathi (*Sasbania grandiflora*) can tolerate seasonal waterlogging. Nutmeg (*Myristica fragrans*) also grows near to water logging cum wet conditions (Aishwath and Lal, 8; Shiva *et al.*, 48).

Table 13. Horticultural crops for waterlogged and wetlands

Sl.No.	*Common Name*	*Botanical Name*	*Conditions*
	Fruit crops		
1	Asian pear	*Pyrus pyrifolia*	Wet land and occasional flood
2	Banana	*Musa paradisiaca*	Intermittently submergence
3	Pond apple/Monkey apple	*Annona glabra*	*Under regular submergence*
4	Water Chestnut	*Trapa natans*	*Under regular submergence*
	Vegetable Crops		
1	Arrow Root	*Colocasia esculenta*	*Under regular submergence*
2	Water spinach	*Ipomea aquatic*	*Under submergence*
	Flower and ornamental plants		
1	Japanese Primrose	*Primula japonica*	*Under submergence*
2	Lotus	*Nelumbo nucifera*	*Under regular submergence*
3	Water Lily	*Nymphaea sps*	*Under regular submergence*
	Spice crops		
1	Anise	*Pimpinella anisum*	Less than a week flood
2	Cardamom	*Elettaria cardamomum*	Best for swampy areas
3	Clove	*Syzygium aromaticum*	*Under submergence*
4	Large Cardamom	*Amomum subulatum*	Intermittently submergence
5	Male Nutmeg	*Myristica fatua*	Intermittently submergence

Sl.No.	Common Name	Botanical Name	Conditions
6	Malabar Nutmeg	*Myristica malabarica*	*Under submergence*
7	Ginger	*Zingiber officinale*	Intermittently submergence
	Medicinal and aromatic plants		
1	Arjun	*Terminalia arjuna*	Submerge conditions
2	Bilimbi/Cucumber Tree	*Averrhoa bilimbi*	*Under regular submergence*
3	Brahmi	*Bacopa monnieri*	*Water bodies surroundings*
4	Bulrush	*Cyperus* spp.	Submerge conditions
5	European Water clover	*Marsilea quadrifolia*	Submerge conditions
6	Kangkong	*Ipomea aquatica*	Submerge conditions
7	Kewda	*Pandanus odoratissimus*	Including see water
8	Mandookaparni	*Centella asiatica*	Wet and submerge
9	Mint	*Mentha arvensis*	Withstand for 8-10 days
10	Sweet flag	*Acorus calamus*	*Water bodies surroundings*
11	Vetiver	*Vetiveria zizanoides*	*Under regular submergence*

Major Constraints and Future Strategies

India has wide horticultural crop diversity yet focused was given towards the selection and breeding for high yield followed by quality also. Hence, the area should be widened by the horticulturists and focus should be shifted for the selection and breeding for adversities like land and soil degradation, climate change, resource efficient cultivars under the supervision of the expert of the area. Unlike conventional breeding, pathologists and entomologist should take lead for the management of insect-pest and diseases of these crops with shifting paradigms. Moreover, selection and breeding is equally essential or even more against the nature's wrath, particularly for changing climate scenario and land degradation, while rejuvenation is needed for damaged other resources by harnessing expertise of soil scientists and environmentalists of the work area.

REFERENCES

1. Aishwath O.P. and Lal, R. 2020. Soil compaction and wetness effects on efflux of greenhouse gases. *J. Soil and Water Conser.* **19** (3): 244-253.
2. Aishwath, O.P. 2004. Growth yield and mineral composition of *Withania somnifera* grown on saline black soil. *Ind. J. Agril. Chem.* **37**(1): 1-8.
3. Aishwath, O.P. 2009. Mineral composition and autecological study of *Plantago ovata* in various soil textural environments of Fluventic Ustochrepts and Vertic Ustochrepts. *Ecol., Environ. and Conserv.* **15**(1): 181-187.
4. Aishwath, O.P. 2018. Spices, medicinal and aromatic plants for sustenance of soil health and water quality in semi-arid regions. In Souvenir of *National Conference on "Managing Natural Resources for Sustainable Agriculture"* held on December 8-9, 2018 at Raja Balwant Singh College, Agra (U.P.) pp 55-62.

5. Aishwath, O.P. 2019. Horticultural crops for degraded land. In: *Handbook of Horticulture* (Eds Verma *et al.,* and Tech. Coordinator Dr K.L. Chadha), Pub. ICAR-DKMA, New Delhi, **1**: 205-215.
6. Aishwath, O.P. and Anwer, M.M. 2010a. Lime induced chlorosis in *Coriandrum sativum*: First report in the world. *ICAR-NRCSS E-Newslet.* **2**(5): 3-6.
7. Aishwath, O.P. and Anwer, M.M. 2010b. Natural resource management through seed spices and aromatic plants. In: *Recent Advances in Seed Spices* (Eds Y.R. babu, R.K. Jaiman, K.D. Patel, N.R. Patel and S.B.S. Tikka). Pub. Daya Publishing House, New Delhi, India. pp 107-118.
8. Aishwath, O.P. and Lal, R. 2016. Resilience of spices, medicinal and aromatic plants with climate change induced abiotic stresses. *Ann. Pl. and Soil Res.* **18**(2): 91-109.
9. Aishwath, O.P. and Nibauria, S.V.S. 2009. Problematic soil resource utilization through aromatic plants. *Ecol., Environ. and Conserv.* **15**(4): 32-41.
10. Aishwath, O.P. and Pal, B. 2000. Performance of palmarosa (*Cymbopogon martinii* var Motia) under various soil texture and salinity levels. *Ind. Perfumer.* **44**: 285-290.
11. Aishwath, O.P. and Pal, B. 2004. Mineral composition of palmarosa (*Cymbopogon martinii*), soil EC and pH as influenced by saline water irrigation in texturally different soils. *J. Gujarat Soc. Agron. and Soil Sci.* **4**: 42-47.
12. Aishwath, O.P. and Tarafdar, J.C. 2007. Role of organic farming in medicinal and aromatic plants. In: *"Organic Farming and Sustainable Agriculture"* (Eds. Tarafdar *et al.,* 2007). Publisher - Scientific Publisher, Jodhpur. pp 135- 185.
13. Aishwath, O.P., Kumar, P., and Jalwania, R. 2019a. Production of seed spices with saline/ alkali soils and water in arid and semi-arid regions. In: *Horticulture in arid and semi-arid regions* (eds Jatav *et al*). Publisher- New India Publishing Agency, New Delhi. pp 127-136.
14. Aishwath, O.P., Kumar, P., Balai, R.C. and Jalwania, R. 2019b. Spices, medicinal and aromatic plants in degraded arid and semi-arid regions. In: *Horticulture in arid and semi-arid regions* (eds Jatav *et al*). Publisher- New India Publishing Agency, New Delhi. pp 137-150.
15. Aishwath, O.P., Meena, S.S., Kant, K., Sharma Y. K. and Anwer, M.M. 2010a. Climate Change and Seed Spices. In: *Challenges of Climate Change-Indian Horticulture* (Eds, H.P. Singh, J.P. Singh, and S.S. Lal). Published by Westville Publishing House, New Delhi. pp 85-89.
16. Aishwath, O.P., Mehta, R.S., Lal, G., Anwer, M.M., Dagar, J.C., Yadav, R.K. and Singh, R. 2010b. Growing seed spices a boon for problematic areas. *Ind. Hort.* **55**(4): 19-21.
17. Aishwath, O.P., Singh, B., Mehta, R.S., Yadav, R.K. and Meena, R.L. 2013. Seed Spices fetches more in saline/sodic soil and with poor quality water. *Ind. Hort.* **58** (6): 33-35.
18. Aishwath, O.P., Vashishtha, B.B. and Malhotra, S.K. 2008. Scope of Production of Aromatic Plants in Arid and Semiarid Regions. **In** *Spices and Aromatic Plants*' (Eds. Z.S. Solanki, G.N. Parihar, B.S. Rathore and S.R. Kumar). Publisher, ZDR, ARS Mandor, Jodhpur, Rajasthan. pp 147-162.

19. Amudha, K., Sakthivel, N. and Mohamed, M.Y. 2011. Rice - A novel food with medicinal value. *Agri. Rev.* **32** (3): 222- 227.

20. Angelone, M, Bini, C. 1992. Trace elements concentrations in soils and plants of Western Europe. In: *Biogeochemistry of trace metals* (ed. Adriano, D.C.). Boca Raton, FL: Lewis Publishers, 19–60.

21. Chadha, K.L. and Pareek, S.K. 1993. Horticulture for wasteland development. *J. Environ. Resour.* **1**(1-2): 58-64.

22. Chaudhari, S.K., Singh, B. and Aishwath, O.P. 2015. A Worldview of Land and Soil Degradation: Extent and Eco-technologies. **In**: Utilization of Degraded Land and Soil through Horticultural Crops for Agricultural Productivity and Environmental Quality (Eds Aishwath *et al.,*). Pub. *ICAR-National Research Centre on Seed Spices,* Ajmer, Rajasthan, pp 1-5.

23. Dagar, J.C. 2008. Cultivation of Medicinal and Aromatic Plants with Poor Quality Water. In: *Diagnosis and Management of poor quality water and salt affected Soils* (eds. Lal *et al*). Central Soil Salinity Research Institute, Karnal. pp 299-302.

24. Dagar, J.C. and Singh, G. 2007. Biodiversity of saline and waterlogged environments: Documentation, utilization and management. National Biodiversity Authority, Chennai, India.

25. Dagar, J.C., Tomar, O.S., Kumar, Y. and Yadav, R.K. 2004. Growing three aromatic grasses in different alkali soils in semi-arid regions of northern India. *Land Degrad. Deve.* **15: 143-151.**

26. Dehne, H.W. (1982) Interaction between vesicular arbuscular mycorrhizal fungi and plant pathogens. *Phytopathol.* **726:** 1115-1119.

27. Eswaran, H., Lal, R. and Reich, P.F. 2001. Land degradation: an overview. In: Bridges, E.M., I.D. Hannam, L.R. Oldeman, F.W.T. Pening de Vries, S.J. Scherr, and S. Sompatpanit (eds.). Responses to Land Degradation. Proc. 2nd International Conference on Land Degradation and Desertification, Khon Kaen, Thailand. Oxford Press, New Delhi, India.

28. Farooqi, A.A. and Sreeramu, B.S. 2001. Ambrette, Chamomile, Dill, Lavender, and Palmarosa. In: *cultivation of medicinal and aromatic plants*. Pub. Univ. Press (India) Ltd. Hyderabad. pp 295-423.

29. Framji, K.K. 1974. International Commission on Irrigation and Drainage bulletin. pp 36-41.

30. Gautam, R.K. and Nayak, A.K. 2008. Screening and Utilization of Salt Tolerant Genotypes of Medicinal Crop under Sodic Soils Conditions. In: *Diagnosis and Management of poor quality water and salt affected Soils* (eds. Lal *et al*). Central Soil Salinity Research Institute, Karnal. pp 296-298.

31. Grill, E., Winnacker, E.L. and Zenk, M.H. 1987. Phytochelatins, a class of heavy metal binding peptides from plants, are functional analogous to metallothioneins. In: *Proc. Nat. Acad. Sci.*, USA. **8:** 439-443.

32. Julie C. (2017) 25 Fruits and vegetables to grow in acidic soil. http: //www. offthegridnews. com/survival-gardening-2/25-fruits-and-vegetables-to-grow-in-acidic – soil/

33. Kothari, S.K., Vijay Kumar, Y., Bhattacharya, A.K. and Ramesh, S. 2004. Harvest management in rose-scented geranium (*Pelargonium sps.*) for higher essential oil yield and superior quality under semi-arid tropics. *Arch. Agron. and Soil Sci.* **50:** 593-600.

34. Lal, R. 1994. Tillage effects on soil degradation, soil resilience, soil quality, and sustainability. *Soil Till. Res.,* **27**: 1–8.

35. Maji, A.K. Reddy, G.P. Obi and Sarkar, D. 2010. Degraded and Wastelands of India: Status and Spatial Distribution (Eds Shashi A Verma). Pub. Project Director, Directorate of Information and Publications of Agriculture, Indian Council of Agricultural Research, Krishi KAB-I, New Delhi-110 012.

36. NRSA. 2005. Wasteland Atlas of India. Ministry of Rural Development and NRSA Publ., NRSA, Hyderabad.

37. Pathak, R.K. 2000. Wasteland, A boon for commercial Horticulture. In: Proc. *Changing scenario in arid horticulture* (A short course). Pub. Central Institute for Arid Horticulture. pp 62.

38. Patra, D.D. and Singh D.V. 1995. Utilization of salt affected soil and saline sodic irrigation water for cultivation of medicinal and aromatic plants. *Curr. Res. Med. and Arom. Pl.* **17:** 378-381.

39. Patra, D.D., Prasad, A., Anwar, M., Singh, D., Chand, S., Ram, B., Katiyar, R.S. and Kumar, S. 2002. Performance of lemongrass cultivars intercropped with chamomile under sodic soil with different levels of gypsum application. *Comm. Soil Sci. and Pl. Analy.*, **33:** 1707-1721.

40. Pitman, M.G. and Läuchli, A. 2002. Global impact of salinity and agricultural ecosystems. In: *Salinity: Environment - Plants - Molecules (Läuchli and Lüttge eds.), Pub. Kluwer Academic Publishers. Printed in the Netherlands.* pp 3–20.

41. Prakasha Rao, E.V.S. 1993 Nutrient management in some important tropical aromatic crops present and future. *Ind. Perfu.* **37:** 35-39.

42. Ramamohan Rao, M.S. 1998. Approaches to management of denudated lands in peninsular, India. In, *Technologies for waste land development* (Eds Abrol, I.P. and Dhurva Narayana, V.V.). Pub. Information Div. ICAR, KAB, New Delhi. pp 232-244.

43. Rao, G.G., Nayak, A.K., Chinchmalatpure, A.R. and Khandelwal, M.K. 2001. Yield of dill (*Anethum graveolens*) on saline black soils of different unirrigated farm sites of Bhal area in Gujarat. *Ind. J. Agril. Sci.* **71:** 711-712.

44. Salt, D.E., Blaylock, M., Kumar, N.P.B.A., Dushenkov, V., Ensley, B.D., Chet, I. and Raskin, I. 1995. Phytoremediation: a novel strategy for the removal of toxic metals from the environment using plants. *Biotechnol.* **13:** 468–474.

45. Schützendübel, A. Schwanz, P. Teichmann, T. Gross, K., LangenfeldHeyser, R., Godbold, D.L. and Polle, A. 2001. Cadmium induced changes in antioxidative systems, H_2O_2 content and differentiation in pine (*Pinus sylvestris*) roots. *Pl. Physiol.* **127:** 887–892.

46. Sehgal, J.L., Saxena, R.K. and Pofali, R.M. 1998. Degraded soil-their mapping through soil surveys. In, *Technologies for waste land development* (Eds Abrol, I.P. and Dhurva Narayana, V.V.). Pub. Information Div. ICAR, KAB, New Delhi pp 1-20.

47. Sharma P.C., Mishra, B., Singh R.K. and Singh, Y.P. 2001. Variability in the response of spinach, fenugreek and coriander to alkalinity and salinity stresses. *Ind. J. Pl. Physiol.* **6:** 329-333.

48. Shiva, M.P. Lehri, A. and Shiva, A. 2002. Introduction, Calamus, Cardamom, Clove Kewada Mint and Vetiver. In. *Aromatic and Medicinal Plants.* Publisher, International Book Distributors, Dehra Dun, India. pp 67-307.

49. Singh, R., Bhargava, G.P., Singh, R. 1995. Response of safflower (*Carthamus tinctorius*) and dill (*Anethum graveolens*) to soil salinity. *Ind. J. Agril. Sci.* **65:** 442-444.

50. Singh, R.P. 1998. Land degradation problems and their management in semi-arid tropics. In, *Technologies for waste land development* (Eds Abrol, I.P. and Dhurva Narayana, V.V.). Pub. Information Div. ICAR, KAB, New Delhi. pp 125-136.

51. Valtcho, D.Z., Lyle E. C. and Baoshan, X. 2006. Effects of Cd, Pb, and Cu on growth and essential oil contents in dill, peppermint, and basil. *Environ. and Exptl. Bot.,* **58:** 9-16.

52. Valtcho, D.Z., Lyle, E. C., Baoshan, X., Niels, E. N. and Andrew, W. 2008. Aromatic plant production on metal contaminated soils. *Sci. Total Environ.* **395:** 51-62.

53. Verma, V. and Chinnamani, S. 1998. Rehabilitation of revine lands. In, *Technologies for waste land development* (Eds Abrol, I.P. and Dhurva Narayana, V.V.). Pub. Information Div. ICAR, KAB, New Delhi. pp 263-272.

54. Zheljazkov, V. and Jekov, D. 1995. Heavy metal content in some essential oils and plant extracts. In: *Proc. International Symposium on Medicinal and Aromatic held on 1 November 1995 at Amherst, USA, Published in Acta Hort. 2006.* pp 426.

Chapter 20

Quality Pomegranate Production in Wastelands of India

R.A. Marathe* and N.V. Singh

ICAR-NRC on Pomegranate, Kegaon, Solapur 413255, Maharashtra

**e-mail: aishwath9@gmail.com*

ABSTRACT

In India, over 32 million ha comes under hot and arid region. Soils of these undulating topographies are characterized by shallow depth and sandy texture with low organic matter emanating poor fertility status. Thus, it is restricting the choice of crops for cultivation. However, by adopting different soil management practices, the drought tolerant pomegranate could be profitably cultivated on these marginal lands. The vast expansion of area (0.283 million ha) under pomegranate cultivation in India has been done on such marginal soils. The hot and dry climate of semi-arid regions favours the pomegranate crop growth but majority of the soils in these areas are on undulating lands having shallow depth, high gravel percent, sandy texture, low organic matter, and low fertility status. It has been observed under field conditions that pomegranate plants raised under poor soils produced lower yield, poor quality fruits and were more susceptible to pests and diseases. On the basis of the investigations related soil management in pomegranate, it can be concluded that, pomegranate can be successfully grown on extremely shallow, rocky soils by digging wide size pits and re-filling it with clay textured soil, provided the filling depth doesn't exceed 60 cm. The performance was equally well on loamy (generally pond soil) soil. If the filling material is of clay texture, then care should be taken towards the supply of sufficient quantity of micro-nutrient, while in case of light soils, application of macro-nutrients should be monitored. For improvement of drainage conditions in clayey soils, use of river sand to the extent of 25 percent is preferred, above which there was drastic reduction in nutrient availability and uptake by the plants, adversely affecting the growth and yield of pomegranate. Incidence and severity of wilt and bacterial blight disease was higher in the plants grown on clayey soils compared to light soils. It is also recommended to plant the pomegranate orchards on trapezoid shaped raised beds of 2 m width and 0.45 m height with sloping sides down on both ends. The standardized and recommended integrated nutrient management practices based on soil fertility status, leaf nutrient analysis and site specific soil related issues need to be adopted at large scale for sustainable pomegranate production on marginal lands.

Keywords: *Pomegranate, Marginal lands, Planting methods, Nutrition.*

INTRODUCTION

Pomegranate (*Punica granatum* L.) is one of the most important commercial fruit crops of the world dry regions of the world. Indian stands tall at *numero uno* position in the world in pomegranate production and acreage with 31.86 lakh tonnes of production from 2.83 lakh ha area. Export of pomegranate has increased from 18.65 thousand MT (Rs. 691 million) in 2010-2011 to 80.55 thousand MT (Rs. 6875 million) in 2019-20. Export from India has also shown significant improvement- 331.9 per cent in quantity, 894.93 per cent in value. Pomegranate cultivation has become a source of livelihood for more than 2.5 lakh rural families in India (Source: http://nhb.gov.in/statistics, http://agriexchange.apeda.gov.in). During last few decades, pomegranate has been evolved as a 'dollar crop' fetching foreign income due to ascending awareness among the consumers for its medicinal and nutraceutical values (Newman and Lanksy, 20). Although native to hot dry regions of Iran, Afghanistan, western Himalayan parts of India and adjoining areas (De Candolle, 7), pomegranate has been widely cultivated in Mediterranean and south east regions of Asia, Africa and Europe. Besides, it is also cultivated in China, Japan, some regions of erstwhile USSR, Mexico and California (USA). Pomegranate has huge potential to replace subsistence farming by virtue of increasing financial benefit due to characteristically short juvenile period of 1-2 years and high returns on investment (Babu, 1).

Pomegranate is grown on diverse soil types ranging from the gravelly soils of Uzbekistan (Saidaliev, 25), vertisols of India (Reddy *et al.*, 24), oil polluted soils of Azerbaijan with amendments (Bairamov *et al.*, 2) and even on poor alkali type soils of California (Day *et al.*, 6). Though, the hot and dry climate of semi-arid regions favours the pomegranate crop growth by creating a congenial atmosphere for the production of this crop. But majority of the soils in these areas are on undulating lands having shallow depth, high gravel percent, sandy texture, low organic matter, and low fertility status. It has been observed under field conditions that pomegranate plants raised under poor soils produced lower yield, poor quality fruits and were more susceptible to pests and diseases.

In India, over 32 million ha comes under hot and arid region. Soils of these undulating topographies are characterized by shallow depth and sandy texture with low organic matter emanating poor fertility status. Thus, it is restricting the choice of crops for cultivation. However, by adopting different soil management practices, the drought tolerant pomegranate could be effectively cultivated on these marginal lands. The vast expansion of area (0.283 million ha) under pomegranate cultivation in India has been done on such marginal soils (Marathe *et al.*, 16). The soils under pomegranate cultivation should be considered as dynamic in nature which varies in their properties due to wide range of geographical and climatic conditions, parent material and distinct and regular manipulation of management practices by pomegranate farmers. Therefore, identification of suitable material, soil suitability and site specific nutrient management are imperative for pomegranate cultivation on marginal lands. However, till date there is an acute dearth of precise information about suitability of soil amendments for plantation of pomegranate on skeletal lands. Under such circumstances, the yield potential of pomegranate has remained unexplored and rather suffered due to crop management issues such as induction of flowering, nutrient uptake, irrigation scheduling *etc.* Thus, effective soil management strategies suitable for pomegranate cultivation have to be adopted.

Major Soil Types of India used for Pomegranate Cultivation

In the Central India, pomegranate growing soils are categorised into three broad soil groups on the basis of depth, texture and plant performance. However, it is also grown on some pockets of Rajasthan on sandy soils which has very poor organic carbon contents and mostly marred with salinity but very well drained. The major soil characteristics and their management practices to be followed are given below.

1. Shallow Coarse Textured Soils

Majority of the soils on marginal land are of this type (Figure 1). The coarse textured or light soils with depth up to 30 cm are grouped under this category. The coarse fragments in these soils are quite high and they are underlain by weathered rock (murrum). Being located on undulating topography and having soil with high percolation rates, the water drained out very rapidly. The hydraulic conductivity is also high, providing an excessive drainage conditions. The water and nutrient holding capacity of these soils are very low. The organic carbon content and soil fertility status are also very poor. Vegetative growth of the plants is also poor owing to shallow rooting depth. Plant can be made under stress by withholding irrigation for a very short duration and usually plant bears heavy flowering. So, one can choose either *Mrig* or *Hasta* flowering and have regular and excessive flowering resulting in higher yield. Owing to low buffering capacity of the soil, plants frequently undergo stress condition as a result productive life of the orchards gets reduced. Disease and pests incidence is also low owing to prevalence of low soil moisture condition most of the time.

Figure 1. Profile of shallow, coarse textured soil.

Intensive management practices are required to be followed on these types of soils. Drip system of irrigation with two laterals placed on both side of plants with minimum four low discharge capacity (up to 4 lph) drippers placed on four direction of the plants need to be used on these soils. Low quantity of irrigation water with frequent interval should be applied; otherwise percolation loss of water below root-zone will occur. During bearing period providing irrigation at irregular or higher interval will adversely affect the fruit yield and quality due to low buffering capacity of the soil. Similarly, fertilizer needs to applied in small quantity and split dozes.

2. Shallow and Heavy Textured Black Soils

The black soils with depth upto 60 cm are categorised under this group (Figure 2). The clay percentage in these soils is high but as it is underlain by murrum or weathered rock, it provides good drainage conditions. Though, the nutrient and water holding capacity of the surface layer is very good, the rooting depth is limited. As such nutrient and water availability is good in this soil. The organic carbon content and soil fertility status of these soils are comparatively better than in shallow coarse textured soils.

Intensive management practices as mentioned above are required to be followed in these soils as well to harvest higher yields. The vegetative growth, flowering and fruit yields are also high in these soils. These are the ideal soils for the plantation of fruit crops like pomegranate.

Figure 2. Profile of shallow, heavy textured black soil.

3. Deep, Heavy Textured Black Soils

The black soils with depth more than 100 cm are categorised under this group (Figure 3). These soils generally occur in low land or intervalley basins. Vegetative growth of the plant is very high because of very high clay content, high water and nutrient holding and supplying capacity and sufficient rooting depth of the soils. In spite of all the favourable conditions, flowering and fruit yield is less owing to very low hydraulic conductivity and poor drainage conditions prevailing in these soils. As the depth of these soils is very high, there is constant supply of water to the plant from the subsurface horizons and it is difficult to impose stress within the plant for flowering induction. Due to lack of proper stress, the plants bear little flowers and resulting in poor yield. During rainy season, waterlogging conditions often arose in the orchards hampering intercultural operations and favours the incidence and spread of diseases. These types of soils are not recommended for the plantation of pomegranate orchards.

Figure 3. Profile of deep, heavy textured black soil.

Different Methods of Pomegranate Plantation

Most of the pomegranate growing areas are characterized by undulating topography with soils having shallow depth, course texture, low water holding capacity and low fertility. While soils of valley or lower basin are very deep, clayey in texture and poor in internal drainage creating waterlogging conditions during the rainy season. The root distribution studied of pomegranate revealed that major portion of the root activity was confined in 0-60 cm radial distance and 0-45 cm vertical distance. So, it is imperative to provide good quality soils at least in this active root-zone for commercial production on sustainable basis. The different methods used for pomegranate plantation are described below.

1. Pit Digging

Pit digging method of plantation is followed in shallow soils on comparatively flat lands. Trapezoidal pits, having base length and width of 1 m, and at surface 1.3 m with 1 m depth are dug at specified interval. The pits should be dug before summer season and exposed to scorching sun so as to have natural solar sterilization of soil. These pits are refilled with good quality pond soil (loamy texture) up to 30 cm above the ground level which subside to ground level after rainy season.

2. Trench Digging

Generally this method of plantation is practised on slopy lands where soil depth is very shallow. With the help of farm machinery trench of 1 cubic meter size are dug on the contour line at specific (row to row distance) interval across the slope. The dugout coarse texture earth materials are dumped beside the trench towards sloping end and refilled with surface soil from upper side of the trench. The trenches may also be filled with pond (loamy texture) soil but it will increase the cost of plantation.

3. Bedding System

The ICAR-NRCP has developed Bedding system of plantation which facilitates better nutrient uptake, higher plant growth and enhances fruit yield. In the areas having shallow soil depth (< 15 cm), bedding system of plantation is followed. This system of plantation comprises of digging 0.60 m wide and 0.30 m deep trenches below the soil surface and filling the trenches with good quality soil which may be from the upper soil layer (area between two rows) of the same field or brought from outside. Similarly, above the ground level, a bed of 2 m wide and 0.45 m height with sloping sides down on both ends are prepared (Marathe *et al.*, 2017). These beds are separated by parallel dead furrows oriented in the direction of land slope. The water drains from the beds into dead furrows, which discharge into a field drain constructed at the lower end of the field and perpendicular to the dead furrows. This system of plantation is recommended by the ICAR-NRCP, Solapur for many merits over other system of plantation (Figure 4). The bedding system ensures greater availability of nutrients to the plants, maintains favourable soil moisture in rhizosphere, hastens the stress physiology required for optimum flowering and plant roots in the bedding system remain well aerated.

Figure 4. Pomegranate plantation on raised bed system.

Innovative Soil Management Technique for Pomegranate Cultivation on Marginal Lands

An experiment was conducted at ICAR-NRCP, Solapur to study the performance of pomegranate cv. Ganesh and cv. Bhagwa grown on different types of soils. The original experimental soil, taxonomically classified as Entisol (Lithic Ustorthent) was characterized by gravelly surface devoid of any soil material and underlained by consolidated materials having montmorillonitic mineralogy. It is classified in Land Capability Classes as Group V land with limitations of soil depth, texture and water holding capacity. Under such soil type, Trapezoidal pits of 1.50 × 1.50 m top and 1.20 × 1.20 m bottom dimension with 0.30 m, 0.60 m, 0.90 m and 1.20 m depth as per various treatments. The pits were re-filled to its full depth with different soil materials that often remain available in such type of marginal lands and texturally classified as gravelly, sandy loam, loam, clayey black soil and only murrum as well.

The results revealed that clayey soil had the highest macro-nutrient supplying capacity (available N: 376.1, P: 47.1 and K: 761.6 kg ha^{-1}) and loamy soil had the highest micro-nutrient supplying capacity (available Fe: 19.2, Cu: 8.1 and Zn: 1.83 ppm). The highest Mn (41.2 ppm) and Zn (22.4 ppm) contents in the leaves from the loam textured soil increased leaf chlorophyll contents and vegetative growth.

The results on nutrient dynamics of soil and plants revealed that the organic carbon and availability of macro- and micro-nutrient contents increased in all types of soil materials over the period of four years. In general, organic carbon content was higher in clayey textured soil (0.95 to 1.39 per cent) as compared to light textured soil (0.73 to 0.98 per cent). Addition of substantial quantity of organic manure in the root zone could be attributed to increase in organic carbon (Bhriguwanshi, 5). Similarly clayey soil material filled up to 1.20 m depth had significantly highest quantity of N (376.1 kg ha^{-1}) and P (47.1 kg ha^{-1}) availability as compared to rest of the treatments. Significantly highest availability of K (761.6 kg ha^{-1}) was observed in clayey soil up to 0.60 m and 1.20 m depth. The micronutrient availability was low in clayey soil material than light textured soil material. The increase of micronutrient contents was several folds higher than initial contents, which might be due to yearly addition of FYM that contains considerable number of micronutrients (Marathe *et al.*, 18). Loam textured soil material showed significantly highest Fe (1.83 ppm), Cu (19.2 ppm) and Zn (8.06 ppm) availability. However, addition of sand (1:1 v/v) in clayey soil significantly reduced micro-nutrient availability. Presence of excessive calcium carbonate in sand (19.24 per cent) and clay (6.31 to 7.77 per cent) might have played vital role in immobilizing the micronutrient availability in clay mixed with sand soil material.

Pooled analysis of leaf nutrient content data revealed that macronutrients' content (N: 1.59-1.67 per cent, K: 1.06-1.09 per cent and Ca: 1.55-1.97 per cent) in the plants grown on clayey textured soil material were higher as compared to light textured soil material (N: 1.45-1.62 per cent, K: 0.84- 0.94 per cent and Ca: 1.67-1.76 per cent). The efficient uptake of macronutrients under clayey soil material could be expected because of higher nutrient and moisture availability in these soils.

Vegetative growth was far better in the plants grown on clayey soil up to 1.20 m depth. Very high water and nutrient holding capacity of clayey soils resulted into better nutrient uptake and growth of the plants than other treatments. However, growth of the plants grown in the pits filled with weathered rock was satisfactory during initials years but it was significantly decreased after four years as compared to other soil types. Earlier Saidaliev (25) reported better performance pomegranate even on gravelly soils in Fergana valley of Uzbekistan. Addition of sand to the extent of 50 per cent (v/v) for the improvement of drainage conditions of heavy soils adversely affected the plant growth. Lower nutrients availability in soil, which reflected in reduced uptake by the plants, might have seized plant growth. Chlorophyll contents in the leaves of the plants significantly varied from 52.28 to 62.54 and was highest in the plants grown in clayey followed by loamy soils.

Coarse textured soils favoured flowering than other soil types. The highest numbers of male flowers (302.5) were induced in gravelly soil while hermaphrodite flowers (144.7) were highest in loamy soil. Soils having poor water holding capacity were prone to deep percolation losses and as such remain dry. Such moisture stress is required to the plants for flower induction (Marathe *et al.,* 15). Availability of sufficient moisture for longer duration in clay soil filled up to 1.20 m depths could not build up sufficient stress and favoured vegetative growth by suppressing reproductive phase in terms of less male (214.8) and hermaphrodite (115.2) flowers.

The highest fruit yield in terms of number and weight was produced on the plants grown on clayey soil and at par in the plants grown on loamy textured soil. Higher nutrient availability and thereby increased uptake by the plants grown in these soils corroborated to optimum growth and higher fruit yield. Further the efficient drainage under these treatments might have played a vital role in increasing the fruit yield. Whereas, drastically reduced fruit yield in plants grown on clayey soil up to 1.20 m depth could be attributed to poor drainage in deep black clay soils as reported earlier in citrus (Marathe *et al.,* 19). As compared to original soil type *i.e.* weathered rock (2.31) cost benefit ratio was significantly high (3.85) in clayey followed by loamy (3.72) soil. Since the variations in recurring cost of cultivation amongst various treatments will be very meagre, third year onwards the net profit increases perpetually.

Serious pathological problem with pomegranate cultivation is the endemic infestation of bacterial blight disease (BBD), which affects all plant parts including fruits (Sharma *et al.,* 28). Disease index, calculated on the basis of incidence and severity on all plant parts, was higher in clayey soil having depth up to 1.20 m as compared to light textured soil material. Although, incidence of wilt disease was also high in clayey soil, it was not observed in the plants grown in clay soil mixed with sand (1:1 v/v) and in weathered rock. High water holding and water retention capacity of clayey soils might have created high humidity in the microclimate of the plants, which proved congenial for the development and increasing disease severity in these plants (Marathe *et al.,* 18).

Integrated Nutrient Management

Balance nutrition is important, both for young, growing and grown up productive trees. However, a bearing tree requires balanced nutrient application for maintaining vegetative growth and sustaining fruit production. In absence of any of the major nutrient elements sharp drop takes place in the performance of pomegranate plants. Apart from fruits, considerable amount of major and micro-nutrients are removed from soil following annual pruning of shoots. Survey of pomegranate orchards in Bijapur district of northern Karnataka, revealed that optimum NPK concentrations in leaves were 0.91-1.66, 0.12-0.18 and 0.61-1.59 per cent, respectively.

Similarly, optimum (available) NPK concentrations in soils having pH 8.1-8.6 were 44-103, 10-20 and 73-115 mg/kg, respectively. Now a days Diagnosis and Recommendation Integrated System (DRIS) and Compositional Nutrient Diagnosis (CND) indices are commonly used for recommendations of nutrients. Using these indices Raghupathi and Bhargava (22) found that yields of pomegranate were limited by more than 2 or 3 nutrients as it is mainly grown on marginal soils with low fertility. They observed highly significant positive correlation between DRIS and CND indices. Three principal components explained 59.5 per cent of the variation in the high yielding population: (N+S+Zn+Fe-Mn-); (N+P+Ca+Zn-); and (K-S+). CND is recommended for diagnosis of nutrient disorders as it expands the DRIS concept from bi-dimensional to multidimensional space and recognizes high-order interactions among nutrients while requiring less computational effort. Furthermore, Gosavi *et al.* (10) established DRIS norms for pomegranate cv. Bhagwa grown in Southwestern Maharashtra, using leaf nutrient analysis and found nutrient sufficiency ranges at 50 per cent flowering as 1.32–2.15 per cent nitrogen, 0.18–0.24 per cent phosphorus, 1.29–1.99 per cent potassium, 0.64–1.20 per cent calcium, 0.23–0.45 per cent magnesium, 0.16–0.26 per cent sulfur, 103.04–149.12 mg kg^{-1} iron, 39.60–72.85 mg kg^{-1} manganese, 15.99–26.10 mg kg^{-1} zinc, 6.16–9.32 mg kg^{-1} copper, 23.38–39.88 mg kg^{-1} boron and 0.29–0.47 mg kg^{-1} molybdenum. The requirement of Fe, Mg, S, Zn and N by the pomegranate plant was found higher at 50 per cent flowering and fruit development stages (Table 1).

Table 1. Recommended fertilizer doses for pomegranate

Age of the Plant (Years)	*FYM (Kg/ tree/ year)*	*Nitrogen (g/tree/year)*			*Phosphorous (g/tree/year)*				*Potassium (g/tree/year)*		
		N Req	*Source*		*P_2O_5 Req*	*Source*			*K_2O Req*	*Source*	
			Urea 46 per cent N	*CAN 25 per cent N*		*TSP 43 per cent P_2O_5*	*SSP 16 per cent P_2O_5*	*DAP 46 per cent P_2O_5*		*MOP 60 per cent K_2O*	*SOP 50 per cent K_2O*
1	10	250	540	1000	125	290	780	271	125	210	250
2	20	250	540	1000	125	290	780	271	125	210	250
3	30	500	1090	2000	125	290	780	271	125	210	250
4	40	500	1090	2000	125	290	780	271	600	1000	1200
5 and above	40	625	1360	2500	250	580	1560	544	600	1000	1200

Note: Depending on availability 1-1.5 kg/plant poultry manure, 1 kg/plant vermicompost and 1-2 kg/plant neem cake DAP also contains 18 per cent N so if used as source of P then amount of nitrogenous fertilizers should be reduced accordingly.

Source: NRCP Nutrient advisory.

Integrated use of organic and inorganic macro and micronutrients sources along with bio-formulations and plant beneficial microbes suffice the nutritional needs of pomegranate plants to produce optimum and sustainable yield. Fertilizer and manures should be applied immediately after harvest, at the time of bud break about a fortnight after defoliation and after fruit set. While integrated supply of organics, in-organics along with microbial formulations were highly beneficial. Application of 10 kg FYM plant^{-1} alone or in combination with N:P:K, poultry manure 5 kg plant^{-1} and bone meal 1 kg with N:P:K were the effective treatments to boost up the all around growth of pomegranate plants (Saraf *et al.,* 26). Application of inadequate or excessive quantity of fertilizers especially micronutrients affect the yield and quality of fruit crops. Foliar sprays of each of $ZnSO_4$ (0.25 to 0.30 per cent), $FeSO_4$ (0.25 to 0.40 per cent), $MnSO_4$ (0.30 per cent) and 0.15 per cent boric acid alone or in various combination significantly increased the fruit yield of pomegranate (Bambal *et al.,* 4; Balkrishnan *et al.,* 3). Individual spray of boric acid effectively decreased the fruit cracking problems in pomegranate. In arid regions of Rajasthan, pomegranate cv. Jalore Seedless with the spraying of 0.2 per cent boron caused least cracking which in turn produced the highest yield (Singh *et al.,* 29). Micronutrients can be managed by giving 3 foliar sprays during crop season (Sharma *et al.,* 27).

Biofertilizers, which plays very important role in pomegranate production, are the preparations containing microorganisms in sufficient numbers helping plant growth and nutrition. They are widely accepted as low cost supplements to chemical fertilizers and have no deleterious effects either on soil health or environment. Inoculation of cuttings with various microorganisms was helpful to increase rooting and survival percentage. Various bioformulations of *Aspergillus niger*, AMF sources (*Glomus* spp.), *Trichoderma* spp., *Penicillium pinophilum, Azospirillum lipoferum, Azospirillum brasilense*, and *Azotobacter sp.* have also been recommended for application in pomegranate twice per year after multiplication or activation on FYM (except AMF) for mobilizing and solubilizing nutrients and facilitating their uptake by plants.

Performance of Pomegranate under Saline-Alkali Conditions

At present majority of soils in arid and semi-arid conditions are available as waste lands with salinity and alkalinity problems, but has potential for pomegranate cultivation. Hardy irrigation water in these areas often exaggerates this problem. However, there are reports that it can be cultivated in the soils of salt content up to 0.4 per cent indicating scope to utilized such lands. Presence of salts such as NaCl, Na_2SO_4 and $CaCl_2$ caused adverse effect in descending order on various growth parameters and Ca uptake (El-Dawwey, 9)

Increase in soil salinity and sodicity levels adversely affected the N, K, P and Ca contents in the leaves, while the chloride, Mg, and Na contents were increased (Patil and Patil, 21; Ram Asrey *et al.,* 23). In arid and semi-arid conditions, poor quality of irrigation water is of serious concern and played important role in increasing soil hazards and limits the performance of pomegranate, though, it is comparatively more tolerant to saline and alkali conditions as compared to other arid and semi-arid fruit crops. The Khog cultivar of pomegranate survived well with irrigation water having EC values of 6.5 mmhos while cv. Jallore seedless could tolerate salinity only up to 4.5 mmhos (Jain and Dass, 13).

Management of Saline Alkali Soils

The researchers advocated the various management practices as replacement of saline-sodic soil with normal soil in trenches and pits, provision of 20 cm thick layer of gravel at the bottom of trenches, frequent irrigation and mulching to maintain a negative salt balance. Secondary salinization in these treatments during subsequent years was prevented by adding small amounts of waste sulphuric acid and iron sulphate. Some reports are available that pomegranate cannot tolerate water stagnation. Further attempts were also made in management of saline alkali soils by Gurbachan Singh *et al.* (11) who tried sulphitation cane filter cake (SCFC) to minimize the ill effects of alkalinity (pH 10.0) in pomegranate and proved much superior in increasing growth and biomass of pomegranate than gypsum and pyrites. Increased concentration of P, K, Ca, Mg and S in the harvested biomass, decreased soil pH, large increase in soil EC, organic carbon, available P and K were recorded in SCFC treatments.

For management of saline alkali soils, several workers suggested soil incorporation of 0.5 to 3.0 per cent bitumen as a soil conditioner which was economical and more effective as compared to surface spray. It was very effective in pomegranate grown on highly calcareous (33.2 per cent $CaCO_3$) sandy clay or highly saline loam soils (29.6 per cent $CaCO_3$). Application of bitumen increased leaf number, leaf area and uptake of N, P, K and Ca (Khattab *et al.*, 14) in pomegranate. Soil condition, bud emergence and plant growth at early stages of growth were also improved. The soil conditioner had probably modified and stabilized soil structure, increased soil aeration, improved hydraulic properties via salt leachability, lowered pH value, dissolved salts in the soil, reduced water loss through evaporation and increased moisture in the root zone. The temperature of the soil, particularly during the colder months was increased and this enhanced root growth and the ability of the trees to absorb water and nutrients from the soil (El Hady *et al.*, 8). The improvement in the fruit quality as indicated by fruit size, total acidity, TSS content and total percentage of sugars, seed weight and juice percentage had also been reported by the use of bitumen (Heikal, 12).

Conclusions

On the basis of the investigations related soil management in pomegranate, it can be concluded that, pomegranate can be successfully grown on extremely shallow, rocky soils by digging wide size pits and re-filling it with clay textured soil, provided the filling depth doesn't exceed 30 cm for cv. Ganesh and 60 cm for cv. Bhagwa. In both varieties drastic reduction in yield was observed in clayey soils if depth exceeds above 60 cm. The performance was equally well on loamy (generally pond soil) soil. If the filling material is of clay texture, then care should be taken towards the supply of sufficient quantity of micro-nutrient, while in case of light soils, application of macro-nutrients should be monitored. For improvement of drainage conditions in clayey soils, use of river sand to the extent of 25 percent is preferred, above which there was drastic reduction in nutrient availability and uptake by the plants, adversely affecting the growth and yield of pomegranate. Incidence and severity of wilt and bacterial blight disease was higher in the plants grown on clayey soils compared to light soils. It is also recommended to plant the pomegranate orchards on trapezoid shaped raised beds of 2 m width and 0.45 m height with sloping sides down on both ends. The standardized and recommended integrated nutrient management practices based on soil fertility status, leaf nutrient analysis and site specific soil related issues need to be adopted at large scale for sustainable pomegranate production on marginal lands.

REFERENCES

1. Babu, K. D. 2010. Floral biology of pomegranate (*Punica granatum* L.). Fruit, vegetable and cereal. *Sci. and Biotech.* **4:** 45-50.

2. Bairamov, Z. R., Orudzhaliev, F. S. and Magrebi, R. A. 1989. Growth and development of forest plantations on derelict land under the influence of organic/mineral fertilizers. *Lesnoe Khozyaistvo,* **10**: 25-27.

3. Balakrishnan K., Venkatesan, K. and Sambandamurthi, S. 1996. Effect of foliar application of Zn, Fe, Mn and B on yield and quality of pomegranate cv. Ganesh. *Orissa J. Hort.* **24**: 33-35.

4. Bambal S. B., Wavhal, K. N. and Nasalkar, S. D. 1991. Effect of foliar application of micro-nutrients on fruit quality and yield of pomegranate (*Punica granatum* L. cv. Ganesh). *Maharashtra J. Hort.* **5**: 32-36.

5. Bhriguwanshi, S.R. 1988. Long-term effect of high doses of farm yard manure on soil properties and crop yield. *J. Indian Soc. Soil Sci.* **36**: 784-786.

6. Day, K. R., Andris, H. L., Klonsky, K. M. and De Moura, R. L. 2005. Sample cost to establish and produce pomegranates. Pamphlet PG-VS-05, University of California Cooperation Extension.

7. De Candolle, A. 1967. *Origin of Cultivated Plants.* Hafner Publishers Co., New York, pp. 468.

8. El Hady, O.A., Khattab, M.M., Hanna, A.H. and Ebtisam, K.M. 1991. Interaction of locally produced bitumen emulsion with calcareous saline soil and the growth response of pomegranate (*Punica granatum* L.) transplants. *Egyptian J. Soil Sci.* **31**: 89-114.

9. El-Dawwey, G.M. 1998. Tolerance of Balady mandarin and Manfalouty pomegranate seedlings to salinity. *Assiut J. Agril. Sci.* **29**: 123-133.

10. Gosavi, A.B., Deshpande, A.N., Maity, A. 2017. Identifying nutrient imbalances in pomegranate (cv. Bhagwa) at different phonological stages by the diagnosis and recommendation integrated system. *J. Plant Nutri.* 40 (13): doi.org/10.1080/01904167.2016.1267209

11. Gurbachan Singh, Soni, M. L., Singh, S. K. and Singh, N. T. 1999. Effect of amendments on pomegranate (*Punica granatum*) calcareous and non-calcareous alkali soils. *J. Indian Soc. Soil Sci.* **47**: 345-352.

12. Heikal, E. K. 1991. Improving fruit growth and development of Manfaloti pomegranate trees grown on sandy soil. *Annals of Agricultural Science,* Moshtohor. **29**: 967-979.

13. Jain, B. L. and Dass, H. C. 1988. Effect of saline water on performance of saplings of jujube (*Ziziphus mauritiana*), Indian cherry (*Cordia dichotoma* var. wallichii) and pomegranate (*Punica granatum*) at nursery stage. *Indian J. Agril Sci.* **58**: 420-421.

14. Khattab, M. M., Hanna, A. H., El Hady, O. A. and Kamal, E. M. 1990. Nutrient uptake by pomegranate (*Punica granatum* L.) transplants grown in highly calcareous soil conditioned with bitumen emulsion. *Acta Horticulturae.* No. 274, 215-222.

15. Marathe R. A., Jyotsana Sharma, Shinde, Y. R. and Chaudhari, D. T. 2016. Standardization of organic manure application in pomegranate (*Punica granatum* L.) orchards grown in semi-arid regions. *Indian J. Agril. Sci.* **86**: 1265-1270.

16. Marathe, R. A., Dhinesh Babu, K. and Shinde, Y. R. 2015. Soil and leaf nutrient constraints in major pomegranate growing states of India. *Agril. Sci. Digest.* **36**: 52-55.

17. Marathe, R. A., Jyotsana Sharma, Dhinesh Babu, K. and Murkute, A. A. 2017. Bedding system: A unique plantation method of pomegranate in arid and semi-arid regions. *Natl. Acad. Sci. Let.* **40**: 249-251.

18. Marathe, R. A., Jyotsana Sharma, Murkute, A. A. and Dhinesh Babu, K. 2016a. Response of nutrient supplementation through organics on growth, yield and quality of pomegranate. *Sci. Hortie.* **214**: 114-121.

19. Marathe, R. A., Mohanty, S. and Shyam Singh. 2003. Soil characterisation in relation to growth and yield of Nagpur mandarin (*Citrus reticulata* Blanco). *J.Indian Soc. Soil Sci.* **51**: 70-73.

20. Newman, R. A. and Lansky, E. P. 2007. Pomegranate - The Most Medicinal Fruit. Basic Health Publication Inc. 28812, Top of the world Drive Laguna Beach, 92651/C, California, USA.

21. Patil P. K. and Patil, V. K. 1982. Effect of soil ESP on the growth and chemical composition of pomegranate (*Punica granatum*). *Progre. Hort.* **14**: 1-5.

22. Raghupathi, H. B. and Bhargava, B. S. 1998. Diagnosis of nutrient imbalance in pomegranate by Diagnosis and Recommendation Integrated System and Compositional Nutrient Diagnosis. *Commu. Soil Sci. and Plant Anal.* **29**: 2881-2892.

23. Ram, Asrey., Singh, R. B. and Shukla, H. S. 2002. Effect of sodicity levels on growth and leaf mineral composition of pomegranate (*Punica granatum* L.). *Annals of Agric. Res.* **23**: 398-401.

24. Reddy, B. G. M., Patil, D. R., Hulamani, N. C. and Patil, S. G. 1998. Performance of pomegranate (*Punica granatum* L.) clones in the Vertisols of Tungabharda command area. *Progressive Horticulture.* **30:** 14-15.

25. Saidaliev, U. V. 1985. Results of variety trials of pomegranate on the gravelly soils of Fergana province. *Trudy Nauchno Issledovatel'skogo Instituta Sadovodstava, Vinogradarstva-i-Vinodeliya Imeni R R Shredera.* **47:** 26-28.

26. Saraf R. K., Samaiya, R. K. and K. C. Shukla. 2004. Effect of different sources of nutrients on growth of pomegranate (*Punica granatum* L.). Proc. Nat. Con. on Biodiversity and Sustainable Utilization of Biological Resources, (eds. Sahu T.R. and P. K. Sahu), March 16-18, 2001, Sagar, Madhya Pradesh, India, pp 203-208.

27. Sharma, J., Chandra, R., Dhinesh Babu., Meshram, D.T., Maity, A., Singh, N.V. and Gaikwad, N.N. 2014. *Pomegranate: cultivation, marketing and utilization.* Technical Bulletin No. NRCP/2014/1.pp.88.

28. Sharma., Jyotsana., Sharma, K. K., Kumar, A., Mondal, K.K., Thalor, S., Maity, A., Gharate, R., Chinchure, S. and Jadhav, V. T. 2017. Pomegranate bacterial blight: symptomatology and rapid inoculation technique for *Xanthomonas axonopodis* pv. *punicae*. *J.Plant Pathology*. **99:** 109-119.

29. Singh D. B., Sharma. B. D. and R. Bhargava. 2003. Effect of boron and GA3 to control fruit cracking in pomegranate (*Punica granatum*). *Current Agriculture*. **27**: 125-127.

Chapter 21

Risk to Mango Production from Newly Emerging Diseases

P.K. Shukla*, Tahseen Fatima, Savita Varma, Nidhi Kumari and Gundappa Baradevanal

Division of Crop Protection, ICAR–Central Institute for Subtropical Horticulture, Rehmankhera, Lucknow 226 101, Uttar Pradesh

**e-mail: pksmush@gmail.com*

ABSTRACT

Mango production has recently been put at risk due to the emergence of new diseases (wilt, decline, root rot, twig/branch drying, sooty blotch, and shoulder browning). During the last 10 years, a gradual increase in the incidence of wilt/decline and twig/branch drying/dieback diseases has been observed. Out of 15 major mango growing states surveyed, the distribution of wilt/decline and twig/branch drying disease has been recorded in 14 and 9 states, respectively. Ceratocystis fimbriata, which causes wilt and decline, and Berkeleyomyces basicola, which causes root rot, are two new pathogens that have recently been reported. It has been confirmed that the Scolytid beetle is associated with Ceratocystis wilt. Specific symptoms have been documented to avoid confusion in the diagnosis of emerging diseases. Excessive intercultural operations that cause root injury, flood irrigation and intercropping have all been identified as major factors contributing to the development of root rot, wilt, and decline diseases, while, an increase in the number of unseasonal rainy days during the fruiting period has been identified as a major cause of sooty blotch and shoulder browning diseases. Sooty blotch and shoulder browning are common during the fruiting season wherever there is heavy rainfall. The management practices for wilt, decline, root rot, sooty blotch and shoulder browning have already been developed however for twig drying, it is yet to be standardized. The efforts have been made to transfer the knowledge generated concerning newly emerged diseases through research publications, extension folders, training programs, seminars, print, electronic and social media; and the agro-advisory service of the ICAR-CISH, Lucknow.

Key words: *Mango, wilt, decline, twig drying, shoulder browning*

INTRODUCTION

Mango (*Mangifera indica* L.) is the world's fifth most important fruit crop. Global fruit production of mango was estimated 50.64 million tonnes from 56.81 million hectares (FAO, 1). In India, the production of mango was estimated at 21.82 million tonnes from a cropped area of 2.258 million hectares during 2017-18 (Hort Stat, 6). Uttar Pradesh has the highest productivity (17.14 tonnes/ha) and production (4.55 million tonnes) of mango in India (Hort Stat, 6). The diseases, *viz.* wilt, decline, root rot, twig/branch drying, sooty blotch, and shoulder browning has emerged as new threats to mango production in India (Shukla *et al.*, 14). Extensive investigations into the causal agents, symptoms, factors influencing disease development, disease distribution, the extent of losses were carried out, and management have been conducted and are currently in progress. To provide immediate relief to farmers and to protect the crop from newly emerged diseases research achievements have been extended using all possible means. The article has summarized account of published findings and the data generated later on regarding the newly emerged issues in mango production along with future needs.

Wilt, Decline and Twig/Branch Drying

The wilting of mango trees has become a serious threat to the mango industry in the recent past. The complete wilting of a healthy-looking tree takes place within days due to infection in the main roots or trunk, which has been termed as sudden wilt. When declining tree dies in months or years due to infection of secondary roots, termed as slow decline (Shukla *et al.*, 15). Wilting or drying of big branches or twigs takes place due to infection in the stem. The incidence of wilt and decline diseases has been observed in the majority of the major mango growing areas of the Rajasthan, Uttar Pradesh, Bihar, Telangana, Jharkhand, West Bengal, Himachal Pradesh, Gujarat, Karnataka, Odisha, Punjab, Tamil Nadu, Maharashtra and Andhra Pradesh. The potential distribution of mango wilt in India was predicted under different climate change scenarios by ecological niche modeling by Max Ent. Study revealed that the most vulnerable areas for the disease occurrence under future climate scenarios in 2050 are parts of Southern Jammu and Kashmir and Western Bihar; whereas in 2070 scenario the highest suitability of disease occurrence was predicted in Uttar Pradesh, Jammu Kashmir, Bihar, Himachal Pradesh and Madhya Pradesh (Shukla *et al.*, 10).

The Pathogen

Several pathogenic fungi (*Botryosphaeria, Ceratocystis, Diplodia, Lasiodiplodia, Fusarium, Fusicoccum, Physalospora, Phytophthora, Pythium, Rhizoctonia, Sclerotium* and *Verticillium*) have been found associated with mango wilt and decline but species of *Ceratocystis* have been most destructive. In India, *Ceratocystis fimbriata* has been identified as the causal agent and found widespread (Shukla *et al.*, 15). The association of *Berkeleyomyces basicola* with mango root rot was also confirmed (Shukla *et al.*, 13) and it has been found to play a significant role in the decline of mango trees. The branch drying has been mostly due to the infection of *C. fimbriata* but twig drying has been incited by *Lasiodiplodia theobromae*. The dispersal of the pathogen takes place by soil, water, air, insects, infected wood, human, tools and planting material. The invasion of *C. fimbriata* mostly occurs through wounds or wood-boring beetles. The soil-borne inoculums has been responsible to cause infection in roots and infection in trunk or branches is caused by scolytid beetles infested with *C. fimbriata* and by contaminated pruning tools. *C. fimbriata* is a slow-growing fungus; the incubation period in a plant varies from 6 months to several years for the expression of symptoms.

Symptoms

Based on initial infection site, symptoms appear as sudden wilt, decline or branch drying (Figure 1). The infection of pathogens causes light grey, brown to dark brown or black staining in the woody portion (Figures 2a, b). Severe exudation of gum from infected stem (Figure 2c) has been recorded particularly at the time of wilting of branches or whole tree (Shukla *et al.*, 15). Leaves of wilted trees stay attached to the tree for long period but gradual defoliation takes place in declining trees. The tree does not express any symptom before wilting but the declining tree exhibit nutrient deficiency syndrome. The presence of *C. fimbriata* in infected tissue has

Figure 1. Symptoms of mango wilt (a), decline (b), branch drying (c) and twig drying (d).

Figure 2. Symptoms of infection in underground trunk (a), aboveground trunk (b), gummosis (c) and beetle infestation (d).

indirectly been confirmed by the presence of minute bores and frass created by Scolytid beetles (Figure 2d). The blackening of bark on branches due to fungal infection followed by gummosis is the primary symptom of twig or branch drying. When infection proceeds into deeper tissues, the drying of branches and twigs takes place.

Factors Affecting Disease Development

C. fimbriata is transmitted by the scolytid beetles, *Hypocryphalus mangifera* and *Xyleborus* species (Gundappa *et al.*, 5); which are attracted to infected tissue and there they create bores in the trunk and branches (Goitia and Rosales, 3). The wood shavings containing fungal propagules pushed outside the stem and the beetles infested with the fungus may cause secondary spread. Deep ploughing and digging the soil under canopy has been found responsible for damage to roots, which facilitates infection in roots by soil-borne inoculum of *C. fimbriata* (Shukla, 8). Pruning of trees enhances infection due to transmission through pruning tools and invasion through wounds (Teviotdale and Harper, 16). Infected wood and sawdust may harbour viable spores for at least 5 years; and thus the pathogen can be transmitted to long distances (Grosclaude *et al.*, 4). Flood irrigation and intercropping of high water demand crops enhances the flow of inoculums and thus incidence of wilt and decline of trees (Malik *et al.*, 7; Shukla *et al.*, 15). The infection of *L. theobromae* in stems mostly occurs during the rainy season due to abundance of inoculums and wetness on the plant surface.

Management

To achieve the management of wilt and decline diseases several experiments were carried out under controlled and field conditions.

In vitro Screening of Fungicides

The fungicides available in the market were evaluated at recommended concentrations against *C. fimbriata* using poison food technique along with *Trichoderma asperellum* isolated from *C. fimbriata* infected mango wood (Table 1). The stock solution of each fungicide was prepared separately in sterilized distilled water and it was added to the PDA medium at 50±2 °C to obtain desired concentrations (Table 1) before pouring into Petri dishes. PDA without fungicide was used for *T. asperellum* and as untreated control. Five mm discs of 20 day-old *C. fimbriata* and *T. asperellum* cultures maintained on potato dextrose agar (PDA) medium were placed either at the centre of the plates or near the sides and incubated at 25 ±2 °C in a BOD incubator. Observations were made regularly and radial growth of the fungus was measured on 14th and 28th days after inoculation (Table 1)

All the fungicides and *T. asperellum* significantly suppressed the radial growth of *C. fimbriata* in comparison to control. The fungicides (thiophanate methyl, carbendazim, propiconazole, hexaconazole, captan + hexaconazole, metalaxyl + mancozeb, carbendazim + mancozeb, mycobutanil, propineb, mancozeb, tetraconazole and difenoconazole) found highly effective at recommended doses (Table 1) were further tested at lower doses (Table 2). Radial growth of the fungus was measured on the 10th and 20th days after inoculation.

Since, *C. fimbriata* is slow growing in comparison to *T. asperellum*, for further confirmation of suppressive effects of *T. asperellum*, *C. fimbriata* was inoculated 10 days before *T. asperellum* after recording the size of *C. fimbriata* colony. The growth of *C. fimbriata* was almost stopped by *T. asperellum* (Figure 3). Efficacy of sterilized culture filtrate of *Trichoderma harzianum* (later characterized as *T. asperellum*) was tested against *C. fimbriata* and it was found effective (Shukla *et al.*, 2012).

Table 1. Evaluation of fungicides and bio-control agent against *C. fimbriata*

Treatments	*Trade Name*	*Radial Growth (mm)*	
		14th day	*28th day*
Metalaxyl 35 per cent WS @ 0.1 per cent	Tagron	6.4[b]	18.2[cd]
Thiophanate methyl 70 WP @ 0.1 per cent	Roko	0.0[a]	0.0[a]
Copper hydroxide 87.5 WP @ 0.3 per cent	Kocide	0.0[a]	0.0[a]
Carbendazim 50 per cent WP @ 0.1 per cent	Bavistin	0.0[a]	0.0[a]
Propiconazole 25 EC @ 0.1 per cent	Result	0.0[a]	0.0[a]
Hexaconazole 5 per cent SC @ 0.1 per cent	Contaf plus	0.0[a]	0.0[a]
Hexaconazole 4 per cent + Zineb 68 per cent WP @ 0.1 per cent	Avtar	0.0[a]	0.0[a]
Propineb 70 WP @ 0.1 per cent	Antracol	0.0[a]	0.0[a]
Captan 70 per cent + Hexaconazole 5 per cent WP @ 0.1 per cent	Taqat	0.0[a]	0.0[a]
Metalaxyl 8 per cent + Mancozeb 64 per cent @ 0.1 per cent	Master	0.0[a]	0.0[a]
Azoxystrobin 23 per cent @ 0.05 per cent	Amistar	6.8[bc]	17.4[c]
Difenoconazole 25 EC @ 0.1 per cent	Score	0.0[a]	0.0[a]
Dinocape 48 per cent EC @ 0.1 per cent	Karathane	1.4[a]	5.8[b]
Tetraconazole 3.8 per cent @ 0.1 per cent	Domark	0.0[a]	0.0[a]
Iprovalicarb 5.5 + Propineb 61.25 @ 0.1 per cent	Melody duo	0.0[a]	0.0[a]
Myclobutanil 10 WP @ 0.1 per cent	Index	0.0[a]	0.0[a]
Chlorothalonil 75 WP @ 0.1 per cent	Kavach	10.2[d]	23.4[e]
Sulphur 80 WP @ 0.2 per cent	Kri-o-sulf	6.6[b]	14.6[c]
Trichoderma asperellum		4.8[b]	7.0[b]
Control		11.6[d]	26.6[f]
CD 0.05		1.92	3.02

Figure 3. Suppression of *C. fimbriata* by *T. asperellum*

Table 2. Radial growth (mm) of *C. fimbriata* at 20 days after inoculation

Doses (per cent)	*Thiophenate Methyl*	*Metalaxyle + Mancozeb*	*Propiconazole*	*Captan + Hexaconazole*	*Difenoconazole*	*Carbendazim + Mancozeb*	*Hexaconazole*	*Carbendazim*	*Propineb*	*Tetraconazole*	*Mycobutanil*	*Mancozeb*
0.10	0^{a}	0^{a}	0^{a}	0^{a}	12.2^{a}	0^{a}	0^{a}	0^{a}	0^{a}	0^{a}	0^{a}	0^{a}
0.05	0^{a}	0^{a}	0^{a}	0^{a}	10.8^{a}	0^{a}	0^{a}	0^{a}	0^{a}	0^{a}	0^{a}	0^{a}
0.02	0^{a}	40.2^{b}	4.8^{b}	0^{a}	18.2^{ab}	0^{a}	0^{a}	0^{a}	34.0^{b}	13.4^{b}	14^{ab}	10.8^{ab}
0.01	0^{a}	37.2^{b}	7.5^{bc}	2.6^{a}	16.4^{ab}	0^{a}	10.8^{b}	0^{a}	63.0^{c}	18.6^{c}	19.2^{ab}	19.6^{b}
0.005	0^{a}	46.2^{bc}	10.4^{cd}	9.2^{b}	17.6^{ab}	0^{a}	18.8^{c}	0^{a}	74.4^{cd}	22.2^{c}	15.2^{bc}	14.2^{b}
0.002	0^{a}	62^{cd}	18^{e}	19.8^{c}	20^{bc}	0^{a}	18.4^{c}	0^{a}	82.6^{cd}	37.4^{d}	27.0^{bc}	38.4^{c}
0.001	0^{a}	56.8^{cd}	21.2^{e}	25.2^{d}	35.6^{d}	0^{a}	28.0^{d}	0^{a}	82.2^{cd}	47.4^{e}	29.4^{bc}	33.8^{c}
0.0	75^{b}	76^{e}	74.2^{f}	72.6^{e}	74.6^{e}	77.6^{b}	77.2^{e}	72.2^{b}	82.6^{cd}	66.6^{f}	75.8^{d}	55^{d}

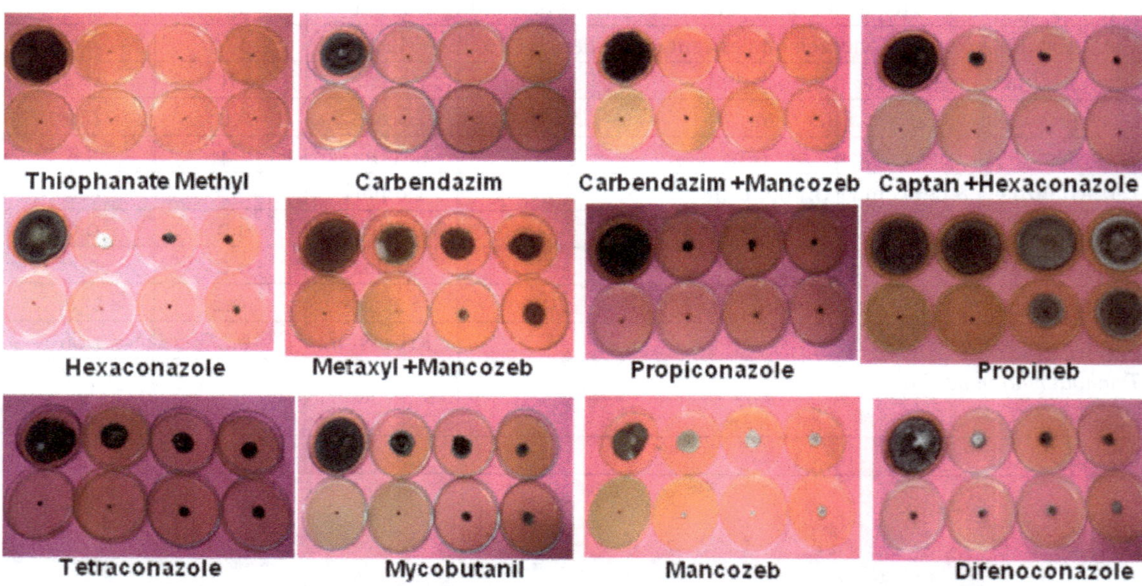

Figure 4. Effect of different fungicides on growth of *C. fimbriata* at different concentrations.

(Left to right - 1- Control, 2- 0.001 per cent, 3- 0.002 per cent, 4- 0.005 per cent, 5- 0.01 per cent, 6- 0.02 per cent, 7- 0.05 per cent, 8- 0.1 per cent)

Complete inhibition of *C. fimbriata* was observed in thiophanate methyl, carbendazim and carbendazim+mancozeb at 0.001 per cent concentration and in captan+hexaconazole and hexaconazole at 0.02 per cent concentration, whereas, metalaxyl + mancozeb, propiconazole, propineb, tetraconazole, mycobutanil and mancozeb were effective at 0.05 per cent concentration (Table 2, Figure 4). Difenoconazole was not found effective even at 0.1 per cent concentration however; it could suppress the growth of fungus to a certain extent.

Multi-Location Trials for the Management of Mango Wilt

Trees exhibiting typical initial symptoms of Ceratocystis wilt or decline were treated with suitable treatments at various locations of disease incidence during 2014 to 2017 (Table 3).

Management Schedules Applied

1. Thiophanate methyl as soil drench @ 150g/tree and spray with propiconazole @ 0.1 per cent.
2. Thiophanate methyl as soil drench @ 150g/tree and spray with carbendazim @ 0.1 per cent.
3. Carbendazim as soil drench @ 150g/tree and spray with propiconazole @ 0.1 per cent.
4. Trunk injection @ 5 g thiophanate methyl dissolved in 5 ml of water/hole along with soil drench @ 150g/tree and spray with carbendazim @ 0.1 per cent.
5. Treatment schedule repeated after 1 year as such in 15 replicates and other 15 trees were applied *Trichoderma asperellum* 100 ml suspension of 10^6cfu/ml + FYM 50-100 kg followed by neem cake 1.0 kg/tree 2 month later.
6. Soil drench treatment of thiophanate methyl @ 50 g repeated after 1 year.

Dried or wilted branches were pruned and cut ends were treated. Treated trees were given 50-100 kg FYM and 2 kg NPK mixture or 1.0 kg di ammonium phosphate and 0.5-1.0 kg of muriate of potash per tree. Treatments were repeated after one year. Observations on recovery were recorded from time to time. Overall, treated trees have shown very good recovery (Table 3).The beetle-infested trunks of wilted trees in experimental orchards were sprayed with chlorpyriphos 20EC @ 0.2 per cent followed by pasting of clay wetted with 0.3 per cent solution of chlorpyriphos and no exit of beetles was recorded from treated trunks.

Table 3. Multi-location trials for the management of mango wilt/decline (2014-2020)

Location	*Age of Orchard (Years)*	*Management Schedule Applied*	*Year of Application/ Repetition*	*No. of Treated Trees*	*No. of Recovered Trees (2019-20)*
Rehmankhera, Lucknow	30	1 and 5	2015	6	4
Pipari, Lucknow	25	1 and 5	2016	9	4
Bilauna, Aligarh	26	1, 2, 3 and 5	2015	30	28
Patarsa, Kanpur nagar	27	1 and 5	2014	3	3
Unchagaon, Bulandshahar	64	4 and 6	2015	3	3
Sitapur	80	1	2017	1	1
Sohawal, Faizabad	35	1	2017	50	46
Sahilamau, Lucknow	28	2	2017	2	1
Total				**104**	**90**

The following conclusions were drawn from multi-location trials:

- ❑ Sudden-wilt affected trees could not be saved, when treated after the leaf droop stage.
- ❑ The trees which were identified as diseased based on early symptoms of wilt or decline and immediately treated were survived.

Treatments Identified Best

- ❑ Thiophanate methyl soil drench @ 50-150g/tree and aerial spray with propiconazole @ 0.1 per cent.
- ❑ Thiophanate methyl soil drench @ 50-150g/tree and aerial spray with carbendazim @ 0.1 per cent.
- ❑ Tree trunk injection @ 5g thiophanate methyl dissolved in 50 ml of water/hole along with soil drench @ 50-150g/tree and aerial spray with carbendazim @ 0.1 per cent.

Efficacy of Newer Fungicides against *Ceratocystis fimbriata* and *Berkeleyomyces basicola*

Keeping in view the possible ban on the recommended fungicides (thiophanate methyl and carbendazim), the efficacy of newer fungicides was tested under laboratory conditions during 2020-21 to keep additional options for the management of *C. fimbriata* and for the management of recently identified mango root rot fungus, *Berkeleyomyces basicola*. The combination product azoxystrobin 11 per cent + tebuconazole 18.3 per cent WW SC @ 100 ppm was found highly effective against *C. fimbriata* (Table 4) and thiophanate methyl 75 per cent WP, propiconazole 25 per cent EC, hexaconazole 5 per cent EC and azoxystrobin 11 per cent + tebuconazole 18.3 per cent WW SC @ 100 ppm were found highly effective against *B. basicola* (Table 5).

Table 4: Radial growth (mm) of *Berkeleyomyces basicola* in PDA treated with various fungicides

Treatments	*Dose 100 ppm*		*Dose 200 ppm*	
	15 day	*30 day*	*15 day*	*30 day*
Control	37.4 [d]	80.8 [f]	37.4 [f]	80.8 [f]
Cholorothalonil 75 per cent WP	51.6 [e]	71.6 [e]	48.8 [g]	70.0 [e]
Bupirimate 26.7 per cent WW EC	36.2 [d]	54.0 [d]	33.6 [e]	48.8 [d]
Kresoxim-Methyl 15 per cent + Cholorothalonil 56 per cent WG	25.6 [c]	27.2 [c]	27.4 [d]	37.4 [c]
Azoxystrobin 18.2 per cent WW+ Difenoconazole 11.4 per cent WW SC	13.8 [b]	19.8 [b]	11.0 [c]	17.6 [b]
Difenoconazole 25 per cent EC	14.4 [b]	19.6 [b]	6.0 [b]	14.0 [b]
Azoxystrobin 11 per cent + Tebuconazole 18.3 per cent WW SC	1.8 [a]	3.0 [a]	1.6 [a]	2.0 [a]
Hexaconazole 5 per cent EC	0 [a]	0 [a]	0 [a]	0 [a]
Propiconazole 25 per cent EC	0 [a]	0 [a]	0 [a]	0 [a]
Thiophanate methyl 75 per cent WP	0 [a]	0 [a]	0 [a]	0 [a]
LSD (0.05)	2.84	4.92	2.95	4.27

Table 5. Radial growth (mm) of *C. fimbriata* in PDA treated with various fungicides

Treatments	*Dose 100 ppm*		*Dose 200 ppm*	
	15 day	*30 day*	*15 day*	*30 day*
Control	54.0[f]	73.0[f]	54.0[d]	73.0[d]
Kresoxim-Methyl 15 per cent + Cholorothalonil 56 per cent WG	19.4[c]	24.4[c]	26.8[c]	36.2[c]
Cholorothalonil 75 per cent WP	45.6[e]	65.4[e]	24.0[c]	30.8[c]
Bupirimate 26.7 per cent WW EC	40.4[d]	53.0[d]	25.2[c]	30.8[c]
Azoxystrobin 18.2 per cent + Difenoconazole 11.4 per cent WW SC	3.6[a]	12.0[b]	6.0[b]	15.0[b]
Azoxystrobin 11 per cent + Tebuconazole 18.3 per cent WW SC	0.0[a]	0.0[a]	0.0[a]	0.0[a]
LSD (0.05)	3.92	5.11	3.81	4.87

Based on results achieved, the integrated management package of practices against mango wilt and decline has been developed (Shukla 8). To avoid damage to roots minimum tillage should be adopted and deep ploughing should completely be avoided under the canopy. Intercrops should not be grown under canopy areas. Wilt affected and adjoining trees should be treated with thiophanate methyl 70WP @ 50-150 g per tree according to the age of trees as soil drench. Cut ends should be pasted with copper sulphate or copper oxychloride 5.0 per cent solution after pruning of infected or wilted branches. The aerial portion of trees may be sprayed with carbendazim 50 WP or propiconazole 25EC @ 0.1 per cent on the appearance of the first symptom. If infestation of Scolytid beetle is observed in an orchard, it should be managed by the spray of chlorpyriphos 20EC @ 0.2 to 0.3 per cent at 15 days interval. Irrigation must be done using a drip system or channels to avoid disease dispersal through water from an infected tree to healthy trees. Recommended doses of manure and fertilizers should be applied in orchards and proper soil moisture should also be maintained regularly.

Shoulder Browning and Sooty Blotch of Mango Fruits

The shoulder browning and sooty blotch of mango fruits have been a serious issue before growers wherever fruiting coincides with rainy season (Giblin *et al.*, 2; Shukla *et al.*, 9). The affected fruit lose their aesthetic value in the market and also suffers from severe rotting during ripening.

The Pathogens

Any of the pathogenic fungi active on mango tree may be found associated with shoulder browning symptoms but *Capnodium mangiferae, Colletotrichum gloeosporioides* and *Alternaria alternata* have been found most commonly associated. *C. mangiferae* use to grow superficially but deposited spores of *C. gloeosporioides* and *A. alternata* germinate during wet weather and cause infection in mature fruits resulting rotting at ripening and storage (Shukla *et al.*, 9).

Symptoms

The symptoms of shoulder browning appear in the form of browning or blackening of the fruits around the shoulder region uniformly or in a tear stain pattern. The first symptom of the disease has been observed to appear in the form of a removable deposit of rain-washed mycelium and spores of sooty mould and other fungi in the shoulder region of fruits after an average of 4.3 rainy days and 57.0 mm rainfall. Similarly, the first symptom of permanent staining of epicarp tissue has been recorded after an average of 6.4 rainy days and 136.8 mm precipitation (Shukla *et al.*, 9). Over 50 per cent of fruit surface area has been observed affected after 15.8 rainy days and 348.5 mm rainfall (Figure 5). Staining of fruits has been observed due to alternate wetting and drying of fruits and deposit of rain-washed fungal propagules and pollutants from the canopy.

Figure 5. Symptoms of shoulder browning.

Factors Affecting Disease Development

The weather has been found to play a key role in the dynamics of shoulder browning incidence. Spatio-temporal variations in rainfall played the most significant role in disease development. The difference in disease dynamics has been recorded among the dates and years because of the difference in time of onset of monsoon, the number of rainy days and cumulative precipitation. Disease severity has been recorded uniform on different cultivars of mango growing under similar conditions. Variations in sooty mould severity on the canopy of mango have been found responsible for variation in severity of the disease among the orchards (Shukla *et al.*, 10).

Incidence and severity of the disease had a significant negative correlation with maximum temperature due to a gradual decrease in maximum temperature during the monsoon period. The correlation of disease with relative humidity was observed positive and significant. Thus, high relative humidity certainly had a positive effect on disease development. Temperature and relative humidity fluctuate according to the distribution and frequency of rains and thus, disease dynamics have also been observed accordingly.

Management

Shoulder browning disease management has great significance in minimizing postharvest rot and for maintaining the aesthetic value of fruits. The efficacy of fungicides was evaluated against shoulder browning and difenoconazole (0.05 per cent) was the most effective (Figure 6) treatment followed by Metalaxyl + mancozeb (0.2 per cent), myclobutanil (0.1 per cent) and propiconazole (0.1 per cent) respectively (Shukla *et al.*, 11). Based on its Codex MRL value of

0.07 mg/kg in mango, the pre-harvest interval for difenoconazole of 35 days was suggested after spraying at 0.5 ml/l of water dose. Besides chemical spray, bagging of fruits was also found useful in minimizing shoulder browning as well as postharvest rot. The limitation in management of shoulder browning is its incidence during monsoon period, therefore, bagging of fruits or one prophylactic spray with fungicide before the onset of monsoon has been recommended.

Figure 6. Management of shoulder browning (a. Bagged fruits. b. Bagged fruits at harvest, c. Difenoconazole, d. Untreated).

Extension of Technologies

The efforts have been made to transfer the knowledge generated about newly emerged diseases through research publications, extension folders, training programmes, seminars, print, electronic and social media; and agro-advisory service of the ICAR-CISH, Lucknow. Many farmers throughout the country have received a correct diagnosis and proper advice by searching the name of the expert and mobile number from the Institute Website, followed by sending photographs of diseased plants. Many farmers, however, continue to rely on the lateral expansion of technologies.

Conclusion

Due attention has been paid on the newly emerged diseases of mango and management measures have been developed. However, the late manifestation of symptoms of wilt disease, when protection of trees becomes almost impossible, is a major constraint in its management, therefore, efforts have been initiated to develop primers for the early detection of pathogens in orchard soil to avoid severe losses. Future research should be done to refine wilt, decline, and root rot management technology by getting additional effective treatments while reducing input costs. Efforts must also be made to develop a cost-effective solution for twig drying disease.

REFERENCES

1. FAO. 2017. FAOSTAT online database at http: //www.fao.org/default.htm.
2. Giblin, F.R., Coates, L.M. andIrwin, J.A.G. 2010. Pathogenic diversity of avocado and mango isolates of *Colletotrichum gloeosporioides* causing anthracnose and pepper spot in Australia. *Aus. Pl. Pathol.*, 39: 50-62.

3. Goitia, W. and Rosales, C.J. 2001. Relacion entre la incidencia de escolitidos y la necrosis del cacao en Aragua, Venezuela. *Manejo Integrado de Plagas*, 62: 65-71.
4. Grosclaude, C., Olivier, R. and Romiti, C. 1995. Chancre colore du platane. Comment l'agent responsable peut survivre dans le sol. *Phytoma*, 479: 41-42.
5. Gundappa, Adak, T. and Shukla, P.K. 2016. Status of mango wilt vector, *Hypocryphalus mangiferae* in India. *Indian Phytopath.* 69 (4s): 361-363.
6. Horticultural Statistics at a glance. 2018. Ministry of Agriculture. agricrop.nic. in>sites>default>files
7. Malik, M.T., Dasti, A.A. and Khan, S.M. 2004. Some manageable predisposing factors of collar/stem rot of mango (*Mangifera indica* Linn.). *Pak. J. Phytopathol.*, 16: 37-42.
8. Shukla, P.K. 2017. Mango wilt disease management. Extension folder, ICAR-CISH, Lucknow publication, pp. 1-4.
9. Shukla, P.K., Adak, T., Misra, A.K., Singh, A. 2016. Appraisal of shoulder browning disease of mango (*Mangifera indica* L.) caused by fungus complex in subtropical region of India. *J. Mycol. Pl. Pathol.* 46: 38-46.
10. Shukla, P.K., Baradevanal, G., Rajan, S. and Fatima, T. 2020. MaxEnt Prediction for potential risk of mango wilt caused by *Ceratocystis fimbriata* Ellis and Halst under different climate change scenario in India. *Journal of Plant Pathology* https: //doi.org/10.1007/s42161-020-00502-9
11. Shukla, P.K., Bhattacharjee, A.K. and Dixit, A. 2018. Field efficacy of difenoconazole against shoulder browning disease of mango (*Mangifera indica* L.) and its residue analysis for safety evaluation in mango fruit. *Indian Phytopathology* 71(1): 147-151. DOI 10.1007/s42360-018-0008-0.
12. Shukla, P.K., Fatima, T. and Gundappa. 2019. Efficacy of sterilized culture filtrate of *Trichoderma harzianum* Rifai against mango wilt pathogen, *Ceratocystis fimbriata* Ellis and Halst. *J. Eco-friendly Agriculture*, 14(2): 69-71.
13. Shukla, P.K.,Fatima, T. and Kumari, N. 2020. First report of *Berkeleyomyces basicola* causing mango root rot and decline in India. *Plant Disease* 105(4): 1214, doi: 10.1094/PDIS-10-20-2133-PDN.
14. Shukla, P.K., Fatima, T. and Misra, A.K. 2019. Recent trends in management of major diseases of mango. In: *Recent Trends in Plant Disease Management*, Sheela, Chourasia, H.K., Sahu, K.K., Jha, R.N. and Prasad, G. (Eds.), Today and Tomorrow's Printers and Publishers, New Delhi, ISBN 10: 81-7019-644-5/ISBN 9788170196440, pp. 35-68.
15. Shukla, P.K., Varma, S., Fatima, T., Mishra, R., Misra, A.K., Bajpai, A., Gundappa and Muthukumar, M. 2018. First report on wilt disease of mango caused by *Ceratocystis fimbriata* in Uttar Pradesh. *Indian Phytopathology* 71(1): 135-142. DOI 10.1007/s42360-018-0021-3.
16. Teviotdale, B.L. and Harper, D.H. 1991. Infection of pruning and small bark wounds in almond by *Ceratocystis fimbriata*. *Plant Disease*, 75(10): 1026-1030.

Chapter 22

New Development in Prolonging Storage of Fruits and Vegetables

Surajit Mitra

Department of Post Harvest Technology, Bidhan Chandra Krishi Viswavidyalaya (State Agricultural University), Mohanpur 741252, Nadia, West Bengal

e-mail: drsurajitmitra@yahoo.co.in, mitra.surajit@bckv.edu.in

ABSTRACT

Desirable dietary intake of nutrients as per the requirement is imperative to maintain good health. Both the macronutrients like carbohydrates, proteins and fats, and micronutrients like vitamins and minerals along with some phytochemicals are required for formulating a balanced diet to lead a healthy and active life. Some of the fruits and vegetables rich in protein, vitamins, minerals and anti-oxidants, and low in carbohydrates, fats and anti-nutrients seem to help in reducing detrimental effects of high fat and high sugar diets to the hippocampus, a major component of the brain. Being protective food, different fruits and vegetables are gaining popularity in almost all over the world as nutrition-security crop to fight oxidative stress, improve general cognition, reduce cholesterol levels and help in maintaining healthy blood pressure levels. The plant foods including fruits and vegetables are only source of dietary fibre, which is essential for bowel movement, growth of gut bacteria and for controlling blood sugar and lipids. Fruits and vegetables having high β-carotene and anthocyanins are also gaining importance as antioxidant foods to tackle the problem of vitamin A deficiency, improving visual acuity and liver function, and to prevent the obesity, heart disease and diabetes. But high moisture content and continuation of metabolic activities as living commodities even after harvest cause huge postharvest losses of fruits and vegetables. In spite of huge production of about 300 million tonnes of fruits and vegetables in India, there is a deficit of 20-25 per cent in per capita availability due to postharvest losses from farm to fork. Globally, about one-third of all food that is produced each year is still wasted, and approximately, 45 per cent of all fruits and vegetables grown are not consumed, causing loss of billions of dollars mainly due to inefficient value chains in developing countries. In this paper, an attempt has been made to discuss a brief account of various modern postharvest management systems including typical value chains for perishables, measures to address the inevitable problem from the base, handling operations starting from harvesting to consumption taking pre-harvest factors into account, use of postharvest tools, nanoparticles, active packaging, improved storage techniques, non-thermal innovative processing technologies including cold plasma technology for microbial decontamination without affecting nutritional and sensory characteristics of foods, pulsed electric fields, ultrasonication and high pressure processing for prolonging shelf life of fruits and vegetables to reduce the loss of harvested produce and thereby to boost farmers' income for lifting rural economy, and to increase the affordability of fruits and vegetables for alleviating global malnutrition. Gene editing to extend shelf-life without loss of quality can also be a promising new avenue for reducing postharvest loss of nutrient-dense fruits and vegetables. Extending storage life of fruits and vegetables with the implementation of improved postharvest technologies and proper monitoring will not only help in cutting food waste and loss, but will be helpful in preventing wastage of natural resources used by the harvested produce, and also in reversing global warming by reducing greenhouse gas emissions due to decomposition of lost and wasted food.

***Keywords**: Fruits and vegetables, Improved techniques, Prolonged storage, Loss reduction.*

Fresh fruits and vegetables as major sources of essential vitamins and minerals constitute important parts of the human diet. As the animal foods do not contain any fibre, we solely depend on plant foods, especially different fruits, vegetables and tubers for dietary fibre, mostly the undigested cellulose which is essential for bowel movement, and some are prebiotic which promote growth of gut bacteria and help in controlling blood sugar and lipids. Dietary Allowance should be based on the energy requirement of an individual in terms of the level of energy intake from a food that balances energy expenditure, and bioavailability of a given nutrient from a diet. Inclusion of nutrient-dense fruits and vegetables in regular diet is required for human wellbeing. The attributes of quality of fruits and vegetables like attractive skin colour, excellent flavour, taste, texture, and nutritional properties are associated with consumer acceptance. Maintenance of the quality of harvested fruits and vegetables is very challenging because of their high perishability and short shelf life due to high moisture content, soft texture, high physiological activities and susceptibility to many postharvest diseases and pests. Postharvest deterioration of horticultural produce including fruits and vegetables starts immediately after harvest because of the continuation of physiological processes like transpiration and respiration at the cost of stored water and food. Unlike the plants in the field, compensation of water lost by transpiration, and food degraded by respiration is not possible in the harvested produce. Rapid loss of water after harvest is one of the most serious postharvest problems in maintaining freshness of fruits and vegetables. Huge quantities of the harvested fruits and vegetables are lost due to high perishability on account of high moisture content, and limited application of postharvest technology in developing countries.

Considering the importance of extension of shelf life of fruits and vegetables for reducing huge postharvest losses, in this paper, an attempt has been made to highlight major causative factors like mechanical damage, physiological deterioration and microbial spoilage of fruits and vegetables with proper postharvest handling systems including improved storage techniques like controlled atmosphere and hypobaric storage, irradiation, emerging technologies like smart packaging including active and intelligent packaging, nanotechnology, and non-thermal innovative processing technologies including cold plasma technology, pulsed electric fields, ultrasonication and high pressure processing for reducing loss and maintaining quality of harvested produce, and to make them available all the year round either in fresh or processed form, enhance food and nutrition security, and grasp the opportunity with both hands for economic development. Gene editing to extend shelf-life without loss of quality is also being tried as a promising new avenue for reducing postharvest loss of nutrient-dense fruits and vegetables. Promotion and proper implementation of new technologies for extending storage life of fruits and vegetables will not only help in reducing postharvest losses, but will also be helpful in preventing wastage of natural resources used by the harvested produce. In addition, if food loss and waste is checked, the harmful effect of greenhouse gas in contributing to global warming can be reduced by retarding its emissions due to the decomposition of lost and wasted food.

Extent of Postharvest Losses

It is estimated that globally the food loss and waste in horticulture sector accounts for 40 to 50 per cent of the total production, and some 2 billion people, about 30 percent of the planet's population suffer from micronutrient deficiencies or other effects of inadequate diet (FAO, 16). In general, the food losses are higher in horticultural crops, illustrated by the post-harvest losses of 52 per cent in cauliflower, 46 per cent in tomato, 20 per cent in mandarin, 29 per cent in banana and 38 per cent in mango which are produced in the SAARC nations (FAO, 15). India is the world's second largest producer of fruits and vegetables next only to China with a total production of about 300 million tonnes in 2019-2020. Yet agriculture, which makes up over 55 per cent of the workforce, generates just 15-16 per cent of GDP. In India, improper postharvest handling, inadequate cold supply chain, lack of cold storage facilities and insufficient food processing units lead to 20-40 per cent losses of harvested fruits and vegetables causing a monetary loss of about Rs. 2 lakh Crores every year. Although, food processing sector is one of the largest sectors in India in terms of production, growth, consumption, and export with a total value of all exported processed foods at about Rs. 37,000 Crores (USD 5000 Millions) including Rs. 11,375 Crores from processed fruit and vegetable products in 2020-21 (APEDA, 4), fruit and vegetable processing industry are still in rudimentary stage with an estimated use of 7-9 million tonnes of fruits and vegetables comprising only 2.5-3.0 per cent of the total production. In spite of the second largest producer of fruits and vegetables in the world, India is far behind from the countries like Malaysia (83 per cent), Philippines (78 per cent), Brazil (70 per cent), USA (65 per cent), Thailand (30 per cent) and China (23 per cent) in value addition and shelf life extension of fruits and vegetables. In the financial year 2019-20, 8.31 million tonnes of fruits and vegetables were processed in India, which is expected to reach 16.39 million tonnes by 2024, expanding at a CAGR of 14.84 per cent during 2020- 2024 period (Business Wire, 9). So, there is a huge scope for expansion of cold chains and food processing industry in India for extending shelf life and reducing postharvest losses of fruits and vegetables.

Types of Postharvest Loss

Food Loss and Food Waste

Food losses refer to the decrease in edible food mass available for human consumption throughout different segments of the supply chain. In addition to quantitative losses, food products can also face a deterioration of quality, leading to a loss of economic and nutritional value. Food waste refers to food losses resulting from decisions to discard food that still has value. Food waste is most often associated with the behaviour of retailers, the food service sectors and consumers, but food waste and losses take place all along food supply chains (FAO, 14). Though wasted food refers to the discarded food due to expiry of date or other reasons, especially in developed countries, it is an integral part of total food loss, and the terms food waste and food loss are interchangeably used. Food loss does not only mean the loss of physical foods, rather it accounts for the waste of water, land, fertilizer, chemicals, labour, money, time, energy and other resources involved in the production of the lost food (Mitra, 29).

Among different kinds of postharvest loss, some important losses are -

1. **Qualitative loss:** Qualitative loss includes loss of caloric and nutritive value, loss of acceptability by consumers, and loss of edibility of the produce which is more difficult to measure than quantitative loss of fresh commodity.
2. **Quantitative loss:** Quantitative loss refers to the physical loss of a produce that can be measured in terms of percentage or any other units. This loss is a higher priority than qualitative loss in developing countries.
3. **Psychological loss:** This loss is due to non-acceptance of consumers to new kinds or varieties of crops due to a feeling of repugnance towards a particular crop with a desire to avoid. Human aversion or disapproval and religious taboos accentuate this type of losses.

Losses at different Points

Postharvest loss of fruits and vegetables can occur at different points starting at harvest in the field with mechanical injuries, at packinghouse during preparation for packing, during transport due to loading and unloading, compression, decay, bruising and deformation, during processing due to removal of skin, seeds, core and pomace after extraction of juice, during storage due to chilling injuries, decay, peel disorders, attack of pest and rodents, at market due to wilting, discolouration, over ripening, softening and decay, and finally losses at consumers' home due to wilting, over ripening, softening and decay of produce purchased more than the requirement of a family.

Causes of Postharvest Deterioration and Losses

In most of the developing countries, postharvest losses of highly perishable fruits and vegetables due to high moisture content generally ranges from 30- 40 per cent, and in some cases, it goes up to 80 per cent as against 5-20 per cent in developed countries depending upon the crop, cultivar, season and handling systems. Not only producers but also different players or actors involved in postharvest handling systems of horticultural produce must understand that harvested fruits, vegetables and other horticultural crops are living commodities and continue their metabolic activities causing deterioration in quality, which leads to senescence, the ultimate death of cells leaving the produce inconsumable or unusable. Along with the physiological processes, the microbial and environmental factors involved in deterioration also need to be understood clearly. Knowing the causes of postharvest deterioration of horticultural crops, use of appropriate postharvest techniques can delay this deterioration and extend the shelf-life of harvested produce maintaining desirable qualities. Maintenance of quality of fruits and vegetables has also direct relation with the food safety. Food safety has become a new concern in today's world due to awareness among the consumers, and active participation of various regulatory agencies (Boindala and Lewis, 7). The agents responsible for food borne diseases include bacteria, viruses, parasites, toxins, various chemical substances including pesticide residues and adulterants. Food safety cannot be achieved by responsibility of single industry or regulatory agency but the role of everyone is important as shown in Figure 1.

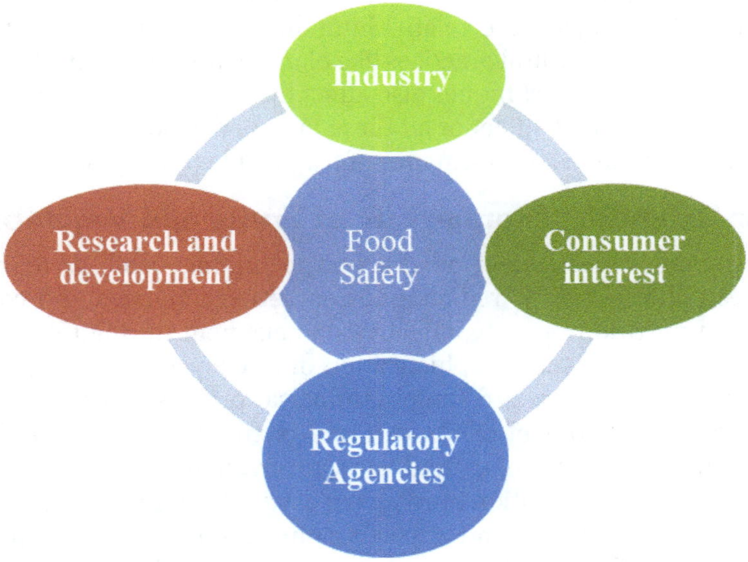

Figure 1. Inter-relationship of different components effecting food safety.

Factors Responsible for Loss of Fruits and Vegetables

1. **Physiological or internal factors:** Causes of deterioration include respiration rate, transpiration, ethylene production and action, rates of compositional changes affecting colour, texture, flavour, and nutritive value, mechanical injuries, water stress, sprouting and rooting, and other physiological disorders. The rate of physiological deterioration also depends on several environmental factors.

2. **Microbial or biological factors:** Different microorganisms, mainly bacteria and fungi including moulds and yeasts deteriorate the quality and market value of horticultural crops by causing various postharvest diseases. Apart from microbes, some pests and rodents are also responsible for postharvest losses.

3. **Environmental or external factors:** These causes include temperature, relative humidity, air velocity, atmospheric composition comprising concentrations of oxygen, carbon dioxide and ethylene, and sanitation procedures.
4. **Socioeconomic factors:** Although the internal and external causes of postharvest losses are well understood and many technologies have been developed to reduce these losses, they have not been implemented properly due to some socioeconomic factors like inadequate facilities for packaging, transportation, storage and marketing.

Apart from the physiological, microbial and environmental causes, some of the important reasons responsible for postharvest losses are lack of knowledge of the growers and handlers about postharvest management systems, improper packaging, inadequate transport, inefficient marketing, insufficient cold chains, lack of government regulations, inadequate supply of tools and equipment for harvesting and postharvest handling of fresh commodities and negligence in postharvest research and development. Furthermore, some pre-harvest factors such as environmental factors like temperature, relative humidity, rainfall, light, frost *etc.*; genetic factors like type of crop and cultivar; cultural practices like planting time, planting density, fertilization, irrigation and plant protection techniques also affect the postharvest quality and shelf life of a produce. Actually, the combined effect of all these factors decides the rate of deterioration in quality, spoilage and postharvest performance of fruits and vegetables.

Strategies for Extending Storage Life of Fruits and Vegetables

For achieving main objectives of postharvest technology such as maintaining quality of produce, protecting food safety, reducing losses between harvest and consumption, and promoting food and nutrition security, effective management during the postharvest period is more important than the level of sophistication of any given technology. One of the aims of postharvest technology is to retard the physiological processes for extending shelf-life of harvested produce. Proper control of the external factors that influence the physiological processes by reducing mechanical injury and temperature, maintaining high relative humidity, manipulating atmospheric composition by reducing oxygen level and increasing carbon dioxide, reducing accumulation of ethylene, and imparting physical barrier by waxing or using plastic films can delay the tissue disintegration or senescence phase and extend the shelf-life of fruits and vegetables. Proper control of both the physiological and external factors having direct effect on the ripening, that occurs at the cellular level involving a series of anabolic and catalytic reactions including degradation of chlorophyll, production of carotenoids and anthocyanins, conversion of starch to sugars, changes in organic acids, proteins and fats, development of flavours, reduction in tannins and fungistatic compounds, softening of cell wall, and finally collapse or death of cells, can retard the rate of deterioration and prolong the storage of fruits and vegetables.

Steps of Postharvest System

There are many interacting steps involved in any postharvest system. Harvested commodities are often handled by many people, transported and stored repeatedly between harvest and consumption. While particular practice and the sequence of operations vary for each crop depending on the nature of the crop and intended use, there is a general series of steps in postharvest handling system as shown in Figure 2.

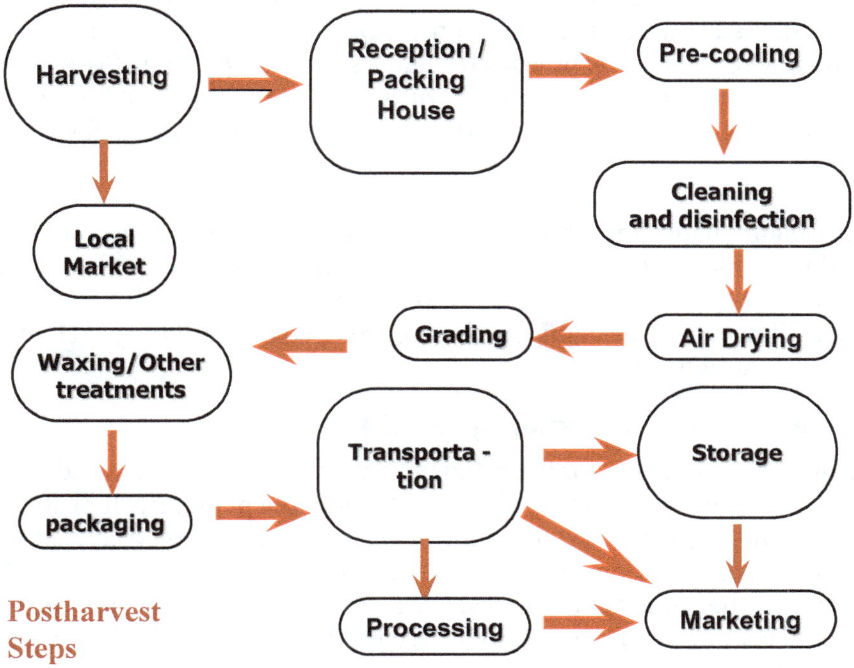

Figure 2. General steps in postharvest handling system. (*Source*: Mitra, 29).

Harvesting

For harvesting a quality produce, care must be taken right from the selection of good genetic materials for planting to proper harvesting of the crop including good agricultural practices (GAP) in-between. If the quality of any crop is not good at the time of harvest, there is no way to improve its quality after harvesting. Harvesting of fruits and vegetables at proper stage of maturity is very important for its subsequent quality, storage and marketable life. Different maturity indices standardised based on different methods like physical (skin colour, shape, size, firmness and aroma), chemical (sugar, starch and acidity), physiological (rate of respiration), computation (heat unit or degree days), and electronic and other methods (electronic colour sorter, and acoustic and vibration test) should be used for judging proper maturity of fruits and vegetables. During harvesting, care should also be taken to avoid mechanical injury by following suitable methods of harvesting either by hands or machines or robots (Mitra, 29).

Precooling

Precooling is a very important step in the postharvest handling system of perishables for rapid removal of field heat *i.e.* the difference in temperature between the harvested crop and the optimal storage temperature of that produce. Precooled fruits and vegetables exhibit relatively slower loss in weight and decline in firmness as compared to the non-precooled. This may be due to the fact that loss of moisture from pre-cooled commodities is slower if higher relative humidity is maintained in the storage atmosphere. Doshi and Sutar (13) reported an additional extension of five days in shelf life of precooled banana stored at 13.5°C and 80-85 per cent relative humidity. To reduce the field heat, harvested fruits and vegetables should be kept under shade in the field immediately after harvest until they are transported from the field.

Packaging

Proper packaging plays important roles in containment, protection, communication and convenience during movement from one place to another, storage and marketing of fresh commodities as well as processed products. In addition to various functions of traditional packaging, many innovative packaging methods developed for extending shelf life of fruits and vegetables are discussed below.

Modified Atmospheric Packaging (MAP)

With the increase in demand for fresh and natural products without the addition of harmful chemicals, modified atmospheric packaging (MAP) is gaining popularity for extending the shelf life of fresh fruits and vegetables without much affecting their fresh-like appearance. Modified atmosphere packaging works on the concept of reducing the level of oxygen (O_2) inside the packing bag and at the same time increases the concentration of carbon dioxide (Veljoviæ *et al.*, 40). Edible coatings composed of polysaccharides, proteins, lipids and composites are also applied directly on the food surface by dipping, spraying or brushing to create a modified atmosphere as alternative to the protective waxy coatings to provide a barrier to moisture, oxygen, and solute movement for prolonging storage life of whole as well as minimally processed fruits and vegetables.

The atmosphere inside a package can be modified either by passive or active means. In case of passive modification, the rate of change and the final gas composition in the package depend largely on both the packaged product and the permeability of the packaging material. Packaged fruits and vegetables consume the oxygen of the surrounding environment and release carbon dioxide during respiration resulting in reduced oxygen concentration over time, which may cause anaerobic respiration and fermentation, deteriorating the quality of packaged produce. Considering the limitations of the passive atmosphere modification in achieving the desired atmosphere very slowly and chances of uncontrolled levels of oxygen, carbon dioxide or ethylene with a detrimental effect on the quality of the packaged produce, active modification of the atmosphere by 'smart packaging' is gaining importance to provide a solution to this problem.

Smart Packaging

It is an emerging technology in the food packaging industry, which incorporates both active and intelligent packaging. Smart packaging can meet the demands for fresh and safe food in better way than traditional packaging by transforming packaging from a passive to active state. Smart packaging is a collective term used to group some emerging food packaging technologies referred to as 'active packaging', and 'intelligent packaging'. The purpose of adding active packaging functionality to a packaging system is to improve the protective functional value *e.g.*, antimicrobial materials or oxygen absorbers. The purpose of adding intelligent packaging functionality to a packaging system is to improve its epistemic value providing additional information about the product *e.g.*, time temperature indicators or ripeness indication (Robertson, 33). One significant area of technological overlap with active and intelligent packaging is that of nanotechnology applied to the food packaging industry. Nanotechnology is a platform technology and can be utilized in packaging to achieve the aims of smart packaging but has potential to pose health risks (Alfadul and Elneshwy, 2). It must also be noted that the use of nanotechnology in food processing or packaging has the potential to cause significant consumer concerns (Siegrist *et al.*, 36). The main concepts and types of active and intelligent food packaging, focusing on recent progress and new trends using biodegradable and biobased polymers were also described by Salgado *et al.* (34). For commercial success, food products utilizing innovative active or intelligent packaging technologies must be accepted by consumers. Different innovative packing systems are shown in Figure 3.

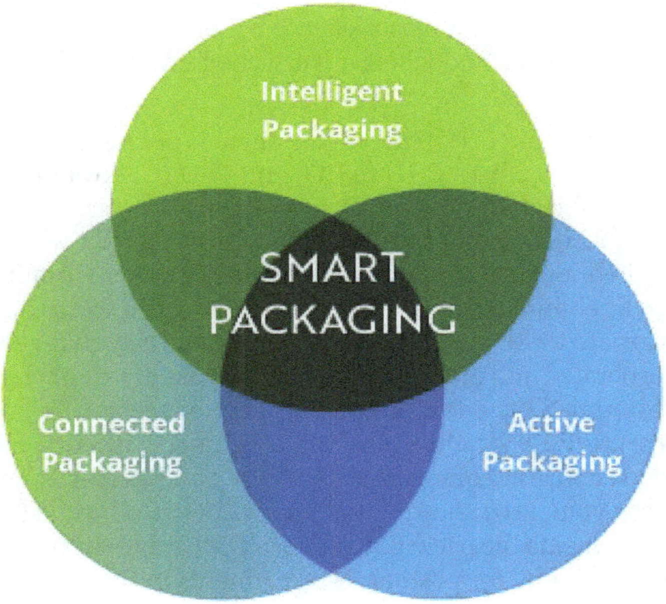

Figure 3. Innovative packaging systems.

Active Packaging

In active packaging, atmospheric modification is usually accomplished by creating a vacuum and then incorporating the desired gas mixture in the package. Compared to the passive method, active atmospheric modification is an instant method and takes place at the beginning of storage. The atmosphere then remain practically unchanged, provided that the proper barrier material is used and there is no leakage through pinholes or poor seals. Three main gases used in MAP are nitrogen (N_2), oxygen (O_2), and carbon dioxide (CO_2). Nitrogen is primarily used to displace oxygen and prevent package collapse. A low oxygen concentration is maintained for the packaged fruits and vegetables in order to sustain their basic process of aerobic respiration. And carbon dioxide is used for its bacteriostatic effect and the ability to slow down the respiration of many products. However, the optimum level of each gas for each commodity must be determined and used in order to maximize the positive effects. In some cases, additional gases are used in combination with the above mentioned gases. Carbon monoxide (CO) is sometimes added to inhibit microbial growth but due to toxicity, its application is limited. Similarly, sulfur dioxide (SO_2) may be used to prevent oxidative browning and to control the growth of bacteria and moulds. Ethanol has also been used to enhance the firmness of tomatoes, improve flavour, and reduce fungal activity. Active packaging techniques for extending shelf life, maintaining quality and safety of foods can be divided into two categories; absorbing system (scavengers) and releasing system (releasers). Absorbing (scavenging) systems remove undesired compounds such as oxygen, carbon dioxide, ethylene, excessive water, taints *etc.* Releasing systems actively add or emit compounds like carbon dioxide, antioxidants and preservatives to the packaged food or into the head-space of the package. Depending on the physical form of active packaging systems, absorbers and releasers can be a sachet, label or film type. Sachets are placed freely in the head-space of the package and labels are attached into the lid of the package avoiding direct contact with food.

The most common ethylene (C_2H_4) scavenging agent is potassium permanganate ($KMnO_4$) which in presence of water, oxidizes C_2H_4 to CO_2 and H_2O (Gaikwad *et al.,* 18). Warsiki (41) developed chitosan films activated with $KMnO_4$ to wrap tomatoes that managed to prolong their shelf-life. Nevertheless, $KMnO_4$ cannot be utilized in direct contact with foods due to its high toxicity and has a limited long-term efficacy in high-moisture environments (Yildirim *et al.,* 43). Ethylene can also be eliminated by physical adsorption on active surfaces such as activated carbon, carbon impregnated with metal catalysts or activated clays like zeolite, vermiculite and montmorillonite incorporated into the packaging film. The use of 1-methylcyclopropane (1-MCP) is another option to diminish the effect of C_2H_4. Its action mechanism implies its irreversible union to the C_2H_4 receptor in fruit, thus blocking C_2H_4 effects by competitive inhibition. It is encapsulated in cyclodextrins in commercial products and releases when the formulation is mixed with water. Ortiz *et al.* (30) showed that pillows prepared with soybean proteins could use the relative humidity inside the package to release 1-MCP and delay tomatoes softening and suggested that it could be useful for postharvest treatments in the own package.

Intelligent Packaging

Intelligent packaging is one type of smart packaging that may include sensors for detecting the freshness or temperature of the packaged contents. Freshness or time/temperature indicators or sensors can be integrated inside or outside the packaging to show retailers and customers if an important limit has been exceeded. An intelligent package can carry out many intelligent functions, like detecting, registering, locating, communicating and applying scientific logics to help in decision-making, extending shelf-life, maintaining quality, improving safety, providing information and indicating possible problems with the help of sensors. Unlike active packaging, intelligent packaging is not aimed at releasing components onto the food or headspace to act directly for extending shelf-life of food, instead, it conveys information about food quality to manufacturers, handlers, retailers, and consumers.

The components, which are attached as labels or incorporated or printed on the food packaging material, offer better possibilities to check product quality, track critical items and provide more detailed information during all the food supply chain including storage, transport, distribution and sale (Lee *et al.,* 25). They can also inform about product history such as storage conditions, composition of headspace, microbial growth, *etc.* (Realini and Marcos, 32). Halonen *et al.* (21) described the development of biobased antimicrobial films and sensors to be used as indirect indicators of food spoilage.

Time–Temperature Indicators (TTIs)

TTIs are typically small self-adhesives for attaching, or small devices for incorporating into the package that provide a visual indication, most commonly by changing colour or moving colour front, as a response to temperature history during distribution and storage. These indicators may be placed in the shipping container or in the individual consumer package. There are three basic types of TTIs- full history (those that start integration immediately once activated), partial history (those that start integration when a certain temperature threshold is achieved), and critical temperature or abuse indicators (that show a change if a certain temperature is achieved and whose response does not depend on the time the product has been packaged). The major mechanisms on which the TTIs are based include enzymatic reaction, polymerisation, melting point and diffusion of a substance. In the first case, once the indicator is activated, an enzymatic reaction takes place causing a pH change and a consequent colour change. Indicators that work based on polymerisation have a polymer coating that gets darker at a rate that depends on the temperature. The colour of the polymer is compared to a colour reference surface printed on the indicator. These indicators are usually self-activated, thus requiring deep low temperature storage before use. Three commercially available TTIs namely MonitorMark, CheckPoint and Fresh-Check work on the principle of diffusion, enzymatic reaction and solid state polymerisation respectively. The TTIs in the fruits and vegetables sector have an interesting application in minimally processed products, such as fresh-cut fruits, vegetables and salads, since these fresh-cut products require a very good temperature control for maintaining quality and safety.

Connected Packaging

It is an innovative packaging, finds applications in almost all retail product categories, including toys, cosmetics, games and clothing. It allows consumers to interact with a product through a code on the package that can be activated with a mobile device. Connected packaging generally used by brands to deliver exclusive content to its customers, can also be used for high value fruits and vegetables to transform simple packaging into digital experience, through a mobile phone tap on a QR code.

Nanotechnology in Packaging

Application of **nanotechnology** in postharvest management of horticultural crops by nanocomposite packaging, nanoencapsulation and nanoemulsion coating of fruits and vegetables can be useful in extending shelf-life of perishables by delaying ripening and/or inhibiting the growth of pathogenic microorganisms. Nanocomposites are materials that incorporate nanoparticles into a matrix of base material. Polyvinyl alcohol (PVA), polyethylene terephthalate (PET), polystyrene (PS) and polyethylene (PE) are some of the polymer matrices generally used in preparation of nanocomposites. The polymer materials are reinforced with nano sized particles of silver (Ag), titanium dioxide (TiO_2), silicon dioxide (SiO_2) also known as silica, and clay or layered silicate to develop nanocomposites. The size of the reinforced material has direct impact on the property of the nanocomposites, as extremely high surface to volume ratio of nanoparticles dramatically changes their properties. Among various applications of nanocomposites, one use as oxygen and gas barrier is very important, and in food packaging, these composite materials are used for blocking CO_2 and water vapour, and removal of ethylene from the packaging material to enhance the shelf life of fruits and vegetables. Electrostatic interactions between nanocomposite and bacterial cell wall or fungal cell membrane destroy the cell structure and increase the cell membrane permeability leading to the leakage of intracellular substances and impairment of vital microbial activities. Nanoparticle based packaging has many advantages over conventional packaging in terms of antimicrobial properties, barrier properties and greater strength. Nano-titanium dioxide (TiO_2) with photocatalytic capabilities can oxidize ethylene into carbon dioxide and water, thus preventing loss due to excessive ripening. Sami *et al.* (35) reported that the chitosan/nano-silica with the addition of nisin coating extended shelf-life of fresh-cut cantaloupe up to 8 days at 4°C by forming semi-films against microorganisms and enzymes. Nano-silicon oxide (SiO_2) coating has also been used in apples for extending shelf life.

Transportation

An ongoing effort has to be made by both public and private sectors to implement innovative measures for adding value to the transportation sector and making it capable of appropriately responding to the growers' demands. Due importance should be given to transport systems for transportation of fruits and vegetables to local market, packing house, processing units, store house, long distant market, and also for export by air and sea transport. Refrigerated containers and trailers need to be customarily used for long distance transportation by sea, rail or truck in extending shelf life and preserving the quality of fresh fruits and vegetables. In spite of high initial investment and operating cost, only cold supply chains can provide basic requirements including proper control of temperature, humidity and ventilation during transportation. In addition, fresh produce should be properly packed and stacked using pallets to avoid compression damage due to excessive movement or vibration. Cold chain model needs to be designed using refrigeration components including pre cooling chiller, cold room, and truck refrigeration. Considering the importance of cold supply chains in prolonging shelf life and reducing postharvest losses of horticultural crops, investment in cold chain is increasing in India. It is estimated that India's investment in cold chain will be over Rs. 11,100 Crores (about USD 15 billion) over the next five years.

Storage

In developing countries, especially in tropical countries, long term storage of fruits and vegetables is a major problem where sufficient cold stores are not available. Along with the microbial spoilage of fruits and vegetables, sprouting in tubers and bulbous crops is a major problem during storage in conventional low cost storage structures. Therefore, proper storage facilities with the provisions for modification of temperature, relative humidity, gaseous concentration and pressure as per the nature of commodity need to be created for prolonging storage of fruits and vegetables. Several improved storage structures have been developed with the manipulation of temperature, humidity, gaseous concentration and atmospheric pressure for shelf life extension of fruits and vegetables.

Cold Storage

Cold storage or low-temperature storage, the enclosed chamber with electric refrigeration unit is the most commonly adopted improved technique for long-term storage of fruits and vegetables. Most of the cold storages are equipped with full control of temperature, and can be adjusted ranging from 0–10 °C or more for a number of commodities.

Controlled Atmosphere (CA) Storage

In CA storage, the concentration of oxygen and carbon dioxide is manipulated and constantly regulated along with the maintenance of low temperature and humidity during storage. This is an advance method of storage in which atmospheric composition is different than the normal atmosphere where oxygen level is generally kept below 8 per cent and CO2 level is kept above 1 per cent combined with refrigeration. Despite the fact that the reduced concentration of oxygen slows down metabolism, it also affects the ethylene production, since oxygen is necessary for transforming 1-amino cyclopropane -1-carboxylic acid to ethylene. In the horticultural industry, CA storage is most successful invention to prolong the shelf life of fresh produce (Bodbodak and Moshfeghifar, 6), which was first reported by Kidd and West (23) with its commercial application in pome fruits. Several studies showed the positive effects of low oxygen and elevated level of carbon dioxide on shelf life and quality of fruits stored in CA storage (Table 1).

Table 1. Effect of various CA storage conditions on quality of fruits

Crop	*Concentrations*		*Observations*	*References*
	O_2	*CO_2*		
Apple (Granny Smith)	<0.5 per cent	1 per cent	Retained firmness for a long time	Mditshwa *et al.* (28)
Strawberry	5 kPa	15 kPa	Increased strawberry shelf-life by 3 days (based on disease incidence)	Alamar *et al.*, 1
Acovado	2.5 and 5 kPa	5 and 7.5 kPa	Optimized shelf life and fruit quality	Basuki *et al.*, 5
Litchi	1 per cent	5 per cent	Reduced weight loss	Ali *et al.*, 3
Mango (cv. 'Kensington Pride' and 'R2E2'	3 per cent	6 per cent	Extended the storage life with good quality up to 6 weeks	Singh (37)
	At 13°C			
Pomegranate	10 kPa	5 kPa	Reduced weight loss	Pareek *et al.* (31)
	5 kPa	5 kPa	Minimized fungal decay	
Plum (cv. Laetitia)	1-2 kPa	1-3 kPa	Delayed ripening	Steffens (38)
Mango (cv. Delta R2E2)	3 per cent	6 per cent	Reduced respiration rate, and delayed de-greening of the skin and fruit softening. Increases TSS, total sugars content and taste of the ripe fruit	Lalel *et al.* (24)

Hypobaric Storage

Hypobaric storage, also known as low pressure storage is a system of storing harvested commodities while ventilating with air at less than atmospheric pressure. This condition helps in increasing shelf life, decay incidence and maintain firmness of the commodity by reducing metabolic activities. Storage of harvested produce by application of reduced pressure was first used by Workman *et al.*, 42, when they stored tomato fruits and recorded reduced respiration rate at 20°C temperature under 88 mm Hg storage as compared to those stored at 20°C under atmospheric pressure. It is very essential for storing crop in hypobaric storage that storage atmosphere is constantly being changed in order to maintain required oxygen level, since the produce in hypobaric storage is continuously respiring. This is achieved by vacuum pump evacuating the air from the store and the store atmosphere is continuously being replaced from the outside. The inlet and outlet of air from the store should be balanced to achieve the desired pressure in the store for a particular commodity. The reduced pressure inside the store can result in rapid water loss from the crop, to overcome this problem, the relative humidity inside it must be cent per cent or near to the saturation point of air (Thompson, 39). Hypobaric storage is effective in reducing decay incidence, controlling browning and imparting better post harvest life in the fruits and vegetables. Effect of hypobaric storage on the quality and shelf life of some fruits is shown in Table 2.

Table 2. Effect of hypobaric storage on the quality and shelf life of some fruits

Crop	*Storage Pressure*	*Observations*	*References*
Strawberries	380 mm Hg, at 20°C or 5°C	Increased the activity of defence related enzymes, and delayed rot development	Hashmi *et al.* (22)
Jujube cv. Lizao	20.3, 50.7 or 101.3 kPa	Delayed the loss of ascorbic acid, and prolonged the postharvest life	(Cui, 11)
Loquat	40-50 mm Hg	Reduced decay and fruit browning	Gao *et al.*, 19
Avocado	61–81 mm Hg	Increased fruit firmness up to 30 days	Burg, 8

Irradiation

Food irradiation is a physical means of food preservation that involves exposure of pre-packaged or bulk foodstuffs to ionizing radiations including gamma rays, X-rays, or electrons, which kills most of the spoilage causing organisms, inactivates the enzymes responsible for browning, and also delay the ripening of fruits and sprouting of vegetables. With an objective to reduce the use of chemicals in preserving fruits and vegetables, the non-residual feature of ionizing radiation has gained popularity in many countries. Ionizing radiation from cobalt-60 or caesium-137, or machine-generated electron beams are used as a source of irradiation for extending the shelf life of fresh produce. The energy from the radiation breaks the bonds in the DNA of microbes, causing defects in the genetic makeup and making them unable to reproduce. Ferrier (17) reported that irradiation is used for the control of potato and onion sprouting, and strawberry decay commercially. While much of the focus of irradiation use on fresh fruits and vegetables for extending shelf life and reducing decay, it has been considered as an effective means for killing, sterilizing or preventing further spread of foreign invasive pests of quarantine importance across international borders. Gamma irradiation has also been employed for decontamination or sterilization of dehydrated fruits and vegetables. However, the irradiation being an expensive method, is generally used for the storage of high-value fresh commodities or for export.

Emerging Processing Technologies

Although conventional processing techniques such as drying, canning, bottling, pickling and freezing are often used for extending shelf life, and to add value to fruits and vegetables, nowadays, considering the loss of nutritional quality in thermal processing, a number of non-thermal novel processing techniques including cold plasma, pulsed electric field (PEF), high-pressure processing (HPP) and ultrasonication are available to extend the shelf life of fruits and vegetables.

Cold Plasma (CP) Technology

Cold plasma technology is one of the emerging non-thermal processing techniques attracting more interest from researchers and food processors to prolong the shelf life of fruits and vegetables. Cold plasma technique can be used for inactivation of microorganisms, removal of toxins, degradation of pesticides and dyes, and packaging modification. Plasma, the fourth state of matter exists as an ionized gas containing electrons, free radicals, ions, *etc.* possessing a net neutral charge. It is induced under varying temperatures and pressures by energizing a neutral gas, and classified as thermal and non-thermal plasma. As the thermal technology has disadvantage of altering food quality, postharvest treatment with cold plasma, a gas-derived mix of atoms, excited molecules and charged particles can extend the shelf life without any adverse effect on fresh produce. It is reported that atmospheric plasma at 18 kV can kill *Salmonella spp.* on strawberry samples by 1.7 - 2.3 log CFU/sample. Likewise, using dielectric barrier discharge (DBD) plasma at 80 kV for 5 min successfully decontaminated cherry tomatoes containing *Escherichia coli, Salmonella typhimurium,* and *Listeria monocytogenes* by up to 3.5, 3.8, and 4.2 log CFU/tomato respectively. Therefore, sterilization using cold plasma at low cost with no or minimal impacts on the quality attributes like colour, texture, sensory and nutritional value of treated products can be an effective alternative to heat-based processing techniques.

Pulsed Electric Fields (PEF)

Though the thermal processing is generally used for microbial decontamination of food, it is not a preferred method due to loss of nutritional quality and sensory properties of food. Therefore, in addition to nanotechnology, irradiation and cold plasma technology, pulsed electric fields (PEF), another novel non-thermal decontamination technology is now being considered as an effective tool for food decontamination, reducing food borne illness and shelf life extension of food by applying short electric pulses with very little or no detrimental effect on quality attributes of food. Application of pulsed electric fields of high intensity, created from high voltage pulses generated by high voltage pulse generator for a very short duration from a few microseconds to milliseconds causes permeabilization or puncturing of cell membranes triggering death of microorganisms.

Food products with no air bubbles and low electrical conductivity are best suited for processing by pulsed electric fields technology. In addition to various liquid foods like milk, yogurt, soups *etc.*, different fruit juices such as apple juice, orange juice, tomato juice, carrot juice, and some other foods like apple sauce, pea soups and salad dressings have been found to maintain sensory attributes and nutritional quality for an extended period after processing by this novel method using pulsed electric fields without application of heat or any preservative. The effects of pulsed electric fields (PEF) on biomembranes have been thoroughly studied by scientists since the use of PEF has attracted great interest in several scientific areas such as cell biology, biotechnology, medicine and food technology.

High Pressure Processing

The growing concern about food borne diseases and intense use of preservatives and chemicals has made the consumers to shift towards natural and healthy foods such as fruits and vegetables and their products in terms of natural flavour and taste, and free from additives and preservatives. This demand has triggered the need for the development of a number of non-thermal processing techniques, of which high-pressure processing technology has proven to be very valuable. High pressure processing commonly known as HPP technology, and sometimes termed as hydrostatic pressure processing (HPP) is a cold pasteurization or non-thermal processing technique in which food, sealed in flexible and water-resistant package is subjected to a high pressure of different ranges transmitted by water known as hydrostatic pressure usually from 400 MPa to 600 MPa (megapascal) in metric system or 58000 psi to 87,000 psi (pounds per square inch) in the imperial system for a few seconds to a few minutes, usually one to five minutes depending on the nature of food and the target. The food product to be treated at high pressure is packaged in a flexible container and placed inside a pressure vessel submerged in water or a liquid medium, which transmits the pressure. These high pressures to packaged fruit and vegetable products in high pressure processing equipment kill most of the harmful microorganisms, such as salmonella, E. coli and Listeria monocytogenes by damaging their cell components without altering colour, flavour, texture or nutritional value of the product. As in this pressure processing method, no chemical preservative is used, it is also referred as natural pressure cold pasteurization. HPP is an effective alternative to heat processing for fruit and vegetable products such as juice, jam, puree, sauce *etc.* in extending shelf life and preserving nutritional and sensory quality. Several commercial products, including fruit juices of mandarin orange, grapefruit, apple and sweet orange, and carrot juice and broccoli-apple juice mixture treated with HHP are currently available in different markets. But, the main limitation in using this technique with high pressure treatment is to safely and repeatedly generate high pressure in a vessel, solid foods have to be processed in batch or semi continuous equipment. In spite of relatively high capital cost of equipment, the use of high-pressure treatments and the range of products on which the treatments are applied continue to grow every year (Gopal, *et al.*, 20).

Ultrasonication

Ultrasonication is one of the emerging novel technologies, now used in the field of food science and technology. Ultrasound being non-toxic and environment friendly is considered as green technology, which saves lot of energy and maximizes intended output. Among diverse applications, ultrasound has been employed in studying composition of fruits, vegetables, and dairy products, and detecting microbial contamination in dairy and canned foods. Ultrasonication has also gained huge application in the food industry such as processing, extraction, emulsification, homogenization and preservation. It is used to maintain post-harvest quality attributes in fresh fruits and vegetables including sensory characteristics, nutritional value and microbial safety, and can be a substitute for washing of fruit and vegetable in food industry. Use of ultrasonication in guava juice, orange juice and tomato juice has been reported to maintain fresh quality, nutritional value, and microbiological safety. Ultrasonication cleaners (20–400 kHz) have been efficiently used to produce fruits and vegetables free of contamination (Lin and Erel, 26) and at 40 kHz, it has been applied on strawberry fruits in which decay and infection was considerably reduced along with maintenance of quality (Cao *et al.*, 10).

Conclusion

Along with securing food for all, food safety also needs to be focused, as it is directly related with human health. Although many techniques have been developed, no method is equally suitable for all the commodities due to wide variations in nature and texture of fruits and vegetables, so recommendations should be crop, variety and location specific. Efforts should also be made for identifying and developing varieties having long shelf life and good processing quality. Coordinated activities by growers, handlers, storage operators, processors, wholesalers and retailers are required to maintain quality and reduce postharvest loss of perishable commodities. Due to multiple uncertainties along the supply chain, microbial contamination may occur leading to spoilage at any stage in the chain from farm to consumer. Therefore, proper harvesting, handling, post-harvest treatments, cold chain management including control of storage temperature, relative humidity and gaseous composition are essential to minimize physiological activities, microbial spoilage and reduction in risk of foodborne illness. Along with the improvement in conventional postharvest technologies, prioritization of cold supply chains, nanotechnology, smart packaging, irradiation, improved storage techniques, and non-thermal novel processing technologies including cold plasma technique, pulsed electric fields, ultrasonication and high pressure processing can be helpful in microbial decontamination without affecting nutritional and sensory characteristics of foods, and in prolonging storage life of fruits and vegetables to reduce the loss of harvested produce, which in turn will boost farmers' income, lift rural economy, and increase the affordability of fruits and vegetables for alleviating global malnutrition. Initiatives from both public and private sectors in investing more on infrastructure development especially cold chains, the most crucial element in prolonging shelf life of perishables, and processing units; allocating more funds for postharvest research areas, and emphasising more on capacity building in postharvest management, and food standards for complying with food safety standards can reduce postharvest losses remarkably.

REFERENCES

1. Alamar, M. C., Collings, E., Cools, K. and Terry, L. A. 2017. Impact of controlled atmosphere scheduling on strawberry and imported avocado fruit. *Postharvest Biol. Technol.*, **134:** 76-86.

2. Alfadul, S. M. and Elneshwy, A. A. 2010. Use of nanotechnology in food processing, packaging and safety- review. *African J. Food, Agric. Nutr. Dev.* **10** (6): 1-21.

3. Ali, S., Khan, A. S., Malik, A. U. and Shahid, M. 2016. Effect of controlled atmosphere storage on pericarp browning, bioactive compounds and antioxidant enzymes of litchi fruits, *Food Chem.*, **206**: 18–29.
4. APEDA. 2021. India export of agro food products. Product group report- processed fruits and vegetables; other processed foods. Agricultural and Processed Food Products Export Development Authority, India.
5. Basuki, E., McGlasson, W. B. and Skurry, G. 2016. Polyamines association in textural changes of avocado during controlled atmosphere storage. *Imp. J. Interdiscip. Res.*, **10** (2): 1716-1721.
6. Bodbodak, S. and Moshfeghifar, M. 2016. Advances in controlled atmosphere storage of fruits and vegetables. **In:** *Eco-Friendly Technology for Post harvest Produce Quality*, Academic Press, Elsevier, pp. 39-76.
7. Boindala S and Lewis J. 2019. The grand challenge of regulating health foods in India. Indian *J. Med. Res.* **150** (3): 248-253.
8. Burg, S. P. 2004. *Postharvest Physiology and Hypobaric Storage of Fresh Produce.* CABI, Wallingford, UK.
9. Business Wire. 2020. Outlook on the fruit and vegetable processing industry in India to 2024- Featuring Dabur India, ITC and Freshtrop Fruits among others - ResearchAndMarkets. com, April 24, 2020, Dublin.
10. Cao, S., Hu, Z., Pang, B., Wang, H., Xie, H., and Wu, F. (2010). Effect of ultrasound treatment on fruit decay and quality maintenance in strawberry after harvest. *Food Control,* **21**: 529–532.
11. Cui, Y. 2008. Effect of hypobaric condition on physiological and biochemical changes of Lizao jujube. *J. Anhui Agric. Sci.,* **36** (29): 12900-12901.
12. de Sao Jose, J. F. B., de Andrade, N. J., Ramos, A. M., Vanetti, M. C. D., Stringheta, P. C. and Chaves, J. B. P. 2014. Decontamination by ultrasound application in fresh fruits and vegetables. *Food Control,* **45:** 36–50.
13. Doshi, J. S. and Sutar, R. F. 2010. Studies on pre-cooling and storage banana for extension of shelf life. *J. Agric. Eng.*, **47** (2): 14-19.
14. FAO. 2012. The role of producer organizations in reducing food loss and waste. International Year of Cooperatives, Issue Brief Series, Rome.
15. FAO. 2014. Better nutrition, better lives. Second International Conference on Nutrition, 19-21 November 2014, FAO Headquarters, Rome, Italy
16. FAO. 2015. Reduction of post-harvest losses in horticultural chains in SAARC Countries. Regional Technical Cooperation Programme (TCP) Project Report, Rome, Italy.
17. Ferrier, P. 2010. Irradiation as a quarantine treatment. *Food Policy.* **35**: 548–555.
18. Gaikwad, K. K., Singh, S., and Negi, Y. S. 2020. Ethylene scavengers for active packaging of fresh food produce. *Environ. Chem. Lett.* **18**: 1–16.

19. Gao, H. Y., H. J. Chen, W. X. Chen, Y. T. Yang, L. L. Song, Jiang, Y. M. and Zheng, Y.H. (2006). Effect of hypobaric storage on physiological and quality attributes of loquat fruit at low temperature. *Acta Hortic,* **712**: 269–274.

20. Gopal, K.R., Kalla, A.M. and Srikanth, K. 2017. High Pressure Processing of Fruits and Vegetable Products: A Review. *Int. J. Pure App. Biosci.* **5** (5): 680-692.

21. Halonen, N. J., Palvölgyi, P. S., Bassani, A., Fiorentini, C., Nair, R., Spigno, G., *et al.,* 2020. Bio-based smart materials for food packaging and sensors–a review. *Front. Mater,* **7**: 82. doi: 10.3389/fmats.2020.00082.

22. Hashmi, M. S., East, A. R., Palmer, J. S. and Heyes, J. A. 2013. Pre-storage hypobaric treatments delay fungal decay of strawberries. *Postharvest Biol. Technol.,* **77**: 75–79.

23. Kidd, F. and West, C. 1927. A relation between the concentration of oxygen and carbon dioxide in the atmosphere, rate of respiration, and length of storage of apples. *Food Investigation Bd. Rpt.* London, pp.41-42.

24. Lalel, H. J. D., Singh, Z. and Tan, S. C. 2005. Controlled atmosphere storage affects fruit ripening and quality of 'Delta R2E2'mango. *J. Hortic. Sci.,* **80** (5): 551-556.

25. Lee, S. Y., Lee, S. J., Choi, D. S., and Hur, S. J. 2015. Current topics in active and intelligent food packaging for preservation of fresh foods. *J. Sci. Food Agric.* **95**: 2799–2810.

26. Lin, I., and Erel, D. 1992. Dynamic ultrasonic cleaning and disinfecting device and method. US Patent No. 5113881A. Washington, DC: U.S. Patent and Trademark Office.

27. Maged E.A. M. and Ayman, H. A. E. 2012. Pulsed Electric Fields for Food Processing Technology. **In**: *Structure and Function of Food Engineering.* A. A. Eissa (Ed,), Intech Open.

28. Mditshwa, A., Fawole, O. A. and Opara, U. L. 2018. Recent developments on dynamic controlled atmosphere storage of apples- A review. *Food Packag. Shelf Life,* **16**: 59-68.

29. Mitra, S. 2021. Postharvest technology-An overview. **In**: *Postharvest Management of Horticultural Crops.* Surajit Mitra (Ed.), Jaya Publishing House, New Delhi-110089, India, pp. 1-16.

30. Ortiz, C. M., Mauri, A. N., and Vicente, A. R. 2013. Use of soy protein based 1-methylcyclopropene-releasing pads to extend the shelf life of tomato (*Solanum lycopersicum* L.) fruit. *Innov. Food Sci. Emerg. Technol,* **20**: 281–287.

31. Pareek, S., Valero, D. and Serrano, M. 2015. Postharvest biology and technology of pomegranate. *J. Sci. Food Agric.,* **95** (12): 2360–2379.

32. Realini, C. E., and Marcos, B. 2014. Active and intelligent packaging systems for a modern society. *Meat Sci.* **98**: 404–419.

33. Robertson, G. L. 2012. *Food Packaging: Principles and Practice.* CRC Press, Massachusetts, MA.

34. Salgado, P. R., Di Giorgio L., Musso, Y. S. and Mauri, A. N. 2021. Recent developments in smart food packaging- focused on biobased and biodegradable polymers. *Front. Sustain. Food Syst.* **5**: 630393. doi: 10.3389/fsufs.2021.630393.

35. Sami, R., Almatrafi, M., Elhakem, A., Alharbi, M., Benajiba, N. and Helal, M. 2021. Effect of nano silicon dioxide coating films on the quality characteristics of fresh-cut cantaloupe. *Membranes,* **11** (2): 140, doi: 10.3390/membranes11020140.

36. Siegrist, M., Cousin, M. E., Kastenholz, H. and Wiek, A. 2007. Public acceptance of nanotechnology foods and food packaging: The influence of affect and trust. *Appetite*, **49** (2): 459-466.

37. Singh, Z. 2015. Controlled atmosphere storage of mango fruit: Challenges and thrusts and its implications in international mango trade. *Acta Hortic.*, **1066**: 179-191

38. Steffens, C. A., Amarante, C. V., Alves, E. O. and Brackmann, A. 2014. Fruit quality preservation of 'Laetitia' plums under controlled atmosphere storage. *An. Acad. Bras. Cienc*, **86** (1): 485-494.

39. Thompson, A. K. 2016. *Fruit and vegetables-harvesting, handling and storage*, 3rd ed. Oxford, UK: Wiley Blackwell.

40. Veljoviæ, S., Boonsiri, K., Maina, M., Semavor, E., and Mitra, S. 2017. Infection capacity of the pathogens *Penicillium italicum* and *P. expansum* in orange during storage. *Food and Feed*, **44** (1): 31-38.

41. Warsiki, E. 2018. Application of chitosan as biomaterial for active packaging of ethylene absorber. *IOP Conf. Ser. Earth Environ. Sci.*, **141**: 12036.

42. Workman, M., Pratt, H. K. and Morris, L. L. 1957. Studies on the physiology of tomato fruit. I. Respiration and ripening behaviour at 20 °C as related to date of harvest. *Proc. Am. Soc. Hortic. Sci*, **69**: 352–365.

43. Yildirim, S., Röcker, B., Pettersen, M. K., Nilsen-Nygaard, J., Ayhan, Z., Rutkaite, R. 2018. Active packaging applications for food. *Compr. Rev. Food Sci. Food Saf.*, **17:** 165–199.

Chapter 23

Rising India's Share in Global Fruit and Vegetable Markets

M. Angamuthu*

Agricultural and Processed Food Products Export Development Authority, August Kranti Marg, New Delhi 110012

Current address: Visakhapatnam Port Authority, Visakhapatnam 530035, Andhra Pradesh

e-mail: angamuthuias@yahoo.com

ABSTRACT

The horticulture sector in India possesses unique advantage because of availability of varied climatic conditions for production of large number of horticultural crops. Besides, there is huge varietal diversity in most of the cultivated fruits, vegetables, flowers, plantation and spices which increase the harvesting season and more revenues to the farmers. Horticulture is the key driver in the agriculture trade with wide range product basket, high value crops and international demand. India's agrarian culture and varied agro-climatic conditions are conducive for cultivation of wide range of horticulture crops around the year. Horticulture sector is considered as the sunrise industry in our Country. The production and supply of quality planting materials, implementation of INM, IDM and IPM, multi cropping system, integrated farming system (IFS), pollinators, pre and post-harvest management, forward and backward linkages in marketing are instrumental for successful cultivation of horticultural crops. Advancement of the horticulture sector in India has been of significant priority in the country's overall development strategy. An important sub-sector to agriculture, horticulture has grown rapidly, contributing over 30 percent to the GDP of agriculture in India. There has been an increased emphasis from various government initiatives to consider the horticulture sector at par with the industrial sector, in order to ensure substantial income from farming. This vast production base coupled with logistical advantages to the Middle East and South East Asia offers India tremendous opportunities for global export. There is a huge market potential for India to become the global hub for important horticultural crops because of growing demand of worldwide consumers for safe and healthy foods, increased urbanization of societies, the growth in scale and influence of supermarkets, and packaged foods, corporatization of farming, and adoption of sustainable farming practices.

Keywords: *Diversity, Export potential, Processed products, Global scenario.*

The Economic Survey 2020-21 identified agriculture as the bright spot amid the economic crisis triggered by the Covid-19 pandemic. The sector faced several adversities as the harvesting season clashing with the lockdown period; this led to a shortfall of working hands due to reverse migration, and a shortage of logistic support. Despite all this, agriculture sector clocked a growth of 3.4 per cent in 2020-21, becoming the only sector that grew during the pandemic crisis. A growing domestic market does provide decent returns on investment, but overseas markets and linkages with the global value chains can lift farmers' incomes substantially and thus fulfilling the vision of Hon'ble PM, Shri Narendra Modi.

Currently, the top 10 export commodities from agriculture and allied sectors are rice, marine products, buffalo meat, spices, cotton, oil meals, sugar, castor oil, and tea. The list does not include horticulture products like fruits, vegetables, or dairy products for which India is among the top two producers. The growth story of horticulture has been an interesting journey. Post independence, the main focus was on cereals for food security which led to the green revolution. Subsequently the domestic and export demand resulted in crop diversification towards oilseeds, sugarcane and horticulture crops during late 1980s due to higher returns. During this period, there was consolidation of efforts and institutional support for development of the horticulture sector. Economic reforms and policies of 1990s gave thrust to the growth of the horticulture sector.

The advent of new technology from production to post harvest, policy initiatives, and infrastructure development including cold storage, quality assurance and institutional mechanisms to provide market linkages changed the paradigms for horticulture development. The infrastructure development, strengthening the supply chain, cool chain management, refer corridors for perishables have helped in reducing post harvest losses while maintaining quality of the produce for exports.

Agriculture is one of the most important sectors of India Economy and accounts for 20 per cent of India's GDP. Agriculture with its allied sectors is the largest source of livelihoods in India. India's agriculture exports grew 17.34 per cent to $41.25 billion in 2020-21, jumping after remaining almost stagnant in the previous three years. There is a huge growth in the export of cereals, non-basmati rice, wheat, millets, maize, and other coarse grains. The trade data of the first six months of 2021-22 revealed the highest merchandise exports in two quarter at $198 billion, which was 57.5 per cent and 24.3 per cent higher than the volumes registered in the same period of 2020-21 and 2019-20. The data underlined high growth percentages recorded in exports of rice, other cereals, spices, cashew nuts, fruits, and vegetables in six months. The trend in export growth prompted the government to announce an ambitious target of $400 billion exports for 2021-22. India is a leading producer of many agriculture commodities such as dairy products, cereals, spices, fruits, vegetables, rice, wheat, and cotton. But it still has a moderate 2.2 per cent share in global agriculture exports. There is scope to augment its share in view of the huge potential.

Horticulture is the key driver in the agriculture trade with wide range product basket, high value crops and international demand. India's agrarian culture and varied agro climatic conditions are conducive for cultivation of wide range of horticulture crops around the year. Horticulture sector is the sunrise industry in our Country; Most of them are SMEs and we need to make linkages within the country and outside the country to harness the benefits of value addition and higher returns. In India, Horticulture products (Fresh Fruits and Vegetables, processed fruits and vegetables, Floriculture, Plantation Crops including tea, Coffee) contributes around 22 per cent of export share in agriculture.

India's Horticulture Export Scenario: India well Placed to be a Major Player

The development achieved in the horticulture sector is indicative of the fact that there is growing demand for horticulture produce. Demand for horticultural crops across the globe is increasing. Horticulture Sector has immense export potential having a high value and diversified product basket. The product basket includes fruits, vegetables, flowers, medicinal and aromatic plants, tea, coffee, cocoa products, coconut products, processed fruits and vegetables *etc.* Organic certified products from horticulture bases such as processed fruits (mango pulp, puree), medicinal plants, herbal teas, dried and desiccated coconuts *etc.* leads to a new dimension for expansion. Among the registered GI Agricultural products in India, majority come under the horticulture products (fresh and processed). GI products give an added advantage of niche product branding being inherent product of the Country. With the Agriculture Export Policy 2018 looked to harness the export potential of Indian agriculture by promoting organic, ethnic, and indigenous products with a focus on high value-added produce and perishables to achieve food security and better income for farmers. The policy stressed the need for diversification into new products and perishables as traditional farms produce like rice, wheat and marine products accounted for more than 50 per cent of the total farm exports from India.

For holistic growth of the horticulture sector in the country, the Ministry of Agriculture and Farmers Welfare has been implementing 'Mission for Integrated Development of Horticulture' since 2014-15. The Government intervention in the horticulture sector has led to the situation wherein horticulture production has surpassed the agriculture production in the country. Total Horticulture production in 2020-21 is estimated at a record 331.05 MT (fruits 103 MT and vegetables 197.2 MT) against the foodgrains production of 308.64 MT. The horticultural crops production has surpassed foodgrains production since 2016-17.

Indian horticulture produce is exported to over 70 countries. The key importing countries include the Middle East, Southeast and South Asia, the UK, European Union countries and the US. India's top three export destinations are United Arab Emirates (16 per cent), Bangladesh (13 per cent) and Netherlands (9 per cent). In the recent years, the horticultural products exports to African countries have increased. The horticultural exports to some markets like the US witness fluctuating trend.

For expanding the horticultural products exports footprint, we need to develop adequate infrastructure such cold chains, connectivity as well as promoting production of quality produce which meets the global standard, across states.

The stress should be on production of uniform quality of horticultural produce in large volumes and create organic and pest free zones. The supply chain must be integrated and barriers to import of plant varieties and seeds needs to taken up for discussion with respective countries.

Cluster Development Approach

In May 2021, Union Minister of Agriculture and Farmers' Welfare Shri Narendra Singh Tomar had launched the Horticulture Cluster Development Programme (CDP). In a pilot phase, the programme is being implemented in 12 horticulture clusters out of the total 53 clusters selected for the programme. A central sector programme implemented by the National Horticulture Board. The CDP aims at growing and developing identified horticulture clusters to make them globally competitive. The clusters of the pilot phase include Shopian (Jammu and Kashmir) and Kinnaur (Himachal Pradesh) for Apple, Lucknow (Uttar Pradesh), Kutch (Gujarat) and Mahbubnagar (Telangana) for Mango, Anantpur (Andhra Pradesh) and Theni (Tamil Nadu) for Banana, Nasik (Maharashtra) for Grapes, Siphahijala (Tripura) for Pineapple, Solapur (Maharashtra) and Chitradurga (Karnataka) for Pomegranate and West Jaintia Hills (Meghalaya) for Turmeric. It is anticipated that CDP is expected to attract an estimated investment of Rs 10,000 crore when it is implemented in all the 53 clusters.

Another key thrust area has been the promotion of Exotic and Indigenous Crops. The government has identified 10 globally popular exotic fruit crops of commercial importance. These indigenous fruit crops have high nutritional and nutraceutical properties. The Exotic Fruit crops include Avocado, Blue berry, Dragon fruit (dragon fruit is also referred to as Kamalam), Fig, Kiwi, Mangosteen, Persimmon, Passion fruit, Rambutan, Strawberry. There are a number of Niche Fruit crops like Aonla, Karonda, Seabuckthorn, Garcinia, Jamun, Hanuman Phal (Soursop), Bael, Tamarind, Phalsa, Jackfruit.

Recently, for the first time, facilitated exports of 'dragon fruit', sourced from farmers of Gujarat, West Bengal and Maharashtra, to the United Kingdom, Kingdom of Bahrain and United Arab Emirates. APEDA also supported exports of the first consignment of bananas sourced from farmers of Lakhimpur district of Uttar Pradesh, to Iran via sea route while Maharashtra and Gujarat have been key exporters of bananas.

APEDA has also have initiated registration of pack-houses for horticulture products for meeting the quality requirements of the international market. Registration of export units for peanut shelling and grading and processing units, for instance, is to ensure quality adherence for the EU and non-EU countries.

Despite the challenges posed by COVID19 pandemic, in 2020-21, India exported fruits and vegetables valued at Rs 11,019 crore compared to Rs 10,114 crore worth of exports in 2019-20, which is an increase of close to 9 per cent.

India's Horticulture Exports – At a Glance

- India's total Agri Export during 2020-21 was 41.2 Billion USD (Source: DGCIS)

- Share of Horticulture products in Agri Exports (per cent)
- Total share of export of major product Categories under Horticulture in Agri Exports is around 22 per cent.
- Break up of percent share is given below:

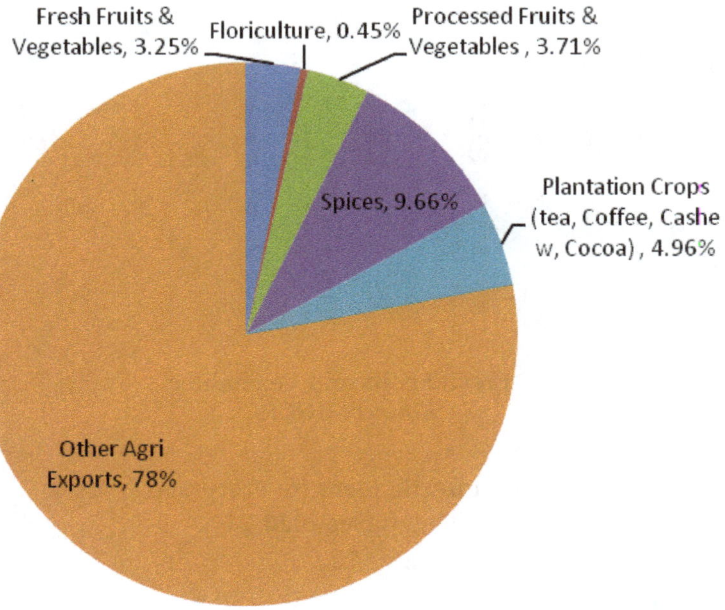

***Source*: DGCIS.**

Export of major product categories under Horticulture are given below:

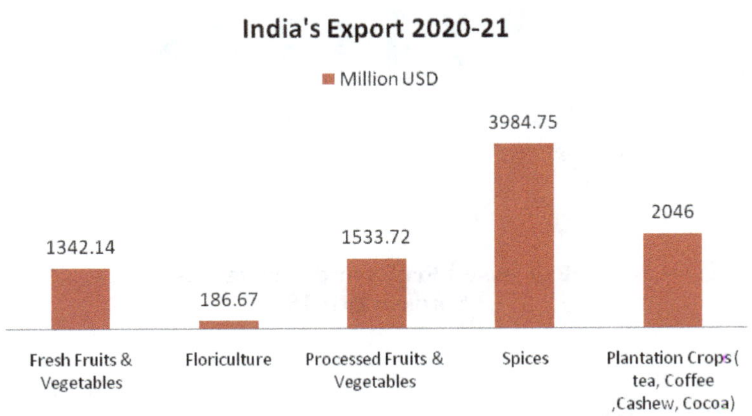

***Source*: DGCIS.**

Major export destinations for Fruits and Vegetables

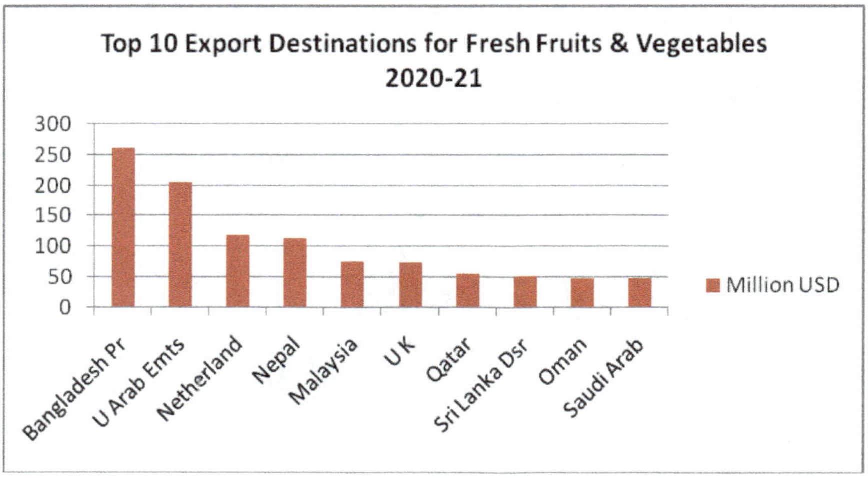

Fresh Fruits and Vegetables
***Source*: DGCIS.**

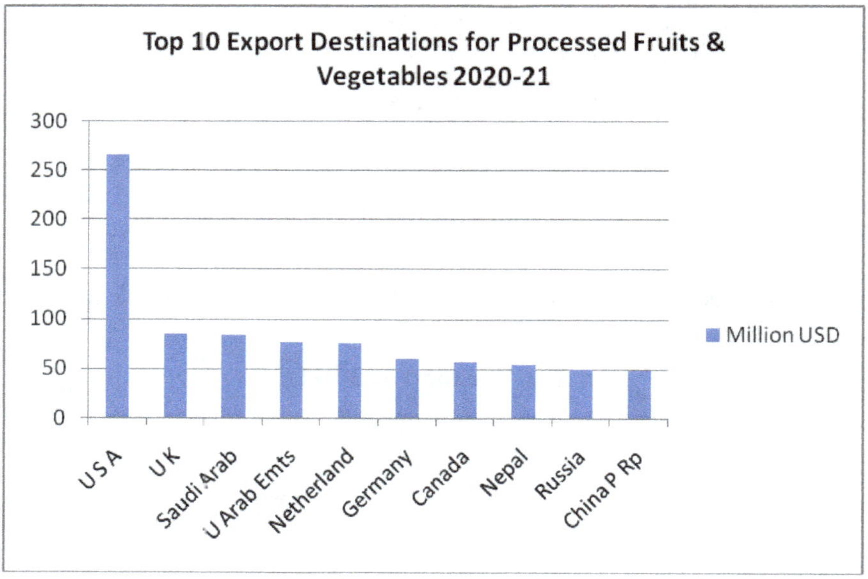

Processed Fruits and Vegetables
***Source*: DGCIS.**

Startups in Horticulture

The startups and farm bills will transform horticulture farmer's income, trade and business. In addition to serving domestic market, these horticulture startups will create spillover effect in exports. Now export policy on horticulture should and will move towards such startups. An Ernst and Young 2020 study pegs the Indian agritech market potential at $24 billion by 2025, of which only 1 per cent has been captured so far. Among various agritech segments, the supply chain technology and output markets have the highest potential, worth $12.1 billion. Currently, it is estimated that there are about 600 to 700 agritech startups in India operating at different levels of agri-value chains. Many of them use artificial intelligence (AI), machine learning (ML), internet of things (IoT), *etc.*, to unlock the potential of big data for greater resource use efficiency, transparency and inclusiveness.

Ninjacart, Dehaat, and Crofarm (Otipy) are a few of the many startups that are redefining the agrifood marketplace. The novelty of startup-led value chain transformation is not limited to empowering farmers but also co-opting local grocery, mom-and-pop, and kirana stores as well as small agrifood businesses that are an integral part of the agrifood ecosystem. At the same time, the startup network is able to leverage the bigger front-end players who demand bulk quality produce and have challenges in directly linking with farmers. This is in contrast to the earlier organised retail (big box) wave that emerged in the mid-2000s, wherein the livelihood of the unorganised retailers and small businesses was perceived to be threatened.

One of the emerging interesting trends is that of multiple models of engagement. Ninjacart started as a farmer-to-consumer (F2C) venture but soon moved onto farmer-to-business (F2B), recognising the need to address the challenges that confront the agrifood ecosystem and achieve greater scale effects. Dehaati Beej Se Baazar Tak is a full-stack agri service start up that engages through B2F (business-to-farmer) and F2B models. It uses data science, agri science and analytics to nurture a thriving ecosystem of farmers, micro-entrepreneurs and institutional buyers. Crofarm is a F2B digital supply chain that manages logistics, inventory and supply of fresh produce directly from farms to retail chains like Big Bazaar, Reliance Retail, BigBasket and Grofers. Crofarm's Otipy is an app-based F2B2C social commerce platform that delivers (in addition to grocery and other household items) fruits and vegetables directly sourced from farmers to consumers through their partner resellers.

The agritech startups have a growing footprint. Dehaat is present in Bihar, West Bengal, Odisha, and Uttar Pradesh, working with 6,50,000 farmers through 1,890 Dehaat Centres. Ninjacart sources fresh produce from farms and supplies to retailers, restaurants, grocery and kirana stores, and small businesses and is operational in nearly 11 cities. With a farmer network of 10,000 plus, Crofarm has served more than 1 lakh consumers and 5,000 businesses. Otipy has emerged as one of the popular app-based platforms with nearly 2 lakh customers and more than 8.25 lakh mobile downloads. It currently works with 10,000-plus resellers in Delhi NCR and also present in UP, Gujarat, and Himachal Pradesh. About 70 per cent of the resellers are women.

The startups have had a demonstrated impact. Ninjacart reduced wastage to 4 per cent compared to up to 25 per cent in traditional chains through demand-driven harvest schedule. Logistics optimisation enabled delivery in less than 12 hours at one-third the cost in traditional chains. Farmers' net incomes are reported to have increased by 20 per cent. Dehaat has enabled up to 50 per cent increase in farmers' income as a result of savings in input costs, increased farm productivity, and better price discovery.

The Global Scenario

Production

In 2018, the world produced a total of 868 million tonnes of fruit and 1089 million tonnes of vegetables (FAOSTAT).The main fruit in order of importance were banana, citrus (orange, tangerine, mandarin, lemon, grapefruit, *etc.*), melons, apple and grape. "Other" fruit accounted for 76 million tonnes, reflecting the wide range of different types of fruit that are grown around the world. The main vegetables were tomato, various alliums (onion, garlic, shallot, leek), brassicas (cabbage, cauliflower, broccoli), and cucumber. But overall, "other" fresh vegetables are more important than any of these individual categories: nearly twice as many "other" vegetables were produced than tomatoes.

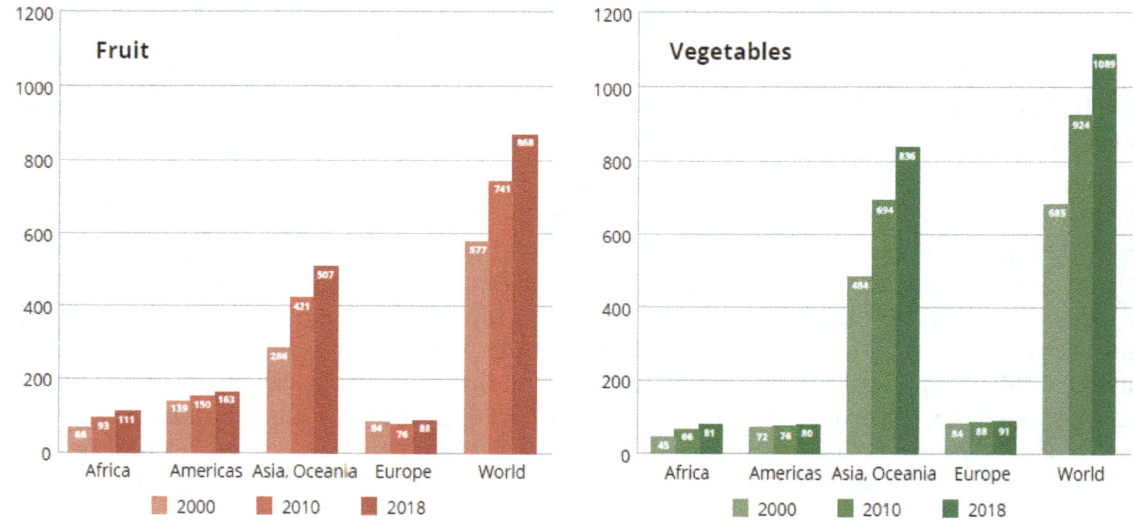

Change in production of fruit and vegetables 2000–2018.
***Source*: FAO Background Paper and FAOSTAT.**

The world's main producing region of both fruit and vegetables is East Asia, followed by South Asia. Other major producing regions are South America, Southeast Asia and Southern Europe (for fruit), and Southeast Asia (for vegetables). Northern and Western Europe produce relatively little, so have to import much of what they consume.

International Trade

International trade of fresh fruit and vegetables represents only around 7–8 percent of total global production (FAOSTAT), but it still ranks among the most valuable crop and livestock commodity groups.

Latin America, the Caribbean and Asia are the major exporting regions of Fresh Fruits and vegetables. The major importers of fresh produce are the European Union, the United States of America (both are also large exporters), China, Canada, Japan and the Russian Federation. Trade agreements, such as the World Trade Organization Agreement on Agriculture and various regional trade agreements, have led to reduced import tariffs (Source FAO background paper on fruits and vegetables-your dietary essentials-2021). The expansion in global trade of fruits and vegetables can be attributed to the growing demand for safe, good-quality, attractively packed fresh produce, a growth in health consciousness, and awareness of the nutritional benefits of fresh fruit and vegetables. The United Nations has declared 2021 as the International Year of Fruits and Vegetables.

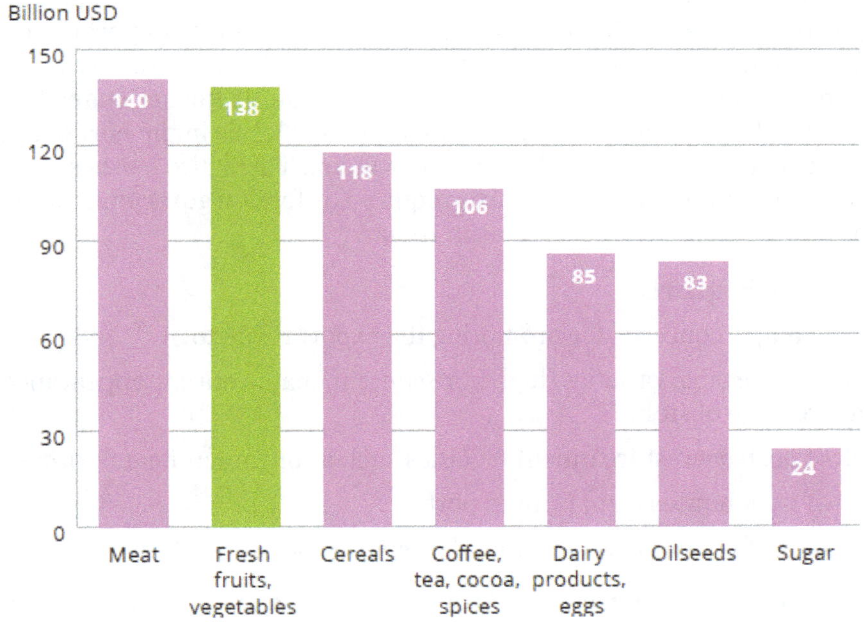

World Export Value of Selected Agricultural Commodities 2018.
***Source*: FAO Background Paper and FAOSTAT.**

Challenges in Export of Horticultural Produce

Although India is a leading fruit producer in the world, the fresh fruit export from India is very small owing to several constraints. These constraints relates to production practices, post-harvest technologies, issues related to supply chain, market access and non-tariff restrictions.

Further share in horticultural crop production is contributed by small holders' farmers as these crops can be grown on small farms. Small and marginal farmers who own less than two hectares of land have a share of 86.2 per cent in total farmers of India.

From Competition at Home to Competing Abroad

As pronounced a paradox as any, India produces about 11 per cent of the world's vegetables and 15 per cent of all fruits, but its share in global exports amount to a meagre 1.7 per cent in vegetables and 0.5 per cent in fruits. One of the reason behind this are International transportation costs are high (20-30 times higher than those faced by other countries) and storage/distribution services are inefficient. A World Bank report says that because of infrastructural and other constraints, a domestic exporter is mostly unable to service destinations beyond 14,000 Kms from the country's borders. For example, it costs $790 to transport one metric tonne of grapes from India to Netherlands - about three times higher than what it takes to transport the same from Chile - which is twice as far from the Netherlands as compared to India. After COVID international transport costs have further increased substantially. Creating an integrated and competitive supply chains for agriculture and improving storage, distribution and transportation network ought to be taken as the two broad priorities of reform. Reports pointed out that insufficient backward linkage in the distribution chain and inadequate access to cold storage facilities lead to seasonal gluts and extreme price variation. It further states that because of the fragmented marketing chains in India average wastage levels in the horticulture sector are quite high as much as 12 per cent of the total production. There also exists a gap between the stringent health, safety and quality standards required by foreign governments and buyers and the standards and assessment mechanisms in India.

Constraints for Export

Some of the major concerns for promoting the export relate to:

- Lack of exportable varieties (high fibre content, inappropriate appearance and texture and large size of stone)
- Lack of post-harvest treatment facilities such as of vapour heat treatment
- Lack of pack houses from farm to port
- High cost of obtaining certification for exports like Eurep-Gap

The issues can be categorized into "**Supply Chain Issues**" which are inherent to the domestic supply chain of food products and "**Market Access Issues**" which comprise of various parameters and factors driven by the requirements of the target countries.

Supply Chain Issues

- Uneconomic scale of operation
- Lack of consistency in supply and quality
- Lack of cost competitiveness due to statutory changes, intermediation and wastages losses
- Inadequate and inappropriate storage and distribution infrastructure
- Lack of technical support for the agro-industrial sector
- High Post Harvest losses
- High freight cost

Market Access Issues

Non-Tariff Barriers: Non-tariff barriers can take various forms. Broadly these can be categorized as under:

- Import Policy Barriers
- Standards, Testing, Labelling and Certification requirements
- Export Subsidies and Domestic Support
- Government procurement
- Short product life cycle
- Lack of brand image

Most exports from India lack scale; the low volume translates into lack of economics in operation and mares exports uncompetitive. Hence, exporters are notable to establish themselves as long term players in the export market, and rely heavily on opportunistic businesses. These factors cumulatively translate into low investments in upgrading skill sets, product innovation, quality improvement and brand building.

Multiple Safety Standards

The emergence of private food safety and quality standards mainly in developed countries is now a well-established fact. These standards operate alongside regulatory systems but in terms of market access and access to the shelves of the leading supermarkets in the rich countries, it become almost mandatory. With these standards becoming a global phenomenon, countries in the developing world face increasing constraints in exporting their food products to markets in the developed countries.

Technological Constraints

- The major technology related constraints contributing to low productivity of horticultural crops and inferior quality of produce are:
- Vast majority of holdings are small and un-irrigated
- Large tracts of low and unproductive plantations needing replacement/rejuvenation.
- Low productivity of crops due to inferior genetic stocks and poor management.
- Inadequate supply of quality' planting materials of improved varieties o High incidence of pests and diseases.
- Heavy post harvest losses and low utilization in processing sector.

Commercial constraints

According to the USDA economic report, the commercial constraints on fruits and vegetables export include:

Trade Barriers

Natural trade barriers include high transportation costs to distant markets, and artificial barriers include legal measures such as protectionist policies. Liberalization of trade through international agreements has been instrumental in relaxing many legal trade barriers by reducing tariffs and by harmonizing the technical barriers to trade.

Scientific Phyto-Sanitary Requirements

Importing countries set the standards that potential trade partners must meet in order to protect human health or prevent the spread of pests and diseases. For instance, Japanese imports of US apples are limited to Red and Golden Delicious apples from Washington and Oregon. The Japanese, who are mainly concerned with the spread of fire blight, impose rigorous and costly import requirements on the US apple shippers. The apples must be subjected to a cold treatment and fumigation with methyl bromide before shipment to Japan, and three inspections of US apple orchards during the production stage. Infestation by fruit flies (Tephritidae: Diptera), common in the tropics, is a major constraint to the production and export of tropical fruits.

Technological Innovations

Countries can increase their competitiveness and world market shares by providing higher quality products and promoting lower prices through technological innovations.

The key reasons why Indian fruits don't reach export markets in substantial quantity are the lack of quality and safety practices in production and packaging. To overcome the challenges, it will need a paradigm shift in approach towards agriculture business, backed by technology.

The strategy of 'farm to fork' transformation needs to be extended to "laboratories to farmland to consumer, some of which are listed below:

Research Room Efforts:

a) Application of tissue culture for clonal propagation of the desired type of pathogen-free plants to improve the quality and nutrition of fruits.

b) Efforts to develop and produce fruits and varieties that are in demand in large overseas markets and to increase their shelf-life.

Plantation Site

a) Proper selection of fruits variety for plantation

b) Using plantation techniques such as High-Density Plantation (HDP).

Productivity

Although the average productivity of fruits in India is closer to the world average, at 11.7 tonne per hectare, it is only half the productivity in the US at 22.2 tonne per hectare, India's growth in the overall production of fruits is more due to increases in the area of cultivation and not enhanced productivity. Thus, a scientific assessment of the nutrient requirement of fruit trees is called for assessing nutrient and irrigation requirements for increasing their productivity.

Training

Technical support and training to the farmers for educating them on adopting best practices to increase their fruit's quality and productivity.

Post-Harvest Infrastructure

1. Adequate cold storage facilities are available for just about 10 per cent of India's horticulture production. As per Economic Survey 2020-21, the lack of modern sorting and grading facilities leads to post-harvest losses of up to 6-18 per cent in fruits. In particular, multi-commodity storage near the farms is required. According to an estimate, in Tamil Nadu, the largest producer state of bananas, the wastage rate is around 30 per cent for lack of post-harvest infrastructure.
2. More facilities for irradiation vapour heat treatment (VHT), and hot water treatment, to maintain quality standards of fruits to increase the export to developed countries.
3. Encouraging and promoting investment including FDI in specialised transport infrastructure and food testing and inspection infrastructure near the farm.
4. Application of approved fruit and vegetable coatings to prevent oxidation, moisture transfer, and pathogen growth in fruits during transportation needs to be encouraged.

Way Forward

In order to increase export competitiveness following inventions are required and India can be a leading global player in horticulture produce exports.

Strengthening the Backward Linkage for Market Driven Production

For linking the farmers to the export market, there is an emergent need to strengthen the supply chain and production efficiency. Developing operation protocols and SOPs based on international standards for each products or group of products, availability of good quality seed and planting materials, approved inputs *etc.* Farmers need to be encouraged and sensitized to upgrade their technical capabilities and production know how for export oriented production. Dissemination of information and technology transfer by the concerned Govt. organizations, research institutions, NGOs and private players for strengthening the backward linkage.

Adherence to the quality parameters

Each exporting Country needs to adhere to the SPS requirements for international trade for quality assurance. Any shortfall in this aspect not only leads to rejection of export consignments but also affect the image of the Country. All concerned stakeholders in the supply chain *i.e.* producers, traders, processors, exporters *etc.* with supervision of the State Governments should ensure SPS adherence and monitoring of usage of pesticides/chemicals, implementation of the export norms related to farm production and traceability at the farm level.

Post Harvest Management and Value Addition

Although India is the 2nd largest producer of fruits and vegetables, we are lacking far behind in processing. Development of horticulture sector should encompass simultaneous growth in the agro processing sector. Processing not only reduces the post harvest losses of fresh produce but also increases the value of the produce multiple times. Volumes saved in post-harvest losses are actually the surpluses generated without additional cost. Contract farming ensures that the product is cultivated as per requirement of the processor with buy back arrangements. Here, the contractor (processor Company) provides the technical knowhow, inputs *etc.* to the farmer. Contract farming of potato for chips, gherkins for export *etc.* are some of the examples. Similarly cultivating the right variety of horticulture crop such as tomato, mango, cucumber *etc.* with buy back arrangements for processing to jam, jelly, pickles *etc.* Development of horticulture based agro processing will help in developing the SMEs making India self reliant towards fulfilling the objectives of Hon'ble Prime Minister Atma Nirbhar Bharat.

Research and Development Activities

The research on horticultural crops needs to be demand driven with technological up gradation, improved variety, pest management *etc.* To enhance exports, there is need for development of new and improved varieties for in demand products and varietal replacement such as in grapes and onions, increasing shelf life of perishables such as Banana, Development of sea protocols for export of perishables by sea for high value products such as Banana, Mango *etc.*, research on increasing yield and productivity *etc.*

Infrastructure and Logistics to Facilitate Agricultural Exports

Perishable horticulture produce need adequate infrastructure such as pack houses, processing unit, cold storages, and refrigerated vans for catering to export requirements in major cities near the production areas. Setting of infrastructure near the exit points such Centre for perishables cargoes (CPC) at the major airports, green channel at sea ports, adequate number of functional container plug in points, Setting up of ICD to handle container cargo, increase rail connectivity for transfer of perishables *etc.* are required to ensure that quality is not compromised across the supply chain from farm to fork.

Market Intelligence, Branding and Promotion

Information about the importing Country's demand, requirements and trade performance helps to make an informed choice of product to be exported and destination. The products for export need to be identified that have a demand in the market in terms of Consumer preference and food habit. Assessing India's strength in exporting those products in terms of taste, quality, special tags such as GI, Organic *etc.* New markets can be explored for products like Grapes, Mangoes *etc.* where India has export strength. Similarly, new products can be introduced in existing markets. Products must confirm with the regulatory regime of the importing countries which can be verified through a quality assurance system, building trust in the consumers. Adequate market promotion, branding, attractive packaging, participation in trade fairs, exploring new platforms for exports through e commerce will help in leveraging India's strength in horticulture exports. Horticulture is a technology driven sector. Crop diversification from staples to horticulture crops needs to be done in a planned manner to harness the maximum benefit and make it economically viable.

Govt Initiatives to Boost Export of Horticultural Produces/Products

The PM's Gati Shakti Yojana

The Prime Minister during his Independence Day speech of 2021 had emphasised that the National Master Plan, GatiShakti would help realize the dreams of crores of our countrymen.

The GatiShakti programme marks a paradigm shift in decision making to break the silos of departmentalism. In the proposed Plan, all the existing and proposed economic zones have been mapped along with the multimodal connectivity infrastructure in a single platform. Individual projects of different line Ministries would be examined and sanctioned in future within the parameters of the overall Plan, leading to synchronisation of efforts. GatiShakti will bring synergy to create a world class, seamless multi-modal transport network in India. The National Master Plan will employ modern technology and the latest IT tools for coordinated planning of infrastructure and will boost trade and commerce in Agri/horti domain.

District as an Export Hub and One District One Product scheme

The scheme adopts the One District One Product (ODOP) approach to reap the benefit of scale in terms of procurement of inputs, availing common services and marketing of products. ODOP for the scheme will provide the framework for value chain development and alignment of support infrastructure. There may be more than one cluster of ODOP products in one district. There may be a cluster of ODOP products consisting of more than one adjacent district in a State.

An illustrative list of such products includes mango, potato, litchi, tomato, tapioca, kinnu, bhujia, petha, papad, pickle, millet-based products, fisheries, poultry, meat as well as animal feed among others. Besides, certain other traditional and innovative products including waste to wealth products could be supported under the Scheme. For example, honey, minor forest products in tribal areas, traditional Indian edible items like turmeric, amla, haldi, *etc.*

Out of 707 ODOP, a total of 403 ODOP are meant for fruits and vegetables which certainly help the farmers to get good price for their produce. It is also expected that under the cluster development programme (CDP) of APEDA, GOI.

Establishment of NABL Accredited Labs to Test Pesticide Residue Analysis

Increasing awareness regarding good nutrition through quality food is creating a rise in the demand for fruits and vegetables.

To increase export of these commodities, concurrent developments in the areas of state-of-the-art cold chain infrastructure and quality assurance measures are required. Ensuring pesticide residue free produce through stringent testing of produce is a key requirement to this end. The trade of fruits, vegetables and other food materials require that the level of pesticide residues in the produce be less than the permissible level (MRL) of a pesticide residue in a particular commodity set by the importing country or an international organization, *viz.* Codex Alimentarius Commission, FAO/WHO. The pesticide residue level should be below the MRL if good agricultural practices (GAP) are followed.

Establishment of Cold Chain Facility in India

It is imperative to establish cold chain facilities right from farm to the folk to maintain the quality, increase shelf life, prevent losses and bring economies of scale. It is important to synergize the efforts of private and public in this direction.

Good Agricultural Practices (IndGAP)

In order to enable farm produce to be internationally competitive innovative farming practices incorporating the concept of globally accepted Good Agricultural Practices (GAP) within the framework of commercial agricultural production for long term improvement and sustainability is essential. GAP in addition to improving the yield and quality of the products, also has environmental and social dimensions. Implementation of GAP would promote optimum utilization of water resources such as pesticides, fertilizers, and water and eco-friendly agriculture. Its social dimension would be to protect the agricultural workers' health from improper use of chemicals and pesticides.

Application of Digital Tools

The challenge to meet food and nutritional requirements of 9.7 billion people by 2050 is a tough one in light of the constraints such as land availability and increased cost of farming inputs. Increasing fruits output could be achieved by deploying advanced digital tools. Besides high yield seeds and fertilisers, digital tools are capable of helping farmers — both big and small — in decision making for more efficient and effective resource use while maximising yields.

It will help in the right management of the soil to meet the needs of the tree such as moisture level, nutrient content, and also for weed and pest control. Smart farming or the application of IoT in the orchard, is the most awaited change agent in India to meet the twin challenges of food security and farmers' welfare.

FTA Negotiations with the US, UK, EU, and Australia

The Indian side needs to continue to flag issues for removing trade barriers and facilitate market access enabling it to integrate into the global value chain of fruits trade. With import norms for fresh fruits becoming stringent regarding the use of agrochemicals and pesticides, raising awareness among producers about these issues is equally important

For the long run, the sector would need to be supported by the government. Import of technologies will also help in enhancing the productivity and skills in the horticulture sector. For instance, in the past, through the Indo-Israel project, Israeli technology "plastic mulching" has improved the yields of tomato farms in Tamil Nadu. Similarly, India should adopt Israel's drip irrigation system which has been proven to be efficient in avoiding water wastage.

Consolidation and Marketing Support

Consolidation of farm holdings is considered a very vital step in boosting demand of horticulture products. Marketing must be supported by integration at the backend. The need of the hour thus, is to consolidate, because it's impossible for farmers to have huge land bank holdings due to the law as well as the division of property when handed down from generation to generation. It is only then that standardization and fair management can be brought into the production process where farmers coming together can take bigger and bolder decisions.

To build scale and sophistication in the value chain, the Centre's Horticulture Cluster Development Programme could act as a game changer. Official estimates expect it to benefit about 10 lakh farmers and move closer to doubling their income as they benefit from access to a common pool of resources and by building last-mile connectivity.

Efforts made by APEDA in each of the above factors has been noteworthy

APEDA has been assisting exporters through its financial assistance under Agriculture and Processed Food Export Promotion Scheme under the Component Infrastructure Development. The Scheme has been beneficial for setting up of modern integrated packhouses having facilities as per requirements of importing countries. In addition, the scheme covers assistance for purchase of refrigerated vehicles to support the supply chain in horticulture. The above two components have largely helped exporters of fresh horticulture produce like mangoes, pomegranates, grapes, bananas, fresh vegetables and floriculture sector. The modern integrated pack houses which are able to meet the requirements of Quality and post-harvest management have transformed the horticulture export sector.

Supply of High Quality Input Material

Government of India has announced Agri Export Policy (AEP) for the first time in the year 2018 in the country. The policy aims at achieving doubling farmer's income and doubling exports with focus at providing adequate resources to farmers to become direct exporters without intermediaries. Under this policy, 14 product clusters spread across in 6 states dedicated to horticulture products have been identified and looking at the success of these clusters, State Governments are approaching APEDA and Ministry of Commerce for approval of more and more product based clusters. Exports has shown increasing trend, due to focused approach adopted in these clusters. Banana exports have increased from cluster of Anantpur and Theni in Tamil Nadu; Jalgaon and Solapur in Maharashtra.

Other successful models have been for grapes from Nashik, vegetables from Uttar Pradesh and Bihar. Yet another farmer centric model of cluster development has taken place in the Varanasi (UP). APEDA has been able to arrange MOU between FPOs/FPCs with exporters for organizing procurement of quality produce directly by FPC for exports.

Market Development

APEDA's Scheme of Market Development helps participation of exporters in the trade fairs. Efforts have been made by organizing product specific and country specific Buyer Seller Meets (BSMs) to help branding of Indian Horticulture Produce with the help of concerned Embassy or Consulate in the major potential countries by APEDA.

Last year, APEDA organized series of Virtual BSMs and B2B meetings. This exercise led to direct exposure of Indian farmers/FPC to the world.

Conclusion

The horticulture sector in India possesses unique advantage because of availability of varied climatic conditions for production of various horticultural crops and several ICAR institutes and SAUs are working for the benefit of farmers and other stakeholder. Besides, there is huge varietal diversity in most of the cultivated fruits, vegetables, flowers, plantation and spices which increase the harvesting season and more revenues to the farmers. The production and supply of quality planting materials, implementation of INM, IDM and IPM, multi cropping system, integrated farming system (IFS), pollinators, pre and post harvest management, forward and backward linkages in marketing are very important for successful cultivation of horticultural crops. The losses in selected fruits and vegetables were found to be in the range of 5.8 per cent - 18.0 per cent. In spite of higher production, the quantity of processed products from the raw materials is low in most of the horticulture crops. Several prediction models have suggested that the area under horticultural crops going to be increasing and cereal crops will be decreasing, hence, its time to breed the climate resilient varieties to with stand the climatic aberrations. Hence, a comprehensive plan covering R and D aspects for all horticultural crops is must and convergence of scientists, policy makers and public is the need of the hour for marching ahead and take the horticulture in to a newer height.

Advancement of the horticulture sector in India has been of significant priority in the country's overall development strategy. An important sub-sector to agriculture, horticulture has grown rapidly, contributing over 30 percent to the GDP of agriculture in India. There has been an increased emphasis from various government initiatives to consider the horticulture sector at par with the industrial sector, in order to ensure substantial income from farming. Doubling farmer's income, building infrastructure, infuse of technology, agriculture export policy and boosting agricultural research and education, are all steps taken envisioning the growth of the agriculture and its subsectors, and farmer welfare.

India's diverse climate gives it the advantage of growing many fresh fruits and vegetables all around the year. This vast production base coupled with logistical advantages to the Middle East and South East Asia offers India tremendous opportunities for global export. In order to capitalize on this market opportunity, India's horticultural revolution has to be able to invest and implement in research and innovation to gain more efficient methods of crop production, refined post-harvest storage and handling methods, efficient supply chain, and innovative methods of knowledge dissemination to cultivators and consumers alike.

With the growing demand of worldwide consumers for safe and healthy foods, increased urbanization of societies, the growth in scale and influence of supermarkets, and packaged foods, corporatization of farming, and adoption of sustainable farming practices, there is a huge market potential for India to become the global hub for fruits.

Chapter 24

Physiology, Quality and Safety of Fresh-Cut Fruits and Vegetables

Sunil Pareek

Department of Agriculture and Environmental Sciences, National Institute of Food Technology Entrepreneurship and Management, Kundli, Sonepat 131028, Haryana

e-mail: sunil_ciah@yahoo.co.in

ABSTRACT

Although fresh-cut fruits and vegetables are growing segments in convenience food, but simultaneously they face many problems. These are very sensitive to microbial attack, browning and other color changes, water loss, nutritional losses, etc. Various chemicals are using by the industry to maintain the color of fresh-cuts. However, safe and organic way of preserving the fresh-cuts is need of the hour. Understanding the exact mechanism of changes in fresh-cuts and application of holistic approach is important. Natural preservatives and edible coatings should be explored to preserve the fresh-cuts. Some newer green chemistry based technologies such as natural plant based oils or coatings, pulsed light treatments, cold plasma, ionizing radiation, nano-coatings, etc. can be standardized for fresh-cut fruits and vegetables. A complete protocol for processing line separately for conventionally grown and organically grown produce is required.

Keywords: *Fruits, Safety, Processing, Enzymatic activities, Packaging.*

Minimally processed fruit and vegetables are defined as any fruit and vegetable that has been subjected to different processing steps (*e.g.*, peeling, trimming, cutting, washing, disinfection, rinsing, *etc.*) to obtain a fully edible product while providing convenience and functionality to consumers and ensuring food safety. The International Fresh-Cut Produce Association (IFPA) defines fresh-cut products as fruit or vegetable that have been trimmed and/or peeled and/or cut into 100 per cent usable product that is bagged or pre-packaged to offer consumers high nutrition, convenience, and flavour while maintaining its freshness (Figure 1).

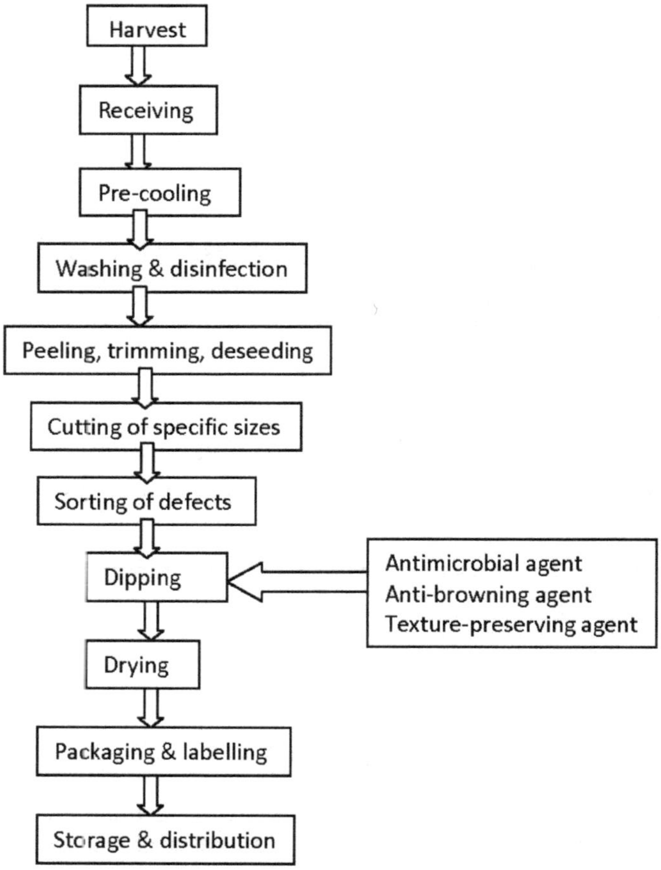

Figure 1. Typical fresh-cut process flow chart for fruits, vegetables and root crops.

In particular, fresh-cut fruits attract consumers because they are fresh, nutritious, low priced, and ready-to-eat. As a consequence, a wide assortment of minimally processed fruits has been developed to meet consumer's need for 'quick' and 'convenient' products, and to benefit from fruit's healthy image. Minimal processing gives additional value to fresh-cut fruits in terms of convenience and time saving, although several hurdles are encountered due to the difficulty in preserving their freshness during prolonged periods. These products, in fact, are characterized by shorter shelf life than their whole counterparts, because of higher susceptibility to microbial spoilage, increased respiration rate and ethylene production, which is stimulated by wounding of the tissue; in fact, the process operations (*i.e.*, cutting, splicing *etc.*) from lesions in the tissue that determine enzymatic browning, texture decay, rapid microbial growth, weight losses and undesirable volatile production, thus reducing, highly the shelf life.

Quality is a term which denotes a degree of excellence, a high standard or value. Quality of foods may be defined as the composite of those characteristics that differentiate individual units of a product, and have significance in determining the degree of acceptability of that unit to the user. Minimal processing of fresh fruits and vegetables, such as trimming, peeling, cutting, slicing and other physical actions, causes injury and damage to tissues, affecting physiological activities and subsequently quality (Pareek, 29). Some problems related to cell disruption are leakage of nutrients, enzymatic reactions, mold growth, lactic acid fermentation, loss of texture, development of off-flavors and off-odors, and appearance defects. All these factors limit the storage and market life of fresh-cut fruits and vegetables.

Now-a-day fresh-cut fruits and vegetables have tremendous market both in developed and developing countries. It is one of the fastest growing sector in developing countries as well. Therefore, it attracts the researchers to develop eco-friendly technologies to maintain the quality and extend the shelf life of fresh-cut products. In this paper, fresh-cut products problems particularly enzymatic browning, its mechanism, and prevailing and emerging technologies, physiology, quality, and safety aspects are discussed.

Mechanism of Enzymatic Browning in Fresh-cut Fruits and Vegetables

Appearance and texture changes are two fundamental characters determining the acceptability of fresh-cut fruits and vegetables. Browning of cut surface is a problem with several products such as apple, eggplant, kiwi, lettuce, mango, peach, pear, pomegranate arils, potato, rose apple, and sweet peppers. The understanding of the processes leading to these changes is essential in developing better approaches to minimize them and, hence, improving quality and shelf life for the consumer. One common issue that much of the biochemistry of appearance and textural changes has been, by-and-large, studied in whole plant or tissue systems. The consequences of the cutting operation and post-cutting processes in fresh-cut products have not been extensively studied. It is estimated that over 50 per cent losses in fruit occur as a result of enzymatic browning. It is one of the most important color reactions that affect horticultural commodities, leading not only to tissue discoloration but also to the development of off-flavors.

While the tissue, or cell structure, remains intact, the enzyme and phenolic compounds are separated and so do not react with one another. It is the breaking down of the cell structure caused by the bruising that initiates the oxidation reaction and so triggers the onset of browning. An enzyme called polyphenol oxidase (PPO) oxidizes the phenolic compounds that are found in the tissue of fruits. The oxidation causes the phenolic compounds to condense into brown spots. Therefore, the enzymatic browning in several fresh-cut fruit and vegetables starts with cellular disruption, causing the release of phenolic compounds stored within the vacuoles in cell wall compartments. Once oxygen penetrates the wounded tissue, the phenolic compounds are the substrate for the enzyme PPO (Yoruk and Marshall, 47). The PPO in the presence of oxygen can catalyse two different reactions: the hydroxylation of monophenols and the oxidation of *o*-Quinones a non-enzymatic polymeraization of quinines occurs, leading to the formation of melanins that are pigments of high molecular mass and dark in color (Figure 2) (Queiroz *et al.*, 30).

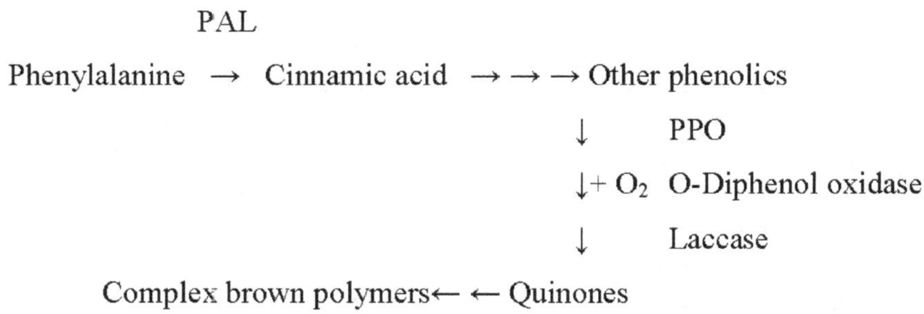

Figure 2. Mechanism of enzymatic browning in fresh-cut fruits and vegetables.

Phenolic compound synthesis is associated with the endoplasmic reticulum (ER). The proteins involved with their synthesis are either incorporated into the ER membrane or are loosely associated with it. Once formed, these compounds are glycosylated and then are extruded within transport vesicles formed from the ER membrane. These vesicles are the vehicle by which the phenolic compounds are transported to the vacuole or into the apoplast/cell wall compartment. There are smaller quantities of phenolic compounds which may be found in chromoplasts, cytoplasm and the mitochondria, but these are normally minute amounts and are associated with specialized metabolic functions. Once a physical stress or deteriorative process (*e.g.*, wounding response or senescence) is initiated, the compartmentalization of the cell begins to fail (Marangoni *et al.*, 25). The consequence of this is the mixing of polyphenol substrates (*e.g.* catechin, polyphenols) with PPO and/or phenolperoxidase (Degl'Innocenti *et al.*, 11). This hypothesis is supported by work with fresh-cut potatoes that clearly shows that browning is not rare limited to by either the enzymes associated with browning or polyphenol substrate concentration (Cantos *et al.*, 7). The authors suggest that membrane stability is potentially a major factor controlling the rate of browning. The peroxidase (POD) activity can also cause browning in the presence of hydrogen peroxide. Hydrogen peroxide concentration is low in fruits and vegetables, so the contribution of POD in the post-cut browning is considered to be lesser than that of PPO (Mayer, 26).

Secondary Browning

Term secondary browning first time used by P.M.A. Toivonen (Toivonen *et al.*, 41). Studies have shown that browning of fresh-cut products is not only caused by broken cells but is also caused by a microbially associated exudation of enzymes, leading to a phenomenon coined as secondary browning (Toivonen, 37). Secondary browning occurs as a result of cross-contamination between outer surfaces of fruit slices with cut surfaces in post-cut solutions but it is not strictly correlated to the microbial counts on the cut surfaces. Secondary browning can be distinguished from cut-edge browning in two ways: (1) secondary browning occurs sometime after cutting, usually towards the expected limit of shelf life, while cut-edge browning occurs within hours of cutting, and (2) secondary browning is localized in nature, whereas cut-edge browning tends to be diffuse across most of the cut edge.

Secondary browning occurs as a consequence of cross contamination between apple slices in post-cut dip solutions. While the first dip solution, after cutting, contains chlorine or some other sanitizer, apple slices very quickly quench these oxidation-based agents (Delaquis *et al.*, 12). As a result, it is extremely difficult to maintain microorganism-free dip solution.

There are many solutions to the secondary browning problem, but often these treatments are not safe for food use or they impart undesirable flavours on fresh-cut produce (Rupasinghe *et al.*, 31). Gaseous, natural anti-microbial agents have been studied, such as hexanal vapour and acetic acid vapour (Sholberg and Randall, 36); however, such materials are difficult to handle in the commercial setting for fresh-cut products. Secondary browning is effectively controlled with the addition of vanillin, a natural food-grade flavoring agent, which is considered GRAS (generally recognized as safe) (Toivonen and Delaquis, 39).

There are very few reports that compare 1-MCP application procedures in processing of fresh-cut fruit. In fresh-cut apple, 1-MCP does not appear to reduce cut-edge browning and can even aggravate incidence of secondary browning (Toivonen, 38). Since secondary browning is associated with microbial activity, it has been suggested that an anti-microbial reagent must be provided if 1-MCP is applied on fresh-cut fruit (Toivonen, 38). A new technology employing a co-release of 1-MCP and yeast-generated volatiles has been shown to control secondary browning in apple slices (Toivonen and Lu, 40). This sachet-based technology reduces secondary browning of fresh-cut products while delaying ripening of the produce.

Technologies to Control Enzymatic Browning and Preserve Fresh-Cut Fruits and Vegetables

Various chemical or physical or combinations of both chemical and physical methods are in use to control or reduce the browning in fresh-cut fruits and vegetables (Table 1). Chemical methods are based on reactions to inhibit the enzymatic activity, remove the substrates which are essential for browning (oxygen and phenolics) or function as preferred substrate. Use of reduced/low temperatures, reduction in oxygen availability, modified atmospheres, high pressure processing (HPP), or gamma irradiation are used as physical means of treatments to inhibit/reduce/control of enzymatic browning.

Table 1. Treatments to control enzymatic browning in fresh-cut fruits and vegetables

Sl.No.	*Treatments to Control Enzymatic Browning*
A.	**Chemical methods**
	Acidulants
	Citric acid
	Reducing agents
	Ascorbic acid
	Cysteine
	Chelating agents
	Ethylenedismine tetra acetic acid (EDTA)
	Sporix

Sl.No.	*Treatments to Control Enzymatic Browning*
	Complexing agents β-cyclodextrin
	Enzyme inhibitors 4-hexylresorcinol
	Other antibrowning agents Sodium chloride Honey Proteases Benzoic acid Cinnamic acid
B.	Physical treatments
	Reducing oxygen availability Modified atmosphere packaging Edible coatings
	Reducing temperature
	Gamma radiation
	Use of other non-thermal technologies High pressure technology Pulsed electric fields Ultrasonic treatments Cold plasma

The testing of different anti-browning agents, and combinations of agents, seems to be the favored approach in the study of color control. Dipping is not only a way for using additives on minimally processed produce, but also a way to rinse enzymes and substrates released by disrupted cells, after cutting, from fruit surfaces. The choice of anti-browning agent should take into account the legislation regarding additives in each country, *e.g.*, Itallian legislation allows the use of ascorbic acid (AA), citric acid (CA), calcium ascorbate and sodium ascorbate as additives in minimally processed fruits and vegetables. In the European Community, government organizations are stressing food safety from not only a microbiological point of view, but also regarding additives and pesticide residues. European Regulation stipulates that any potential allergenic substance must be shown to the consumer in food label specifications (Salcini and Massantini, 32).

Factors Affecting Quality and Safety of Fresh-Cut Fruits and Vegetables

Many factors have a role on the final quality of fresh-cut produce. Genotype, growing conditions, cultural practices, and maturity stage at harvest may greatly influence initial quality of the product to be processed; on the other hand, postharvest handling and storage of raw material and processing conditions, including finished product fate throughout the distribution chain, markedly determine its final quality.

Genotype

Genotype is the very first factor to be considered since post-cutting and particularly susceptibility to browning may widely differ from cultivar to cultivar. Differences in browning potential may be related to the different phenol content and to the rate of enzymatic activities of PPO, PAL, POD, other than to the content of antioxidant compounds (*i.e.* ascorbic acid) which prevent oxidation of phenol compounds (Cornacchia *et al.*, 9). The activity of PPO was found to be cultivar dependent in potato (Cornacchia *et al.*, 9) and apples, where 'Red Delicious' showed an enzymatic activity 10 times higher than 'Elstar' cultivar (Janovitz-Klapp *et al.*, 20). Several studies reported the different browning potential and shelf life for different cultivar of the same crop as for shredded cabbage (Yano and Saijo, 46), apples; Luo and Barbosa-Cánovas, (24) reported that 'Jonagold' cultivars had the lowest browning rate and 'Braeburn' the highest, while Kim *et al.* (21) found that 'Delicious' cultivars had the lowest browning rate whereas 'Monroe' had the highest. Studies on fresh-cut pears also showed a wide browning susceptibility among different cultivars; Gorny *et al.* (18) and Dong *et al.* (14) found 'Bartlett' pears and 'Anjou', respectively, as the cultivars showing less color changes after cutting.

Cultural Practices

The effect of different irrigation regime on quality and storability of fresh-cut 'Iceberg' lettuce and 'Romaine' lettuce were investigated by Luna *et al.* (23). On both species stored in air, browning increased with the increase of irrigation, and was related to an increase of PPO activity, even though no differences were observed in MA packaged samples due to the positive effect of the gas composition. In addition, microbial loads at harvest were higher in whole heads subjected to the highest irrigation regime for fresh-cut 'Romaine' lettuce, although this difference was not detected during storage, while an increase in off-odor was related to the highest irrigation regime for the 'Iceberg' lettuce. On 'Iceberg' lettuce other authors also investigated the effect of salinity applied in floating system. Pre-harvest salinity of 3.8 and 4.8 m S cm^{-1} improved the properties of fresh-cut lettuce, reducing CO_2 production with a subsequent control of the decay process. Salinity treatments also reduced enzymatic oxidase activity and browning phenomena during storage and helped to control the growth of mesophilic bacteria and of moulds and yeasts (Scuderi *et al.*, 34).

Maturity

Maturity and or physiological stage at harvest and time of harvest for plants with long productive cycles may strongly affect post-cutting performances and not only for fruits. For 'Iceberg' lettuce, in fact, maturity stage is significantly related with the browning susceptibility and the development of off odors in the packaging. Immature lettuces after cutting showed less phenolic and less browning than that of more mature lettuce (Couture *et al.*, 10), but another study reported that higher CO_2 concentrations and higher accumulation of ethanol and acetaldehyde were detected in the headspace of MA-packed lettuce from immature heads, compared to over-mature heads. Moreover, differences in respiration rate of the fresh-cut product from heads cultivated during the winter–spring seasons were very high; respiration rate of fresh-cut 'Iceberg' lettuce increased when medium temperature during cultivation increased and this factor need to be taken in account when designing packaging in order to avoid fermentative conditions (Tudela *et al.*, 42). Many studies report the effect of maturity stages which is very important for sensorial quality and shelf life of cut fruits. Honeydew fresh-cut melon with 8.8 per cent soluble solids failed to retain typical flavour and deteriorated more rapidly than those with 13 per cent soluble solids (Watada and Qi, 45). Significant flavor differences were reported for mature-green versus ripe mangoes (Bender *et al.*, 5), as well as from fresh-cut slices obtained from firm-ripe and soft-ripe mangoes (Beaulieu and Lea, 4).

Storage Conditions before Processing

Appropriate ambient conditions should be applied when storing fresh fruit and vegetable before processing, also considering the need to decrease product temperature before cutting in order to delay metabolic tissue responses after wounding. Moreover, for fruits which are normally stored for long time, conditions applied during storage may have a strong impact on shelf life of the final product (Gorny *et al.*, 17; Calderon-Lopez *et al.*, 6). Bartlett pears held at -1°C in a CA of 2 per cent O_2 + 98 per cent N_2 had a longer post-cutting life than those held in air at -1°C for the same duration (Gorny *et al.*, 17). Testing the effect of CO_2-enriched atmosphere and DPA pre-cutting treatment on different apple cultivars, it was found that slice color was lighter when fruits were previously stored in elevated CO_2-atmospheres, but atmosphere and DPA effects varied according to the cultivar and were affected by the storage time before slicing. The optimum CO_2 concentration for raw material in order to inhibit browning with lowest accumulation of fermentative products was 30 per cent for 'McIntosh', 15 per cent for 'Empire' and 45 per cent for 'Delicious'. Storage of apples for 6 weeks prior to slice preparation reduced the firmness and increased the amount of acetaldehyde, ethanol and ethyl acetate produced by cut apples; therefore processing apples close to harvest or maintaining whole apple fruit quality by CA or 1-MCP will produce slices with the best quality and longest shelf life (Amissah *et al.*, 2).

Emerging Sanitation Techniques

Ozone Based Technology

In 2001, the U.S. Food and Drug Administration (FDA) declared ozone as Generally Recognised as Safe (GRAS) substance for commercial use as a disinfectant and sanitizer in foods and food processing, as long as good manufacturing practices are provided. Thus, the application of ozone has received interest from industries in terms of food preservation, surface hygienization, sanitation, water disinfection and wastewater reutilization. However, the susceptibility of microorganisms to ozone depends on the physiological state of the culture, pH of the medium, temperature, humidity and presence of additives like acids, surfactants and sugars (Sharma, 35). Other factors, such as the nature and composition of food surface, type of microbial contaminant and the degree of attachment to or association of microorganisms with food also limit the success of the sanitation process. The organic matter of food may compete with microorganisms for ozone, limiting the accessibility of ozone to food contaminants (Sharma, 35).

Ultraviolet Light C Radiation

UV-C treatments consist on the food exposition to radiation for a certain period of time, under a bank of UV lamps with a maximum peak emission at 254 nm, which is the most efficient wavelength for damaging the DNA of microorganisms. The lamps typically used in UV disinfection consist of the usual low-pressure mercury arc lamps with a monochromatic output at 254 nm (conventional and amalgam types), and the medium-pressure mercury arc lamps and pulsed lamps with a polychromatic output in all UV wavelengths (however concentrated at certain peaks inside of the UV-C region). Conventional low-pressure UV lamps are the most efficient in terms of energy consumption, while amalgam UV lamps offers up to three times the UV-C radiation output compared with conventional types. Lamps that emit polychromatic light continuously have a medium light intensity. Medium-pressure UV lamps produce more spectral lines, being less electrically efficient but produce much greater UV-C radiation intensity (less treatment time is required). Some juices, fruits and vegetables, meat and fish can be processed by UV light aiming at microbial load reduction. The critical parameters to be considered when designing a convenient UV radiation process for controlling microbial contamination, extension of shelf life and maintenance of satisfactory sensory and nutritional quality are the: type of food, type of food surface, food composition, intensity of radiation and exposure time (Guerrero-Beltrán and Barbosa-Cánovas, 19).

Ultrasounds Technology

Ultrasonication has been successfully used to clean equipment or to sterilize liquids. However, it is unlikely that ultrasounds technology will be used for fruits and vegetables preservation, because their tissues have different acoustical properties from those of biological and industrial materials. The ultrasounds effects may cause denaturation of proteins that will result in a decrease of enzymatic activity. However, short bursts of ultrasounds may increase enzyme activity. This may be explained by large molecular structures that break down, making enzymes more accessible to react with substrates. Thus, the resistance of most microorganisms and enzymes to ultrasounds is so high that the intensity of treatment would produce adverse changes in the texture and other important physical properties of the food products (Fellows, 15). Other important limiting factors are the dispersion of void and pores that influence the monitoring and interpretation of data. However, the development of sensors, microprocessors and methods of signal analysis promote the use of ultrasounds. At industrial level, it is necessary to study the appropriate frequency and power requirements for different products and to choose the best equipment and measurement tests. Nevertheless, ultrasounds technology has been used in the food industry, either for analysis or for modification of foods (Knorr *et al.*, 22). Low intensity ultrasounds have been used for non-evasive detection, providing information about physical-chemical properties of food materials and have been used in relaxation phenomena studies (Knorr *et al.*, 22). High intensity ultrasounds have been used to induce physical and chemical changes of many food properties. Processes such as extraction, emulsification, homogenization, separation, extrusion, cell disruption, degassing of liquid foods, viscosity alteration, promotion of chemical oxidation/reduction, enzyme and protein extraction, meat softening and modification of crystallisation processes may be enhanced by ultrasonication (Pareek, 29). The use of ultrasounds as a food preservation process is therefore limited by the food complexity and its protective effect on microorganisms. Nevertheless, there are other processes that can be combined with ultrasounds. For example, sonication has been frequently combined with heat (thermosonication), with pressure (manosonication) or with pressure and heat (manothermosonication) to sanitize fresh products. These hurdle combinations may cause additive, antagonist or synergistic effects, depending on the type of combination and on safety or quality targets. For fruits and vegetables, these combined processes may improve applications, since minor changes in quality characteristics occur, when compared to conventional thermal processes.

High Pressure Processing (HPP)

HPP is becoming a commercially implemented technology throughout the world to process a wide range of different food products. Several pressurized products are already available in the global marketplace, due to some initiatives of food industries and to the positive responses of consumers to these products (Alexandre *et al.*, 1). In fruits and vegetable sector several studies have been developed. Zhou *et al.* (49) studied the effect of high hydrostatic pressure on the microbiological quality and physical–chemical characteristics of pumpkin during refrigerated storage. These authors concluded that 550 MPa for 10 min ensures physicochemical features decontamination of pumpkin. Denoya *et al.* (13) studied the effect of high pressure processing and vacuum packaging on the preservation of fresh-cut peaches. They observed that peaches subjected to this combined processes had lower ethanol content and enzymatic activity, which evidenced that fermentation and browning were successfully inhibited by this combination.

To assure the efficacy of HPP treatment the following must be taken into consideration: pressure applied, come-up time to achieve a given pressure, holding time, decompression time, nature of pressure treatment (cyclic or continuous), initial temperature of food, process temperature, temperature distribution in the pressure vessel due to the adiabatic heating, characteristics of the food (such as pH, composition and water activity), packing material and type of microorganisms present in food products (Venugopal, 43).

Edible Coatings and Films

One of the most important limitations in the application of edible coatings/films to fresh-cut products are related with the fact that coatings/films efficacy is dramatically affected by water availability, so the moisture content of the foodstuff and the relative humidity in the storage environment should be taken into account in order to develop effective films and coatings. Another limitation is related with the use of essential oils; their strong flavor can negatively affect the original taste of the fresh-cut products (Ciolacu and Nicolau, 8). The impact of coatings/films on safety and quality attributes of fresh-cut products has been reported in some works. Oms-Oliu *et al.* (27) applied edible coatings with antibrowning agents to maintain sensory quality and antioxidant properties of fresh-cut pears. Oms-Oliu *et al.* (28) used polysaccharide-based edible coatings to enhance quality and antioxidant properties of fresh-cut melon. Ghidelli *et al.* (16) extended the shelf life of fresh-cut eggplant with a soy protein–cysteine based edible coating and modified atmosphere packaging. Salvia-Trujillo *et al.* (33) used antimicrobial nanoemulsions as edible coatings to improve safety and quality attributes of fresh-cut 'Fuji' apples. Arnon *et al.* (3) developed polysaccharides-based edible coatings for citrus fruits. They observed that the layer-by-layer coatings notably improved all quantified quality parameters, concluding that polysaccharide-based edible coating may offer an alternative to synthetic waxes.

Packing with Gaseous Atmospheres

In the last years, the development, validation, and research of optimum MAP conditions for individual fresh fruits and vegetable have been actively pursued globally (Zhuang *et al.*, 50). Waghmare and Annapure (44) studied the combined effect of chemical treatment and MAP on fresh-cut papaya, concluding that sensory and quality characteristics of papaya were highly retained for 25 days. Zhang *et al.* (48) studied the effect of high oxygen and high carbon dioxide atmosphere packaging on the microbial spoilage and shelf life of fresh-cut honeydew melon. They observed that an atmosphere with 50 per cent O_2 and 50 per cent CO_2 extended the shelf life of the product.

Some Considerations about Fresh-cut Fruit

The fresh-cut fruit industry is projected to grow continue in the forthcoming years. However, there are still several challenges to face on the research and on the processing side in order to assist and increase this expected grow. The first area to exploit is the development of more automated and less invasive technologies to minimize the amount of employed labor (potential source of contamination) and to improve quality of the final product. This specifically applies to the processing phase but need to be extended to the entire processing chain, from the growing fields to consumers ("from farm to fork"). At the same time, attention should be dedicated to investigate new technologies and treatments addressed to extend the microbiological shelf life in order to prevent the growth of pathogenic microorganisms. Little research has been performed on food-borne human pathogens on fresh-cut fruits. This is an area that shall have to be duly investigated in the future. It has been also noticed that irradiation of fresh-cut fruit products may be beneficial in considerably reducing the number of bacteria present on the product. Lastly nontraditional sanitizers and new compounds from natural sources for coating and treatment needs further research, to offer healthier products to the consumers.

All these different steps will give the opportunity to the processors to achieve more stable products while meeting higher hygienic standards. This is the key to reinforce the trust of the consumers on fresh-cut fruit based products. It is recommended that in the future researchers, manufacturer and the industry will work together to overcome barriers that hamper delivery of high quality fresh-cut fruits throughout the year. In fact, as the food service industry and home meal replacement expand, there will be a greater demand for fresh-cut fruits with acceptable shelf life.

REFERENCES

1. Alexandre, E.M.C., Silva, C.L.M. and Brandão, T.R.S. 2014. Traditional and emerging technologies for strawberry processing. In: *Strawberries: cultivation, antioxidant properties and health benefits*, Malone, N. (Ed.), Nova Science Publishers, Inc., New York, USA, pp. 73-92.
2. Amissah, J.G.N., Hotchkiss, J.H. and Watkins, C.B. 2006. Diphenylamine and pre-slicing storage effects on the responses of apple slices to elevated CO_2 atmospheres. *Postharvest Biol. Technol.* **39:** 178-184.

3. Arnon, H., Granit, R., Porat, R. and Poverenov, E. 2015. Development of polysaccharides-based edible coatings for citrus fruits: a layer-by-layer approach. *Food Chem.* **166**: 465-472.
4. Beaulieu, J.C. and Lea, J.M. 2003. Volatile and quality changes in fresh-cut mangos prepared from firm-ripe and soft-ripe fruit, stored in clamshell containers and passive map. *Postharvest Biol. Technol.* **30:** 15-28.
5. Bender, R.J., Brecht, J.K., Baldwin, E.A. and Malundo, T.M.M. 2000. Aroma volatiles of mature-green and tree-ripe 'Tommy Atkins' mangoes after controlled atmosphere vs. air storage. *HortScience.* **35:** 684-686.
6. Calderon-Lopez, B., Bartsch, J.A., Lee, C.Y. and Watkins, C.B. 2005. Cultivar effects on quality of fresh cut apple slices from 1-methylcyclopropene (1-MCP)-treated apple fruit. *J. Food Sci.* **70:** S221-S227.
7. Cantos, E., Tudela, J.A., Gil, M.I. and Espin, J.C. 2002. Phenolic compounds and related enzymes are not rate-limiting in browning development of fresh-cut potatoes. *J. Agric. Food Chem.* **50:** 3015-3023.
8. Ciolacu, L. and Nicolau, A.I. 2014. Edible coatings for fresh and minimally processed fruits and vegetables. In: *Global safety of fresh produce: a handbook of best practice, innovative commercial solutions and case studies*, Hoorfar, J. (Ed.), Woodhead Publishing Limited, Cambridge, UK, pp. 233-244.
9. Cornacchia, R., Cabezas-Serrano, A., Amodio, M. and Colelli, G. 2011. Suitability of 4 potato cultivars (*Solanum tuberosum* L.) to be processed as fresh-cut product: early cultivars. *Am. J. Potato Res.* **88:** 403-412.
10. Couture, R., Cantwell, M.I., Ke, D. and Saltveit, M.E. 1993. Physiological attributes related to quality attributes and storage life of minimally processed lettuce. *HortScience.* **28:** 723-725.
11. Degl'Innocenti, E., Guidi, L., Paradossi, A. and Tognoni, F. 2005. Biochemical study of leaf browning in minimally processed leaves of lettuce (*Lactuca sativa* L. var. *Acephala*). *J. Agric. Food Chem.* **52:** 9980-9984.
12. Delaquis, P., Toivonen, P., Walsh, K., Rivest, K. and Stanich, K. 2004. Chlorine depletion in sanitizing solutions used for apple slice disinfection. *Food Prot. Trends.* **24:** 323–327.
13. Denoya, G.I., Vaudagna, S.R. and Polenta, G. 2015. Effect of high pressure processing and vacuum packaging on the preservation of fresh-cut peaches. *LWT Food Sci. Technol.* **62:** 801-806.
14. Dong, X., Wrolstad, R.E. and Sugar, D. 2000. Extending shelf life of fresh-cut pears. *J. Food Sci.* **65:** 181-186.
15. Fellows, P.J. 2000. *Food processing technology: principles and practice.* Woodhead Publishing Limited and CRC Press, LLC, Cambridge, UK.
16. Ghidelli, C., Mateos, M., Rojas-Argudo, C. and Pérez-Gago, M.B. 2014. Extending the shelf life of fresh-cut eggplant with a soy protein–cysteine based edible coating and modified atmosphere packaging. *Postharvest Biol. Technol.* **95:** 81-87.

17. Gorny, J.R., Cifuentes, R.A., Hess-Pierce, B. and Kader, A.A. 2000. Quality changes in fresh-cut pear slices as affected by cultivar, ripeness stage, fruit size, and storage regime. *J. Food Sci.* **65:** 541-544.

18. Gorny, J.R., Cifuentes, R.A., Hess-Pierce, B. and Kader, A.A. 2000. Quality changes in fresh-cut pear slices as affected by cultivar, ripeness stage, fruit size, and storage regime. *J. Food Sci.* **65:** 541-544.

19. Guerrero-Beltrán, J.A. and Barbosa-Cánovas, G.V. 2004. Advances and limitations on processing foods by UV light. *Food Sci. Technol. Int.* **10:** 137-147.

20. Janovitz-Klapp, A., Richard, F. and Nicolas, J. 1989. Polyphenoloxidase from apple: partial purification and some properties. *Phytochem.* **28:** 2903-2907.

21. Kim, D.M., Smith, N.L. and Lee, C.Y. 1993. Quality of minimally processed apple slices from selected cultivars. *J. Food Sci.* **58:** 1115-1117.

22. Knorr, D., Ade-Omowaye, B.I.O. and Heinz, V. 2002. Nutritional improvement of plant foods by non-thermal processing. *Proc. Nutr. Soc.* **61:** 311-318.

23. Luna, M.C., Tudela, J.A. Martínez-Sánchez, A., Allende, A. and Gil, M.I. 2013. Optimizing water management to control respiration rate and reduce browning and microbial load of fresh-cut Romaine lettuce. *Postharvest Biol. Technol.* **80:** 9-17.

24. Luo, Y. and Barbosa-Cánovas, G.V. 1997. Enzymatic browning and its inhibition in new apple cultivars slices using 4-hexylresorcinol in combination with ascorbic acid. *Food Sci. Technol. Int.* **3:** 195-201.

25. Marangoni, A.G., Palma, T. and Stanley, D.W. 1996. Membrane effects in postharvest physiology. *Postharvest Biol. Technol.* **7:** 193-217.

26. Mayer, A.M. 2006. Polyphenol oxidases in plants and fungi: Going places? A review. *Phytochem.* **67:** 2318-2331.

27. Oms-Oliu, G., Soliva-Fortuny, R. and Martín-Belloso, O. 2008. Edible coatings with antibrowning agents to maintain sensory quality and antioxidant properties of fresh-cut pears. *Postharvest Biol. Technol.* **50:** 87-94.

28. Oms-Oliu, G., Soliva-Fortuny, R. and Martín-Belloso, O. 2008. Using polysaccharide-based edible coatings to enhance quality and antioxidant properties of fresh-cut melon. *LWT Food Sci. Technol.* **41:** 1862-1870.

29. Pareek, S. 2017. Fresh-cut fruits and vegetables: technology, physiology, and safety. CRC Press, Tylor and Francis, Boca Raton, Florida, USA. Pp. 623.

30. Queiroz, C., Mendes Lopes, M.L. Fialho, E., and Valente-Mesquita, M.L. 2008. Polyphenol oxidase: characteristics and mechanisms of browning control. *Food Rev. Int.* **24:** 361-375.

31. Rupasinghe, H.P.V., Boulter-Bitzer, J., Ahn, T. and Odumeru, J.A. 2006. Vanillin inhibits pathogenic and spoilage microorganisms in vitro and aerobic microbial growth in fresh-cut apples. *Food Res. Int.* **39:** 575–580.

32. Salcini, M.C. and Massantini, R. 2005. Minimally processed fruits: an update on browning control. *Stewart Postharv. Rev.* **3:** 1-7.

33. Salvia-Trujillo, L., Rojas-Graü, M.A., Soliva-Fortuny, R. and Martín-Belloso, O. 2015. Use of antimicrobial nanoemulsions as edible coatings: impact on safety and quality attributes of fresh-cut 'Fuji' apples. *Postharvest Biol. Technol.* **105:** 8-16.

34. Scuderi, D., Restuccia, C., Chisari, M., Barbagallo, R.N., Caggia, C. and Giuffrida, F. 2011. Salinity of nutrient solution influences the shelf-life of fresh-cut lettuce grown in floating system. *Postharvest Biol. Technol.* **59:** 132-137.

35. Sharma, R. 2005. Ozone decontamination of fresh fruits and vegetables. In: *Improving the safety of fresh fruit and vegetables*, Jongen, W. (Ed.), Woodhead Publishing Limited, Cambridge, UK, pp. 373-386.

36. Sholberg, P.L. and Randall, P. 2007. Fumigation of stored pome fruit with hexanal reduces blue and grey mold decay. *HortScience.* **42:** 611-616.

37. Toivonen, P.M.A. 2006. Fresh-cut apples: challenges and opportunities for multidisciplinary research. *Can. J. Plant Sci.* **86:** 1361-1368.

38. Toivonen, P.M.A. 2008. Application of 1-MCP in fresh-cut/minimal processing systems. *HortScience.* **43:** 102-105.

39. Toivonen, P.M.A. and Delaquis, P. 2006. Low volume sprays to treat fresh-sliced apples with anti-browning solution. *HortTechnol.* **16:** 257–261.

40. Toivonen, P.M.A. and Lu, C. 2006. Compositions and methods to improve the storage quality of packaged plants. U.S. Patent Application 20060154822.16 pp.

41. Toivonen, P.M.A., Delaquis, P., Cliff, M., Beveridge, T. and Moyls, L. 2001. Development of apple quality standards for slicing and optimization of sanitation procedures. I*n* Apple postharvest research review, 2001 July 26. Washington Tree Fruit Research Commission, Wenatchee, WA. Pp. 112–117.

42. Tudela, J.A., Marín, A., Martínez-Sánchez, A., Luna, M.C. and Gil, M.I. 2013. Preharvest and postharvest factors related to off-odours of fresh-cut Iceberg lettuce. *Postharvest Biol. Technol.* **86:** 463-471.

43. Venugopal, V. 2006. High pressure processing. In: *Seafood processing - adding value through quick freezing, retortable packaging, and cook-chilling*, Venugopal, V. (Ed.), CRC Press, Boca Raton, FL, pp. 319-340.

44. Waghmare, R.B. and Annapure, U.S. 2015. Integrated effect of sodium hypochlorite and modified atmosphere packaging on quality and shelf life of fresh-cut cilantro. *Food Packag. Shelf Life* **3:** 62-69.

45. Watada, A.E. and Qi, L. 1999. Quality of fresh-cut produce. *Postharvest Biol. Technol.* **15:** 201-205.

46. Yano, M. and Saijo, R. 1987. New preservation method for shredded cabbage with special reference to non-browning cultivar *J. Japan. Soc. Cold Preserv. Food* **13:** 11-13.

47. Yoruk, R. and Marshall, M.R. 2003. Physicochemical properties and function of plant polyphenol oxidase: a review. *J. Food Biochem.* **27:** 361-422.

48. Zhang, B., Samapundo, S., Pothakos, V., Sürengil, G. and Devlieghere, F. 2013. Effect of high oxygen and high carbon dioxide atmosphere packaging on the microbial spoilage and shelf-life of fresh-cut honeydew melon. *Int. J. Food Microbiol.* **166:** 378-390.

49. Zhou, C.-L., Liu, W., Zhao, J., Yuan, C., Song, Y., Chen, D., Yuan-Ying, N. and Quan-Hong, L. 2014. The effect of high hydrostatic pressure on the microbiological quality and physical–chemical characteristics of pumpkin (*Cucurbita maxima* Duch.) during refrigerated storage. *Innovative Food Sci. Emerg. Technol.* **21:** 24-34.

50. Zhuang, H., Barth, M.M. and Cisneros-Zevallos, L. 2014. Modified atmosphere packaging of fresh fruits and vegetables. In: *Innovations in food packaging,* Han, J.H. (Ed.), Elsevier Ltd., LLC, pp. 445-474.

Chapter 25

Seed Spices Exports: Constraints and Opportunities

R.K. Kakani[1]*, R.K. Solanki[1] and Balraj Singh[2]

[1]ICAR-Central Arid Zone Research Institute, Jodhpur 342006, Rajasthan

[2]Sri Karan Narendra Agriculture University, Jobner 303329, Rajasthan

**e-mail: kakanirak@gmail.com*

ABSTRACT

Spices are non-leafy parts like bud, fruit, seed, bark, rhizome, bulb of plants used for flavouring or seasoning. They are used to add flavour to foods and beverages and as herbal medicines. It is defined as vegetable products or mixtures thereof, free from extraneous matter, used for imparting aroma in foods. Spices have a profound influence on the course of human civilization. They permeate our lives from birth to death. In everyday life, spices succor us, cure us, relax us, and excite us. Ancient peoples such as the Egyptians, the Arab, and the Roman made extensive uses of spices, not only to add flavour to foods and beverages but as medicines, disinfectants, incenses, stimulants, and even as aphrodisiac agents. No wonder they were sought after in the same manner as gold and precious metals.

***Keywords**: Seed spices, Export, Quality, Growing areas, Strategies.*

Spices Grown in India

India is known as the 'Land of Spices' due to its rich repository of spices. Among the wide varieties of spices crops, there is a distinct group, seed-spices, which constitutes all the annual crops whose dried seeds are used as spices. Out of the 63 spices grown in India, 20 are classified as seed-spices with 36 percent share in area and 17 percent share in the production of the total spice in India. The main crops with a larger area under cultivation are cumin, coriander, fenugreek, and fennel, whereas minor crops covering small acreage are ajwain, dill seed, nigella, anise, and celery. India is largest producer, consumer and exporter of seed spices in the entire world. Seed spices have good international market. These also have golden future as most of the importing countries are from the western developed world.

Seed Spices Production Status

On Indian lands the beauty of seed spices is that, at least one crop is grown in each state of India for seed propose or as leafy vegetable. In some states, they are grown on very marginal land for leafy vegetables, while in others they are grown on a large scale for seed spices. Rajasthan and Gujarat have maximum area under cultivation and accounts as major producer of seed spices. Thus both the states are known as "Seed Spice Bowl". Rajasthan ranks first in coriander, fenugreek and ajwain production and second in cumin, fennel, and dill seed. Gujarat stands first in cumin, dill and fennel production and second in fenugreek and ajwain. Punjab is the largest producer of celery. Uttar Pradesh and Bihar are major producer of nigella. Andhra Pradesh is the major producers of ajwain and coriander. Madhya Pradesh, Assam, Odisha, Bihar and Uttar Pradesh contribute significantly in coriander production. Haryana, West Bengal, Uttarakhand are mainly fenugreek producing states. The seed spices production statistics and details of states growing various seed spice crops in India are given in Table 1 and 2 respectively.

Global Demand and Export Scenario

The change in dietary habits of people and increased use of seed spices in medicines, pharmaceutical, neutraceutical, cosmetic, perfumery and other industries has increased the dimensions of market. This has increased the demand of seed spice, its extracts and products in food processing industry. Various seed spices products like oils, oleoresins, essential oils, organic products, curry powders/mixtures and spices mixes are expanding the seed spices export basket. There is a boom in the export of spice (including seed spices) oils and oleoresins. Most of the seed spices are having centre of origin in Mediterranean region; but seed spices are grown by many non Mediterranean countries (Table 3) among which India is the largest producer.

The global demand of seed spices is more than five lakh tonnes and India fulfills the need by about 70 per cent. India exports raw spices as well value added items to the nearly 70 countries in the world. Seed spices have golden future as most of the importing counties are from the western world. The important export destination of Indian Seed spices are presented in Table 4.

The seed spices export as whole has been increased from 188530 tonnes with a value of Rs. 170593 lakhs during 2011-12 to 349695 tonnes with a value of Rs. 433568 lakhs. This increase is 85.48 per cent in terms of quantity and 154.15 per cent in terms of value is massive. The abrupt rise in seed spices export is indicative of the quality India is producing and competing well with other nations to meet the export standards of developed countries by following improved agricultural practices, role of government bodies, research institutions, exporting agencies, and spice board is also significant.

Table 1. Major seed spices area and production during 2016-17 to 2020-21 (area 000, ha and Prod 000, tonnes)

Seed Spices	*2016-17*		*2017-18*		*2018-19*		*2019-20*		*2020-21*	
	Area	*Production*	*Area*	*Production*	*Area*	*Production*	*Area*	*Production*	*Area*	*Production*
Coriander	673.70	864.37	544.24	721.39	469.98	599.48	528.97	700.81	628.62	822.21
Cumin	780.92	500.36	966.17	689.42	1027.94	700.64	1276.28	912.04	1241.30	856.51
Celery	4.01	5.51	4.01	5.51	4.51	6.38	4.56	6.60	4.27	6.00
Fennel	89.54	148.56	63.81	103.83	90.46	157.48	82.73	139.76	79.84	128.50
Fenugreek	220.67	310.07	149.33	213.34	119.78	183.91	126.29	182.17	133.23	203.36
Ajwan	24.23	13.82	35.41	24.20	35.88	19.62	41.13	28.97	36.07	23.35
Total	**1793.07**	**1842.69**	**1762.97**	**1757.69**	**1748.55**	**1667.51**	**911.31**	**1970.35**	**2123.33**	**2039.93**

Source: Spice Board of India and DASD, India.

Table 2. Major seed spice growing regions

Seed Spice	*Distribution in India*
Coriander	Rajasthan, Madhya Pradesh, Andhra Pradesh., Assam, Tamil Nadu, Gujarat, Karnataka, Uttar Pradesh., Bihar, Meghalaya, Uttrakhand and Haryana
Cumin	Gujarat, Rajasthan and Madhya Pradesh
Fennel	Gujarat, Rajasthan, Madhya Pradesh., Haryana and Uttar Pradesh.
Fenugreek	Rajasthan, Madhya Pradesh., Maharashtra, Haryana, Uttar Pradesh., Uttrakhand and Gujarat
Ajwain	Rajasthan, Andhra Pradesh, Maharashtra and Bihar
Dill	Rajasthan, Gujarat, Jammu and Kashmir, Uttar Pradesh, Orissa, M.P. and Punjab
Caraway	Jammu and Kashmir and hilly areas
Celery	Punjab, Haryana, Uttar Pradesh and Delhi
Nigella	Uttar Pradesh, Bihar, Jharkhand and West Bengal
Aniseed	Punjab, Jammu and Kashmir, Andhra Pradesh and foothills area

Table 3. Distribution of seed spices in world

Crop	*Distribution*
	World
Coriander	India, Bulgaria, Iran, Canada, China, Morocco, Syria, Romania, Egypt, Russia, Mexico, Argentina, China, Japan, Italy
Cumin	India, Syria, Iran, China, Afghanistan, Turkey, Ethiopia, Iran, Egypt, Pakistan, Syria, Italy
Fennel	India, Egypt, China, Bulgaria, Turkey, Romania, Russia
Dill	India, Germany, Hungary, Netherlands, Pakistan, USA
Fenugreek	India, Morocco, Bulgaria

Source: Spice Board of India.

Table 4. Export destinations of different seed spices from India

Seed Spice	*Major Importing Countries with Decreasing Order of per cent Share during 2019-20*
Coriander	Malaysia, UAE, UK, South Africa, Nepal, Saudi Arabia, USA, Qatar, Oman, Australia, Kuwait
Cumin	China, Bangladesh, USA, Afghanistan, Egypt, UAE, Nepal, Morocco, Pakistan, UK, Malaysia, Brazil, Saudi Arab, Mexico, Sri Lanka and Japan
Fennel	Vietnam, USA, Malaysia, Saudi Arabia, China, UAE, UK, Canada, Nepal and Bangladesh
Fenugreek	USA, YAR, UAE, South Africa, Nepal, Saudi Arabia, Bangladesh, Sri Lanka, Germany
Other seeds	Nepal, USA, UK, Sri Lanka, Saudi Arabia, UAE, Pakistan, Canada, Malaysia

Source: Spice Board of India.

Constraints in Export

Good growth is observed in spices consumption due to increasing trends towards eating ethnic foods mainly in developed countries. Consumers in Asia, latin America and Middle eastern developing countries are demanding natural products for seasoning and flavouring of foods, rather than artificial additives. Still the scope to enhance export from India is immense; there are few constraints which are limiting the seed spices exports which needs to be addressed:

1. Lack of proper tracking of production area
2. Lower productivity or inconsistency in production
3. Poor quality of produce both in extrinsic *e.g.* physical purity and cleanliness and intrinsic quality *e.g.* essential oils content, residual toxicity, microbial load, *etc.*
4. Low exploration of value addition potential
5. Lack of quality management in crop production
6. Lack of proper training to the spice grower, traders, processors and exporters
7. Lack of market intelligence
8. Lack of organic seed spices production
9. Competitive prices from other sources

Table 5. Export of seed spices from India (Quantity in tonnes and value in Rs. In lakhs)

Item	*2016-17*		*2017-18*		*2018-19*		*2019-20*		*2020-21*	
	Qty.	*Value*	*Qty.*	*Value*	*Qty.*	*Value*	*Qty.*	*Value*	*Qty.*	*Value*
Coriander	40100	42680	30300	29207	35185	27274	48890	35208	47135	39831
Cumin	97790	153113	119000	196320	143670	241798	180300	288480	214190	332806
Fennel	15320	17239	35150	30875	34550	25906	26250	24412	24220	23161
Fenugreek	33330	23380	34680	18276	29280	12688	27150	13846	26570	15690
Other seed spices	29190	21533	24350	21701	22175	16045	29740	18736	37580	22080
Total	215730	257945	243480	296379	264860	323711	312330	380682	349695	433568

Source: Spice Board of India.

Strategies to Increase Production

It is expected that the total export of seed spices can be further increased if India broaden the production and supply base of minor seed spices *viz.* ajwain, dill, anise, caraway and nigella in addition to major seed spices as this the time where everyone is thinking of immunity; in the post COVID situation herbal sources for boosting immunity have gained high demand and the crops like ajwain, nigella, dill seed have the inherent properties of boosting immunity. In this endevour the prime research institute of India under NARS like ICAR-National Research Centre on Seed Spices has succeeded in development of better production technology of seed spices for high yield and quality. The efforts are being made to acclimatize caraway, anise and celery to semi-arid conditions. For increasing the production of seed spices we have to increase not only the productivity but also to explore new potential area where these crop can be grown successfully with higher productivity.

The foremost task now is to breed varieties of seed spices having higher yield and also resistance to biotic stresses (*Fusarium* wilt and *Alternaria* blight and powdery mildew in cumin, coriander and fenugreek) and abiotic stress like limited moisture and heat tolerance. It's also important to improve the intrinsic quality of all the seed spices for fetching high value from international trade. For this purpose, need is to broaden the genetic resource base by incorporating germplasm from centre of origin, use of mutation breeding and biotechnological approach is also required to generate variation for enhancing production potential. Population improvement programme should also be started by using gene pool concept with a vision to meet the immediate as well as long term needs. New breeding initiatives should also be taken to exploit the heterosis potential by searching genetic stock or by inducing male sterility, use of chemical hybridizing agents (CHAs) or other non-conventional methods can also play significant role in these crops.

Production technology research mandate should be focused to harvest maximum by investing minimum per unit input. Production techniques should be package of methods having maximum water use efficiency, nutrient efficiency, weed management, integrated disease and pest management. Besides this our production technology should be such that it would not use hazardous chemicals not only to meet human bio-safety standards but also to maintain better soil heath and sustainable production environment in long run. We should also develop techniques for different conditions like protected environment to harvest assured and maximum production per unit area.

In changing scenario of competitiveness, quality of produce would be major issue needing a fool-proof quality assurance mechanism. Share in value-added products of spices has to be increased through product diversification. Changing food habits demand product diversification, especially 'clean spices' and 'encapsulated spices'. The seed spice production, processing, value addition and export areas demands for well set action plans. Quality has become the key word and the focal point for spices production, processing, transport and retailing. More intense research is also needed for value added products (essential oils and oleoresins) for meet the export purpose need and to enhance foreign exchange. The exporters of seed spices are of opinion that the produce thus received is not properly cleaned, graded and packed. NRCSS has taken initiative and the farmers will be educated on this issue so that they may get more price for their produce, NRCSS is striving hard for coming up to the expectations of farmers and the Council.

Strategies to Increase Production and Export

1. Production of seed spices with good agricultural practices (GAP) for exportable quality production.
2. Meeting regular and assured supply of seed spices through proper planning for development of new elite varieties, addressing seed replacement rate, area expansion in non-traditional regions and by use of protected cultivation system.

3. By increasing quality through organic seed spices production
4. By assuring quality of the seed material by providing proper threshing, winnowing, grading and storage facilities at the production or collection sites.
5. Seed testing and certification by national referral lab for the quality of the produce for export.
6. Encouraging the small scale processing units (cleaning, grinding, extraction units *etc.*) in major assembling centres.
7. Promoting the value added products like grounded powder, oil and oleoresins.
8. Transfer of modern production and post harvest technology through a new socio economic model having Scientific consultancy for contract farming between farmers and Industry/commerce/exporters to ensure clean production and traceability.
9. Establishment of market intelligence setup and to have production planning as per demand in developed and developing nations
10. Developing production technologies having lower input cost.

Conclusion

Seed spices are an important horticultural commodity. It is low volume, high value and export-oriented crop. Despite of challenges, it is contributing significantly in Indian economy and world market. Recent increase in demand in world market had created a good scope for seed spice cultivation in India. These crops have potential to increase farmers income; in India, these crops are mainly cultivated in the resource limited areas of arid and semi arid zones. In the era of global trade and climate change, these crops have high scope for expansion. The post COVID situations has also increased the demand of natural source for immunity boosting; seed spice are having inherent metabolites which boost immunity. Hence, the future of these crops is very productive for India as its pride of the country to stand as the largest producer and consumer of seed spice globally. Minor seed spice needs to be scaled up for export to maximize the total seed spice export from the country.

Chapter 26

New Initiatives in Establishing Farmers Producer Organizations for Quality Production and Export Promotion in Horticultural Commodities

R.N. Padaria*, Subhashree Sahu and Aishwarya S.

ICAR–Indian Agricultural Research Institute, New Delhi

e-mail: rabi64@gmail.com

ABSTRACT

Atmanirbhar Krishak needs the transformation of agriculture from "a way of life" into a dynamic entity, i.e. agribusiness through FPOs. The farmer will work as an agripreneur, with appropriate business skills, in a collaborative mode with other stakeholders and try to efficiently utilise the available resources, minimise the costa and generate profits. FPOs will foster enhance cost-effective production and productivity and higher net incomes to the members of the FPO. It will also improve rural economy and create job opportunities for rural youths in villages itself. The various issues such as working capital, lack of infrastructure have to be addressed by making policy changes and suggesting the banks to have structured products for lending to FPOs. They need to have strong and sound backward and forward linkages with input companies, technical service providers, marketing/processing companies, retailers. As "AtmanirbharKrishak" takes root, FPOs emerge as the catalysts propelling horticultural growth, rural empowerment, and a thriving economy. Through their collective efforts, FPOs are poised to redefine agriculture, steering it towards a prosperous and self-reliant future for farming communities.

Keywords: *FPOs, Challenges, Horticulture production, Risks, Community development.*

Farmers Producer Organizations (FPOs) indicate a legal entity constituted by a collective of farmers, aimingfor mutual benefit through collective cooperation. They help overcome the constraints faced by individual small-scale farmers like access to resources, infrastructure, markets and thus realise the benefits of economies of scale and gain access to better and efficient services and information. In the long run, they can help in risk mitigation, socio-economic empowerment and policy advocacies at regional, national and international levels. In recent years, their potential to drive agricultural development and thereby food security, poverty reduction and sustainable rural development have been well recognised by governments at national and international level. They can take various forms like cooperative societies, Farmer producer companies, SHGs, community-based organisations, *etc.*

NABARD claims to promote almost 5100 FPOs across different states, majority as Producer Companies and about 10-15 per cent as Co-operative Society. Similarly, SFAC has registered total 897 FPCs as on 31st March 2021. Therefore, we may assume that currently there are about 10,000 registered FPOs operating in India in different forms and supported by different agencies. Although, several of these FPOs might be in dormant stage. Currently, almost 60 per cent of total registered FPOs are located in only four states (Maharashtra, Uttar Pradesh, Haryana, and Bihar), and another 30 per cent in seven states (Tamil Nadu, Andhra Pradesh, West Bengal, Telangana, Odisha, Karnataka, and Madhya Pradesh) (Raju *et al.,* 2; Govil *et al.,* 3).

Horticulture has been a sunrise sector and commodities like fruits, vegetables, flowers and medicinal plants possess significant economic value and contribute substantially to agricultural trade. In the agricultural sector, the FPOs have emerged as promising players driving positive changes in the value chain of horticultural commodities, by enhancing the production quality and boosting the exports. FPOs can serve as game-changers in the areas of value-addition, branding and marketing, capacity-building, access to improved technologies and thus help the farmers realise increased incomes and being competitive in the global market.

Challenges in Horticultural Production and Export

Though the horticultural sector has huge potential to boost the economic growth of a nation, it is marred with certain key challenges, which have been discussed in the following section:

1. Huge post-harvest losses are incurred due to the perishable nature of the produce, unavailability of adequate infrastructure and storage facilities, coupled up with weather extremes and unpredictable seasonal irregularities.
2. Fragmented and small size of the land holdings lead to inefficient and suboptimal utilization of resources and reduced economies of scale.
3. The farmers unaware of the internationally accepted quality control and standards, which makes our produce uncompetitive at the global level. Export requires the players to possess huge volumes of horticultural produce of optimum and consistent quality meeting the international export standards. The farmers need to be aware and adhere to the phytosanitary and food safety requirements of importing countries.
4. Use of obsolete and unscientific technologies, inefficient logistics, inadequate technical information has ledto increased costs and huge losses of horticultural produce.

5. The small-scale farmers face a lot of hurdles and difficulties in accessing export markets due to stringent regulations and competitive barriers.
6. Reduced expenditure towards skill development and capacity building of the small-scale farmers in modern cultivation practices, post-harvest management, processing and value addition activities also limit the potential for higher value exports.
7. The farmers with limited capital availability to invest in improved varieties, plant protection chemicals, plant growth regulators, modern farming techniques, *etc.* pose a serious threat to horticultural development.
8. The Indian market is also faced with policy discrepancies and regulatory challenges. Inconsistent policies, regulations, and bureaucratic hurdles affect the horticultural production and export. Complex export procedures and too much of paperwork hurt the export.
9. The International market demand is largely volatile and ever-changing, which ultimately affects the exportability. So, being dependant on a single or narrow market will make the farmers prone to high risk and vulnerability.

Emerging Role of FPOs for Horticultural Commodities

- **Quality Enhancement:** FPOs are innovative approaches based on collaborative farming models that harness the power of collective action among the participants for quality enhancement. These models promote sharing of knowledge, resources, and best practices and adopt the tried and tested modern cultivation practices to enhance the crop quality and productivity, along with reduced losses, increased market acceptability and opening the scope for export markets.
- **Economies of Scale:** The FPOs facilitate collective or joint procurement of inputs, fertilizers, machines, and packaging materials, which can significantly reduce the costs of production and thus improve the competitiveness of the horticultural commodities in export markets. They can invest in improved infrastructure like cold storage, and transportation networks, thereby reducing post-harvest losses. FPOs can leverage precision agriculture and smart-farming technologies like Remote sensing and GIS mapping for crop-monitoring and pest detection, application of drones for aerial crop monitoring and field assessments, automated irrigation systems and sensor-based technologies to monitor soil moisture, temperature, real-time environmental factors, pest populations and implement targeted control measures to optimize the resource use efficiency.
- **Market Access:** FPOs provide a platform for small-scale horticultural producers to aggregate their produce and collectively access the export markets, which would have been otherwise impossible individually. FPOs can collaborate with the export agencies and traders, explore the ongoing international market trends and demands and update the same to its member farmers. It can build transparency and traceability of the value chain systems, thus facilitating the procurement of certifications that will help ensure a wide consumer base in international markets. The use of Block chain technology can be explored for ensuring the supply chain transparency and quality assurance.

- Through use of digital technology, FPOs can overcome geographical limitations, access real-time market information, streamline transactions, and strengthen their engagement with buyers, resulting in improved bargaining power and increased profitability. FPOs can establish their e-marketplace where the member farmers can directly enter into a seller-buyer link with the customers. They can develop their dedicated App and website, though which the visiting customers can browse through the site and place the order for the produce of the desired quality. Digital platforms will help establish the supply chain transparency and build trust among buyers. Integrating market intelligence features into the digital platforms will provide the member farmers with access to the current market trends, demand, and prices. The transactions can be made smooth by offering digital payment options by tying up with suitable financial institutions. It can also collect and analyse the data gathered through these platforms to make informed decisions about production planning, pricing, and market strategies.
- **Microfinance and credit access for FPOs:** FPOs can effectively access microfinance and credit to expand their activities by exploring government-sponsored schemes that provide subsidized credit, grants, or subsidies for agricultural development. FPOs can undertake group lending models, where the members of the FPO jointly take responsibility for loan repayment. This approach builds social cohesion and reduces default risk. FPOs can establish savings groups or reserve funds to accumulate capital that can be used as collateral or seed money for accessing credit. FPOs can foster partnerships with Financial Institutions like banks, credit unions, and microfinance institutions to develop tailored credit products with favourable interest rates and loan terms, which suit the seasonal and cash flow dynamics of agricultural production. They can also partner with insurance providers, in order to offer the FPOs credit-linked insurance to mitigate risks associated with weather, pests, or market fluctuations.
- **Training and Capacity Building:** FPOs can play a major role in arranging regular programmes and platforms for the capacity-building of the farmer members and the FPO leaders, regarding the improved package of practices, modern technologies, the market requirements, post-harvest management (processing, packaging, and branding), quality standards, and entrepreneurial and management skills for FPO management.
- **Standardization and Certification:** FPOs can collaborate with certification bodies and experts to facilitate the process of obtaining appropriate quality certifications required for export. Examples of some such certifications needed for export markets are Global Good Agricultural Practices (Global GAP) and Organic certification. This will add value to products and opens up premium markets.
- **Negotiation Power:** Export markets often require compliance with stringent quality standards and price negotiations. FPOs empower small-scale farmers with collective bargaining power, enabling them to negotiate better terms and prices with buyers and exporters.
- **Risk Diversification:** Through diversified cultivation and market-oriented strategies, FPOs can reduce the vulnerability of horticultural farmers to market fluctuations, climate risks, and pest outbreaks.

- **Sustainable Practices:** FPOs can encourage the adoption of sustainable farming practices, including organic and environmentally friendly cultivation methods, aligning with international consumer trends and preferences. This will contribute to the conservation of natural resources, improved soil health, and the reduction of negative environmental impacts. The FPOs can implement agro ecological principles that emphasize the integration of diverse crops (through crop rotation, intercropping, and polyculture), use of organic inputs, and mimicry of natural ecosystems. By adopting sustainable and organic farming practices, FPOs will be able to catering to consumers who value healthier and more environmentally-conscious choices, thus providing better economic prospects for FPO members.
- **Community Development:** The success of FPOs contributes to community development by increasing income, employment opportunities, and overall livelihood improvement among rural horticultural communities.

Government Policies and Support for FPOs

The government offers various financial incentives in order to support the formation, capacity-building, and overall development of the FPOs. FPOs receive initial seed capital or start-up funds to cover expenses such as legal registrations, office setup, and training programs. Equity Grants and Revolving Funds are provided to FPOs, helping them secure capital for infrastructure development, training, and operational expenses. FPOs are offered matching grants, where a certain amount is contributed by the government or organization for every rupee raised by the FPO itself. Financial assistance is extended to FPOs for constructing storage facilities, processing units, packaging centres, and market infrastructure. Training and Capacity-building Grants are allocated for conducting training programs, workshops, and skill development sessions for FPO members to enhance their agricultural and entrepreneurial skills. Market Development Funds are earmarked to assist FPOs in participating in agricultural exhibitions, trade fairs, and market linkages, expanding their market reach. NAFED provides market and value chain linkages to the FPOs. There are well defined training institutions like Bankers Institute of Rural Development (BIRD), Lucknow and LaxmanraoInamdar National Academy for Co-operative Research and Development (LINAC), Gurugramfor capacity development and trainings of FPOs.

The Government of India has implemented various policies and schemes to promote the development of Farmers Producer Organizations (FPOs) in the country. Following are some key Indian government policies for FPO formation and development in India:

- Central Sector Scheme titled "**Formation and Promotion of 10,000 Farmer Produce Organizations (FPOs)**" to form and promote 10,000 new FPOs in the country with budgetary provision of Rs 6865 crore. FPOs are to be developed in produce clusters, based on "One District One Product" cluster; to promote specialization and better processing, marketing, branding and export. Under this Scheme, formation and Promotion of FPOs are to be done through the 9 Implementing Agencies (IAs) *viz.* Small Farmers Agri-Business Consortium (SFAC), National Cooperative Development Corporation (NCDC), National Bank for Agriculture and Rural Development (NABARD), National Agricultural Cooperative Marketing Federation of India (NAFED),

North Eastern Regional Agricultural Marketing Corporation Limited (NERAMAC), Tamil Nadu-Small Farmers Agri-Business Consortium (TN-SFAC), Small Farmers Agri-Business Consortium Haryana (SFACH), Watershed Development Department (WDD)- Karnataka and Foundation for Development of Rural Value Chains (FDRVC)- Ministry of Rural Development (MoRD). FPOs will be provided financial assistance uptoRs 18.00 lakh per FPO for a period of 03 years. In addition to this, provision has been made for matching equity grant uptoRs. 2,000 per farmer member of FPO with a limit of Rs. 15.00 lakh per FPO and a credit guarantee facility uptoRs. 2 crore of project loan per FPO from eligible lending institution to ensure institutional credit accessibility to FPOs. During 2020-21, a total of 2200 FPO produce clusters have been allocated for formation of FPOs.

- **SFAC's Equity Grant and Revolving Fund Scheme:** The Small Farmers Agribusiness Consortium (SFAC) provides equity grants and revolving funds to FPOs, offering initial financial support for their establishment and growth.
- **Formation and Promotion of FPOs Scheme by SFAC:** SFAC's scheme offers financial assistance for the formation, promotion, and capacity-building of FPOs, enabling them to organize and function effectively.
- **National Cooperative Development Corporation (NCDC) Support:** NCDC provides financial assistance to FPOs through grants, loans, and subsidies for infrastructure development, capacity-building, and marketing initiatives.
- **National Rural Livelihood Mission (NRLM):** NRLM supports the formation and strengthening of FPOs by offering grants and capacity-building funds to enhance the livelihoods of rural farmers.
- **Integrated Cooperative Development Project (ICDP):** ICDP focuses on strengthening cooperatives, including FPOs, by providing technical assistance, training, and capacity-building.
- **Startup India:** The Start-up India initiative promotes entrepreneurship across sectors, including agriculture. FPOs with innovative ideas can benefit from financial incentives, mentoring, and networking opportunities.
- **Cooperative Development Fund (CDF):** The CDF provides financial support for the development of cooperatives, including FPOs. It helps FPOs strengthen their operations and expand their reach.

Successful Case Studies of FPOs in India

Following are some of the examples of FPOs that have been successful in promoting horticultural production in India:

- **Jeevika Women Agri Producer Company Limited (JWAPCL), Khagaria, Bihar:** started in November 2009 under theCompanies Act 1956 is supportedby Bihar Rural Livelihood Promotion societyunder Govt. of Bihar. The company has been incorporated with amain objective to support community members to enhance their income in terms of providing them betterforward and backward linkages, *i.e.* better market opportunities to sale their agri-produce and get realistic prices.

- **Himalaya Fresh Producers Company Limited, Himachal Pradesh:** This FPO is engaged in horticultural production, specifically apple cultivation. It has adopted modern farming practices, including integrated pest management and controlled atmosphere storage, to enhance fruit quality and extend shelf life. The FPO's collective efforts have resulted in higher profits and improved market access for apple growers.
- **Jyotisagar Farmers Producers Company Limited, Karnataka** specializes in floriculture and ornamental plant cultivation. The FPO has introduced advanced propagation techniques, efficient irrigation systems, and integrated pest management practices. By offering a variety of flowers and plants to the market, the FPO has achieved success and expanded its reach beyond local markets.
- **Kunthunadu Farmers Producers Company Limited, Kerala:** Kunthunadu FPO has focused on promoting horticultural products such as vegetables, fruits, and spices. The FPO has introduced sustainable farming practices, organic cultivation, and value addition. It has established a collection centre and processing unit, adding value to the produce and ensuring fair returns to farmers.
- **Bharatiya Agro Industries Foundation (BAIF)-Shashwat Producer Company Limited, Maharashtra:** BAIF's FPO focuses on promoting horticultural crops and value-added products. The FPO emphasizes training in modern cultivation techniques, organic farming, and post-harvest management. It has successfully established market linkages, enabling farmers to access better prices for their produce.
- These case studies highlight how FPOs in India have played a pivotal role in promoting horticultural production through the adoption of modern techniques, market linkages, value addition, and sustainable practices. These efforts have not only enhanced the livelihoods of farmers but also contributed to the growth and diversification of the agricultural sector in the country.

Innovative Approaches for Promoting Horticulture based FPOs

Different innovative approaches for promoting horticulture based FPOs incorporating elements of collaboration, technology adoption, and market orientation are discussed below:

a) **AMULs SAFE (Save Food Save Farmers) Model**: An innovative approach aimed at enhancing the income of marginal farmers involved in horticultural activities. This model expands AMULs dairy structure to include the marketing and processing of produce. Through this model dairy cooperative societies play a role, in collecting, processing and ensuring the quality and traceability of products. This approach helps farmers in enjoying several benefits such as decreased harvest losses, improved market access and direct interaction with consumers. It demonstrates the potential, for cross-commodity integration and utilizing established cooperative networks to boost the horticulture sector.

Table 1: Examples of Successful FPOs

Name of the FPO/Model	*Crop Focus*	*State*	*Key Features/Objectives*
Cooch Behar Areca nut Growers Cooperative Society	Areca Nut	West Bengal	Scientific cultivation of Areca nut, fair pricing, marketing strategies, strengthening farmers' positions, promotion of environmentally friendly practises
Chittoor Mango Growers Association	Mango	Andhra Pradesh	Training, collective marketing efforts, quality enhancement
Udaipur Floriculture Farmers Producer Company	Flower crops	Rajasthan	Floriculture promotion, training, value addition, tapping into domestic and export markets
Kangra Organic Strawberry Farmers Association	Strawberry	Himachal Pradesh	Organic strawberry cultivation, fair prices, promoting sustainable practices
Kanyakumari Orchid Growers Association	Orchid	Tamil Nadu	Research, training, marketing support for orchid growers
Nilambur Teak Planters Cooperative Society	Teak	Kerala	Collaborative cultivation, sustainable forestry, fair pricing
Kurnool Tomato Growers Association	Tomato	Andhra Pradesh	Tomato cultivation, processing units, supply chain optimization
Darjeeling Green Tea Cooperative Society	Tea	West Bengal	Green tea cultivation, quality control, cooperative marketing
Kohima Organic Spice Growers Association	Spices	Nagaland	Organic spice cultivation, sustainable practices, value addition, export promotion
Puri Betel Leaf Growers Producer Company	Betel Leaf	Odisha	Product diversification, market expansion

b) **APEX Model by IKSL**: The Agri-Horticultural Society of India (IKSL) has introduced the APEX model (Agri-Production, Extension and Marketing) to support horticulture-based FPOs. This model emphasizes creating integrated value chains through partnership agreements. FPOs collaborate with input suppliers, technology providers, research institutions, and marketing agencies. The APEX model aims to improve production practices through training, access to high-quality inputs, and technology infusion. It also focuses on market linkages, helping FPOs negotiate better prices and access to markets.

c) **Jain Irrigation's Farmer Producer Company Model**: Jain Irrigation, an agri-business company, has been involved in promoting FPOs among horticulture farmers. Their approach involves integrating technology, precision agriculture, and market intelligence. The company helps FPOs implement drip irrigation, greenhouse technology, and other precision practices, enhancing productivity and quality. These FPOs are also provided with real-time market information, helping them align production with market demand. This model showcases the integration of private sector expertise to support FPO development.

d) **HOPCOMS in Karnataka, India**: Horticultural Producers' Cooperative Marketing and Processing Society (HOPCOMS) is a successful example of an FPO model in Karnataka, India. HOPCOMS operates a network of outlets that directly link farmers to urban consumers. By eliminating intermediaries, HOPCOMS ensures that farmers receive a fair share of the consumer price. This model not only benefits farmers by providing them with better prices but also ensures that consumers have access to fresh and quality produce.

e) **FFRC Model in Maharashtra**: The Farmer Field Resource Center (FFRC) model is implemented by the Maharashtra State Agricultural Marketing Board (MSAMB). These centers serve as hubs of knowledge and resources for farmers engaged in horticulture. FFRCs provide training on advanced cultivation techniques, post-harvest management, and market linkages. FFRCs empower farmers to enhance their horticultural production and market access by equipping farmers with up-to-date information and skills.

f) **Horti Hub FPO Model in Karnataka**: The Horti Hub model is being implemented by FPOs in Karnataka, facilitated by the State Horticulture Department. Under this model, multiple FPOs collaborate to establish a central hub equipped with facilities like cold storage, processing units, and packaging centers. This allows FPOs to collectively process, package, and store their horticultural produce, ensuring better quality and extended shelf life. The Horti Hub approach emphasizes value addition and efficient supply chain management.

g) **FLAAP FPOs for Large Cardamom in Kerala**: The Federation of Large Cardamom Producer Organizations (FLAAP) is a collective initiative in Kerala, specifically targeted at cardamom farmers. FLAAP assists its member FPOs with training in organic farming, pest management, and market linkages. FLAAP aim to improve the livelihoods of farmers in the Western Ghats region by collaborating on quality control and export endeavours.

Challenges Faced by the FPOs

The FPOs are to be considered as business models and farmers need to have the requisite amount of business acumen to handle it efficiently, which many a times is lacking in the Indian farmers. They have limited knowledge on modern technologies, limited access to real time information, poor entrepreneurial and managerial skills, *etc.* which hampers the functioning of FPOs. FPOs also find a hard time in arranging the required amount of finance. Due to lack of any collateral assets with the FPOs, financial institutions are hesitant to finance the FPOs, unless the latter is well-capitalized (Nikam *et al.,* 1). The FPOs find huge challenge in diversifying their activities and be in full operation round the year. The regulatory compliances like book keeping, accounting, taxation, obligations of directors of the board, *etc.* are highly challenging and exhaustive for the illiterate farmers. At times, the progressive farmers try to dominate the governance and administrativestructure of the FPO, thus sidelining the small farmers and thus, there is a clash between personal interests and theinterest of the farmer-members. Many FPOs lack technical skills, inadequate professional management, inadequate risk mitigation mechanism, and inadequate access to market and infrastructure. Also, the FPOs are not up-to-date with the real-time data on market information and prices.

Extension Strategies for Promoting Horticulture based FPOs

FPOs are increasingly adopting approaches to ensure the quality of their products and facilitate smooth entry, into global markets. This effort requires a range of strategies that combine expertise strategic partnerships, value addition and adherence to strict quality standards. In this section, we discuss a range of extension strategies carefully designed to empower FPOs in their efforts to establish a foundation, for producing quality horticultural commodities and promoting successful exports.

- Provision of Technical Expertise: It is the need of the hour to offer hands-on technical expertise and advisory services directly on-site to the members of FPOs. This will entail comprehensive guidance encompassing contemporary cultivation methodologies, pest and disease management protocols, optimal irrigation techniques, and strategies for augmenting soil health. Different organisations should organise specialized seminars and workshops designed to enhance the knowledge and competencies of FPO members in horticultural practices, post-harvest handling, compliance with quality benchmarks, and adherence to export regulations.
- Elevation of Quality Standards: It is necessary to establish strong quality assurance programs that guarantee careful supervision and maintenance of the highest benchmarks for horticultural products. Therefore, implementing rigorous grading, meticulous sorting procedures, and judicious packaging methodologies in line with the strict standards required for export markets is necessary.
- Facilitated Market Infiltration and Understanding: We need to equip FPOs with actionable insights derived from comprehensive market intelligence. This encompasses consumer preferences, evolving trends, and export opportunities.
- Cultivating Market Networks and Relationships: There is a need to engineer strategic platforms for networking, enabling FPOs to cultivate meaningful relationships with potential buyers, exporters, and international entities. Therefore, participation in trade fairs, exhibitions, and industry events will furnish a conducive environment for showcasing their product portfolios.
- Adherence to Export Regulations and Conformity: FPOs should be equipped with comprehensive knowledge about export regulations, mandatory certifications, and the necessary documentation to ensure the facilitation of smooth cross-border trade. Within the context of international trade, offering focused guidance helps FPOs adeptly navigate the inherent complexities of legal and regulatory frameworks.
- Strategic Financial and Institutional Facilitation: It is essential to create opportunities for FPOs to access financial resources, including credit facilities, grants, subsidies, and government initiatives. This will aid in nurturing a well-organized institutional framework that sustains the long-term viability of FPO initiatives.
- Orchestrating Supply Chain Integration: FPOs must be organized to ensure synchronized coordination along the supply chain by working together with key stakeholders throughout the process, from input suppliers to processors and exporters.

This orchestrated approach ensures a seamless flow of horticultural commodities from FPOs to the ultimate end consumer.

- Expanding Market Horizons: FPOs should be encouraged to explore a multifaceted market spectrum, spanning both domestic and international spheres, thereby mitigating overreliance on a singular market. They need to be enabled to identify emerging markets and niches that align with the unique characteristics of their horticultural products

REFERENCES

1. Nikam, V., Singh, P., Ashok, A. and Kumar, S. 2019. Farmer producer organisations: Innovative institutions for upliftment of small farmers. *Indian Journal of Agricultural Sciences*. **89(9):** 15-24.
2. Raju, K V., Kumar, R., Vikraman, S., Moses Shyam, D., Srikanth, R., Kumara Charyulu, D. and Wani, S. P. 2017. Farmer Producer Organization in Andhra Pradesh: A Scoping Study. RythuKosam Project. Research Report IDC-16. Patancheru 502 324. Telangana, India: International Crops Research Institute for the Semi-Arid Tropics. 160 pp. ISBN 978-92-9066-592-2
3. Govil, R., Neti, A., and Rao, M. R. 2020. Farmer Producer Companies: Past, Present and Future. Azim Premji University, Bangalore.

Chapter 27

Empowering Tribals in Jharkhand through Horticultural Enterprises

Rajeev Ranjan

Professional Assistance for Development Action (PRADAN), Ranchi, Jharkhand
e-mail: rajeevranjan@pradan.net

ABSTRACT

The objective of this paper is to share the experience of PRADAN's horticulture work done with tribal families in Jharkhand. This livelihood prototype has emerged after our long work with these communities and the technical support of ICAR RCER Farming System Research Centre for Hill and Plateau Region, Plandu, Ranch. The work has been done intensively in 12 PRADAN teams of Jharkhand. The PRADAN professionals have directly implemented the prototype with the help of community cadres in the villages.

Keywords: *PRADAN, Responsibility, Tribal area, Empowerment, Success story.*

INTRODUCTION AND BACKGROUND

Jharkhand was carved out of Bihar in 2000. The State is home to nearly a tenth of the country's Scheduled Tribes, who constitute 27 per cent of Jharkhand's population. (Singh *et al.*, 9) More than 70 percent of people in this state live in rural areas (Government of Jharkhand, 4), and around 37 percent of the state's population are classified as living below the poverty line (BPL). Despite possessing a favorable agro-climatic condition, huge forest cover, rich mineral, and human resources; continues to be one of the poorest and backward states of India. The State has a sub-humid climate with an average annual rainfall of 1,100 mm. However, due to the high variability of monsoon rains, limited irrigation coverage (less than 11 per cent of the Net

Sown Area), and undulating terrain, agriculture is fraught with high risks. Agriculture remains the principal source of livelihood for an overwhelming majority of the population, with more than 60 per cent of the main workers (as high as 85 per cent in Gumla) classified as cultivators (Census, 1). About two-thirds of the farmers are classified as marginal farmers, owning less than a hectare of land. Rice paddy is the principal crop of the State. Maize, minor millets, and pulses are the other crops. Yields of all the crops are at abysmally low levels. Low productivity of agriculture, one-dimensional nature of the farm economy, and lack of development in non-farm livelihood sectors make food security, hunger, and subsistence the primary concern of farmers (Sharma, 8). The average farm size among the tribal population is less than 2 hectares. The livelihood is mainly rain-fed agriculture-based and the prevalence of subsistence economy leaves with little produce for markets. Mono-cropping is prevalent across regions, leading to increased food insecurity and reduced prices of produce (Cornish *et al.*, 2). According to the 70th round of the National Sample Survey, the average monthly income of agricultural households in Jharkhand was less than Rs. 5000 (Government of India, 3).

PRADAN's Response to the Situation

Professional Assistance for Development Action (PRADAN) was founded in 1983 by two young professionals Deep Joshi and Vijay Mahajan inspired by the belief that well-educated people, with empathy towards the poor, must work directly with them at the grassroots to alleviate mass poverty in rural India. They realized that the critical gap in creating change was the absence of capable people, not material resources. PRADAN seeks to realize its vision of creating a just and equitable society where everyone lives and works with dignity by building robust collectives of women that will strive for large-scale change in the human condition. At present, PRADAN works in the 7 poorest states of India with nearly 8.9 lakh smallholder women farmer's families to improve their livelihood. In Jharkhand the SHG outreach is nearly 3.61 lakhs spread across 9 districts. PRADAN was taken as conscious decision to work with women farmers.

Smallholder farmers are affected by four key issues namely low productivity, negative/ marginal capital formation, low income, and alienation of youth from agriculture. The issues need to be solved by intervening at the farm economy level especially with smallholder farmers. To address this issue PRADAN has developed several livelihood prototypes for its participating families. Based on the People, Institutions and Resources framework; livelihood prototypes are being developed and selected for the families *e.g.* if the families are habitant near forest fringe area goat rearing, TASAR, or lac cultivation is promoted while if they are having irrigation facilities around the year cash crop is promoted as livelihood option for them. PRADAN has a twin focus in agriculture promotion, first to ensure food sufficiency from her own and second to give additional income. So while doing agriculture planning with families, cereals, pulses, oilseeds, and vegetables are included in the crop basket to meet the food, nutrition, and income need of the families.

Unavailability of timely and quality input, lack of agro advisory services, lack of basic infrastructure, and lack of output marketing linkages is common feature of this area. The reasons for poverty are multidimensional as well as structural. This creates a complex and challenging situation. To address this situation, PRADAN is adopting the following strategies:

- Organizing women farmer's into producer groups at the hamlet level
- Promoting Agriculture Production Cluster (APC) in those area. APC is coordinated market-linked farming, leveraging resources, skillsets and market opportunities by enhancing the capacities of the farming ecosystem.
- Identifying crops based on smallholder suitability, agroecological compatibility and market attractiveness.
- Providing handholding support during standing crops
- Promoting Farmer producer organization to provide marketing services to farmer
- Partner with resource organizations, market players, and government to bring knowledge, technology, and investment for smallholder farmers.

Some Achievements in Agriculture Livelihood

PRADAN in Jharkhand currently working with 2.41 lakhs HHs in agriculture out of which 1.15 lakhs are doing vegetable cultivation. The crops are Paddy, Pulses, Oilseeds, and vegetables. In the case of vegetables, the crops are creepers, Tomatoes, Cauliflower, Cabbage, Green peas and watermelon, *etc.* 160 AEs are providing various services like Input, Saplings, Mechanization and output marketing. PRADAN has promoted 13 women FPOs in this state to help these women farmers negotiate with bigger market forces and influence the stakeholder in their favor. The FPOs are helping member farmer in cultivation crops and selling their produce in the distant market *e.g.* The Gumla Mahila Kisan Sawbkamban Trust (FPO) has sold 189.5 MT of Tomato, Green peas, cabbage and cauliflower and 110MT of mangoes in the FY 2020-21 while Torpa Mahila Krishi Bagwani Swablambi Sahkari Samity Ltd has sold 274 MT of watermelon during the last summer 2021.

Challenges Faced in Agriculture

- United infrastructure *e.g.* irrigation and post harvest processing infrastructure restricting farmers to engage with her land and market for limited time period and not round the year.
- The area under vegetable cultivation sharply declines after Kharif. The engagement of farmers with the market is for a limited period leading to poor market connection.
- Bacterial wilt is causing 20-30 per cent plant mortality especially during Kharif.
- The state has been witnessing huge migration of male folk (Rao and Mitra, 6). Resultantly, the pressure of agriculture is falling on women. To reduce the drudgery, suitable farm tools and implements need to be made for women farmer. The penetration of technology in the tribal area is far away.
- Availability of quality and timely input at four prices.
- The climate is changing very fast, quite unpredictable, and is affecting crop cultivation. Need to come up with crops/varieties which can perform better in these changing climatic situation.

Experience with Mango based Intercropping Model

Reason for choosing Mango with Intercrop as livelihood option:

- The majority of the families with whom we work are tribal families and their worldview revolves around "Jal-jungle-Jameen"[1]. Plantation as an livelihood option fits in the view.
- Availability of upland. These families have an average landholding of 2.5-5 acres. They are ready to take plantation in 0.5-1 acre of their upland,
- Mango is a relatively high yielding and high-value cash crop that has a robust domestic market; it has potentially high returns for smallholders,
- This is the intervention which in addition to giving cash income also takes care of the environment by supplying fresh air and building soil carbon through sequestration,
- In comparison to the other crop, mango plantation seems to be very remunerative for smallholders in Jharkhand. An acre of the orchard has the potential to make a family 'Lakhpati[2] family',
- Support from Government programs helped sealed this activity.
- The presence of technical resource organizations ICAR RCER Farming System Research Centre for Hill and Plateau Region, Plandu, Ranchi, Palandu Ranchi, and their scientists had helped in developing scientific know-how to establish the activity. They are also helping us in reviving older orchards in the area, making these commercially remunerative for smallholder farmers,
- The mango plantation has been taken as a 'Pension scheme' by the families as they see the plantation as regular income source giving money at every season.

Key Elements of PRADAN's Implementation Model Include

The key elements are–building community interest and perspective, ensuring selection of suitable farmers and areas, careful consideration of suitable varieties based on smallholder suitability and market attractiveness, developing locally tailored prototypes (including intercrop), building community-level capabilities and resource persons, completing detailed planning, and ensuring family level ownership through weekly review meetings, Linkages with markets and government institution and unlike them with FPOs.

---

1. 'Jal Jangal Jameen' the popular slogan of Adivasi movements, was first given by Komaram Bheem (sourc e: http://adivasiresurgence.com/2016/10/16/komaram-bheem-a-forgotten-adivasi-leader-who-gave-the-slogan-jal-jangal-jameen/)

2 Lakh means 100,000. Lakhpati means individual / household, owning Rs. 100,000 worth of wealth.

Brief Description of PRADAN's Efforts Towards Promotion of Mango Cultivation among Tribal

As one of the different livelihood options, PRADAN started promoting mango-based farming in 1997 in the Purulia district of West Bengal. In Jharkhand, it was taken during 2004-05 as a demonstration. This was demonstrated in 4 blocks of the Gumla district namely Palkot, Raidihh, Ghagara, and Gumla (PRADAN, 5). Around 38 families in 22 acres have taken the cultivation. In the 1st year the mortality was very high ranging from 8-30 per cent in different blocks mentioned above.

The reasons were technical, social, economics and environmental in nature. With these practical learnings from 2006-07, we started scaling up under various government programs like Special Central Assistance to Tribal Sub-plan Areas (SCA to TSP) prototype schorte, NABARD's scheme for orchard Promotion in Tribal Areas (WADI yojna), and Special Swarnajayanti Gram Swarojgar Yojna (Special SGSY Project, a flagship initiative of the Ministry of Rural Development, GoI) in Jharkhand. We are thankful to ICAR-RCER, Palandu, Ranchi as they provided technical support to us while implementing these seherds with women farmers. Dr Sibendra Kumar and Dr Bikas Das has helped in redesigning the Package and Practices (PoP) based on the local context and provided us with technology for orchard management. Central Opening and pheromone traps were the major interventions suggested by the ICARRCER for our project areas. Their support had enhanced our technical understanding of raising orchards and most of the orchards came in shape. Jalkund and Hydrogel were introduced in the orchard through ICAERRCER to combat water scarcity during the summer months. Central Opening and pheromone traps were quickly adopted by the orchard growers across locations and these have been integrated with the package of practices.

In July 2009, the first harvest came. The total production was 9 MT. On the first day, the farmers brought nearly 1MT of mango from the project areas (from 22 acres of demonstration sites) in the Gumla market. We were quite excited about it but the market didn't responded equally. The first day was very disappointing as the sale was meager. People thought that this was a wild variety as *Amarpali* was smaller in size compared to the established table variety as *Langra* is the most preferred variety and Amarpali was competing with langra in marketing starteagies was langra so we can't compete with it. So from the next day, we did two important things, one to educate the consumer that these mangoes are carbide-free and the second lowered the rate compared to langra. The women farmer also made slices of mangoes and given the consumer to taste them. It was our strategy to build the product in the existing market. The taste of Amrapali mango was liked by the local people in Gumla and suddenly demand for Amrapali went up to attract consumer. Stalls were put up in blocks, districts and even in the state capital Ranchi. Stickers were fixed on mangoes and the farmers were encouraged to bring their produce in crates to give the harvest an attractive look. Posters and banners were raised in the local markets mentioning that the Mango was produced in the local villages and the same was free from carbide. The demands for *Amrapali* started growing in the local areas and the variety acquired its identity and market share in Gumla and later on in the Ranchi market.

On average, a good family earned Rs. INR 8,000-10,000 from mango and another 15,000-20,000 from intercrop so a total of 23,000-30,000 from an acre of Mango in the 1st harvest itself. That was well beyond the aspirations of the local communities. Seeing the success of horticulture, from 2012 onwards, the local community had started adding more plants in their existing orchards even planting new orchards from Government progrms that or even planting new mangoes from their contribution. It was during 2016-17, we discussed with MGNREGS Jharkhand cell to include Mango as a model/scheme in the shelf of activities under MGNREGS. Amrapali, being a regular bearing variety was at the core of the orchard model under the MGNREGS. This opened up a window for the smallholder farmers to access public funding to raise Mango under MGNREGS (Sabhikhi, 7). This was a welcoming step of the Jharkhand government.

Currently, PRADAN in Jharkhand is supporting 17,666 households to raise mango orchards in 4600 Ha. in 9 districts of Jharkhand namely Lohardaga, Gumla, Godda, Khunti, Dumka, Bokaro, Hazaribagh, Ramgarh, and West Singhbhum. Data from the field shows an immense impact on the income gain from the mango based intercropping model. The family applying three times fertilizer and pesticide the yield per ha is 6.5-8.5 MT/ha making the family lakhpati from an 8-10 year orchard. These kinds of farmers are earning Rs. 65,000-75,000 from one acre of land and another Rs. 25,000-35,000 from Intercrops. It is worth sharing the story of Mahamuni Oraon of Ghunti village, Ghagara block in this context. Her family was a migrant labourer before she planted mangoes. She is a hard-working tribal farmer who used to follow the prescribed POP stringently and earns more than 1.0 lakhs of rupees every year from 2018 onwards from mango based intercropping model.

The plantation is being taken with the funding support of MGNREGS, SCA to TSP, Special SGSY, NABARD-WADI, and individual farmer contributions.

Strategies/Steps done for Promoting Mango Orchards

The strategies which worked well are:

Successful Demonstration

At Gumla, we did a successful demonstration. We organized demonstrations in the selected pocket (4 villages in 4 blocks). We learned from our successes as well as from our failures. We learned how to implement the mango plantation *i.e.* What to do and what not to do. We also learned the criteria for the selection of farmers who could potentially succeed in orchard promotion. The demonstrations also helped us in deciding about the optimum size of orchard per household as also standardized the month-wise Package of Practices.

Cluster Approach

The prototype is taken in a cluster approach. One cluster should have a minimum of 60 acres of land to attain economy is scale and organize requisite services around orchard growers.

Identification of Cultivators, Farmer, Land According to the Context

Cultivars with low gestation periods, such as Amarpali and Mallika were chosen. Since orchard establishment is a labor-intensive activity and the labor clashes with Paddy transplantation, hence families with extra working hands were encouraged to join the program.

Plots were selected seeing the availability of irrigation sources nearby. Disputed land, plots with less soil depth, shady/waterlogged plots were avoided.

Size of the Land for Plantation

The size of the land was determined by the optimum number of plants to be grown by individual households. In the context of smallholder farmers, it was found that 50-100 plants per household were the optimum scale per household. Less than 50 plants did not make a good commercial proposition to retain the interest of the farmers, resulting in low survival and growth. On the other hand, more than 100 plants would become cumbersome for a nuclear family to manage. That means, dependence on hired laborers to tender the orchard, which in most cases failed to take off.

Preparing Community for Plantation

Exposure visits of potential farmer are done in nearby older orchards to convince them and build their vision to older orchards before finalizing their selection for orchard growing. Farmer producer groups were promoted after the exposure of families. The visioning exercise was done on a mandatory basis with all farmers. Both male and female members were invited to this event.

Quality Inputs

Inputs like Saplings, vermicomposts, bone meal, and neem cakes were also bought from a reputed vendor. Communities were taught techniques for checking the quality of inputs like vermicompost and saplings to ensure quality standards.

Fencing before Plantation

Dry as well as live fencing is mandatory. Only dry fencing did not serve the purpose and it broke into pieces after the rainy season. It is a norm that if one wants plantation, she has to complete fencing first. Pad yatra or Rallies against free grazing helped ensure social fencing.

Transplantation

From "truck to field": Planting of mango sapling on the same day of plant arrival. Even if it was too late, the farmer had to transplant the sapling on the same day else it resulted in higher mortality.

Intercropping is Mandatory

Intercropping was made mandatory so the program was renamed **"Mango with Intercrop"**. During the gestation period, the intercrop kept motivated farmers to visit her field and so the focus remained on the orchard. In addition, intercrop gave additional income to the families.

Ensuring Monthly POPs

Month wise activity calendar was prepared with support from ICAR RCER Farming System Research Centre for Hill and Plateau Region, Plandu, Ranchi and cadres were promoted to ensure this in the field. Producer group meeting was done in orchards as also cross-field visit were promoted for cross learning and ownership building for the community. Community cadres called Horticulture Community Service Providers (CSP) were promoted to provide handholding support to the farmers. Fortnightly review and planning meetings of CSPs were conducted by PRADAN professionals.

Marketing

Stalls, stickers, banners, and posters were used as part of a marketing campaign for the promotion of *Amarpali* and *Mallika* mango in the market. Selling mango in the local market had created more space for *Amarpali* and *Mallika* than going outside distant markets. Farmer's fair during rabi season helped in the dissemination of knowledge and best practices among farmers. Buyer–seller meeting was held in scale.

Policy Advocacy

A lot of exposure visits of senior state beurocrats, Resource institutions, MLAs, MPs, and Ministers were done to showcase the good work to mobilize the state force for taking the mango into scale. Finally, it got included in MGNREGS.

Current Challenges with the Program

1. With the increasing scale of cultivation every year, the availability of quality input is becoming a concern now, an issue also highlighted in other studies such as Vijay *et al.* (10). We need to think of how to make it available or do we need to search for other alternatives.
2. This summer, the wholesale rate of mango was INR 15-20 kg. Covid Pandemic has also impacted the rate. Meanwhile, there is a lot of demand for mango cultivation in the community. It seems *Amrapali* has saturated the Jharkhand market. Now it's time to bring in other cultivators of mango and also demonstrate mixed fruit orchards.
3. Some of the production-related technology has been reached to the farmer but we are far lagging in post-harvest technologies. Now it's high time to grasp processing opportunities.
4. Blackening of mango is a serious concern. During the harvest season, we are facing rain sometimes, which, increases the intensity of blackening. The quality of mango is going down and thus impacting the income of the farmer. Need solutions around this. Currently control opening and united paper bagging is done. What other alternatives are available?

Story of Change through Mango Plantation

Ghasni Oraon is residents of Chota Pasanga, a village in Fori of Gumla block. They were a wage earner and migrants. With their income, they were somehow able to fulfill their daily needs. With the traditional agriculture practices, her family was able to grow food only for six months from her land.

On the 14th of March, 2006, a PRADAN convinced this village to form women SHG. After a lot of hesitation and resistance, the women were mobilized in Chala Swayam Sahayata Samuh Ghasni Devi was chosen as the accountant of this SHG. Some more SHGs were added in the village. PRADAN then conducted training on improved paddy cultivation and vegetable cultivation. In this training, Ghasni Devi participated enthusiastically. After the training, she decided to grow Kharif vegetables and made an income of Rs 25000. Motivated by the result, she decided to do summer and Rabi crops that year. She adopted the SRI technique for paddy cultivation and with improved practices did cash crops like tomato, Chilli, Cauliflower, Cabbage, Green Peas, *etc.* The same fields which were anyhow not able to fulfill the needs of her family now start giving two-three times the better output of paddy, and handsome cash from vegetables. This was just the beginning of their development story.

In 2007 under special SGSY scheme, PRADAN started forming Tola Sabha in the village for INRM work. Many of the families constructed small water harvesting tanks on their low land. This not only conserved the runoff but also save the paddy during the dry spell. With the help of a water harvesting tank, Ghasni Devi can bring more area into rabi cultivation. Her family is food secure now. Now her next goal was to earn some additional cash to focus on her child's education, health, nutrition, wellbeing, *etc.*

In 2007 with the financial support from a special SGSY scheme, She started mango cultivation. Earlier the villagers were skeptical about mango cultivation but Ghasni Devi decided to plant 125 plants in her uplands. Following her, other farmers also decided to go for a mango plantation. Along with the mango plantation, she also got an irrigation facility on her land. In 2011, her mango plantation started bearing fruits. In 1st year Ghasni Devi earned INR 30,000 from her orchards. From 2018 onwards the family is earning nearly a lakh from mango and intercrop. This year, they had sold 8587 kg of mangoes to their FPO and got INR 98,469. The family have constructed a pond last year and is rearing fish to increase their income further.

Today Ghasni Devi is living a happy and prosperous life with her family. She managed to have all the basic necessary items that are required for a happy life. He children is now studying in good school and above all of it, the self-confidence of Ghasni Devi has increased a lot. She says "I can face any situation now".

Success Story of Lalita Bhenga

Lalita Bhengra is a resident of Kumang village of Torpa block of Khunti district of Jharkhand. She is an inspirational farmer in the Torpa block. Despite available land, her situation was not good. She had upland available but was not yielding that much return which can provide a livelihood for her family. The story begins by joining her to SHG named Roshini Mahila Mandal. She was helped by PRADAN in taking mango orchard under MGNREGS. With some initial hesitation, she agreed for taking the plantation in 1.25 acres of land. Her family member supported her in the activity. The whole plot was dug into small pits of 1m*1m*1m depth at

a spacing of 5m*5m spacing. In the same rainy season, the upland got quickly converted into orchards. The family was also motivated to take intercrops between the row spaces to tide over the gestation period of mango as well as return from the existing land. Growing intercrops also helps the main crops by providing irrigation and fertilizer.

The plantation was done during 2017 and she was capacitated to grow crops like Tomato, Chilly during Kharif and Cauliflower, and Green pea during Rabi. She also grew Cucumber during the summer season. In addition, she grew leafy vegetables too. This made the vegetable available at her family's plate around the year as well as she started earning handsome earning from selling it. She earned INR 13,200 from vegetable cultivation during the first year. In addition, since the scheme was under MGNREGS, the family completed 100 days and have earned INR 16,800.

With this increase in income, she repaired her old house and purchased a pump set for irrigating her land.

She has been able to manage the agriculture work very nicely and learned to work in sync with the market. Her plot is now an exposure site for others. The fellow farmer draw inspiration from her work. This year she is expecting an income of 70,000 from mango only despite this being the first year of bearing. It has happened only because of her hard work that made the plant more robust and yields to the fullest.

Year	*Income from Intercropped Vegetable (INR)*	*Income from Mango (INR)*
2017	13,200	Not to be taken
2018	32,700	Not to be taken
2019	62,500	Not to be taken
2020	20,000	35,000
2021	30,000	40,000

In her language, she says "The dream of my Orchard has come true"

REFERENCES

1. Census of India. 2011. Primary Census Abstracts: Registrar General of India. Government of India: Ministry of Home Affairs.
2. Cornish, P. S., Karmakar, D., Kumar, A., Das, S., Croke. B. 2015. Improving crop production for food security and improved livelihoods on the East India Plateau I. Rainfall-related risks with rice and opportunities for improved cropping systems. Agricultural Systems. **137**: 166-179.
3. Government of India. 2013. Income, Expenditure, Productive Assets and Indebtedness of Agricultural Households in India. Ministry of Statistics and Programme Implementation. National Sample Survey Office.
4. Government of Jharkhand. 2017. Jharkhand Economic Survey 2016–17. Ranchi: Planning-cum-Finance Department, Finance Division
5. PRADAN. 2011. Mango Horticulture: A livelihood opportunity for the poor: A workshop report extract. *News reach: The livelihoods and development monthly*. **12(4)**: 25-30
6. Rao, N., and Mitra, A. 2013. Migration, representations, and social relations: Experiences of Jharkhand labor to Western Uttar Pradesh. *Journal of Development Studies*, **49(6)**: 846–860.
7. Sabhikhi, I. A. 2017. A mango plantation in Jharkhand shows how MGNREGA can empower rural families. Scroll. in. https: //scroll.in/article/834458/a-mango-plantation-in-jharkhand-shows-how-mgnrega-can-really-empower-rural-families

8. Sharma, K. 2019. Hunger in Jharkhand: dimensions of poverty and food security in palamu district. S. Asia Res. https: //journals.sagepub.com/doi/abs/10.1177/0262728018816407
9. Singh, K. M., Meena, M. S., Singh, R. K. P., Kumar, A. and Kumar, A. 2012. Rural poverty in Jharkhand: An empirical study based on panel data. Retrieved from SSRN: http: //ssrn.com/abstract=2330844 or https: //doi.org/10.2139/ssrn.2330844
10. Vijay, K.V., Rahaman, S.M., Wadhwani, M.K., Kumari, S., Kumari, M., Kumar, S., Homa, F. Sengupta, S., and Bairwa, S.L. 2019. Diagnosis on strengths and threats of mango in the context of Bihar: *A case study. J Crop and Weed.* **15(2)**: 94-9.

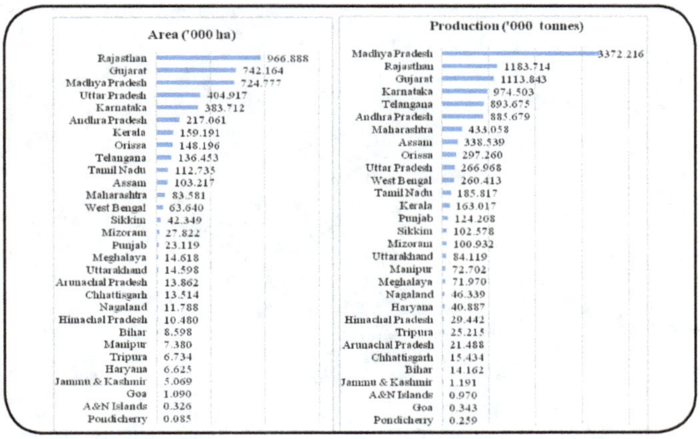

Figure 1. State wise area and production of total spices. Source: Directorate of Arecanut and Spices Development (DASD), Ministry of Agriculture and Farmers Welfare, Government of India. (pg.119)

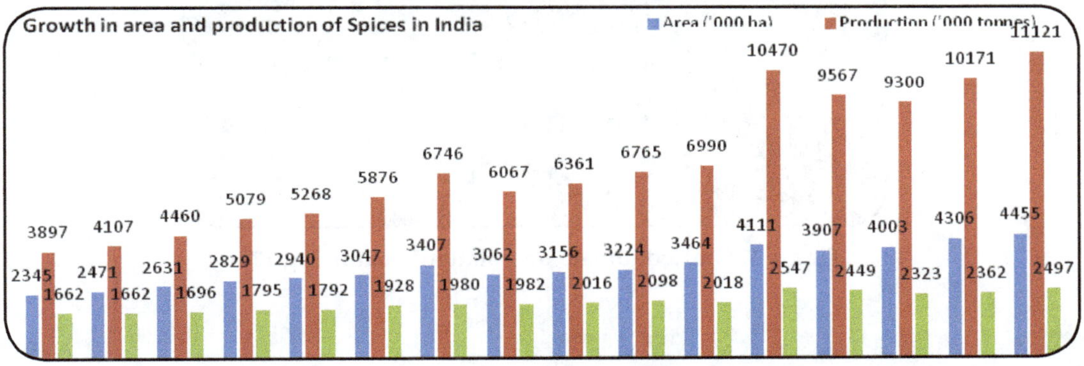

Figure 2. Growth in area and production of spices in India during 2005- 2020. Source: Directorate of Arecanut and Spices Development (DASD), Ministry of Agriculture and Farmers Welfare, Government of India. (pg.119)

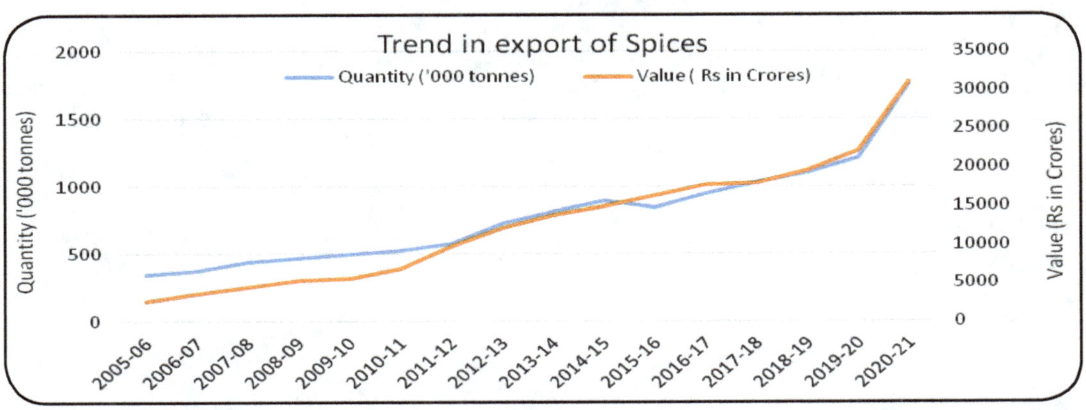

Figure 3. Growth in export of spices from India. Source: Directorate of Arecanut and Spices Development (DASD), Ministry of Agriculture and Farmers Welfare, Government of India. (pg.120)

Figure 1. Grafted tomato on potato (Pomato). (pg.177)

Figure 2. Dual grafting of brinjal and tomato (Brimato). (pg.178)

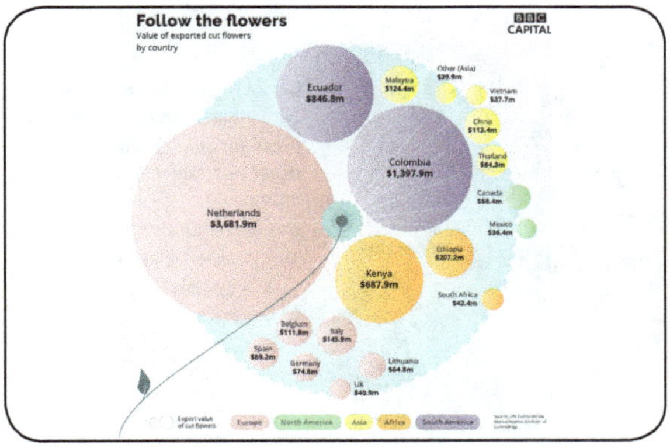

Figure 1. Major players in the global export of flowers (2020). (pg.196)

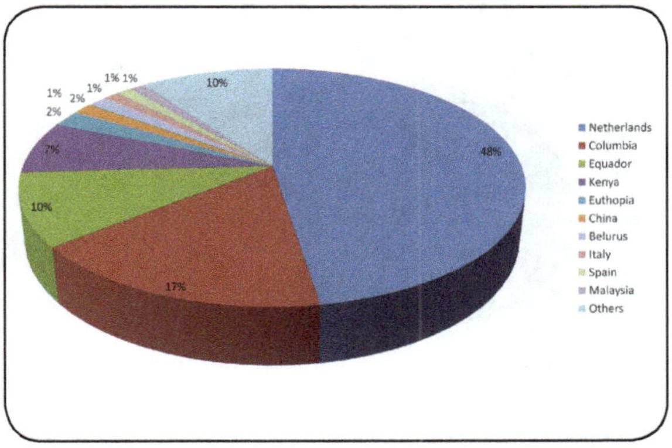

Figure 2. Share of different countries in global exports. (pg.197)

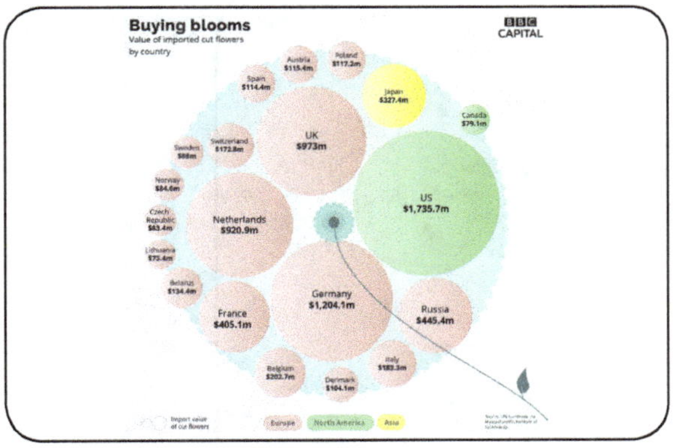

Figure 3. Major importers of floricultural produce. (pg.198)

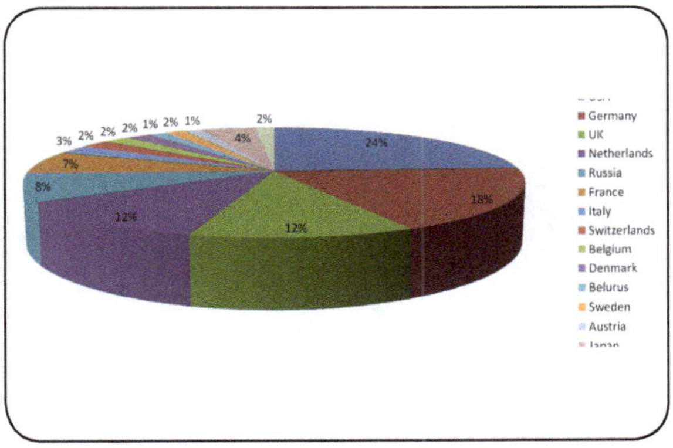

Figure 4. Leading importers of floricultural produce. (pg.199)

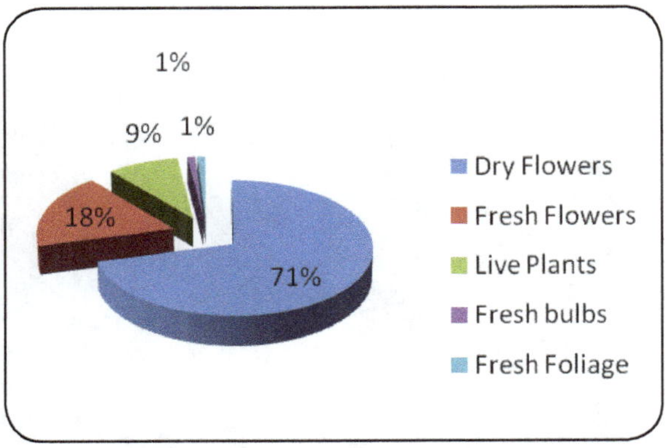

Figure 5. Composition of Indian floriculture exports. (pg.203)

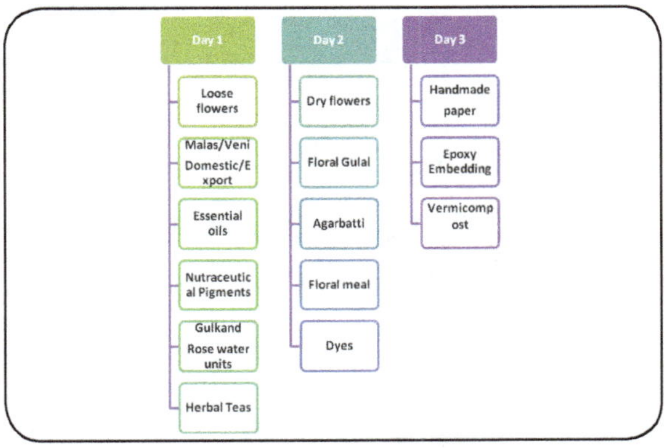

Figure 6. Model system for value addition in integrated markets. (pg.207)

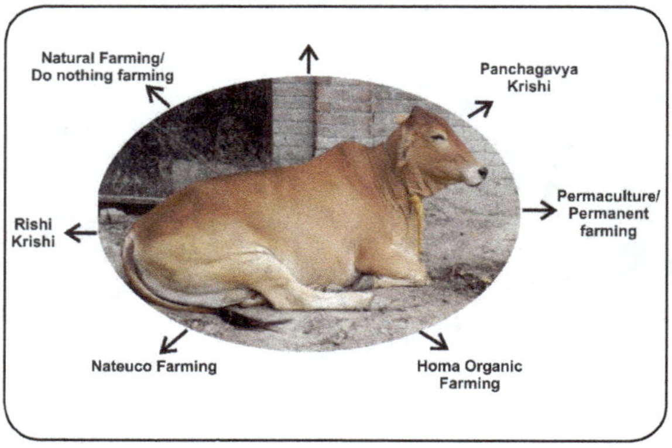

Figure 1. Cow based organic farming systems. (pg.227)

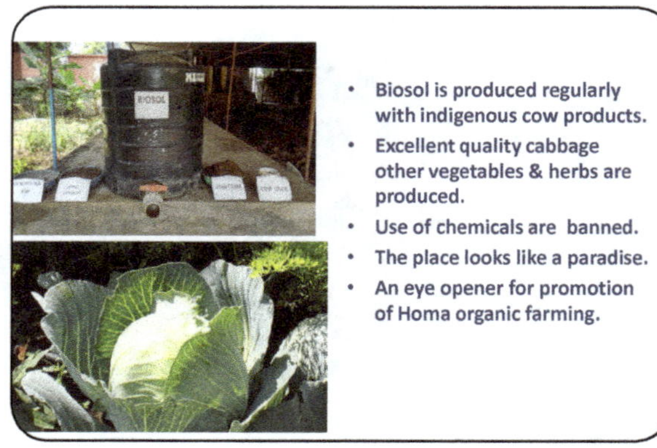

Figure 2. View of preparation of Biosol and its impact in cabbage (pg.239)

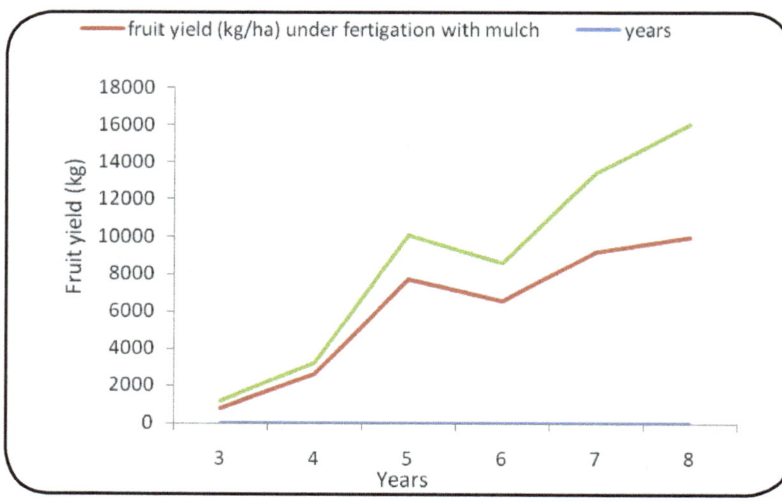

Figure 1. Fruit yield (kg tree–1) under fertigation with mulch. (pg.260)

Figure 1. Profile of shallow, coarse textured soil. (pg.293)

Figure 2. Profile of shallow, heavy textured black soil. (pg.294)

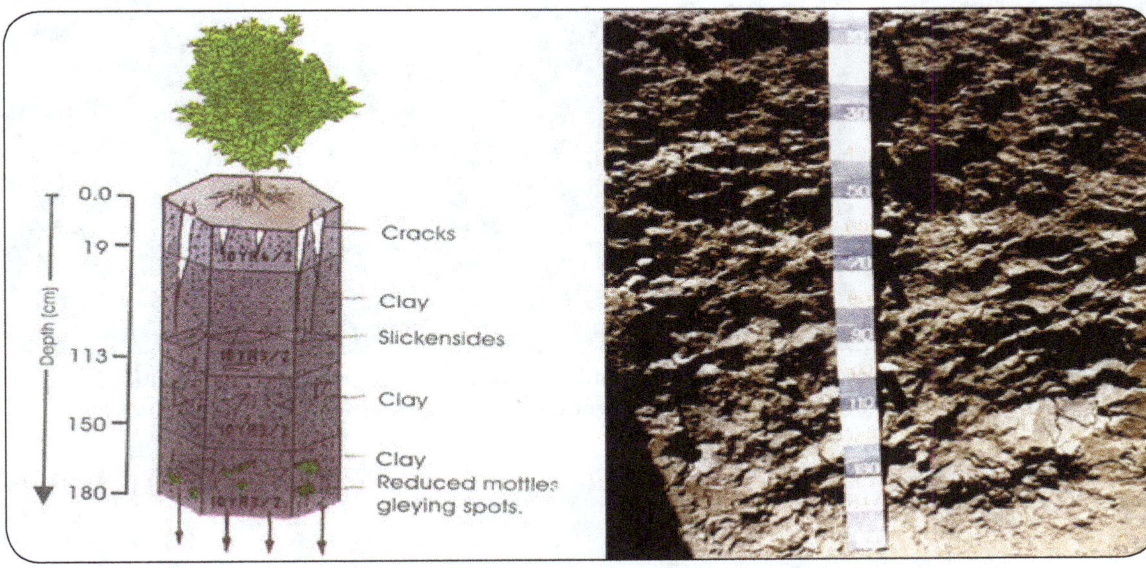

Figure 3. Profile of deep, heavy textured black soil. (pg.295)

Figure 4. Pomegranate plantation on raised bed system. (pg.296)

Figure 1. Symptoms of mango wilt (a), decline (b), branch drying (c) and twig drying (d). (pg.307)

Figure 2. Symptoms of infection in underground trunk (a), aboveground trunk (b), gummosis (c) and beetle infestation (d). (pg.307)

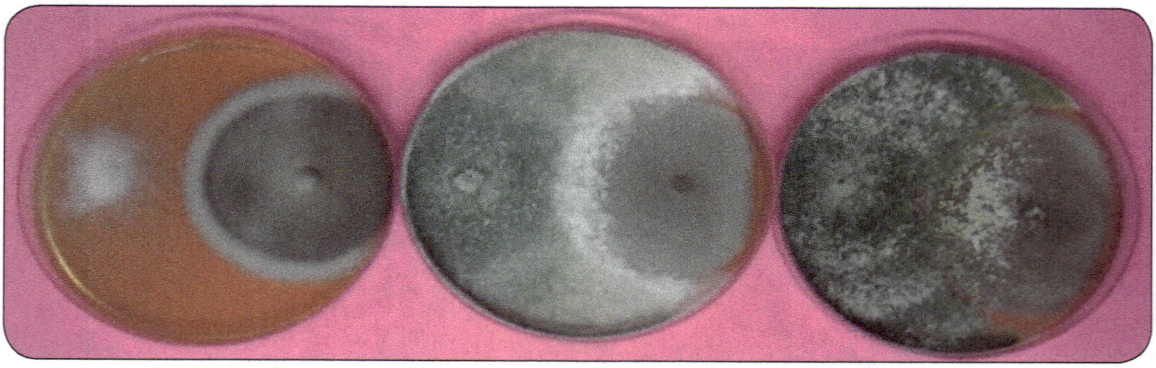

Figure 3. Suppression of C. fimbriata by T. asperellum (pg.309)

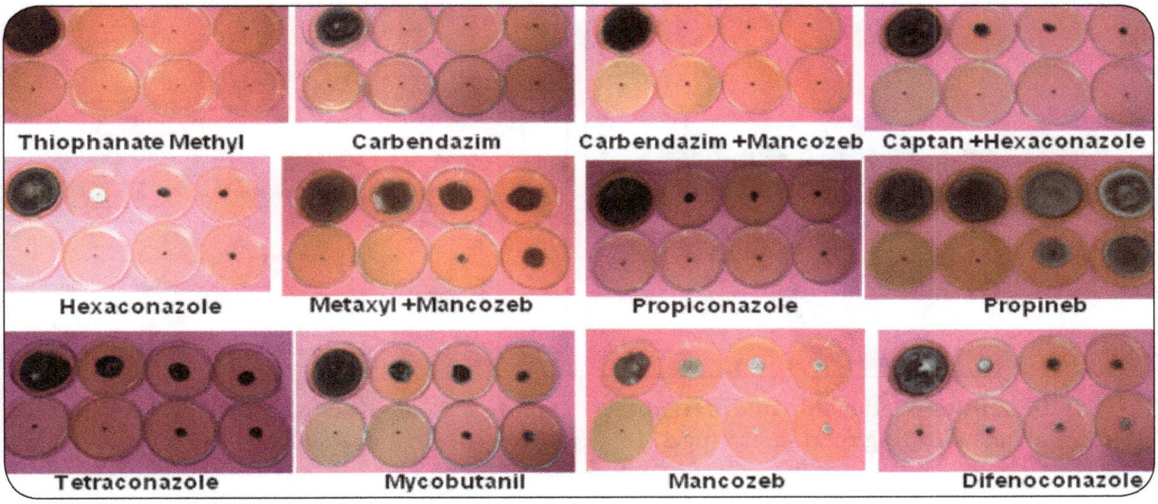

Figure 4. Effect of different fungicides on growth of C. fimbriata at different concentrations. (pg.310)

Figure 5. Symptoms of shoulder browning. (pg.314)

Figure 6. Management of shoulder browning (a. Bagged fruits. b. Bagged fruits at harvest, c. Difenoconazole, d. Untreated). (pg.315)

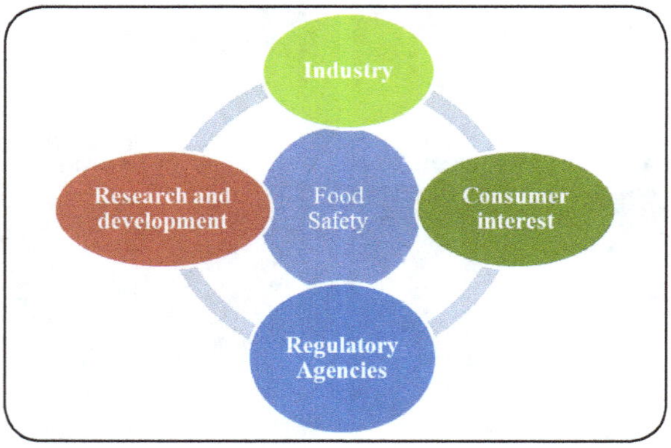

Figure 1. Inter-relationship of different components effecting food safety. (pg.321)

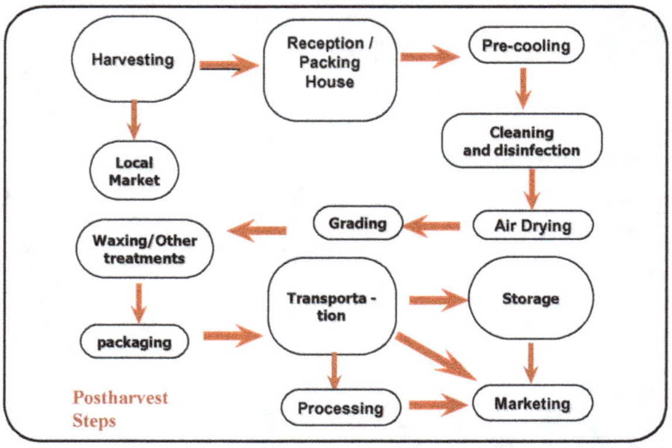

Figure 2. General steps in postharvest handling system. (pg.323)

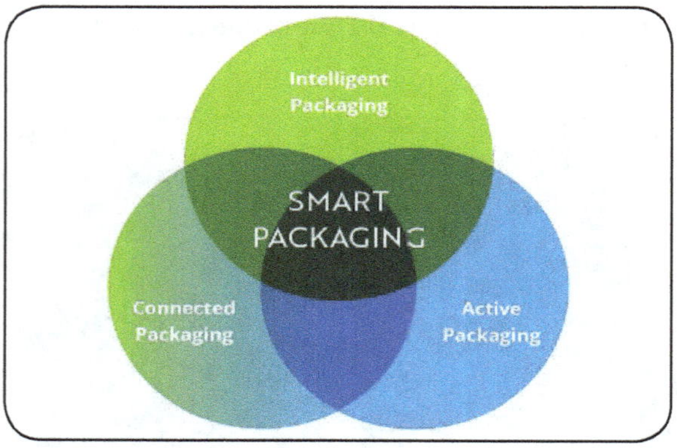

Figure 3. Innovative packaging systems. (pg.325)

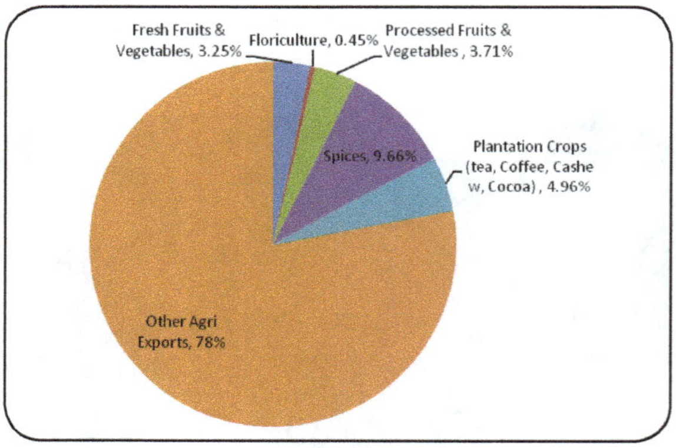

Source: DGCIS. (pg.341)

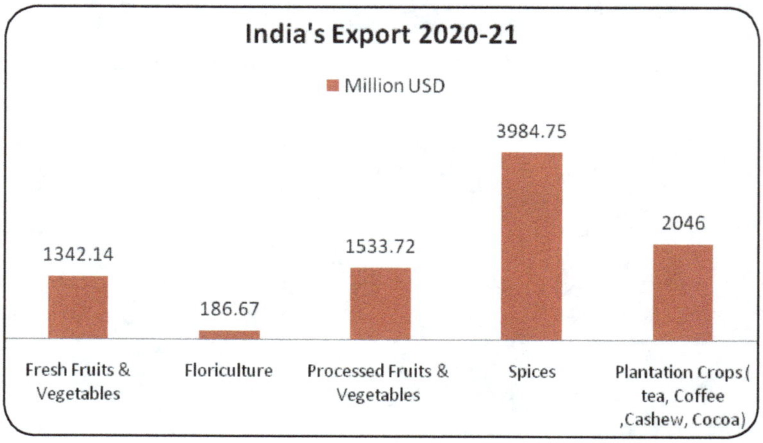

Source: DGCIS. (pg.341)

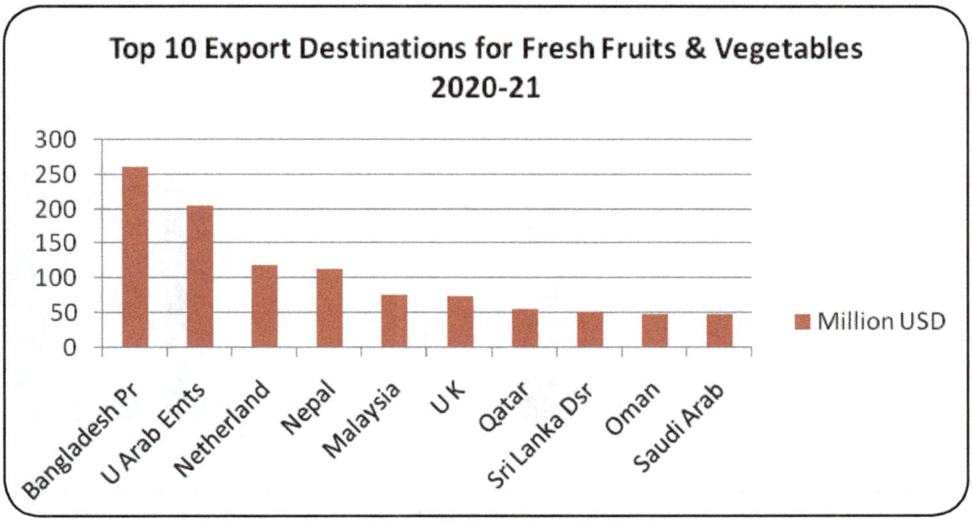

Fresh Fruits and Vegetables
(pg.342)

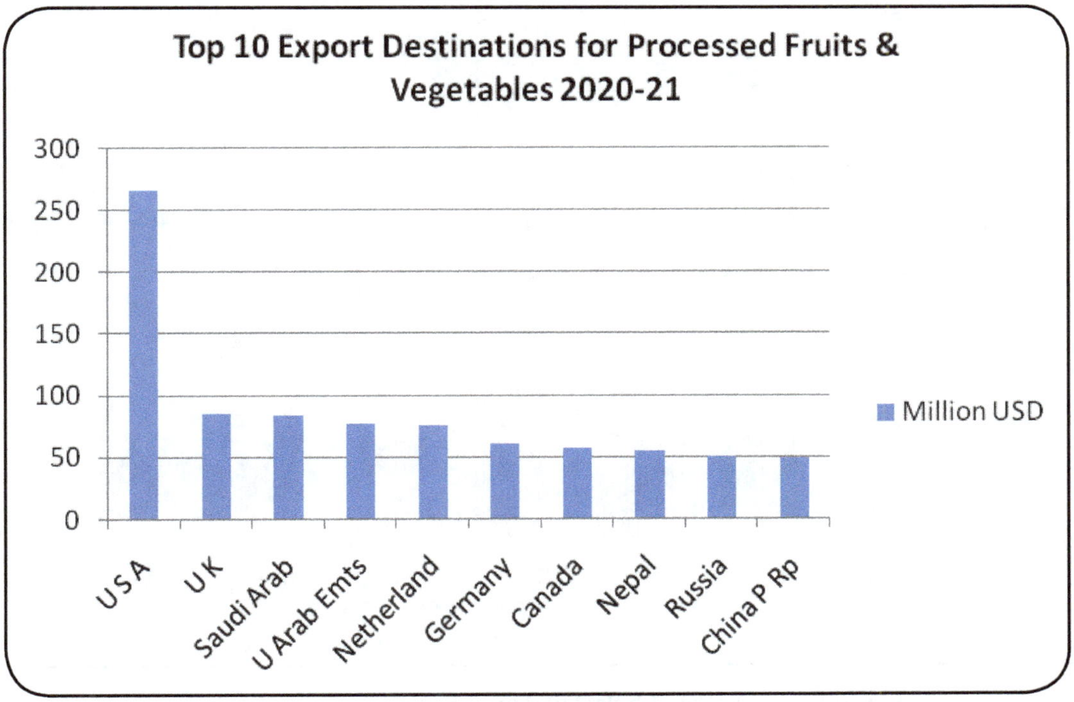

Processed Fruits and Vegetables
(pg.342)

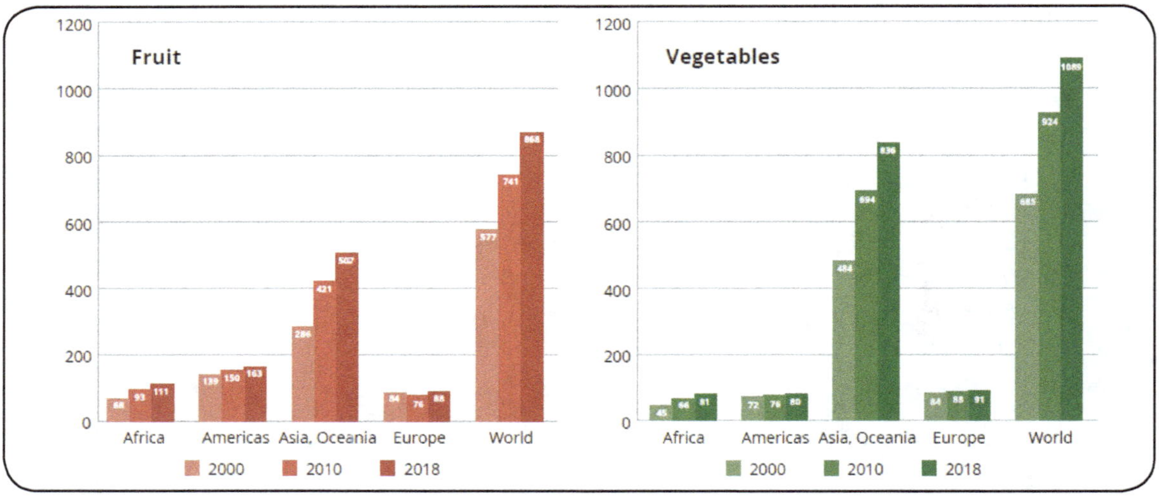

Change in production of fruit and vegetables 2000–2018.
Source: FAO Background Paper and FAOSTAT.
(pg.344)

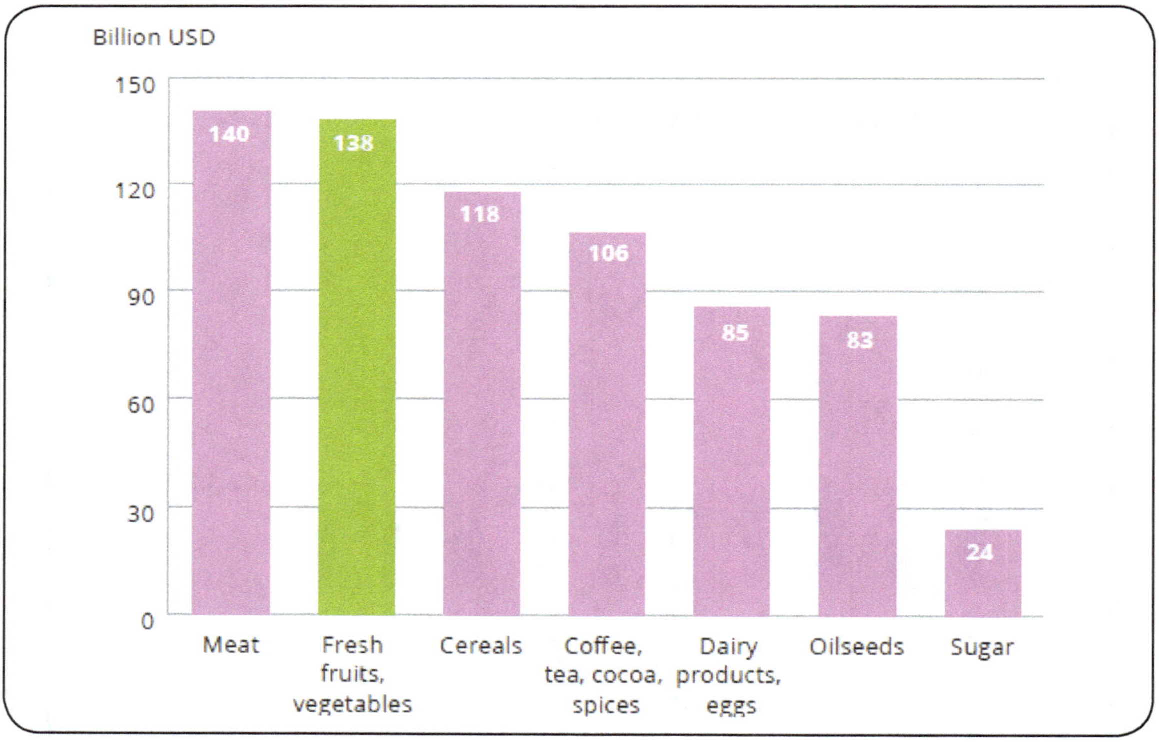

World Export Value of Selected Agricultural Commodities 2018.
Source: FAO Background Paper and FAOSTAT.
(pg.345)

pg.399

pg.399

pg.400

pg.401

pg.401

Index

www.ingramcontent.com/pod-product-compliance
Lightning Source LLC
LaVergne TN
LVHW062214200926
848484LV00006B/1395